Statistical Design and Analysis of Industrial Experiments

STATISTICS: Textbooks and Monographs

A Series Edited by

D. B. Owen, Coordinating Editor
Department of Statistics
Southern Methodist University
Dallas, Texas

R. G. Cornell, Associate Editor
for Biostatistics
University of Michigan

W. J. Kennedy, Associate Editor
for Statistical Computing
Iowa State University

A. M. Kshirsagar, Associate Editor
for Multivariate Analysis and
Experimental Design
University of Michigan

E. G. Schilling, Associate Editor
for Statistical Quality Control
Rochester Institute of Technology

ADDITIONAL VOLUMES IN PREPARATION

Statistical Design and Analysis of Industrial Experiments

edited by

SUBIR GHOSH

The University of California, Riverside
Riverside, California

MARCEL DEKKER, INC.　　　　　New York and Basel

Library of Congress Cataloging-in-Publication Data

Statistical design and analysis of industrial experiments / edited by
  Subir Ghosh.
    p. cm.-(Statistical, textbooks and monographs; vol. 109)
  Includes bibliographical references.
  ISBN 0-8247-8251-8
    1. Experimental design. 2. Research, Industrial--Statistical
  methods. I. Ghosh, Subir. II. Series: Statistical, textbooks
  and monographs; vol. 109.
  T57.37.S69 1990
  607'.2--dc20
                                                                90-3125
                                                                CIP

This book is printed on acid-free paper.

MARCEL DEKKER, INC.
270 Madison Avenue, New York, New York 10016

Current printing (last digit):
10 9 8 7 6 5 4 3 2 1

PRINTED IN THE UNITED STATES OF AMERICA

Preface

Quality improvement using statistical planning and design of experiments is a national concern and a major goal in American industries. The name of Dr. Genichi Taguchi and his contributions to quality improvement through experimental planning and designs are well known in industries across the nation. Many quality improvement centers have mushroomed in the last half-decade in industries and at universities in the United States. The need for understanding and implementing statistical planning and design of experiments is voiced loudly from these centers.

This book is an attempt to present and explain the scope of statistical planning in industries; the contributed chapters present techniques now available for application. The authors are experts and eminent statisticians throughout the United States and abroad. The topics covered include factorial and fractional factorial experiments, response surface methods, mixture experiments, crossover designs, design and analysis for quality control studies (including Taguchi methods), expert systems for designs, and designs for biotechnologies and health studies. The book is divided into two sections, Applications and Methods. Each chapter is fortified with examples demonstrating the application of the methods presented.

This book will be indispensable for everybody concerned with experimentation in industries and serve as a guide for students, instructors, and researchers at colleges and universities. It has a wealth of detailed information on available methods in planning and designing industrial experiments. It is also intended to provide future directions in planning, designing, and analyzing industrial experiments.

I would like to thank all contributors, whose cooperation, effort and valuable chapters made this project possible. My sincere thanks go to Don Owen and two excellent people at Marcel Dekker, Inc., Maria Allegra and Walter Brownfield. I would also like to thank all the people in the Department of Statistics at the University of California, Riverside, particularly Sarah DiMenna. A special thanks also goes to Louise De Hayes.

The chapters in this volume have all been refereed. The following persons have served as referees. The four referees who are not also contributors to this book are designated with an asterisk. We are grateful to all these colleagues for their cooperation and help.

Chand Chauhan*	Karen Kafadar
Ching-Shui Cheng	André I. Khuri
Virginia A. Clark*	Eric S. Lagergren
John A. Cornell	Thomas J. Lorenzen
Subir Ghosh	Raymond H. Myers
Raghu Kacker*	Harji Patel*

I want to express my appreciation to Professor Jagdish Srivastava, who approximately two decades ago introduced me to statistical planning and inference in scientific experiments.

I am thankful to my wife, Susnata, and daughter, Malancha, for their patience, help, and understanding. I would also like to express my gratitude to my parents, Subimal and Padma Renu Ghosh. In 1965 my parents dreamed of me as an engineer while I pursued my own dream of becoming a statistician. I believe that this book will bridge the gap between the two dreams.

Subir Ghosh

Contents

Contributors

Ching-Shui Cheng   Department of Statistics, University of California, Berkeley, Berkeley, California

John A. Cornell*   Department of Statistics, Colorado State University, Fort Collins, Colorado

Angela M. Dean   Department of Statistics, The Ohio State University, Columbus, Ohio

Subir Ghosh   Department of Statistics, University of California, Riverside, Riverside, California

Diane I. Gibbons   Mathematics Department, General Motors Research Laboratories, Warren, Michigan

Perry D. Haaland   Physics Department, Becton Dickinson Research Center, Becton Dickinson and Company, Research Triangle Park, North Carolina

---

Current affiliation: Department of Statistics, University of Florida, Gainesville, Florida

A. S. Hedayat Department of Mathematics, Statistics, and Computer Science, University of Illinois at Chicago, Chicago, Illinois

Mike Jacroux Department of Pure and Applied Mathematics, Washington State University, Pullman, Washington

Paul E. Johnson Department of Information and Decision Sciences, Curtis L. Carlson School of Management, University of Minnesota, Minneapolis, Minnesota

Karen Kafadar Measurement Systems Department, Hewlett-Packard Laboratories, Palo Alto, California

André I. Khuri Department of Statistics, University of Florida, Gainesville, Florida

Kenneth D. Kotnour Statistical Consulting, 3M Information Systems and Data Processing, 3M Company, St. Paul, Minnesota

Eric S. Lagergren Department of Statistics, University of California, Riverside, Riverside, California

Thomas J. Lorenzen Mathematics Department, General Motors Research Laboratories, Warren, Michigan

F. S. Ma Department of Mathematics, Tianjin University, Tianjin, People's Republic of China

S. S. Mao Department of Mathematical Statistics, East China Normal University, Shanghai, People's Republic of China

Gary C. McDonald Mathematics Department, General Motors Research Laboratories, Warren, Michigan

Ruth K. Meyer Department of Business Computer Information Systems, St. Cloud State University, St. Cloud, Minnesota

Raymond H. Myers Department of Statistics, Virginia Polytechnic Institute and State University, Blacksburg, Virginia

Christopher J. Nachtsheim Department of Operations and Management Science, Curtis L. Carlson School of Management, University of Minnesota, Minneapolis, Minnesota

Vijay N. Nair Statistics and Data Analysis Research Department, AT&T Bell Laboratories, Murray Hill, New Jersey

Damaraju Raghavarao   Department of Statistics, School of Business and Management, Temple University, Philadelphia, Pennsylvania

Anne C. Shoemaker   Quality Theory and Technology Department, AT&T Bell Laboratories, Holmdel, New Jersey

Jagdish N. Srivastava   Department of Statistics, Colorado State University, Fort Collins, Colorado

Genichi Taguchi   American Supplier Institute, Inc., Dearborn, Michigan

C. F. Jeff Wu   Department of Statistical and Actuarial Sciences, University of Waterloo, Waterloo, Ontario, Canada

Imran A. Zualkernan   Department of Computer Science, University of Minnesota, Minneapolis, Minnesota

Statistical Design and Analysis
of Industrial Experiments

I
APPLICATIONS

1

Experimental Design for Product Design

Genichi Taguchi   American Supplier Institute, Inc., Dearborn, Michigan

1   QUALITY DESIGN

1.1   Introduction

In the primary stage of designing a new product, design of experiments is effectively used for quality and cost improvement. It has been traditional to utilize technological expertise in specialized fields for functional design. But in order to design a product with little functional variation at a low cost, it is equally important to utilize the design of experiments as a universal technique.

The purpose of this paper is to discuss the role of design of experiments for quality engineering: there are three major sections: (1) Quality Engineering, (2) Experimental Design for Parameter Design, and (3) Design of Experiments for Dynamic Functions. Because of space limitations, discussion is limited to the functional quality of products.

1.2   What Is Functional Quality?

Manufacturers sell their products to make a living. A marketable product meets the needs of consumers. Need is called utility in economics. For

1

example, the utility of an alcoholic beverage is demonstrated by its intoxicating characteristic as well as by its flavor and taste. Utility is function. Product planning is forecasting the demand of a product with a particular function at a certain price after determining the different kinds of function the product should have.

A big question that once plagued people concerned with economics was: "Why is water, an absolute requirement to sustain life, priced at only 150 yen a ton, while diamonds, which have nothing to do with life, cost several million yen a carat?" It was a question directed toward the value of a product. Now we have an answer: "Price is determined by marginal utility." People feel that a product is more or less valuable than its price according to their subjective point of view. Since the value of function is subjective, product planning is a matter of humanity, not a problem of science and technology.

The author has defined quality as follows: "Quality is determined in terms of the losses a product imparts to society from the time the product is shipped." The losses include what are caused by variation from the target functions and what are caused by the unintended effects of the product. For example, in the case of medicine, thalidomide had a good target function, sleep, but its side effects were harmful and resulted in great losses.

A motor manufacturer has advertised in a TV commercial that "Our motors always run correctly." It would be a reliable motor if it ran at a constant speed under any condition. It would be unreliable and of poor quality if the revolutions varied with changes in input voltage, temperature, or humidity; if the lube oil evaporated; or if a component part deteriorated. Everybody agrees that the Shinkansen (the bullet train in Japan) has fantastic functions. However, it slows down in heavy snow and stops because of earthquakes. Thus, its functional quality varies. In addition, its excessive vibration and noise demonstrate poor quality on harmful-effect items. Therefore we conclude that while the functions of the Shinkansen are a breakthrough in technology, its quality is rather poor.

A good-quality product has no functional variations, which means that losses caused by cost and harmful-effect items should be small. If cost control is defined as taking countermeasures to reduce all types of losses before the product is shipped, then quality technology is meant to deal with these countermeasures.

Function itself, however, is not a problem of quality technology regardless of the extent to which the function may cause losses to society.

For example, alcoholic beverages have functions of taste, flavor, and intoxication. Many people suffer losses caused by drunken accidents or quarrels. If someone produced an alcoholic beverage that does not intoxicate, the function would be changed. What kind of products with what kind of functions should be allowed in society is a matter for culture or the law to decide, not a problem of technology. Even for those products accompanied by known harmful side effects, a highly cultured nation allows their existence, thus satisfying the freedom and preferences of individual people. Thus, to discuss the utility and value of a product is to discuss culture, not technology. It would be acceptable for an engineer to discuss these subjects as a private person but, again, it must be emphasized that he or she is discussing a cultural problem, not a quality problem.

Sometimes children watch too much TV and thus may suffer a personal loss, such as being unable to enter a good school or to acquire vital skills. Adults may say that it would have been better if the TV screen had been out of focus. But from the standpoint of quality control, it is desirable to have a clear screen and no trouble with the TV set. Whether one *should* watch TV is a problem of individual freedom. Therefore, when we discuss quality problems, we should limit our criteria to the losses caused by functional variations and to *unintended* harmful effects that have nothing to do with function. Otherwise, we would stray from technology and engineering and into cultural problems.

In the following sections, known harmful effects, such as pollution or the cost of the product, that are not related to functional quality items will not be discussed. Instead, our subject will be limited to quality design: the functional quality variation.

1.3   Error and Their Countermeasures by Design

I call the variables that disturb the function of a product a "noise or error factor." An example is the brightness of a fluorescent lamp, which varies with input voltage or from deterioration. Noise variables are classified into the following three types:

1. Outer noise, such as temperature, humidity, voltage, dust, or individual difference. Operating conditions or environmental conditions that affect the functions of a product are called outer noises.

2. Inner noise or deteriorating noise, that is, property change or variation of a product which occurs after usage, storage or abrasion. These are also called deterioration noises.
3. Variation noise or between-product noise, primarily, variations among products manufactured from the same specification.

For example, a series of specifications or drawings are prepared for the manufacture of a product with a certain function. Among the manufactured products, some have the target function while others don't. Here is a functional variation caused by "between-product noise." These products may function well at first and then be functionally damaged by deterioration, or inner noise. The product may function well under normal conditions but may fail at a high temperature, high humidity, when the power voltage is 20% off, or in other conditions of outer noise.

Good functional quality, then, means less functional variation resulting from inner or outer noise, and it means that the product always functions correctly during its lifetime under a wide range of conditions. The functional quality of a product would be at its best if the product functioned normally when the environmental conditions or outer noises such as temperature, humidity or power voltage varied, functioned normally even when deterioration or abrasion occurred after a long period of use and functioned normally when there were variations among products. Quality, in terms of target function, is measured by the degree of variation from the target (the nominal value or ideal function) in the specification.

Let's take as an example of outer noise an experiment for the manufacture of caramels. In Figure 1, A_1 shows the chewing plasticity of

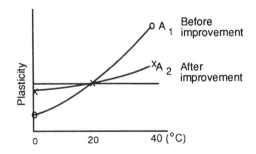

Figure 1   Room temperature and plasticity.

caramels before the experiment. The plasticity varied greatly with room temperature. In other words, we wanted to produce caramels with a flat "temperature-plasticity curve." There were more than 10 raw materials used to make caramels. Many variables were cited, and the experiment was conducted using orthogonal tables. Although it was not a complete success, we could improve the quality and obtain a curve much like A_2 in Figure 1. Through this experiment, we could improve the reliability of the caramels by increasing their resistance to outer noise.

Inner noise is the variation in target function stemming from the variation in what should be the inner constants of a product. Broadly, we can define two types of inner noise: functional variation resulting from deterioration (or timewise noise), and the noise between products (or spacewise noise).

The target function of a TV power circuit is to convert 100 volts of alternating current input into 115 volts of direct current output. If the power circuit maintains 115 volts anytime and anywhere, the quality of power circuit is perfect as far as voltage is concerned. The variables causing voltage to move away from the target of 115 volts are called noise variables and are classified as:

1. Outer noise: the variation of environmental conditions such as temperature, humidity, dust, input voltage, and so on.
2. Inner noise: the deterioration of elements or materials constituting the circuit. For example, the resistance of a resistor increases 10% in ten years, or the parameter of a transistor, such as h_{FE}, changes with time.
3. Variational noise: the variation among products that causes different output voltages from the same input voltage.

In order to minimize the effects caused by these three noise sources, some countermeasures may be considered. Of course, the most important is the countermeasure by design, an off-line countermeasure. It consists of the following three steps:

1. Concept (Primary) Design: functional design including redundancy. Apply the technology of specialized fields.
2. Parameter (Secondary) Design: cost reduction and quality improvement. Design of experiment is used.
3. Allowance (Tertiary) Design: a method to reduce or eliminate causes. Design of experiment is applied.

1.4   Concept (Primary or Functional) Design

In this step, the knowledge of a specialized field is applied. In the case of electronics, the type of circuit to convert alternating current into direct current is researched. In the chemical industry, a chemical reaction to establish a manufacturing process for a product is sought.

An automatic controlling system or redundant design may be included in system design if necessary. For example, an output voltage is measured continuously. When there is a deviation from the target voltage, such as 115 V, a parameter in the circuit, such as the resistance of a variable resistor, is automatically changed. However, it is difficult to control the deterioration or variation of the automatic controlling system itself. There is also an additional cost for the automatic control mechanism. Therefore, putting an automatic controlling system in the circuit does not necessarily mean designing a stable circuit without a cost increase.

Although conceptual design is very important, we cannot afford to research all concepts. Therefore, the research is limited to a few concepts, selected on the basis of past experience or as an informed guess.

There are many steps in new product development. In developing TV sets, there are:

1. Design of system
2. Design of subsystem
3. Design of units or component parts
4. Development of elements
5. Development of materials

Any of the above steps includes concept design, parameter design, and allowance design. System development is embodied in the forecasting of a designer, which is necessarily uncertain. Also, in concept design, the technology of a special field plays a major role. The areas in which the design of experiments plays its role are parameter design and allowance design.

1.5   Parameter (Secondary) Design

Once a system design is completed, the next step is to determine the optimum level of individual parameters of the system. Such research has been neglected, especially in developing countries. In these countries, researchers look for information from literature or copy the technology of developed countries, then design a power circuit that "seems to be the best." A prototype is made, then charged with 100 volts AC current. If the output voltage equals the target value, it is labeled a success, and if the result is unsatisfactory, the value of a parameter that seemed to be effective to tune the target value. This approach is dated, reminiscent of the dyer of old, who has a color sample and tries to create a dye to match. If the color is identical to the sample, he thinks he has succeeded. And if not, he changes either the mixing ratio of the dyes or the dyeing conditions and tries again.

This trial-and-error approach is not a design but merely tuning work, called a modification or a calibration. The so-called designs in developing countries are similar to the dyer's method; their engineers proceed along the same line. For example, the h_{FE} of a transistor A and an element resistor B in a power circuit affect the output voltage as shown in Figure 2.

Suppose a prototype of a power circuit system is made, charged with 100 V AC power, and only 90 V output is obtained instead of 115 V. In order to obtain the target voltage, assume that a transistor in the circuit, h_{FE}, that is the most influential parameter regarding output voltage is changed to 100, 300, and 500. The output voltage may be obtained either by calculating from circuit theory or by experimentation when no theory is available. The left side of Figure 2 shows the relationship. According to the curve, a designing engineer might take $A' = 200$ as the value of h_{FE}, since the curve meets 115 V at $A' = 200$. From the standpoint of quality technology, this is really poor methodology.

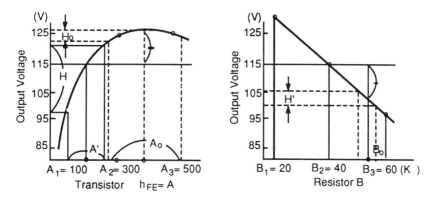

Figure 2   Parameter levels and output voltage.

The purpose of the design is to obtain a circuit using inexpensive component parts without output variation. In this example, only output voltage is discussed as the objective characteristic, although there are others such as current. In general, the elements such as transistor with small variation or small deterioration are expensive. The cheapest grade costs less than one-tenth of the best. Let's assume we selected the cheapest third-grade element. In 10 years of the product life, the h_{FE} value of the transistor either varies or deteriorates $\pm 30\%$ from its nominal value. When $A' = 200$ is selected, this cheap component part varies within $20 \pm 20 \times 0.3$, or within 140–260. As a result, the output voltage varies within 98–121 V, or a 23-V variation, as seen from the figure.

If A_0 is selected as the value of h_{FE}, instead of $A' = 200$, and assuming the varying range of h_{FE} is also 30% around 400, or from 280 to 420, the output voltage varies from 122 to 127 V, or only 5 V. Therefore, we can reduce the varying range of the output voltage when even a cheap component part is used. Instead, the resulting output voltage becomes about 10 V higher than the objective value. In order to adjust this, a component part that linearly affects the output voltage may be selected, as shown on the right side of Figure 2.

In this example, we have the parameter h_{FE}, which affects the output voltage nonlinearly. When the top of the curve, which is A_0, is selected, the influence of its own variation can be reduced. This is a very sophisticated design skill, and reducing the cost also improves the quality. However, it would not be appropriate if such a design affected the increase of the variation of other parameters or component parts. It is important to select a design research path by considering the influences of the variations of environmental conditions or other parameters. That is why orthogonal tables and accumulating analysis are useful. These will be described in the next section.

Parameter design is the most important step in design procedures. It is required to obtain a stable and reliable product or a manufacturing process. It is the technique of utilizing nonlinearity "unconsciously." So it is important in the application of the design of experiments, it is therefore important to select an objective characteristic for such a purpose. This is the most important quality technology countermeasure in developing countries. The secondary design technique is badly needed as the ideal quality technology for all three types of noise sources.

Secondary, or parameter, design is the heart of experimental design applications. In India, engineers are very familiar with all kinds of literature, and many industries are cooperating technically with leading foreign companies. However, the product quality often is not good, because Indian engineers tend simply to swallow whole the literatures or the parameter values specified by their partner companies.

1.6 Allowance (Tertiary) Design

Once the system design is completed, and the midvalues of the factors constituting the system are determined, the next step is to determine the tolerances of these factors. To make this determination, we must consider environmental conditions as well as system elements. The midvalues and varying ranges of these factors and conditions are considered as noise factors and are arranged in orthogonal tables so that the magnitude of their influences to the final output characteristics may be determined. A narrower allowance will be given to those noise factors having a large influence on the output. Cost considerations determine the allowance. Different from secondary design, this step results in "cost-up" by controlling noise in narrower ranges; therefore, quality-controlling countermeasures should be best achieved through secondary design. If this is not an option, the tertiary design is the only countermeasure left that will allow for influencing factors within a narrow range. In such cases, it is important to evaluate the loss caused by variation.

For a simple example, the power circuit of a TV set is again used to study its loss by its functional variation. The power circuit of TV sets converts 100 V AC to 115 V DC. The reason why 115 V was used as the output voltage should be determined in total system design. Once the target value is set at 115 V, no matter how rationally it was determined, other parts of the system, such as the receiving circuit or the Brown tubes, will be designed following that voltage.

When the output voltage of the power source deviates from 115 V, a loss occurs. Let the output voltage be y and the loss caused by the deviation from the target value of 115 V be $L(y)$. Since there is no loss when the voltage is 115, its loss function $L(y)$ would be

$$L(115) = 0 \qquad (1)$$

The loss function becomes minimum when $y = 115$; therefore, its differential value becomes zero.

$$L'(115) = 0 \qquad (2)$$

An approximate equation of the loss function can be derived using the Taylor expansion with the target value at $y = 115$.

$$L(y) = L(115 + y - 115)$$
$$= L(115) + \frac{L'(115)}{1!}(y - 115) + \frac{L''(115)}{2!} \times (y - 115)^2 + \cdots \quad (3)$$

Putting Eqs. (1) and (2) together, the above equation would be:

$$L(y) = \frac{L''(115)}{2!}(y - 115)^2 + \cdots \quad (4)$$

It is seen that the first term of the loss function is given by the square term of deviation from the target value. Letting the coefficient of $(y - 115)^2$ be k, the following approximate equation is obtained after omitting the higher-order terms

$$L(y) \approx k(y - 115)^2 \quad (5)$$

Figure 3 shows the relationship. The constant in the approximate equation (5) is determined in the following way. The loss increases when the output voltage deviates from the target value. Suppose a household bought a TV set, and the output voltage dropped from 115 to 90 after several years usage. When the screen becomes darker or less contrasty, the TV set must be repaired or replaced. Assume the average loss to such customers, denoted by D, to be $200.

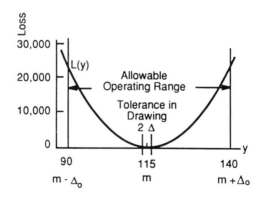

Figure 3   Output voltage and loss function $L(y)$.

When the voltage is too high, repair or replacement is also needed because of such troubles as damaged component parts or too much contrast. Generally, the tolerance limit of the upper side is different from the lower side, but assume for simplicity that these limits are the same. Therefore,

Tolerance limits of consumers $= 115 \pm 25\mathrm{V}$ (6)

Average loss of consumers beyond the tolerance limits: $D = \$200$ (7)

Since the tolerance limits may be different form one person to another, assume that 50% of the consumers either repair or replace their TV sets at this point. Loss function is very complicated since it is the average relationship of different applying conditions with different probabilities, but we approximate the relationship by using Eq. (5). This is called the law of the square of losses.

In Figure 3, when the output voltage y becomes either higher than 140 V or lower than 90 V, 50% of the consumers cannot tolerate watching such TV sets and spend an average of \$200 either to repair or to replace them. Putting \$200 on the left side of Eq. (5) and 25 to the absolute value of $(y - 115)$ on the right side,

$$200 = k \times 25^2 \tag{8}$$

Solving the equation,

$$k = \frac{200}{25^2} = 0.32 \ (\$\sqrt{\mathrm{V}}) \tag{9}$$

The loss function is approximately given by

$$L = 0.32(y - 115)^2 \tag{10}$$

At the time of product and process design, predicted mean square error σ^2 caused by all noises is important.

$$\sigma^2 = E(y - 150)^2 \tag{11}$$

Then,

$$L = 0.32\sigma^2 \tag{12}$$

The role of product design is to reduce the above loss.

2   EXPERIMENTAL DESIGN FOR PARAMETER DESIGN

2.1 R, L Circuit

This section illustrates how to conduct the parameter design. We will also discuss the difference between the parameter design and the usual design of experiments method.

The output current y of an AC circuit with resistance R and self-inductance L is given by the equation below when it is assumed that

V: Input AC voltage (V)
R: Resistance (Ω)
f: Frequency of input AC (Hz)
L: Self-inductance (H)

and its unit is the ampere, A,

$$y = \frac{V}{\sqrt{R^2 + (2\pi f L)^2}} \tag{13}$$

The target of output current y is 10 (A); if there is a shift of at least 4 (A) during use by the consumer, it no longer functions, and it will be regarded that loss to society, D, would be \$100. This is the value obtained when functional trouble occurs in the marketplace (the destination of the user of the product) when the current that is the output characteristic shifts 4 (A) from the target. If trouble occurs when the lower limit is 6 (A) and trouble occurs when the upper limit is 16 (A), one may also represent the functional tolerance as

$$10^{-4}_{+6} \text{ (A)} \tag{14}$$

Since oftentimes such a tolerance, which differs at the top and bottom, is troublesome, once the target value, 10 (A), has been decided on, one frequently resorts to a drastic method, using a symmetrical tolerance even in the case of (14)

$$10 \pm 4 \text{ (A)} \tag{15}$$

Once one has found the functional tolerance for the output characteristic and the mean loss D_0 when it exceeds the tolerance, one finally enters the stage of design. There are three steps to the design as stated before:

1. Concept, or primary, design, system selection

2. Parameter, or secondary, design
3. Tolerance, or allowance (tertiary) design

Concept development is the step by which one searches and determines what combination of circuit and parts possesses the target function and, in the example given above, the circuit has already been given. That means that system selection has already been completed. Many innovations are made in system development, and the core of research is in concept design. It is best to develop, to the extent possible, new systems not found in the literature. The system becomes known to competitors immediately, once the product is sold, and thus becomes a matter for the law, so that the protection of a patent may be applied for.

Tolerance design is deciding on the tolerance ahead of and behind the central value of the parameter. Trading off between loss due to dispersion and cost is necessary for this method of determining the tolerance, and it is wise to use the loss function. However, this method involves the problem of separately deciding on the tolerance for the individual systems and not on distributing the tolerance of the target characteristic value among the system elements. It is in parameter design, not tolerance design, that we consider the cause of dispersion comprehensively. We will consider the problem of quality design for the circuit of Eq. (13) as a step in parameter design.

2.2   Classification of Factors, Control Factor, and Noise Factor

Parameters in the equation giving the output, whose central value and levels can be decided on by freely by the designer, are the control factors.

The voltage of the input source is AC 100 (V) in the case of a household source and, since the designer cannot alter this, it is not a control factor. Nor is the frequency f a control factor.

Only R and L are control factors. Not only can the resistance value be 1 or 3 or 9 (Ω) as specified by the designer, but it must necessarily be decided on. It is the same with self-inductance L. It is logical that since the target value of the output current is 10 (A) in this case, once one decides on the central value of resistance R, the central value of inductance L will be determined by the equation.

In this case, there are only two control factors, and the target value has already been decided. But when there are at least three control factors, it is best not to consider such constraints. The reason for this is that it is more advantageous to take measures for correction toward the target value by changing the levels of control factors that have little influence on the stability among the factors, after optimum design for stability. Factors that are favorable for adjustment are those that change the output to the target value. Such factors for adjustment are sometimes also termed tuning factors. It is one of the major characteristics of the design of experiments method for technical research not to consider restricting conditions but to perform parallel analysis toward the signal-to-noise (SN) ratio and sensitivity, which are measures of the stability, and to adjust later.

Since it is possible to consider restricting conditions among control factors later as long as it relates to the objective, one does not consider the target value or target characteristic curve but changes the control factor independently. Essentially, in this case, we decide on the three levels of R and L independently, as follows:

$$R_1 = 0.5 \ (\Omega) \qquad R_2 = 5.0 \ (\Omega) \qquad R_3 = 9.5 \ (\Omega)$$
$$L_1 = 0.010 \ (H) \qquad L_2 = 0.020 \ (H) \qquad L_3 = 0.030 \ (H)$$

An orthogonal array is used if the number of control factors is more numerous, but since there are only two here, we assign this a two-way layout. An orthogonal array to which control factors have been assigned is termed an inner orthogonal array, or control orthogonal array. We use as many control factors as possible and take the range of levels wide, and we assign only the main effects.

Next, we take up the noise factors. "Noise factor" is a summary term for a cause that disperses the target characteristic from the ideal value, and it can be categorized into the following three types.

1. Dispersion of environmental conditions; this is termed outer noise.
2. Dispersion due to degradation; this is a type of inner noise.
3. Dispersion among products; this refers to differences from product to product.

Noise factor signifies the whole of the cause of dispersion but, in parameter design, the above three types may be regarded as the same. This is because environmental differences change the characteristics of the parts, materials, and elements used for that product and because, as a result, they cause the target characteristics to disperse. Degradation of the product is the dispersion of function due to degradation of the parts, materials, and elements, etc., that are used for the product in question. This is the same as changing the characteristic values of the parts, materials, and elements, etc., that are used for the product in question.

Dispersion of the environment consists of the following two parameters in the case of this example, and their levels are taken as follows:

Voltage of input source V 90   100   110   (V)

Frequency f 50   60   (Hz)

The environment temperature too, changes the resistance value of the resistance and coil inductance L, although only slightly. However, changes due to degradation of resistance and coil are greater than this. Having considered changes due to dispersion of the initial period value, degradation, and environment, let us assign to the changes of resistance R and changes of coil inductance L the following three levels:

First level: Nominal value × 0.9

Second level: Nominal value

Third level: Nominal value × 1.1

Therefore, the levels of the noise factors are as given in Table 1. It should be noted that a prime symbol has been affixed to the noise factors. R and L, in this instance, indicating that there is no error "noise" with respect to noise factors V, R', and L' when the level is the second level. Frequency f is 50 and 60 (Hz); this difference depends on the location in Japan. If one wishes to develop a product that can be used in the whole of Japan, both in eastern and western Japan, it is best to so design it that the target value will be obtained by an output of 55 (Hz), which is midway between the two. Sometimes, one uses an intermediate value that has been weighted by the population ratio. Here, let us assume that 55 (Hz) is to be used. Therefore, since every one of V, R', f, and L' will be adjusted later

Table 1   Levels of Noise Factors

	First Level	Second Level	Third Level
V	90	100	100 (V)
R'	−10	0	+10 (%)
f	50	55	60 (Hz)
L'	−10	0	+10 (%)

so that the output will become exactly 10 (A) when it is at its central value, namely, 100 (V), nominal value, 55 (Hz), and nominal value, it is indicated that as far as the noise factors are concerned, it suffices to consider only the two levels at both ends and assign them.

But, now, as is evident from formula (13), the output becomes minimum at $V_1 R'_3 f_2 L'_3$ and the output becomes maximum at $V_3 R'_1 f_1 L'_1$. When, with respect to the influence of the noise factors, the qualitative influence on the output is clear, the whole of the noise factors is compounded into a single factor. If the compound factor is expressed as N, it is at two levels in this case and is

$$N_1 = V_1 R'_3 f_2 L'_3 \qquad \text{Minus side worst condition}$$
$$N_2 = V_3 R'_1 f_1 L'_1 \qquad \text{Plus side worst condition}$$

When the qualitative tendencies of the noise factors (tendencies such as which level causes the measured characteristic to become smaller) are determined a good method is to compound to obtain one factor at two levels, or to compound into one factor at three levels, including the central level combination. The result is a single compound factor, no matter how many factors there are whose qualitative tendencies one can discern. Since one can determine the qualitative tendencies of all four noise factors in this example, essentially they have been converted into a compound noise factor N at two levels. It means that one need merely investigate the value of the outgoing current at just its two levels.

Such compounding of noise factors is even more important when there is no theoretical equation. If there is a theoretical equation, usually calculation time is within several seconds, even if the noise factors are assigned separately to the columns of an orthogonal array, and little efficiency of design calculation is gained, even when their combinations are rendered into the two levels of minimum and maximum, as explained. Today, computers have rendered this unnecessary. Even so, when there are too many noise factors, the orthogonal array becomes large and much CPU time is necessary. When this happens, compounding noises becomes useful.

2.3   Method of Parameter Design

Design calculations are performed (experiments are performed when there is no theoretical equation) at all combinations of the inner orthogonal array into which control factors R and L have been assigned (two-way layout here), and noise factor assignment (two-level compounded noise factor here). This is the logical method for parameter design. Therefore the assignment is as presented in the left half of Table 2.

Since the design conditions (level combination of the parameters) for No. 1 is $R_1 = 0.5$ (Ω) and $L_1 = 0.01$ (H) in Table 2, the level of the two-level noise factor N_1 and N_2 is as follows:

$$N_1 = V_1 R'_3 f_3 L'_3 = 90 \text{ (V)}, \ 0.55 \ (\Omega), \ 60 \text{ (Hz)}, \ 0.011 \text{ (H)}$$

$$N_2 = V_3 R'_1 f_1 L'_1 = 110 \text{ (V)}, \ 0.45 \ (\Omega), \ 50 \text{ (Hz)}, \ 0.009 \text{ (H)}$$

R'_3 and L'_3 are $+10\%$ of 0.5 (Ω) and 0.01 (H) and are therefore 0.55 (Ω) and 0.011 (H); and R'_1 and L'_1 are -10%, 0.45 and 0.009 (H). Therefore, from formula (13), the values of the current value y at N_1 and N_2, are as follows:

$$y_1 = \frac{90}{\sqrt{0.55^2 + (2\pi \times 60 \times 0.011)^2}} = 21.5 \tag{16}$$

$$y_2 = \frac{110}{\sqrt{0.45^2 + (2\pi \times 50 \times 0.009)^2}} = 38.5 \tag{17}$$

These are the data for N_1 and N_2 of No. 1. What is important is that calculations (and experiments) have been performed without regard to the target value, 10 (A). What raises the efficiency of design study is allowing constraints to be ignored. For Nos. 2–9, similarly, the values of the output

Table 2 Assignment and Data

No.	R	L	Data N_1	N_2	SN Ratio η	Sensitivity S
1	1	1	21.5	38.5	7.6	29.2
2	1	2	10.8	19.4	7.5	23.2
3	1	3	7.2	13.0	7.4	19.7
4	2	1	13.1	20.7	9.7	24.3
5	2	2	9.0	15.2	8.5	21.4
6	2	3	6.6	11.5	8.0	18.8
7	3	1	8.0	12.2	10.4	20.0
8	3	2	6.8	10.7	9.6	18.6
9	3	3	5.5	9.1	8.9	17.0

current at N_1 and N_2 were obtained. These constitute the two values in the data column of Table 2.

What constitutes a measure of stability (robustness) in the case of a nominally best characteristic is the ratio of the mean value m to the standard deviation σ provided nonnegative value never possible.

The square of this ratio is termed the SN ratio, η.

$$\eta = \frac{m^2}{\sigma^2} \tag{18}$$

The reciprocal of the square of the relative error (also termed coefficient of variation) is the measure of stability (robustness), SN ratio. Although the calculation equation for the SN ratio varies depending on the type of characteristic value, all have the same properties, and when the SN ratio becomes ten times as great, loss due to dispersion decreases to one-tenth. The SN ratio is a measure possessing rationality economically. The question, then is how to estimate m^2 and σ^2. When there are n items of data, as $y_1, y_2, \ldots, y_n$, the mean value m and variance σ^2 are estimated by

$$\hat{m} = \frac{y_1 + \cdots y_n}{n} \tag{19}$$

$$\hat{\sigma}^2 = \frac{1}{n-1} \sum (y_i - \bar{y})^2 \tag{20}$$

When there are only two data, one estimates the variance by

$$V_e = y_1^2 + y_2^2 - \frac{(y_1 + y_2)^2}{2} \tag{21}$$

On the other hand, since mean value

$$\bar{y} = \frac{y_1 + y_2 + \cdots + y_n}{n}$$

contains error, if one estimates m^2 by $(\bar{y})^2$, it becomes a little excessive as the mean. When there are n data, if we write

$$S_m = \frac{(y_1 + y_2 + \cdots + y_n)^2}{n} = n(\bar{y})^2 \tag{22}$$

The expected value of this is

$$E(S_m) = \sigma^2 + nm^2 \tag{23}$$

as is well known. Thus, the following statistic:

$$\hat{m}^2 = \frac{1}{n}(S_m - V_e) \tag{24}$$

is more desirable than $(\bar{y})^2$ as the estimated value of m^2. The $\hat{m}^2$ in Eq. (24) is the estimated value of the sensitivity S.

Therefore, we estimate η and S, which are the SN ratio and sensitivity, by the following equations:

$$\eta = \frac{\frac{1}{2}(S_m - V_e)}{V_e} \tag{25}$$

$$S = \frac{1}{2}(S_m - V_e) \tag{26}$$

where

$$S_m = \frac{(y_1 + y_2)^2}{2} \tag{27}$$

$$V_e = \frac{(y_1 - y_2)^2}{2} \tag{28}$$

We use decibel (dB) values, which are ten times the common logarithm values of (25) and (26), to analyze the effects of the control factors.

In the case of the data of No. 1, $y_1 = 21.5$ and $y_2 = 38.5$.

$$S_m = \frac{(21.5 + 38.5)^2}{2} = 1800.00$$

$$V_e = \frac{(21.5 - 39.5)^2}{2} = 144.5$$

Therefore, the decibel values of SN ratio η and sensitivity S are

$$10\log\frac{\frac{1}{2}(1800 - 144.5)}{144.5} = 7.6 \text{ (dB)} \tag{29}$$

$$10\log\frac{1}{2}(1800 - 144.5) = 29.2 \text{ (dB)} \tag{30}$$

It was in this way that the values of SN ratio η and sensitivity S in Table 2 were found. SN ratio η is a measure for optimum robust design and sensitivity S, in order to select one (sometimes two or more) factor(s) by which to match the mean value of the output with the target value later when necessary.

For SN ratio η and sensitivity S, we construct a table of mean values for comparisons of control factors classed by level. For example, in the case of R_1,

Mean of R_1, of SN ratio η:

$$\bar{R} = \frac{1}{3}(7.6 + 7.5 + 7.4) = 7.5 \tag{31}$$

Mean of R_1, of sensitivity S:

$$\bar{R} = \frac{1}{3}(29.2 + 23.2 + 19.7) = 24.0 \tag{32}$$

In Table 3, the mean values were found for the other levels as well. We direct our attention toward the measure of stability, SN ratio, in Table 3. The optimum level of R is R_1 and the optimum level of L is L_1 and therefore the optimum design is $R_3 L_3$. Confirmatory calculation (trial) is performed at R_3, L_1, and one finds the mean value of the output. If there is no difference between this mean value and the target value, one may leave matters as is; but if there is a difference, one compares the influence on the SN ratio η and the influence on the sensitivity S, and one adjusts the output to do away with this difference by using a control factor whose effect on the sensitivity is great compared with the effect on the SN ratio. In this instance, under the optimum design $R_3 \times L_1$ the calculated output value is No. 7 in Table 2. The current is $y_1 = 8.0$ and $y_2 = 12.2$; the mean value is 10.1 and, coincidentally, there is no difference from the target value. If the difference were found to be great and, for example, if the mean value were 15 (A), it would be necessary to lower the sensitivity by just the difference in decibel value relative to the target value,

$$10\log 10^2 - 10\log 15^2 = -3.5 \text{ (dB)} \tag{33}$$

Table 3   Table of Factorial Effects

	η	S		η	S
R_1	7.5	24.0	L_1	9.2	24.5
R_2	8.7	21.5	L_2	8.5	21.1
R_3	9.6	18.5	L_3	8.1	18.5

Since, from Table 3, the effect on S is greater in the case of L than R, we correct by using L. Such adjustment is easy when conducted on numerous control factors, one often finds a control factor that has great influence on the sensitivity, unrelated to the SN ratio.

If the initial value considered by the designer had been R_2L_2, at R_3L_1 loss is rendered into $1/1.5$ in terms of quality improvement by gaining 1.6 dB. When there are two or more target characteristics, design research such as described above is carried out for each. For correction to the target value, factors for adjustment that influence each of the sensitivities become necessary in the same number as the target characteristics.

The method of minimizing the percent change of Eq. (13) is more or less the same as the method used here. In the case of the percent change, it is assumed that the initial value can be adjusted to the target value. If searching the factors for adjustment in such analysis is conducted sequentially for the initial value, this too is equivalent to analysis of the sensitivity in parameter design.

3   EXPERIMENTAL DESIGN FOR DYNAMIC CHARACTERISTICS

3.1   Design Experiment for Dynamic Characteristics

The layout techniques of experimental design and variance calculation are useful especially for the evaluation of dynamic characteristics in design research. Let's define a dynamic characteristic first. The controllability of an airplane or a car; the functions of machine tools, robots, and sporting equipment; and the athletic ability of humans are dynamic characteristics. Dynamic characteristics have the following form:

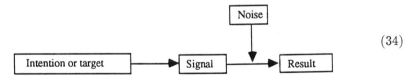

$$(34)$$

Using skiing as an example, when we want to ski straight ahead or turn to the right, we put our weight to the center or to the left, respectively. When there is an intention or a target value, we must dispatch a controllable variable (called a signal) to actualize the intention. Since we want to actualize a specific state, such as turning a certain degree to the right, the extent of using our weight must be changed accordingly. For this reason, the term dynamic characteristic is used.

The skis used for jumping require a straightforward characteristic only; that characteristic may be improved by making a few straight grooves on the bottom surface. In this case, a dynamic characteristic is not necessary.

However, in the case of a raw material with good rollability, it must be rolled to many different thicknesses. If we want to roll it to 1 mm thick, it easily becomes 1 mm, and if we want it to be 0.8 mm, it becomes 0.8 mm. A raw material with a good dynamic characteristic meets our demand, no matter how our target changes. The demand of rolling a material to 1 or 0.8 mm thick is performed by a signal such as the pressure or the number of pressing times.

In skiing, the signal for making a curve is the weight shift or edge control. In a car the steering wheel is used as a medium to express the intention to turn; its turning angle is the signal. Acceleration and braking are the signals of speed. When we make a turn by using the same turning angle, the turning radius varies according to road condition and type of tires. When the turning direction varies by various other conditions (error factors), even if its turning effect is good, it is an unstable car. We need to design a car whose effects caused by signal factors are certain and the effects caused by error are small.

In the case of dyeing cloth, a good dyeing means we can get the color and shade of our choice. If the quantity of dye is used as the signal for changing darkness, the quantity of dye is then a signal factor. If darkness varies according to yarn count or the thickness of dyed material, or varies by temperature or agitation, even if the same quantity is used, it means that darkness is changed by factors other than the signal factor, so that the dyeing is not good.

A dye with a good dyeing ability changes darkness consistently when the signal level changes but will not be much affected by nonsignal factors (error factors or noise factors). The former is the signal factor effect, and the later is error factor effect. SN ratio is obtained by dividing the former by the latter.

If the effect of a signal is very large, a slight change in the applied amount makes a difference in darkness; hence the dye would appear to be inferior, i.e., too sensitive. It is true that the dyeing ability, including the ease of application, cannot be expressed by SN ratio alone.

With this kind of dye, a small quantity error greatly affects the darkness after dyeing and so the application must be done very carefully. It is dye that is difficult to handle. Let M be a signal factor, and the darkness after dyeing be y.

$$y = f(M) \tag{35}$$

If a slight change in application quantity greatly affects the darkness after dyeing, it means that the differential coefficient of Eq. (35), or

$$y' = f'(M) \tag{36}$$

is large. Here, $f'(M)$ is sensitivity. However, no matter how large the sensitivity may be, it can be easily adjusted. For example, by diluting the dye or by adding filler, the sensitivity can be infinitely adjusted. A sensitive dye is not desirable, but the solution is easy. The adjustment of sensitivity is not a substantial technological improvement. What we desire is to find a condition with the largest SN ratio and then determine the sensitivity.

A power circuit may also be deemed to have a dynamic characteristic, since we need the output to be changed according to the input or signal. For the design of such a circuit, we should not only aim at stability as a target. This is because even after the parameter design is completed, the variation among the manufactured products cannot become zero. A 100% of the products will not give exactly 115 V DC when 100 V AC is input. If the specification of the output voltage is 115 ± 1.4 V, a circuit giving 112 V cannot pass inspection. In many cases, the circuit is repaired (instead of the whole product being discarded) by exchanging an element in the circuit. The variable or the element for such exchanging must have a sufficient influencing effect. Such a variable is called a signal factor; its sensitivity must be great.

In the case of machine tools, there is an objective characteristic when material is processed. When the objective characteristic deviates from the target value, it is desirable that the calibration be efficiently done.

When there are three objective characteristics, it is necessary to have at least three independent variables or three signal factors for calibration. In addition, it is desirable that one specific variable calibrate only one specific character. In other words, it is desirable to design in such a way that a signal factor affects only one objective characteristic and, at the same time, dampens the effects caused by other signal factors. Thus, it becomes easier to calculate the amount of calibration when a deviation from the target value is found.

Automatic control or an automatic adjusting mechanism may also have a dynamic characteristic nature. The variable for calibrating the deviation from the target value is a signal factor.

3.2   How to Evaluate a Dynamic Function

Product designers have been designing their products to perform their required function at various conditions as long as possible. The methods designers used are as follows:

1. Analytical method.
2. Operating window method. (This method was proposed by Don Clausing when he worked at Xerox, allowing us to use any design parameters as a continuous measure to evaluate a digital type of function such as an on-off switching system.)

Hitachi used the analytical method to develop their electron microscope around the end of World War II. The ideal function of the microscope is

$$y = \beta M \tag{37}$$

where y is output image and M input image. The coefficient called sensitivity is the parameter we want to be as large as possible. However, an actual product has no such ideal relation. Let us denote the actual relation by

$$y = f(M, x_1, x_2, \ldots, x_n) \tag{38}$$

where x_1, x_2, $\ldots$, x_n are various noise variables that affect y.

Hitachi used the following signal-to-noise ratio as the functional quality measure to be optimized at design stage.

$$\eta = 10 \log \frac{\left(\frac{\partial f}{\partial M}\right)^2}{\left(\frac{\partial f}{\partial x_1}\right)\sigma_{x_1}^2 + \cdots + \left(\frac{\partial f}{\partial x_n}\right)^2 \sigma_{x_n}^2} \tag{39}$$

where σ_{x_1}, $\ldots$, σ_{x_n} are standard deviation of those noises of x_1, $\ldots$, x_n. Hitachi had a good math model, which was used for the calculation of formula (39). However, the SN ratio above has the following weakness:

1. The most important quality problem is usually the nonlinearity of M as signal M changes. Formula (39) fails to evaluate the error called distortion in engineering fields.

2. The noise effect measured by the following:

$$\sigma_N^2 = \left(\frac{\partial f}{\partial x_1}\right)^2 \partial_{x_1}^2 + \cdots + \left(\frac{\partial f}{\partial x_n}\right)^2 \partial_{x_n}^2 \qquad (40)$$

considers only the first order of the effects of those noises and indicates a lack of consideration of higher-order effects of noises, especially interactions with signal M.

3. It is not easy to calculate the η_G, even when a mathematical model is available. It is practically impossible to conduct a series of experiments to obtain the value of the effects of noises, σ_N^2, when a math model is not available.

How to obtain a signal-to-noise ratio considering both distortion and total effect of noises in an efficient way? Finally the writer reached a conclusion. Analysis of variance technique developed in experimental design by R. A. Fisher may be used by changing its theoretical standpoint from statistical theory to quadratic expansion theory.

3.3   The Role of Orthogonal Array

Orthogonal arrays are used to find optimum design, assigning mostly main effects. The strategy of assigning main effects only requires the assumption that all interactions be zero. The existence of significant interaction is the result of poor scientific knowledge of the response function. Usually, engineers don't have enough knowledge of the nature of the response function, and they should conduct experiments to check their knowledge. Most of the direct quality characteristics, such as yield, percent defective, and reliability index, are useless without an appropriate calibration (adjustment) because main effects tend to be interactive. Engineers know that all functions are energy transformations. Orthogonal arrays are considered as testing devices to check the existence or nonexistence of significant interactions, detecting poor experimental design resulting from lack of engineering knowledge. Inspection is the method used to detect a poor-quality product. If the inspection cannot detect defective products, it is a useless inspection.

The use of orthogonal arrays in engineering fields is not fractional replication. It is essential to conduct experiments so that main effects will not be confounded with interactions. If significant interactions exist, optimization fails. The object of using an orthogonal array is not to allow poor design to keep going downstream. The writer often recommends that engineers use only orthogonal arrays L_{18}, L_{12}, L_{36} so that they can't assign interactions.

The following subsection introduces signal-to-noise ratios for dynamic function and presents a case study using an orthogonal array.

3.4   Signal-to-Noise Ratio Using Quadratic Expansion

To obtain an appropriate signal-to-noise ratio, we have to complete the following five steps:

1. Divide the system selected into modules so that each module has no feedback element. This is the most important step for dynamic function, requiring a deep engineering knowledge.
2. Identify the ideal function for each module. For example, the following simple function:

$$y = \beta M \qquad (41)$$

where M is input signal and y is output, is the ideal function in many engineering fields, such as information processing. This step includes the range of signal and of noise spaces to be studied.
3. Select several points from the signal-noise space at which data on measure selection step 4 are obtained, whether by computing or by conducting experiments.
4. Select the most effective design parameter, including output, y, when it is a continuous variable, to get the data.
5. Calculate an appropriate signal-to-noise ratio from the data.

These steps include engineering knowledge concerning each function of each module. There is no general way to discuss them. Therefore, we can discuss only one case study.

3.5   An Example of Parameter Design: Air Gauge Design

Consider an air gauge that is used to measure some dimension M without touching the specimen.

Such an air gauge has been used in many manufacturing processes. Before this design, an air gauge could measure a clearance up to only 220 μm. An air gauge that can measure a wider range of clearances was wanted for controlling dimensions having variation bigger than 220 μm.

An air gauge (see Fig. 4) to be used for a clearance two times wider was studied using experimental designs.

The height of the float y (cm) is calculated using aerodynamic theory as below. The terms are defined in Fig. 4.

$$G = \frac{CdA_2\sqrt{C_1 P_2 r_2}\left[(P_1/P_2^{K-1/K} - 1\right]}{\sqrt{1-(P_2/P_1)^{3/K}(A_2/A_1)^2}} \tag{42}$$

$$r = P/r_0 t \tag{43}$$

$$\Delta P = C_2 \frac{L}{r}\left(\frac{uG^7}{D^{19}}\right)^{0.25} \tag{44}$$

$$A_t = \left[c_4(\pi(D_t + 2R)M) - C_3 + \left(\frac{\pi D_1^2}{4}\right)^{-C_3}\right]^{-1/C_3} \tag{45}$$

$$A_c = (D_r^2(y) - D_c^2) \times \frac{\pi}{4} \tag{46}$$

$$D_{r(y)} = \sqrt{D_c^2 + V_y} \tag{47}$$

$$G = G_c + C_g = G_i + G = 1 \tag{48}$$

$$P_\alpha = \frac{P_\phi}{1 + C_5 G + C_6 G^2} \tag{49}$$

Eleven control factors (design parameters), all having three levels, were picked up, as shown in Table 4.

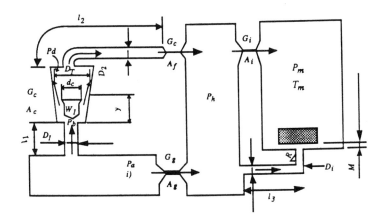

Figure 4   Air gauge diagram.

Table 4   Design Parameters

	Parameter	Unit	1st	2nd	3rd
A	Pressure P_ϕ	gf/cm^2	1733	1933	2533
B	Diameter D_1	cm	0.3	0.4	0.5
C	Diameter D_2	cm	0.4	0.6	0.8
D	Diameter D_3	cm	0.4	0.6	0.8
E	Diameter D_l	cm	0.2	0.3	0.4
F	Diameter D_e	cm	0.9	1.3	1.8
G	Weight W_f	gf	0.35	0.7	1.4
H	Shape Coeff. V	cm	0.007	0.014	0.028
I	Size A_f	cm^2	0.025	0.05	0.1
J	Curvature R	cm	0.01	0.02	0.04
K	Height M_0	cm	0.005	0.01	0.02

The signal factor, i.e., clearance M (cm), to be estimated by observing the height y was chosen to have five levels

$$M_1 = M_0, \qquad M_2 = M_0 + 0.011, \qquad M_3 = M_0 + 0.022,$$
$$M_4 = M_0 + 0,033, \qquad M_5 = M_0 + 0,044$$

Two noise factors, pressure variations N (g/cm^2) and room temperature R (K), are chosen as the most significant environmental conditions.

$$N_1 = -2\%, \qquad N_2 = \text{Nominal}, \qquad N_3 = +2\%,$$
$$R_1 = 290, \qquad R_2 = 293, \qquad R_3 = 296$$

The control factors are assigned to an orthogonal array L_{36}, and the signal and these noise factors are assigned to an outside three-way layout.

Measurements of height y were obtained by the aerodynamic theory for all $36 \times 5 \times 3 \times 3 = 1620$ combinations.

The data for the No. 1 combination of L_{36} are shown in Table 5.

The signal-to-noise ratio SN was obtained from the following analysis.

$$S_T = 18.22^2 + 25.99^2 + \cdots + 41.17^2 - CF \tag{50}$$
$$= 2873.1016, \qquad (f = 44)$$
$$S_\beta = \text{Linear effect of signal } M$$

Table 5   Orthogonal Array No. 1 Data

	M_1	M_2	M_3	M_4	M_5
$N_1 R_1$	18.22	25.99	32.58	37.20	39.80
R_2	18.22	25.99	32.57	37.19	37.78
R_3	18.21	25.98	32.56	37.17	39.76
$N_2 R_1$	18.51	26.41	33.13	37.85	40.52
R_2	18.77	26.40	33.12	37.84	40.50
R_3	18.77	26.39	33.11	37.82	40.49
$N_3 R_1$	18.78	26.79	33.64	38.47	41.20
R_2	18.77	16.79	33.63	38.45	41.19
R_3	18.77	16.78	33.62	38.44	41.17
Total	166.48	237.52	297.96	340.43	364.41

$$= \frac{(-2 \times 166.48 - 237.52 + 340.43 + 2 \times 364.41)^2}{9 \times 10}$$

$$= 2764.1279, \qquad (f = 1) \tag{51}$$

$$S_e = S_T - S_\beta = 108.9737, \qquad (f = 43) \tag{52}$$

$$V_e = \frac{108.9737}{43} = 2.5342 \tag{53}$$

Then, the SN ratio η was obtained from formula (54)

$$\eta = \frac{\frac{1}{\text{r.s.h.}^2}(S_\beta - V_e)}{V_e} \tag{54}$$

$$= \frac{\frac{1}{9 \times 10 \times 0.011^2}(27.64.1279 - 2.5342)}{2.5342}$$

$$= 100067 \tag{55}$$

$$= 50.1 \text{ (dB)} \tag{56}$$

Signal-to-noise ratios for all combinations of control orthogonal arrays (OA) are shown in Table 6. However, there were ten combinations where calibration was impossible as error variance became bigger than S_β.

So, SN ratios are classified into five categories:

1. $\eta \geq 59.0$
2. $59.0 \geq \eta \geq 54.0$

Table 6 Control OA and SN Ratios

	A 1	B 2	C 3	D 4	E 5	F 6	G 7	H 8	I 9	J 10	K 11	e 12	e 13	η_G	Categories
1	1	1	1	1	1	1	1	1	1	1	1	1	1	50.1	(3)
2	2	2	2	2	2	2	2	2	2	2	2	2	1	55.6	(2)
3	3	3	3	3	3	3	3	3	3	3	3	3	1	59.7	(1)
4	1	1	1	1	2	2	2	2	3	3	3	3	1	$-\infty$	(5)
5	2	2	2	2	3	3	3	3	1	1	1	1	1	$-\infty$	(5)
6	3	3	3	3	1	1	1	1	2	2	2	2	1	48.9	(4)
7	1	1	2	3	1	2	3	3	1	2	2	3	1	$-\infty$	(5)
8	2	2	3	1	2	3	1	1	2	3	3	1	1	$-\infty$	(5)
9	3	3	1	2	3	1	2	2	3	1	1	2	1	64.9	(1)
10	1	1	3	2	1	3	2	3	2	1	3	2	1	$-\infty$	(5)
11	2	2	1	3	2	1	3	1	3	2	1	3	1	59.2	(1)
12	3	3	2	1	3	2	1	2	1	3	2	1	1	47.2	(4)
13	1	2	3	1	3	2	1	3	3	2	1	2	2	47.9	(4)
14	2	3	1	2	1	3	2	1	1	3	2	3	2	46.2	(4)
15	3	1	2	3	2	1	3	2	2	1	3	1	2	54.7	(2)
16	1	2	3	2	1	1	3	2	3	3	2	1	2	45.9	(4)
17	2	3	1	3	2	2	1	3	1	1	3	2	2	$-\infty$	(5)
18	3	1	2	1	3	3	2	1	2	2	1	3	2	50.8	(3)
19	1	2	1	3	3	3	1	2	2	1	2	3	2	60.8	(1)
20	2	3	2	1	1	1	2	3	3	2	3	1	2	$-\infty$	(5)
21	3	1	3	2	2	2	3	1	1	3	1	2	2	57.3	(2)
22	1	2	2	3	3	1	2	1	1	3	3	2	2	55.6	(2)
23	2	3	3	1	1	2	3	2	2	1	1	3	2	50.1	(3)
24	3	1	1	2	2	3	1	3	3	2	2	1	2	59.2	(1)
25	1	3	2	1	2	3	3	1	3	1	2	2	3	49.3	(3)
26	2	1	3	2	3	1	1	2	1	2	3	3	3	53.4	(3)
27	3	2	1	3	1	2	2	3	2	3	1	1	3	47.8	(4)
28	1	3	2	2	2	1	1	3	2	3	1	3	3	51.9	(3)
29	2	1	3	3	3	2	2	1	3	1	2	1	3	61.2	(1)
30	3	2	1	1	1	3	3	2	1	2	3	2	3	$-\infty$	(5)
31	1	3	3	3	2	3	2	2	1	2	1	1	3	57.2	(2)
32	2	1	1	1	3	1	3	3	2	3	2	2	3	$-\infty$	(5)
33	3	2	2	2	1	2	1	1	3	1	3	3	3	47.0	(4)
34	1	3	1	2	3	2	3	1	2	2	3	1	3	55.5	(2)
35	2	1	2	3	1	3	1	2	3	3	1	2	3	48.4	(4)
36	3	2	3	1	2	1	2	3	1	1	2	3	3	$-\infty$	(5)

Table 7 Summary Data

Factor	Level	(1)	(2)	(3)	(4)	(5)	Total
$P_\phi A$	1	1	3	3	2	3	12
	2	2	1	2	2	5	12
	3	3	2	1	4	2	12
$D_1 B$	1	2	2	3	1	4	12
	2	2	2	0	4	4	12
	3	2	2	3	3	2	12
$D_2 C$	1	4	1	1	2	4	12
	2	0	3	3	3	3	12
	3	2	2	2	3	3	12
$D_3 D$	1	0	0	4	2	6	12
	2	2	3	2	3	2	12
	3	4	3	0	3	2	12
$D_l E$	1	0	0	2	6	4	12
	2	2	4	2	0	4	12
	3	4	2	2	2	2	12
$D_e F$	1	2	2	3	2	3	12
	2	1	3	1	4	3	12
	3	3	1	2	2	4	12
$W_f G$	1	2	0	3	3	2	12
	2	2	3	1	2	4	12
	3	2	3	2	1	4	12
VH	1	2	3	3	3	1	12
	2	2	3	2	3	2	12
	3	2	0	1	2	7	12
$A_f I$	1	0	3	2	2	5	12
	2	1	3	3	2	3	12
	3	5	0	1	4	2	12
RJ	1	3	1	3	1	4	12
	2	2	3	2	2	3	12
	3	1	2	1	5	3	12
$M_0 K$	1	2	2	4	3	1	12
	2	3	1	1	4	3	12
	3	1	3	1	1	6	12
Total		6	6	8	8	10	36

Table 8 Confirmation

	M_1	M_2	M_3	M_4	M_5
$N_1 R_1$	18.50	24.19	29.75	35.13	40.39
R_2	18.47	24.17	29.66	35.10	40.32
R_3	18.44	24.12	29.62	35.05	40.27
$N_2 R_1$	18.64	24.34	29.98	35.44	40.72
R_2	18.77	24.38	29.91	35.39	40.66
R_3	18.77	24.34	29.85	35.33	40.59
$N_3 R_1$	18.84	24.58	30.23	35.72	40.98
R_2	18.81	24.58	30.15	35.66	40.94
R_3	18.78	24.54	30.10	35.61	40.88
Total	167.73	219.24	269.25	318.43	365.75

3. $54.0 \geq \eta > 49.0$
4. $49.0 > \eta$
5. $\eta = -\infty$

The effects of control factors on the categories were estimated as shown in Table 7.

From Table 7, the optimum combination was chosen as A_3, B_3, C_1, D_3, F_3, G_3, I_3, J_3, K_1, and the confirmation calculation was done as shown in Table 8. The results show that the gain from choosing the optimum is $(69.24-55.6) = 14.62$ dB better than the one proposed first by the engineers and better by 4.34 dB than the best one among the 36 combinations.

4 CONCLUSION

Analysis variance and orthogonal arrays are two major techniques used in experimental designs. Parameter design has been used in the communication industry. In order to marry both methods, it was necessary to liberate the mathematical standpoint from mathematical statistics to quadratic expansion theory related to spectral analysis, naturally leading to use of signal-to-noise ratios as a functional quality measure. Traditional experimental designs consider that an orthogonal array may be used as fractional factorial experiments where no interaction exists. Quality engineering uses the array to check the reproducibility of the optimum obtained at laboratory conditions to downstream conditions.

Those two major points are the most significant differences from transitional experimental design. All other methodologies are not essential and were not mentioned here. See Reference 1.

REFERENCES

[1] Taguchi, G. *System of Experimental Design*, Unipub/Kraus International Publications, White Plains, NY and American Supplier Institute, Dearborn, MI, 1987.

[2] Phadke, M. S. and Taguchi G. "Selection of Quality Characteristics and S/N Ratios for Robust Design." *Conference Record, GLOBECOM '84 Meeting*, IEEE Communications Society, Atlanta, GA (Nov. 1984), pp. 1106–1113.

2

Designing Experiments in Research and Development: Four Case Studies

Karen Kafadar   Hewlett-Packard Laboratories, Palo Alto, California

1   INTRODUCTION

The title of this chapter will seem perfectly obvious to some and will greatly surprise others. Most statisticians recognize the value of experimental design as an efficient means of collecting data. These data serve to refine the hypothesized models of the measurement process and suggest further experiments, leading to an iteration in learning between induction and deduction as delineated by Box, Hunter, and Hunter (1978, p. 2). Production engineers in industry are learning to use experimental design to achieve quality control in the manufacturing process. But relatively few engineers see how experimental design can aid in what they believe is the most creative phase in the product cycle, namely the research and development process. Applications of these methods in R&D have not been published to anywhere near the extent that they have in the area of process improvement.

   This chapter strives to convince such nonbelievers by discussing four case studies that arose in connection with the author's consulting in an R&D laboratory at Hewlett-Packard Company. The first example, basic by intention, shows how decisions about instrument design result from efficient data collection. In the second example, instrument calibration is

35

facilitated by the use of factorial experiments. Graphical displays of main effects and interactions of the factors in the experiment lead to model refinements and thus greater accuracy of the calibration. Circuit design is the subject of the third and fourth examples. The simultaneous goals of minimizing variation and achieving target output levels, advocated by Taguchi and described by various authors (Hunter 1985; Kackar 1985; Taguchi and Wu 1980), assist in printed circuit board design in one case and integrated circuit manufacturing in the other. These examples provide convincing evidence that statistical methods and experimental design can, and should, be applied early in the product development phase to achieve better performance and tighter specifications.

2   POWER METER MEASUREMENTS

The first case study is a very straightforward example of how statistics can be used to save engineers many hours of potentially unfruitful design time. The goal of the data collection is to better understand the measurement process in the power meter, particularly as it is affected by short-term variation and long-term drift. Averaging reduces the effects of random noise, but it may introduce bias if the measurement process drifts. The optimal choice of averaging balances these two errors and is an important design consideration for high-precision instruments.

A design engineer for the HP 437A Power Meter needed a very accurate characterization of the instrument's measurement process to determine the feasibility of implementing 'user-friendly' features. One such feature under consideration was a 'user-defined uncertainty.' When a user measures the power in a signal which is subject to noise from various sources, different readings arise depending on the level of noise in relation to the signal strength (signal-to-noise ratio, or SNR). The user generally is interested not in the noise level but only in reducing the uncertainty of the signal power reading. When the noise can be characterized as *random*, as opposed to, say $1/f$ noise (Keshner 1982), the average of n measurements reduces its effect on the overall uncertainty in the final reading by the factor $\sqrt{n}$. Knowing the noise level, the user can achieve the desired uncertainty if the power meter has an option to vary this number n. But, since the main concern is a specified uncertainty, the design engineer wanted to offer an option which would allow the user to select the desired *uncertainty*, say U. Then the estimation of the noise could be incorporated into the software and the instrument would choose n accordingly. Can such a feature be implemented practically?

To answer this question, some estimates of short-term error and long-term drift are needed. The most likely source of these estimates is from a small sample of measurements. In the presence of only short-term random noise, the variance of the mean of a set of readings, say $X_1, \ldots, X_n$, is $\mathrm{Var}(\bar{X}) = \sigma_e^2/n$. But if the measurement process drifts, then there is a component of variation *between* sample means, e.g.,

$$\mathrm{Var}(\bar{X}) = \sigma_b^2 + \sigma_e^2/n. \tag{1}$$

(We will derive shortly the form of σ_b^2 in the presence of a constant drift.) In this case there is a 'law of diminishing return' in additional averaging, and there is a value of n, say n_b, at which σ_b^2 dominates the variation (making further averaging ineffective in reducing the uncertainty in the result). In addition, about twice the standard deviation of $\bar{X}$ should be less than the specified uncertainty U. So the ideal number of measurements, n^*, satisfies

$$\begin{aligned} n^* &< n_b && \text{(for low bias in } \bar{X}) \\ n^* &\geq \sigma_e^2/(U^2/4 - \sigma_b^2) && \text{(for low variance in } \bar{X}). \end{aligned} \tag{2}$$

Estimates of σ_e^2 and σ_b^2 or, more conservatively, upper and lower confidence limits, would replace the parameters in (2) for the determination of n^*. This optimal choice of n is probably the simplest instance of the familiar variance versus bias trade-off in problems of estimation. Other instances are the degree of smoothing an ordered sequence, the bandwidth parameter in density estimation, and the type of filtering in spectrum estimation.

Since the meter must be able to measure power in incoming signals of various strengths, the instrument is designed to operate in one of five ranges depending on their absolute power levels. To measure low power (range 1), signals must be amplified, which of course amplifies the noise as well. Thus the uncertainty in the reading tends to be greater at lower power, but not necessarily, since, for the same power, different signals could have different levels of noise. Therefore, the components of the noise would have to be estimated for each and every incoming signal, regardless of the range which should be used to measure its power.

Because of hardware differences in the different ranges, the between-sample variance component may depend on the range. Within a given range, estimates of σ_e^2 and σ_b^2 can be obtained by taking N samples of measurements of size n. Let X_{ij} be the ith measurement ($i = 1, \ldots, n$) in the jth sample ($j = 1, \ldots, N$), $\bar{X}_j$ be the sample mean of the sample mean of the jth sample, and $\bar{X}$ be the overall sample mean. Then σ_e^2 can be estimated as the pooled within-sample variance

$$\hat{\sigma}_e^2 = s_w^2 = \sum_{j=1}^{N}\sum_{i=1}^{n}(X_{ij} - \bar{X}_j)^2/[N(n-1)] \tag{3}$$

and σ_b^2 as the difference between the between-sample variance

$$s_b^2 = \sum_{j=1}^{N}(\bar{X}_j - \bar{X})^2/(N-1)$$

and s_w^2/n; i.e.,

$$\hat{\sigma}_b^2 = s_b^2 - s_w^2/n. \tag{4}$$

In the presence of random noise only, $\sigma_b^2 \approx 0$, and so the ratio ns_b^2/s_w^2 should be about 1. (A formal test based on the F-distribution optimistically assumes underlying Gaussian noise; often mere inspection suffices.)

If this ratio greatly exceeds 1, the next step is to quantify, with some confidence, this between-sample variance component. Using the estimator in (4), an optimistic (Gaussian) variance of it can be derived from:

$$\mathrm{Var}(s_w^2) = 2\sigma_e^4/[N(n-1)]$$
$$\mathrm{Var}(s_b^2) = 2(\sigma_e^2/n)^2/(N-1)$$
$$\mathrm{Var}(\hat{\sigma}_b^2) = \mathrm{Var}(s_b^2) + \mathrm{Var}(s_w^2/n) - 2\,\mathrm{Cov}(s_b^2, s_w^2/n) \tag{5}$$
$$= 2(\sigma_e^2/n)^2/(N-1) + 2\sigma_e^4/[n^2(n-1)N]$$
$$\approx [2\sigma_e^4/n^2(N-1)][1 + 1/(n-1)] = 2\sigma_e^4/[n(n-1)(N-1)].$$

(Further properties of this estimator, and other estimators of the variance component σ_b^2, are found in Searle 1970, Sections 9.5, 11.1.)

The estimation of the within and between variance components was carried out on the most sensitive (1) and least sensitive (5) ranges; for added replication, five sets of 100 samples were taken at each sample size n for range 1. (Hardware design considerations encourage values of n that are powers of 2; here, $\log_2 n = 2, 3, \ldots, 9$). Figure 1 plots the ratios ns_b^2/s_w^2 in range 1, which strongly suggest that σ_b^2 is far from zero at all sample sizes. Fortunately, the estimates for both σ_e^2 and σ_b^2 remain fairly constant

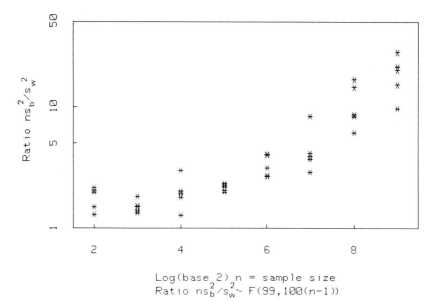

Figure 1 Ratio of ns_b^2/s_e^2, between- to within-sample variances, from 100 samples of size n for a signal measured with the HP 437A Power Meter (range 1). The departure of these ratios from unity indicates the presence of variation between sample means beyond that due to random noise.

for these values of $n = 4, 8, \ldots, 512$ (Figure 2). (Another set of estimates of σ_b^2 and σ_e^2 may be obtained as the intercept and slope of the fitted line to the pairs of points $(1/n, s_b^2)$. These estimates may be less variable than those in (3) and (4) if in fact neither σ_b^2 nor σ_e^2 depends on the sample size n; otherwise, the estimates will be based on untenable assumptions.)

What kind of model for the measurement process is consistent with these observations? The simplest model assume that it drifts at a rate of B units per unit measurement time; i.e.,

$$X_{ij} = A + Bt_{ij} + e_{ij}$$

where the measurement X_{ij} occurs at time $t_{ij} = i + n(j-1)$, and e_{ij} is the associated random noise component having mean 0. Under such a model, s_w^2, the average of the within-sample variances, is expected to increase as

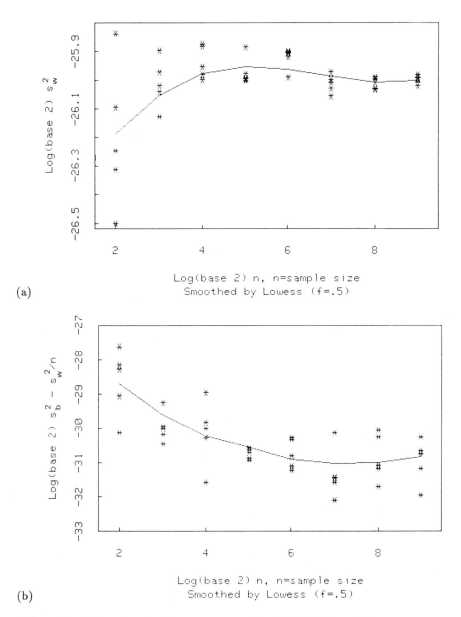

(a)

(b)

Figure 2   Estimates of variance components in HP 437A Power Meter measurement process. Measurements in 100 samples of size n, taken on range 1 (most sensitive, 0.001–0.01 mW). (a) s_w^2 = Averaged (over 100 samples) within-sample variance ($\log_2$ scale). (b) Between-sample variance component = $s_b^2 - s_w^2/n$ ($\log_2$ scale)

the sample size n increases, since

$$E\left[\sum_{i=1}^{n}(X_{ij} - \bar{X}_j)^2/(n-1)\right]$$

$$= E\left[\sum_{i=1}^{n}B^2(t_{ij} - \bar{t}_j)^2\right]\Big/(n-1)$$

$$+2BE\left[\sum_{i=1}^{n}(t_{ij} - \bar{t}_j)(e_{ij} - \bar{e}_j)\right]\Big/(n-1)$$

$$+E\left[\sum_{i=1}^{n}(e_{ij} - \bar{e})^2\right]\Big/(n-1)$$

$$= [B^2/(n-1)]\left[\sum_{i=1}^{n}i^2 - n((n+1)/2)^2\right] + 0 + \sigma_e^2$$

$$= n(n+1)B^2/12 + \sigma_e^2.$$

If the drift parameter B is very small, then the first term will be dominated by σ_e^2. But B may be noticeable in the between-variance component, because the expected value of the between-sample variance s_b^2 is

$$E(s_b^2) = n^2 B^2 N(N+1)/12 + \sigma_e^2/n,$$

so

$$E(\hat{\sigma}_b^2) \approx (nNB)^2/12.$$

Then a plot of $\log_2 \hat{\sigma}_b^2$ versus $\log_2 n$ should be approximately linear with slope $= 2$, and the drift parameter B could be estimated via a multiple of the intercept.

Unfortunately, Figure 2b is not consistent with the constant-drift model: it appears that σ_b^2 actually *decreases* as n increases to about 16 and then remains fairly constant. This suggests that the drift is not constant from the first sample onward but rather diminishes over time.

Furthermore, the within-sample variances are not consistent with Gaussian variation. The sample variance of the five estimates of σ_e^2 at a given sample size is somewhat more than that predicted by the Gaussian variation (5) (see Table 1).

These results offer both good and disappointing news to the engineer. First, range 1 frequently has the lowest signal-to-noise ratio, so the absence

Table 1 Comparison of Actual and Gaussian
variance of s_w^2

n	Gaussian* (2.5)	Actual* (5 reps)	Ratio (Actual/Gaussian)
4	1.153	5.377	4.66
8	0.627	0.745	1.19
16	0.323	0.375	1.16
32	0.149	0.272	1.83
64	0.0784	0.161	2.05
128	0.0344	0.107	3.11
256	0.0172	0.0388	2.26
512	0.0087	0.0254	2.92

*All variances multiplied by 10^{10}.

of a long-term drift is encouraging. However, the time to achieve this steady state may be unreasonably long. In addition, the variability of the estimates above and beyond that predicted by Gaussian variation implies that the optimistic Gaussian variance (5) cannot assure the degree of confidence in σ_e^2 which is needed to determine n^* for the required uncertainty (2).

The engineer repeated the measurement process on a signal in power range 5 (generally higher signal-to-noise ratio). Only one replication at each sample size was run, and so standard errors on the estimates $\hat{\sigma}_e^2$ are estimated using the optimistic Gaussian assumption. These estimates of the components of variation at each sample size are shown in Figure 3. In panel a, the estimates $\hat{\sigma}_e^2$ actually *increase*, suggesting a drift may be affecting the measurement process. However, because the between-sample variance component $\hat{\sigma}_b^2$ *decreases* with increasing n (panel b), the drift may not be constant, but may slowly level off in this range as well. For unexplained reasons, between-sample variation increases dramatically when $n = 256$ or $n = 512$.

The analysis of these data suggests that the uncertainty in the resulting averaged power reading depends upon the nature of the drift which affects the measurement process. Time plots of the data reveal that the noise level on the signal, the range on which it is measured, and environmental conditions all contribute to this drift. Because the variation in the process may be substantially different in the different ranges, too many measurements would be required to estimate it precisely to guarantee the desired level of uncertainty. The recognition of the impracticality of the feature

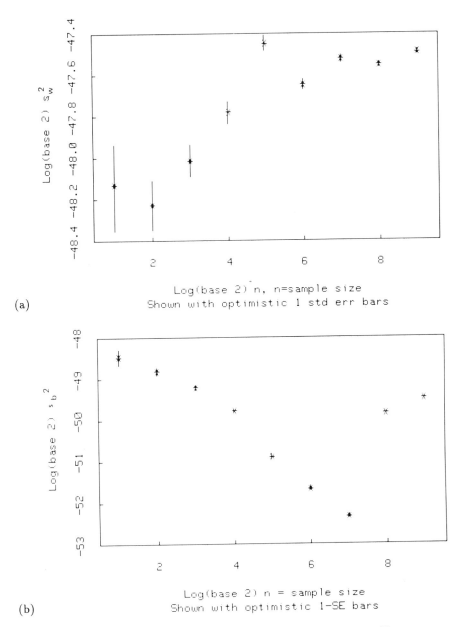

Figure 3 Estimates of variance components in HP 437A Power Meter measurement process. Measurements in 100 samples of size n, taken on range 5 (maximum power levels, 10–100 milliwatts). Single estimate at each sample size n is shown with optimistic (Gaussian) one standard error bars. (a) s_w^2 = Averaged (over 100 samples) within-sample variance ($\log_2$ scale). (b) Between-sample variance component = $s_b^2 - s_w^2/n$ ($\log_2$ scale).

saved the engineer many hours of fruitless design time. Rather, the design effort concentrated on reporting associated standard errors with average values, and on educating the customer on the appropriate selection of n to suit the application.

3   MODEL CONFIRMATION AND ANALYSIS

An electrical engineer typically relies on a mathematical model to predict the performance of a circuit design. Factorial experiments are invaluable for confirmation of the model and often suggest improvements in the form of more accurate and simpler designs.

Such experiments proved useful in calibrating the HP 8981A Vector Modulation Analyzer. This instrument acts as both a vector signal generator and a demodulator in the following way. Consider an ideal signal at a specified frequency ω_0:

$$
\begin{aligned}
s(t) &= A\cos(\omega_0 t - \phi_t)\\
&= A\cos(\phi_t)\cos(\omega_0 t) + A\sin(\phi_t)\sin(\omega_0 t)\\
&\equiv I_t\cos(\omega_0 t) + Q_t\sin(\omega_0 t).
\end{aligned}
$$

The HP 8981A then 'demodulates' the signal into two components, I_t and Q_t, respectively known as the 'in-phase' and 'quadrature' components. The generation of this signal is called vector modulation, because at any instant it can be represented as a point (I_t, Q_t) in a Cartesian coordinate system, and the part of the instrument which generates the components is called a vector demodulator or analyzer. For the signal $s(t)$ above, a plot of Q_t versus I_t for varying times t is a series of points around a circle of radius A.

For purposes of testing satellite receivers, radar transmitters, or other electronic equipment, sometimes vector modulation is distorted intentionally by errors in the demodulation; e.g.,

$$
\begin{aligned}
s(t) &= [A\cos(\phi_t) + I_0]\cos(\omega_0 t) + [A\sin(\phi_t + \phi_0) + Q_0]\sin(\omega_0 t)\\
&= (I_t + I_0)\cos(\omega_0 t) + (\sin\phi_0 I_t + \cos\phi_0 Q_t + Q_0)\sin(\omega_0 t)
\end{aligned}
$$

where $I_t = A\cos(\omega t)$, $Q_t = A\sin(\omega t)$ as above. Here, I_0 and Q_0 represent offsets in the channel (signals at 0 frequency), and ϕ_0 is the error in perfect $90°$ demodulation ('quadrature error'). Then the demodulated components can be expressed via matrices in terms of the ideal demodulation as:

$$\begin{pmatrix} I_t^* \\ Q_t^* \end{pmatrix} = \begin{pmatrix} 1 & 0 \\ \sin\phi_0 & \cos\phi_0 \end{pmatrix} \begin{pmatrix} I_t \\ Q_t \end{pmatrix} + \begin{pmatrix} I_0 \\ Q_0 \end{pmatrix}. \tag{6}$$

A plot of Q_t^* versus I_t^* is now an ellipse centered at the offsets (I_0, Q_0). In fact, when the instrument is *not* calibrated for zero demodulation error, such a pattern in the vector components will be evident.

To increase the speed in signal generation, it is more efficient to design the instrument digitally and convert the digital counts to analog voltages by means of a digital-to-analog converter (dac) which has sufficient resolution (here, 12 bits, allowing 4096 counts of resolution, for the offset dacs, and 11 bits for the quadrature dac). Naturally the demodulation errors are specified in analog units of offset voltage and degrees of quadrature error. The question then becomes: What is the relationship between the offset voltages (or degrees of quadrature error) and their corresponding dacs, so that the instrument can be calibrated accurately for zero demodulation error?

This problem relies on a combination of both linear regression and experimental design. The need for linear regression is as follows: Suppose the ideal signal is generated with a given frequency so that the 'points' (I_t, Q_t) along a circle are easily identified. (For example, the phase ϕ_t can take on one of the eight values $0°, 45°, \ldots, 315°$, so that (I_t, Q_t) lie in one of eight equally spaced places on the circle, $45°$ apart. They are shown as 'x's in Figure 4.) The dacs corresponding to the I and Q channel offsets and the quadrature error are set to a certain value. Then, the measured (I_t^*, Q_t^*) in the two channels should be related to the ideal points (I_t, Q_t) via the equations in (6). Regressing I_t^* on I_t and Q_t^* on I_t and Q_t yields estimates of I_0, Q_0, and ϕ_0. (Notice that $\cos\phi_0$ can be factored from the matrix in (6), yielding an estimate of $\tan\phi_0$. The scale difference can be modelled into the relationship between dac counts and regression coefficients.) In addition, the residuals from these regressions can be examined for possible systematic departures from the model, indicating differences between the theoretical model and the hardware designed to implement it.

Now, this procedure can be repeated for various settings of the three dacs corresponding to I_0, Q_0, and ϕ_0, and here is where the concepts of experimental design arise. Ideally, there is some monotonic relationship between the dac counts and the regression coefficient for that dac. By varying the dac settings factorially, this relationship can be identified, and

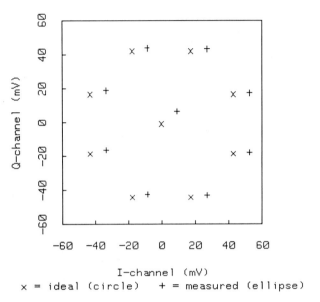

Figure 4   Demodulated (in-phase I, and quadrature Q) components of an eight (equally spaced) phase modulated signal, both without ($\times$) and with ($+$) errors in demodulation.

possible interactions among the settings can be explored. Each of the three dacs is set at one of, say, k settings, and the regression coefficients corresponding to (I_0, Q_0, ϕ_0) from each of the k^3 regressions are related to them.

Ideally, the relationship between each dac setting and its corresponding regression coefficient is linear, but this assumption may be optimistic. Thus, $k = 4$ or 5 settings on each of the three dacs were selected for two separate experiments: one in which the dac settings are fairly close together (local settings, within 10% of the calibrated settings), and the other in which the dac settings span nearly their entire range (global settings, within 80% of the calibrated settings).

Figure 4 shows eight measured pairs (I_t^*, Q_t^*), $t = 1, \ldots, 8$. (Actually, I_t^* and Q_t^* are the averages of five replications each, but only their averages are reported.) These measured pairs lie on an off-centered ellipse and are denoted with '$+$', and the ideal pairs on the circle are denoted with '$\times$'. Linear regression of I_t^* on I_t and Q_t^* on I_t and Q_t as described above yields the following estimates:

$$\hat{I}_0 = 9.182 \text{ mV}, \qquad \hat{Q}_0 = 6.591 \text{ mV}, \qquad \hat{\phi}_0 = -15.28°.$$

The actual dac settings corresponding to these parameters when the data were measured are:

$$I\text{-channel offset} = 3200, \qquad Q\text{-channel offset} = 3200,$$

$$\text{quadrature} = 1600.$$

By varying these dac settings factorially, one may determine the relationship between dac setting and the coefficient which is specified in more familiar units.

To establish this relationship, regression estimates were obtained for each combination of the following dac settings:

dac	'Local' (5 levels)	'Global' (4 levels)
I-offset	1449(100)1849	800(800)3200
Q-offset	1543(100)1943	800(800)3200
quadrature	831(50)1031	400(400)1600

Figure 5a is a plot of the regression coefficients I_0 as a function of the dac setting corresponding to the offset in the I channel. While there is some variability in the regression coefficients, there is a nearly linear relationship between the coefficient (in millivolts) and dac setting. The same is true for the regression coefficients corresponding to offset in the Q channel and quadrature (panels b,c).

A pairwise plot of the regression coefficients (Figure 6) gives an idea of the variability in the measured data, as manifested in the variability of the regression coefficients. When the demodulation errors are relatively small (panel a, 'local settings') the regression coefficients are fairly consistent, as shown by the relatively 'clean' patterns in the pairwise plots. The patterns are not nearly so distinct when the demodulation errors are large (panel b, 'global settings'), presumably because of the increased variability in the measurements which are taken on a highly noncalibrated instrument.

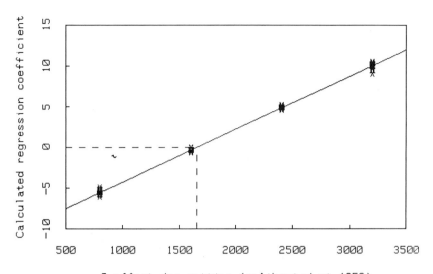

(a)

I offset dac setting (calibrated at 1652)

Regression coefficient ~ -l0.664 + .006456(I dac setting)

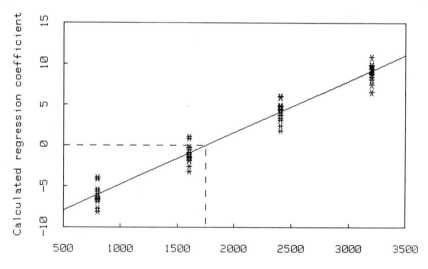

(b)

Q offset dac setting (calibrated at 1750)

Regression coefficient ~ -11.07 + .00632(Q dac setting)

Figure 5 Relationship between state settings of digital-to-analog converters (dac) and corresponding regression coefficients in model for HP 8981A Vector Modulation Analyzer; global dac settings (a) offset in I channel: *regression coefficient* $\approx -10.66 + 0.00646 \times I$-*offset dac setting*. Calibrated setting = 1652. (b) offset in Q channel: *regression coefficient* $\approx -11.07 + 0.00632 \times Q$-*offset dac setting*. Calibrated setting = 1750. (c) quadrature offset: *regression coefficient* $\approx -15.94 - 0.01706 \times quadrature\ dac\ setting$. Calibrated setting = 933.

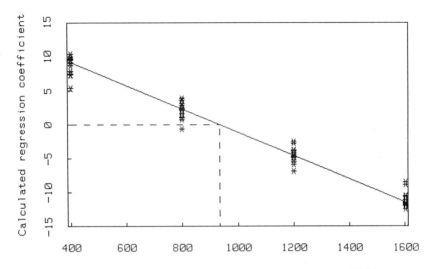

Quadrature dac setting (calibrated at 933)

(c)   Regression coefficient ~ -15.939 - .01706(quad dac setting)

Another benefit of the factorial design is the assessment of relationships *among* the coefficients themselves. For all k^2 pairs of offset dac settings (I and Q) and a single quadrature dac setting (e.g., 831), the k^2 computed coefficients for $\hat{\phi}_0$ should all be approximately the same. Then a perspective plot of (I dac setting, Q dac setting, $\hat{\phi}_0$), where $\hat{\phi}_0$ is the height of the surface, should be relatively flat. Two such plots are shown in Figure 7, one for the 'local' experiment (panel a) and one for the 'global' experiment (panel b). The data at $I = 1943$, $Q = 1449$ in Figure 7a result in highly unusual coefficients, not only for $\hat{\phi}_0$ shown here but also for $\hat{I}_0$ and $\hat{Q}_0$ (not shown). Apart from this outlier, there appears to be some nonlinear trend in this surface when the quadrature dac setting is *below* its calibrated setting. Specifically, the regression coefficients for $\hat{\phi}_0$ are *smaller* when I- and Q-offset dac settings are both *above* or both *below* their calibrated settings, and *larger* when one setting is *above* and the other is *below* its calibrated setting. This nonlinearity is more obvious from the results of the 'global' experiment, as in Figure 7b. Depending on the magnitude of the effect of this nonlinearity on instrument performance, corrections may be investigated in the hardware (using different circuitry) or the software.

This model provided an accurate description of instrument operation, leading to greater accuracy in its calibration. This in turn is passed on to

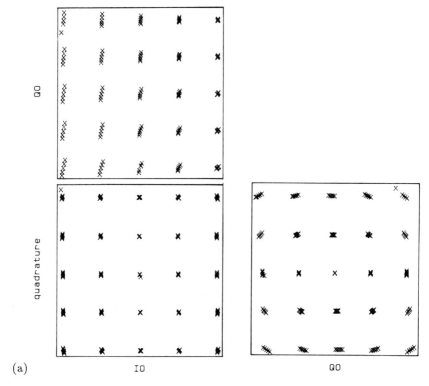

Figure 6 Pairwise plots of regression coefficients corresponding to demodulation errors in model for HP 8981A Vector Modulation Analyzer. (*a*) Dac settings within 10% of calibrated state. (*b*) Dac settings within 80% of calibrated state.

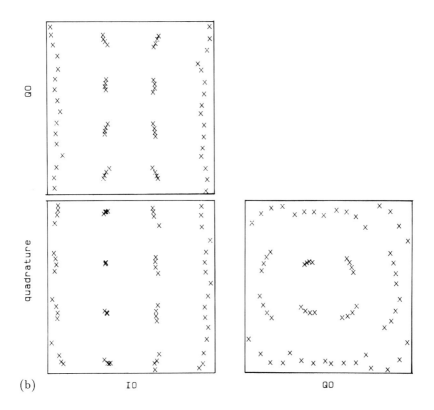

(b)

Local data

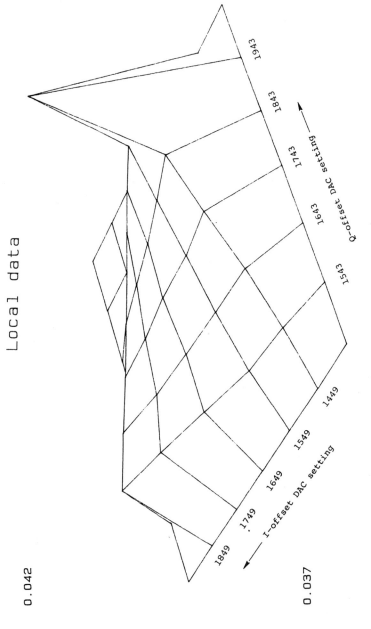

1943

1843

1743

Q-offset DAC setting

1643

1543

25 quadrature estimates when quad DAC set at 831

I-offset DAC setting

1449

1549

1649

1749

1849

0.042

0.037

(a)

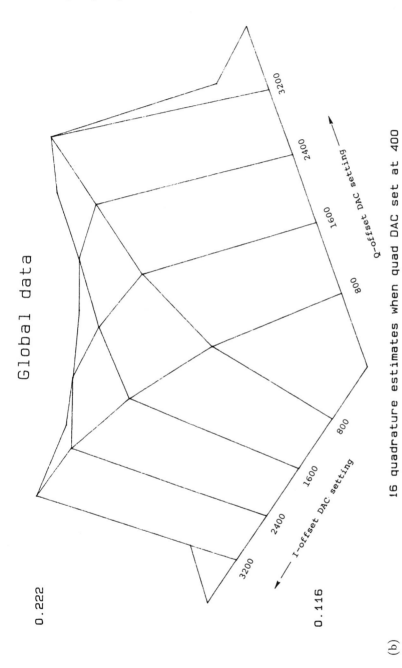

Global data

0.222

0.116

(b)   16 quadrature estimates when quad DAC set at 400

Figure 7 Perspective plots of regression coefficients corresponding to offset demodulation errors for a given quadrature error. Ideally, regardless of settings in offset dacs, all regression coefficients corresponding to the (single) quadrature error should be the same. (*a*) Offset dacs settings within 10% of calibrated state, quadrature dac setting = 831 (< 933 when calibrated). (*b*) Offset dacs settings within 80% of calibrated state, quadrature dac setting = 400 (≪ 933 when calibrated).

the users who rely on it to test their electronic systems. In practice, the HP 8981A calibration algorithm was complicated by the need for nonlinear regression and additional parameters; some of these aspects are described in Kafadar 1988.

4 PRINTED CIRCUIT BOARD DESIGN

Optimal design of printed circuit boards, used in instruments, computers, and most modern electronic equipment, involve many considerations such as target output levels, meeting specifications, and manufacturability. Several authors (e.g., Godfrey 1986, Gunter 1987) have credited Taguchi with proposing that such considerations be taken into account when the product is in its earliest design stage, as only limited improvements can be achieved once it is in production. The circuit designer is faced with the problem of choosing the design parameters for certain circuits to meet these goals. Often there are several choices, all of which may meet target output levels. But some choices may be preferable when issues of variability and manufacturability are considered.

Such a situation arises in the design of the HP 86792A Agile Upconverter, a fast-switching synthesizer that can upconvert complex broadband modulation anywhere in the 10–3000 MHz range. The upconverter includes many printed circuit boards which were designed to meet the simultaneous goals of prescribed target levels and minimum variation. Figure 8 shows a typical path of power through a series of circuits for one of the circuit boards.

While there are many different power paths in this complex instrument, Figure 8 provides an example on which the basic methodology for choosing parameters is illustrated. Input power is received from a previous power path and is split into two paths, one of which is shown here to include an amplifier, a switch, and a low-pass filter. Since this final switch is used to drive another circuit (a mixer), its output level is critical: if it is too high, spurious power in this mixer degrades instrument performance; if it is too low, then the signal power will not be sufficiently high relative to the noise (low signal-to-noise ratio). Optimally, the output power should be within a 3 dB window (± 1.5 dB), regardless of output frequency of the signal, temperature, or components used to assemble the circuit.

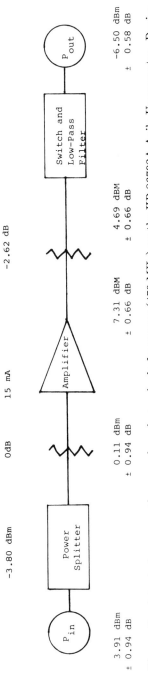

Figure 8 A typical power path on one board at a single frequency (470 MHz) on the HP 86792A Agile Upconverter. Device parameters shown above circuits; nominal powers levels ± one standard deviation shown between and below circuits. For amplifier and switch response curves, see Figures 10 and 11.

Some circuits in Figure 8, such as the power splitter and low-pass filters, involve few if any design choices. The response of the amplifier, however, is governed by (1) gain, or how much additional power is in the output per dB of input power, and (2) compression, or maximum output power of the amplifier. An equivalent characterization (2) is often given by (3) point of compression, or the input power level at which the amplifier's output deviates from linearity (due to reduced gain) by 1 dB. Figure 9 shows three typical amplifier curves which illustrate these two parameters. In addition to these two parameters, various values may be assigned to pads in the circuitry to absorb excess power.

It is clear that more than one combination of parameter choices will meet the target output level. Previously, convenient component values were assigned, and if the switch output did not meet its target power level, then the pads before and/or after the circuit were adjusted accordingly. These adjustments are very costly in production, since there are many checkpoints on many boards for each instrument. A better approach is to choose a single set of design parameters, one for which the output variation due to incoming power variation is minimized. This output variation can be controlled through intelligent selection of design parameters. Then the bulk of the 3 dB window of uncertainty will be due to noise that is beyond our control (temperature, components, etc.).

The simple principles of transmission of error through a circuit provides the greatest leverage in reducing overall output power. Figure 10 shows a response curve from a typical switch, $S(P)$, as a function of incoming power. Clearly the output variation transmitted around P_0, where the slope is relatively steep, is larger than that around P_1, where the curve is quite flat [i.e., $S'(P_0) > S'(P_1)$]. Taylor's theorem provides the basis for this simplest of the 'laws of propagation of error' (Ku 1966):

$$P_{out} = S(P_{in}) \approx S(P_0) + (P_{in} - P_0) \times S'(P_0)$$

$$\implies standard\ deviation\ (P_{out}) \approx |S'(P_0)| \times standard\ deviation\ (P_{in}).$$

Examples indicate that approximations to the standard deviation such as those used above are in fact quite satisfactory (Tukey 1961).

For other engineering reasons, the switch in this particular power path is constrained to operate in the linear (nonconstant) region of the response curve, so $S'(P_0)$ is fixed. The amplifier, on the other hand, has no such constraint on its region of operation. However, only a certain level of

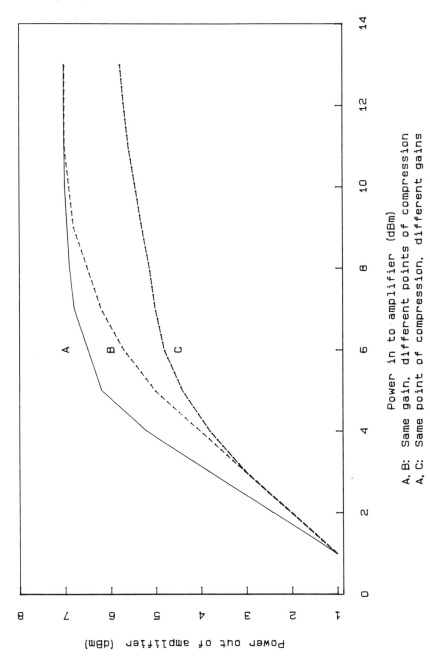

Figure 9 Response curves from three 'typical' amplifiers. B, C have the same gain, but different points of compression. A, C have the same point of compression, but different gains.

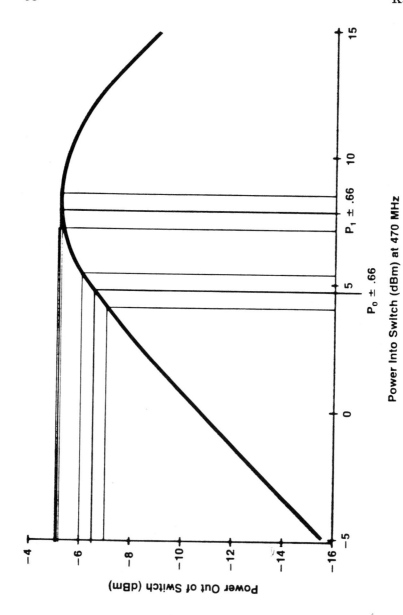

Figure 10 Typical switch response curve at 470 MHz, showing input power at $P_0 \pm 0.66$ to achieve target output of -6.5 ± 0.7 dBm. Copyright © 1988 Hewlett-Packard Journal; reprinted by permission.

input power is available to it due to losses in power from the preceding circuits. Referring to Figure 8, the maximum input power to the amplifier is 0.11 dBm $\pm$ 0.94 dB. The target switch output is $P_{out} = -6.5$ dBm $= S$ (4.69 dBm) with a 99% confidence interval of ± 1.5 dB. This ± 1.5 dB variation translates into $1\sigma = 1.5$ dB/2.576 $= 0.58$ dB and must account for variability due to switch components (estimated as 0.3 dB) as well as that due to power ($\sigma_{P_{out\ of\ S}} = 0.50 = [(0.58)^2 - (0.30)^2]^{1/2}$, using an assumption of independence between the errors in the components and in the input power levels). Since the slope at the target switch output $S'(4.69) = 0.76$,

$$\sigma_{P_{out\ of\ S}} = 0.76\sigma_{P_{into\ S}} \Longrightarrow \sigma_{P_{into\ S}} = 0.50/0.76 = 0.66 \text{ dB} = \sigma_{P_{out\ of\ A}}$$

where $A(\cdot)$ represents the response of the amplifier. Now, for the amplifier curve,

$$\sigma_{P_{out\ of\ A}} = |A'(P_{in\ amp})| \times \sigma_{P_{into\ A}}$$
$$\Longrightarrow 0.66 \text{ dB} = |A'(P_{in\ amp})| \times 0.94 \text{ dB}$$
$$\Longrightarrow |A'(P_{in\ amp})| \leq 0.70.$$

This puts a limit on the maximum slope around the region of operation for the amplifier.

Figure 11 shows a family of amplifier curves with maximum allowable gain and three values of the collector current which have the effect of varying the point of compression. The switch needs at least 4.69 dBm to achieve its target output of -6.5 dBm. (If the amplifier output exceeds 4.69 dBm, the loss of the pad before the switch can be increased to absorb the excess power.) In addition to meeting the target power, the previous paragraph puts a limit on the slope of the amplifier curve not to exceed 0.70. Only the lowest (15 mA) curve satisfies both the incoming power constraint and the maximum slope constraint. Then:

$$P_{in} = 0.11 \text{ dBm} \pm 0.95 \text{ dB}$$
$$\longrightarrow \text{ Amplifier } A(P_{in}) = 7.31 \text{ dBm} \pm 0.66 \text{ dB}$$
$$\longrightarrow \text{ Pad Loss} = -2.62 \text{ dBm} \longrightarrow 4.69 \text{ dBm} \pm 0.66 \text{ dB}$$
$$\longrightarrow \text{ Switch } S(4.69) = -6.5 \text{ dBm} \pm 0.58 \text{ dB}.$$

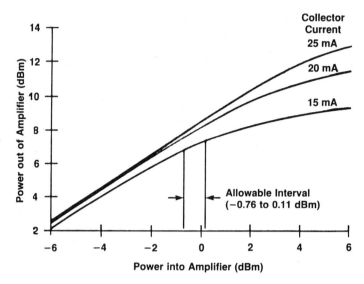

Figure 11 Amplifier response curves for three values of collector current at 470 MHz. Copyright © 1988 Hewlett-Packard Journal; reprinted by permission.

This same analysis is applied to the power paths on several boards during the prototype phase of instrument design.

The effect of component variation on these circuits can be assessed in one of three ways: (1) manufacturer specification (the believer); (2) actual measured data (the skeptic); and (3) simulation (the pragmatist). The present project applied both (1) and (2) (reference was made earlier to component variation estimated at 0.3 dB) and has been discussed more fully in Kafadar and Plouse 1988. Various authors confirm Taguchi's recommendation of using simulation (e.g., Phadke 1986); modern computational tools make such a computer-intensive analysis feasible. Realistic ranges for the component variation must be assigned, and then the circuit model with the previously determined parameters can be simulated with component variability as a noise factor. Examples of this approach are given in Taguchi and Wu 1980, Phadke 1986, and Kackar 1985.

Circuit performance on this prototype design stage compared favorably with performance of units in the first production run. The result of applying this methodology to these circuit boards is the elimination of switch adjustments. Because of a more reliable and cost-effective design, printed circuit board manufacturing is simplified, and assembly time in production is greatly reduced.

5   INTEGRATED CIRCUIT MANUFACTURING PROCESS

The final example arises in integrated circuit design. Since these circuits have extremely small dimensions and tight tolerances, parameter specification is critical. The manufacturing process for these circuits involves photoresist deposit, mask of pattern, film exposure, and final etching of many circuits on a single wafer. The wafer is subjected to various processing conditions (temperatures, pressures, feed rates), or 'noise factors,' which contribute to variation in circuit performance beyond that caused by the circuit parameters themselves. Values of those parameters which affect the signal-to-noise ratio can be selected to maximize this ratio, and values of those parameters which affect the 'signal' can be adjusted to attain target output levels.

Figure 12 is a diagram of an active circuit which is used to simulate a resistance in integrated circuit manufacturing. The physical situation which corresponds to this process involves depositing layers of chemicals (e.g., boron, phosphorous) having defined resistance (e.g., R_s, R_d) onto a substrate layer (e.g., silicon). The interface between the two surface layers has length L and width W. The resistivity of this interface affects the action of the electrons from the upper to the lower surfaces and is governed by the parameter β (which itself is expressed in terms of another parameter r). R_s and R_d are the resistances of the upper and lower surfaces, respectively, and R_c is the output resistance of the circuit. The goal of this analysis is the optimal choices for R_s, R_d, L, and r (W is fixed by other constraints) so that R_c is small and has minimum variation.

This preliminary analysis was conducted completely via simulation. Three levels of each of the four design parameters were considered:

R_s	30 ± 5	ohms/square
R_d	5 ± 1	ohms/square
r	150 ± 50	ohms
L	0.5 ± 0.1	microns
W	1.5	microns

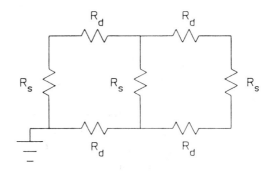

Output:

$$R_c = \frac{2R_sR_d + (R_s^2 + R_d^2)\cosh(\beta L)}{\beta W(R_s + R_d)\sinh(\beta L)} \quad,$$

where:

$$\beta = \sqrt{(R_s + R_d)/\rho}$$

Physical situation:

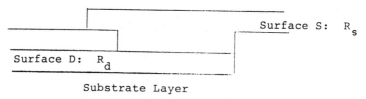

Parameters and Initial values:

R_s	resisitance of S surface	30 ± 5 ohms per square
R_d	resisitance of D surface	5 ± 1 ohms per square
L	Length of interface between surfaces S and D	0.5 ± 0.1 microns
r	Resistance across interface	150 ± 50 ohms
W	Width of interface between surfaces S and D	1.5 microns (fixed)
R_c	output resistance of circuit	ohms per square

Figure 12 Circuit design used to simulate a resistance in integrated circuit manufacturing.

The resulting resistance is given in terms of these parameters via the equation

$$R_c = [2R_s R_d + (R_s^2 + R_d^2) \cosh(\beta L)]/[\beta W (R_s + R_d) \sinh(\beta L)],$$
$$\beta = [(R_s + R_d)/\rho]^{1/2}.$$

This equation is calculated when each parameter is at one of its nominal values *plus* some error (presumably due to component variation or other noise source). The ideal combination of parameters is that for which the variation in R_c due to the error is minimized and R_c itself is minimized.

What should be the error distributions on each of these parameters? Resistors are generally specified as a percentage of their nominal values, and any resistor not falling within, say, 10%, is simply sold as a resistor having a different nominal value. Thus, the distribution around this nominal value is more or less constant, so the simulated distributions for R_s and R_d were uniform, with limits of ±10% (= ±1.732σ_{unif}). In the absence of additional knowledge about r and L, the simulated distributions for these parameters were Gaussian with mean equal to the nominal value and standard deviation equal to 5% of the nominal value. The simulated distribution for W was also Gaussian but with standard deviation equal to 0.5% of the nominal 1.5 microns (because the tolerance on W is considerably tighter). Instead of simulating these noises using Taguchi's proposed 'outer array' (Phadke 1986), here a random sample of size 20 from the respective distributions provided the error in the parameter. Then a sample mean and sample variance of the 20 resulting values of R_c in each of the 81 situations were calculated. Taguchi then recommends that those factors influencing both the logarithm of the appropriate signal-to-noise ratios and the means be investigated.

Following the data analysis strategy outlined by Nair and Pregibon 1986, Figure 13a is a plot of the 81 sample variances versus the 81 sample means on a log-log scale. Since the slope of the line is very nearly 2, both Nair and Pregibon (1986) and Box (1988) recommend instead that the data be transformed via logarithms. Then the 81 means and variances can be analyzed directly for main effects and interactions, rather than constructing an appropriate signal-to-noise ratio. Restarting the simulation, this time calculating the mean and standard deviation of the 20 values of $\log_{10} R_c$, the mean-variance plot is now without structure (Figure 13b). As further confirmation of the success of this transformation, Figure 14 compares the 81 standard deviations on the data, where it is seen that the ratio of the

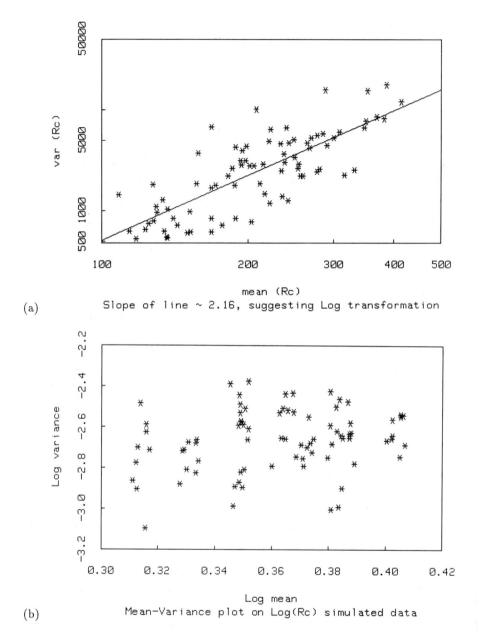

(a) Slope of line ~ 2.16, suggesting Log transformation

(b) Mean-Variance plot on Log(Rc) simulated data

Figure 13 Mean-variance (log-log) plot of simulated R_c values corresponding to circuit in Figure 12. (a) R_c: Slope of plot = 2.16, suggesting logarithmic transformation. (b) $\text{Log}_{10} R_c$: Lack of structure indicates correct scale for analysis.

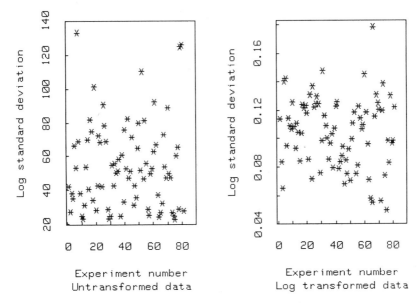

Figure 14   Plots of logarithms of the standard deviations in each of the 81 simulated IC experiment runs. Wider range of standard deviations on untransformed R_c (*left*) versus that on transformed Log R_c (*right*) further confirms logarithmic transformation.

largest to smallest standard deviation for the log-transformed data is about half that for the untransformed data.

Table 2 shows the analysis of variance for the 81 (logged) variances. Variable r is the main contributor to the variation in $\log R_c$. (The $R_s \times R_d$ effect is due mainly to a low variance when $R_s = 25$ and $R_d = 4$.) Having two degrees of freedom each, the sums of squares for the main effects can be decomposed into their linear and quadratic (single degree of freedom) sums of squares. Again following Nair and Pregibon 1986, a half-normal probability plot of these sums of squares shows that the effect of r on the variation is quadratic (Figure 15). The plot of the main effects on the variance at the three levels, with limits of one standard error, are shown in Figure 16. It can be seen that the variation in $\log R_c$ is smallest when $r = 100$.

Table 2　Analysis of Variance for Simulated 3^4 Integrated Circuit Experiment

Source	df	Mean($\log R_c$) Sum of Squares ($\times 10^4$)	Mean Square ($\times 10^4$)	$\log \text{Var}(\log R_c)$ Sum of Squares	Mean Square
R_s	2	0.1811	0.0906	0.07536	0.03760
linear	1	0.0301	0.0300	0.04728	0.04728
quadratic	1	0.1510	0.1510	0.02808	0.02808
R_d	2	1.1660	0.5830	0.07427	0.03713
linear	1	0.0210	0.0210	0.02614	0.02614
quadratic	1	1.1450	1.1450	0.04812	0.04812
r	2	11938.6	5969.3	0.10839	0.05419
linear	1	11820.4	11820.4	0.03857	0.03857
quadratic	1	118.2	118.2	0.06981	0.06981
L	2	4100.4	2050.2	0.03841	0.03841
linear	1	4092.5	4092.5	0.03452	0.03452
quadratic	1	7.9	7.9	0.00389	0.00389
$R_s \times R_d$	4	0.8696	0.2174	0.07042	0.17604
$R_s \times r$	4	5.2295	1.3074	0.15048	0.02637
$R_s \times L$	4	4.0688	1.0172	0.02708	0.00677
$R_d \times r$	4	3.6721	0.9180	0.03469	0.00867
$R_d \times L$	4	3.0623	0.7656	0.1134	0.02834
$r \times L$	4	3.5549	0.8887	0.10566	0.02642
$R_s \times R_d \times r$	8	7.3683	0.9210	0.30028	0.03754
$R_s \times R_d \times L$	8	15.0168	1.8771	0.18319	0.02290
$R_d \times r \times L$	8	10.0839	1.0922	0.14670	0.01834
Error	24	27.527	1.1470	0.44398	0.01850
Total	80	16121.1		1.82730	

Table 2 also gives the corresponding analysis of variance for the 81 means. Both parameters r and L affect the mean level of $\log R_c$. The half-normal probability plot of the sums of squares for the single degree of freedom contrasts is shown in Figure 17; the linear components of both L and r have the greatest impact on the average level of $\log R_c$. Plots of the

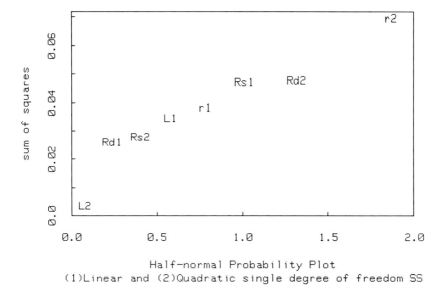

Half-normal Probability Plot
(1)Linear and (2)Quadratic single degree of freedom SS

Figure 15 Half-normal probability plot of sums of squares corresponding to single degree of freedom contrasts for main effects in analysis of variance for logarithm of $\mathrm{Var}(\mathrm{Log}\, R_c)$. Linear and quadratic contrasts denoted by 1 and 2 respectively.

main effects (Figure 18) confirm the value of $r = 100$ to minimize R_c and indicate the optimal value of L is 0.6 micron.

Having identified influential parameters and their initial values for the design, the next stage is to build a prototype circuit with the selected values. Several measurements on the circuit can be made by subjecting it to various environments. Ultimately a highly robust design with minimal variation will go into production.

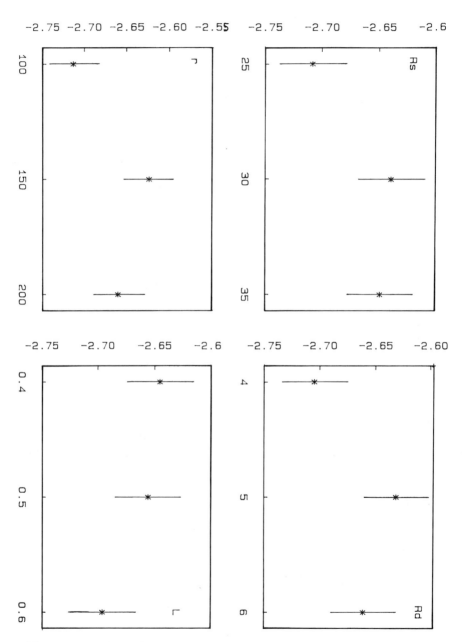

Figure 16 Plots of main effects in analysis of variance for logarithm of $\text{Var}(\text{Log}\, R_c)$. Shown with one standard error bars.

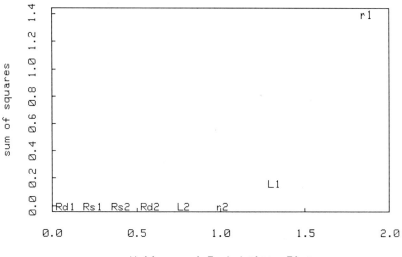

Half-normal Probability Plot
Linear (1) and Quadratic (2) single degree of freedom SS

Figure 17   Half-normal probability plot of sums of squares corresponding to single degree of freedom contrasts for main effects in analysis of variance for $\text{Mean}(\text{Log } R_c)$. Linear and quadratic contrasts denoted by 1 and 2 respectively.

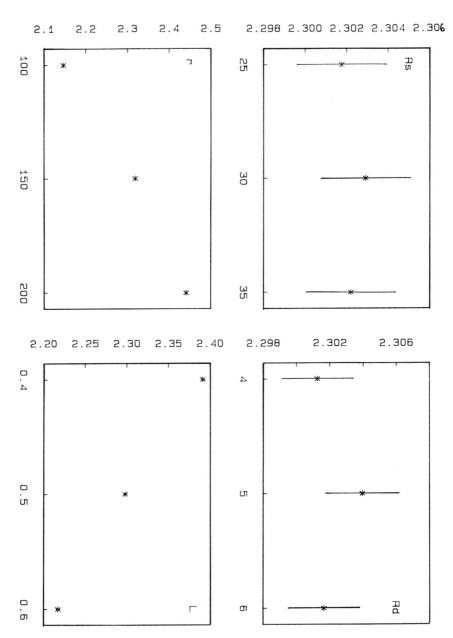

Figure 18 Plots of main effects in analysis of variance for Mean(Log R_c). Shown with one standard error bars.

6 SUMMARY

Some very basic strategies of experimental design can lead to enormous gains in product performance in the research and development stage of the product cycle. These gains include an increased understanding of product characteristics, model confirmation of functional performance, and the achievement of tight specifications with manufacturing considerations in mind. By encouraging the use of statistical methods early in the R&D phase, subsequent problems with manufacturing and warranty failures can be avoided.

ACKNOWLEDGMENTS

The author wishes to thank Paul Zander, Paul Stafford, Lynn Plouse, and Paul Marcoux for providing the data from their respective projects used in this paper, and Sherry Read and Tim Read for their comments on earlier drafts of this chapter.

REFERENCES

Box, G. E. P., Hunter, W. G., and Hunter, J. S. (1978), *Statistics for Experimenters*, Wiley, New York.

Box, G. E. P., (1988), Signal to noise ratios, performance criteria, and transformations (with discussion). *Technometrics 30(1)*, pp. 28–35.

Godfrey, A. Blanton (1986), The history of evolution and quality at AT&T. *AT&T Technical Journal 65*, 9–20.

Gunter, Berton (1987), A perspective on the Taguchi methods. *Quality Progress 20(6)*, 44–50.

Hunter, J. Stuart (1985), Statistical design applied to product design. *Journal of Quality Technology 17(4)*, 210–221.

Kackar, R. N. (1985), Off-line quality control, parameter design, and the Taguchi method. *Journal of Quality Technology 17(4)*, 176–188.

Kafadar, Karen (1988), Statistical calibration of a vector demodulator. *Hewlett-Packard Journal 2*, 18–25.

Kafadar, Karen and Plouse, Lynn (1988), Circuit design using statistical data analysis. *Hewlett-Packard Journal 2*, 12–17.

Keshner, Marvin (1982), 1/f noise. *Proceedings of the IEEE (3)*, 212–218.

Ku, H. H. (1966), Notes on the use of propagation of error formulas. Reprinted in H. H. Ku. ed., *Precision Measurement and Calibration: Statistical Concepts and Procedures*, U. S. National Bureau of Standards Special Publication 300, 1969, 331–341.

Nair, Vijay N. and Pregibon, Daryl (1986), A data analysis strategy for quality engineering experiments. *AT&T Technical Journal 65(3)*, 73–84.

Phadke, Madhav S. (1986), Design optimization case studies. *AT&T Technical Journal 65(3)*, 51–84.

Searle, S. R. (1970). *Linear Models*, Wiley, New York.

Taguchi, Genechi, and Wu, Yuin (1980), Introduction to off-line quality control. *Central Japan Quality Control Association.*

Tukey, J. W. (1961), Statistical and quantitative methodology. Reprinted in L. V. Jones, ed., *The Collected Works of John Tukey, Volume III*, Wadsworth, Belmont, California, pp. 143–185.

3

Biotechnology Experimental Design

Perry D. Haaland   Becton Dickinson Company, Research Triangle Park, North Carolina

1 INTRODUCTION

For most bioprocess technologies there are no theoretical models which can be used to explain process performance. Consequently, successful research is characterized by effective empirical problem solving. Typically, the problem solving process is governed by limitations on time and resources. Therefore, research productivity is a critical concern. Statistical problem solving provides a set of powerful tools which can be used to maximize the efficiency and productivity of empirical problem solving.

1.1 Collecting Information-Rich Data

Research in biotechnology generates great quantities of data. The recent spread of computers in the research environment has greatly increased our ability to collect and manage these data. However, although computers increase the amount of data we can create and manage, they do not necessarily increase the information in the data. Since there are limited time and resources available to generate and understand these data, it is important that the data be information-rich. Statistical design is one way to increase the amount of information-rich data we gather.

Purpose of Collecting Data   Data are collected to solve empirical problems. Data serve as a basis both for understanding and action. Some typical reasons for collecting data for empirical problem solving are as follows:

- determine which few out of many factors significantly affect process performance
- determine how the factor settings should be changed in order to improve the process performance
- determine the optimal performance level and specify what actions must be taken to achieve this level.

These reasons for collecting data have long been a part of statistical methodology. Methods for their use were described, for example, in the books *Statistics for Experimenters* by George E. P. Box, William Hunter, and J. Stuart Hunter (1978), *Applications of Statistics to Industrial Experimentation* by Cuthbert Daniel (1976), and *Practical Experimental Designs* by William Diamond (1981). In this article, we use the power of real examples to show how these statistical problem solving methods can be applied to the collection and analysis of data from biotechnology experiments.

How Should Data Be Collected?   Every experiment has a design; some designs are better than others. Since data must be collected anyway, the use of statistically designed experiments for data collection adds only incrementally to its cost. However, well-designed experiments significantly increase the information content of the data.

Since data serve as the basis of understanding and action, it is essential to have correct data in order to make correct decisions. In his book *Guide to Quality Control*, K. Ishikawa (1976) outlined the following considerations in collecting data:

- "Will the data determine the facts?"
- "Are the data collected, analyzed and compared in such a way as to reveal the facts?"

In order to ensure that the data will determine the facts, we must use an appropriate experimental design. In order to reveal the facts contained in the data, we can use statistical graphics and analysis. The better the experimental design, the simpler the analysis of the data. However, data which are not collected properly are usually difficult to analyze and understand, no matter what methods of analysis are applied.

The first step in the proper collection of data is to clarify the objectives of the experiment. Next, the experimenter must determine what data to collect, how to measure it, and how the data relate to process performance and the experimental objectives. The experimenter must insure that the data collected are representative of the process so that the data will lead to correct conclusions. Finally, an experimental design must be chosen which will reveal the facts as they relate to the experimental objectives.

What Is an Experimental Design? An experimental design consists of a specific number of measurements made at predetermined settings of the experimental conditions. We usually call each experimental condition a run and each measurement an observation. For example, if an investigator is studying a process which depends on the acidity of the buffer (pH) and on incubation time, a possible experimental design is as follows:

- run 1: pH = 6.7, time = 20 minutes
- run 2: pH = 6.7, time = 30 minutes
- run 3: pH = 7.2, time = 20 minutes
- run 4: pH = 7.2, time = 30 minutes

At each of these runs, process performance would be observed. The resulting data set should reveal how pH and time affect process performance and what can be done to improve process performance.

Progress on the Learning Curve Empirical problem solving is difficult because the problems are complex and progress along the learning curve is often slow and difficult. The use of statistical experimental designs speeds progress along the learning curve because these experiments answer questions. (See Figure 1.) This sounds simple, but anyone who has carried out an experiment, examined the results, and then was unable to provide any clear answers can appreciate the fundamental importance of this idea.

1.2 Separating Signals from the Noise

Experimental results consist of signals and noise. Both aspects are important to the problem solving process. A signal is a measure of the effect that a process variable has on process performance. The noise is all of the other variation in the data. A signal which is large in comparison to the noise identifies a process variable which has an important effect on process per-

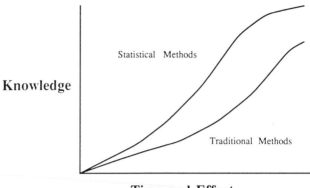

Knowledge

Statistical Methods

Traditional Methods

Time and Effort

Figure 1 A learning curve measures how much time and effort it takes to acquire knowledge about the problem being studied. Statistical methods provide an advantage over traditional problem solving methods because they focus on small, well-designed experiments which answer questions about the process being studied. This results in a steeper learning curve and faster progress toward the problem solution.

formance. Unimportant factors will have signals which fall below the noise level. Statistical methods can be used to separate signals from the noise.

Noise   If we repeat a biological experiment, we don't expect to get exactly the same results as in the first experiment. This is true even if we are very careful to run each experiment under controlled conditions. The differences between experiments contribute to the uncertainty about the true results. Statistical methods allow us to reason in the presence of this uncertainty.

Uncertainty stems from variability in measurements of process performance. Variability can be thought of in terms of two components; namely, experimental error and measurement error. Measurement error is the variability we get when remeasuring the same experimental outcome. Experimental error is the difference in outcomes when an experiment is independently repeated under controlled conditions. Experimental error is usually much larger then measurement error. When we refer to noise, we are usually referring to the sum of both experimental and measurement error.

Clear Signal Designs   Experimenters can often identify many factors which may affect a process, but they usually have limited resources to carry out their experiments. Therefore, the use of small experiments is especially attractive. However, special care must be taken in using a small experiment under these conditions so that meaningful results will be achieved.

For example, suppose that in formulating a new assay buffer, it is known that only the pH and the ionic strength of the buffer affect the stability of the assay components. In this case, the experimenter needs to be confident that the effects of these two factors can be clearly distinguished. However, if the process if observed first with pH and ionic strength both at low levels and then with both factors at high levels, the effects of these two factors cannot be separated.

If we study only pH and ionic strength, we can observe all four combinations of their high and low levels. This experiment clearly separates the effects of the two factors. In general, there will be many more than just two factors of interest so efficiency requires that we study as many experimental factors as we can with the fewest number of observations. As the number of factors increases, it becomes more difficult to choose combinations of their levels which clearly separate their effects yet at the same time minimize the size of the experiment.

In the Design Digest provided by Haaland (1989), a collection of efficient experimental designs is presented which allows many factors to be investigated in small experiments. These designs are part of a class of designs which J. Stuart Hunter characterizes as "clear signal designs" (Hunter, 1987). That is, they

- separate the signals from the noise and
- clearly distinguish signals from each other.

These designs form the basis for efficient, effective statistical problem solving.

The Pareto Principle   The statistical analysis provides estimates of how each experimental factor affects process performance. These estimates reveal which factors are most important and how changing their settings affects process performance. An interesting way to compare the relative importance of the estimated effects of the factors is by means of a Pareto chart (Ishikawa, 1976).

The Pareto chart in Figure 2 graphically displays the magnitudes of the effects from a typical experiment. The effects are sorted from largest to smallest. The Pareto chart clearly shows that the first effect is by far the most important one. Whether or not the second effect is also above the noise level is less clear. We use statistical methods to answer such

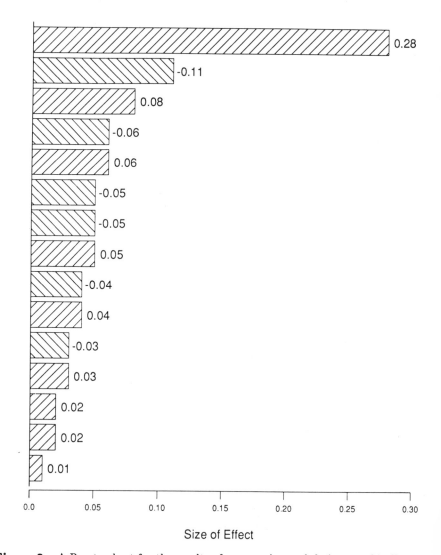

Figure 2 A Pareto chart for the results of an experimental design graphically depicts the relative magnitudes of the effects of each of the experimental factors. On the y-axis, we list the experimental factors (and possibly their interactions). On the x-axis we measure the absolute magnitude of the effect of each factor as determined by the statistical analysis. The Pareto principle or factor sparsity assumption suggests that most of the improvement in the process performance can be obtained by changes in only a few of the experimental factors.

questions. To improve process performance, it is obvious that we should begin by adjusting the factor which had the biggest effect.

J. M. Juran, in his book *Managerial Breakthrough* (1964), described the Pareto principle as "the vital few and the trivial many." That is, improvements is process performance come from paying attention to the few really important factors which affect the process. In order to find these "vital few," we use a statistical experimental design which gives us clear estimates of the effects of each factor which may affect the process. Then we collect the data which allow us to estimate the effects. A Pareto chart should point out clearly the "vital few" effects with which we should concern ourselves. The statistical analysis can help us identify factors with signals which rise less clearly above the noise level.

The Pareto principle enters more formally into statistical analysis as Box and Meyer's (1986) assumption of "factor sparsity." Their assumption is that "in relation to the noise only a small proportion of the factors have effects that are large"; that is, a few factors are "active" and the rest are "inert." In empirical problem solving, factor sparsity suggests that out of the long list of factors which may affect process performance, only a few actually have an important effect. The experimental designs we discuss in this chapter are especially good for finding the few most important factors.

2 SOLVING BIOTECHNOLOGY PROBLEMS

Although statistical experimental design has been widely used in many areas of science and industry (Box, Hunter, and Hunter, 1978; Daniel, 1976; Diamond, 1981; and Taguchi, 1986), it has not yet been widely adopted in the biological sciences. We believe this is partly because of the lack of a clear explanation of the methods using biotechnology examples. Haaland (1989) and this chapter address this need by showing examples of how statistical problem solving is used effectively in real biotechnology applications.

Biotechnology applications require effective problem solving methods because they involve

- many factors
- no theoretical model
- data which may include high levels of noise
- possible interactions among variables.

This description fits many problems in the development of bioprocess technologies. Indeed, the reader can easily see the general applicability of these methods to scientific research.

2.1 Examples

The researchers with whom we work use statistical problem solving successfully in many projects. In fact, their experiences form the basis of this article. Some real-life examples (Haaland, 1989) illustrating the use of statistical methods are as follows:

- formulate stability-enhancing storage buffers for an enzyme-immunoassays
- configure an enzyme-immunoassay so as to maximize its signal-to-noise ratio
- identify treatment factors which increase the activity of a bioactive compound coated on a polymer surface
- optimize production of monoclonal antibodies harvested from ascites-producing mice
- identify important factors affecting the yield of monoclonal antibodies produced by a cell culture system.

In all of these examples, statistical problem solving methods were successfully used to find practical solutions. The statistical methods were learned and used by the researchers themselves. The use of statistical experimental design increased the efficiency and productivity of the research experiments. The statistical analysis provided a powerful framework within which to ask and answer questions about possible solutions. Finally, the systematic problem solving strategy allowed the scientists to effectively integrate all of these tools into their research.

The statistical approach to solving empirical problems provides a common language which scientists, engineers, and managers can use to communicate in planning, carrying out, and evaluating experimental programs. Even more importantly, it provides a means by which problem solvers can speed up their progress along the learning curve. Real examples make these powerful tools more readily accessible to biotechnology researchers.

2.2 Comparison of Strategies for Problem Solving

Successful problem solving consists of

- asking good questions
- collecting data which can answer the questions
- analyzing the data to reveal the answers.

Problem solving can be made more productive by using small, efficient experimental designs when collecting data and powerful statistical graphics and analysis to understand the results (Figure 3).

One-at-a-Time and Matrix Methods   Two methods which are sometimes used as alternatives to statistical experimental design and analysis are "one-at-a-time" and "matrix" experimentation. Each of these methods can be used for problem solving, but neither of them is economical or efficient.

The "one-variable-at-a-time" approach is to fix all of the variables except one and then study the behavior of the system at several levels of that variable. For each variable the best value is found, and then the process is repeated for the remaining variables until all variables have been considered. Figure 4a illustrates this approach. This method may be effective in some situations, but it is very inefficient. It just takes too many experiments to come up with an answer. If there are interactions among the variables, the "one-at-a-time" method may miss the optimal settings because it doesn't thoroughly explore the space of possible solutions.

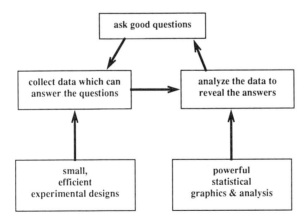

Figure 3   Problem solving consists of a cycle of asking questions, collecting data which can answer the questions, and analyzing the data to reveal the answers. The effectiveness of this process can be greatly increased by using statistical experimental design to collect better data and statistical analysis to more clearly reveal the answers.

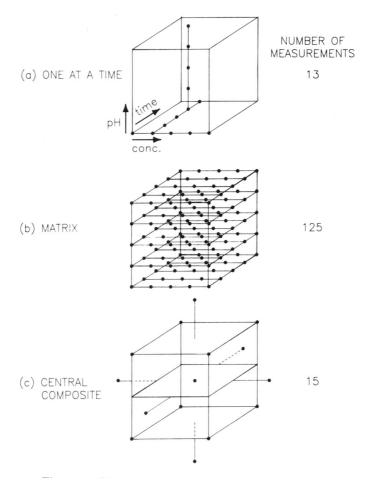

NUMBER OF
MEASUREMENTS

(a) ONE AT A TIME   13

(b) MATRIX   125

(c) CENTRAL
    COMPOSITE   15

Figure 4   Three possible approaches to experimentation in solving empirical problems include the one-at-a-time approach (a), the matrix method (b) and the statistical design approach (c). The one-at-a-time method (a) doesn't require many measurements, but it doesn't explore the experimental space very well so that it may miss the solution. The matrix approach (b) is effective but inefficient because it requires too many measurements. The statistical design method is efficient and effective because it provides good coverage of the experimental space with as few measurements as possible.

A second traditional approach to empirical problem solving is to lay out a matrix of all interesting combinations of the variables being investigated. Then all of the combinations in the matrix are investigated until the solution is found. This method is illustrated in Figure 4b. The matrix method has the advantage of thoroughly exploring the experimental space, but it requires an unnecessarily large number of measurements. Even with only four or five variables, the matrix is too large to be explored in a realistic amount of time.

Statistical Design The statistical problem solving approach uses a series of small, carefully designed experiments. Each experiment carefully explores the experimental space while studying many variables and using a small number of observations. This method is illustrated in Figure 4c. For each small experiment, well-defined questions are asked, and simple statistical methods are used to provide answers. A clear strategy is used to insure efficient progress toward a solution.

2.3 Iterative Problem Solving Strategy

We sometimes call the statistical design approach "strategic experimentation" because it couples statistical experimental design and analysis with a well-defined problem solving strategy. An important characteristic of this approach is that it is iterative. By this we mean that we use a series of small experiments to solve difficult problems. We also call this the "stop, look, and listen" approach to experimentation. That is, we do a small experiment, learn from the results, and then plan the next experiment. (See Figure 5.) This iterative procedure refines and improves our questions to take advantage of what we have learned so far.

When planning an experiment, we have different objectives depending on what is known so far about the solution to the problem. If we are just beginning, there may be a long list of possible variables or factors which may affect the process. We may only have a preliminary idea of the reasonable ranges for the factor values. On the other hand, if we are close to finding a solution, there may only be a few factors which are known to be important. Finally, at the conclusion of an investigation, the results are confirmed in a follow-up experiment. These three stages of experimentation are called, respectively, screening, optimization, and verification.

A useful analogy to the idea of iterative problem solving is climbing a mountain. For example, we don't want to expend most of our resources

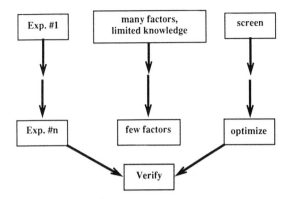

Figure 5 An iterative approach to solving problems involves thinking carefully about the problem, designing an experiment, collecting the data, and analyzing the results. This process may be repeated as necessary until the problem is solved.

just getting to the base camp. At timely intervals, we want to evaluate whether or not the current direction is leading us toward the summit. Different climbing strategies are required for the final ascent to the summit than for earlier stages of the climb. In this context, screening experiments help us locate the summit, optimization experiments are the ascent to the summit, and verification experiments prove we were there. Thus, the iterative nature of strategic experimentation allows us to adapt our methods to the changing nature of our objectives.

Using Small Experiments   If we are going to do several experiments, then each experiment has to be fairly small. In fact, given fixed resources, it is usually better to do several small experiments rather than one large one because the information from one experiment can improve the design of the next experiment. A valuable rule of thumb is to not use more than 25% of our resources for the first experiment (Box, Hunter, and Hunter, 1978).

It is usually a good idea for a first experiment to be as small as possible. For one reason, initial guesses of the best settings of the factors may not be very close to their best settings. We may even find that some of the experimental conditions are not practical or simply don't work. After inspecting the data, we may find that an important factor was left out of the design. On the other hand, the initial guesses for the important factors and their values may have been good enough to suggest that we should go directly to an optimization experiment.

Screening–Optimization–Verification   One purpose of doing a series of small experiments is that, as the investigation progresses, we can change the kind of experiment we do in order to reflect our changing objectives. We classify these different experiments as screening, optimization, and verification experiments. Figure 6 provides a brief description of the conditions under which we use these three types of experiments.

Screening experiments are small experiments which include many variables. They play an important role in the early stages of an investigation. Because they are small and include many factors, they don't have as much information per factor. Their objective is problem reduction; that is, to focus on the important variables and to find out more about their best settings.

The purpose of an optimization experiment is to build a mathematical model which can be used to predict the behavior of the process being investigated. This generally requires a lot of information about each factor so that optimization experiments usually only include a few factors but are fairly large compared to screening experiments. Their objective is to produce specific optimal values for the experimental factors.

	OBJECTIVES	DESCRIPTION
Screening	identify important factors	many factors imprecise knowledge
Optimization	build predictive model	few factors in region of optimum
Verification	confirm results	at predicted best settings

Figure 6   Screening and optimization experiments have different objectives. Screening experiments are used early on in an investigation to narrow the focus of the problem. Optimization experiments are used at the end of an investigation to build a predictive model which can be used to provide specific information about the solution. A verification experiment concludes the investigation.

The simplest type of verification experiment shows that the predicted optimal process performance can be reproduced in a second experiment. This may involve production runs or further laboratory experiments. A second, larger type of verification experiment may be designed to verify that, over a given range, a set important factors has the predicted effect on the process being studied.

Building a Solution   One way to think of empirical problem solving is that we are trying to build a solution to the problem (Figure 7); that is, we lay a solid foundation and then systematically add layers of further information. We start by thinking about which factors may affect process performance, which factors contribute to process variability, and how we can utilize this information to get the process working at a basic level. This information provides the basis for moving to the screening stage in which we try to identify which of the (possibly many) factors are important and to discover how we can change the factor settings to improve process

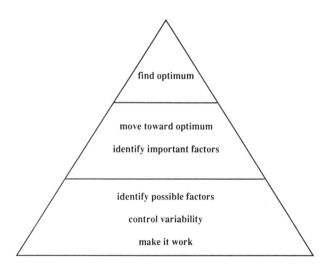

Figure 7   Solving an empirical problem involves building on earlier knowledge by systematically adding layers of further information. At the foundation of this structure is our scientific knowledge about the problem. From there we add the information obtained from screening experiments and then the information obtained from optimization experiments. The culmination of this structure is an experiment which verifies that we found the optimum process performance.

performance. The culmination of the problem solving process is finding the optimal settings of the most important factors.

3   ASCITES PRODUCTION EXAMPLE

In this section, we present a real example which illustrates the statistical problem solving methodology. The problem involves the optimization of in vivo production of monoclonal antibodies as described by Adrion et al. (1984). The researchers who carried out this investigation used statistical methods for their efficiency, power, and productivity.

3.1   Problem Description

In this production process, large quantities of monoclonal antibodies may be prepared by growing antibody-secreting cells in mice and then collecting an antibody-rich fluid called ascites. The antibody-secreting cells used are all identical twins of a single cell which was originally selected using monoclonal antibody technology. This original cell and all of its twins (clones) produce identical antibody molecules, so the antibodies are called monoclonal. Once the cell line is established, animals (e.g., mice) may be used as hosts to support further growth of the cell line. The antibodies can then be harvested by collecting the fluid (ascites) which is continuously secreted by the cells.

Problem-Solving Approach   In early stages of monoclonal antibody development, it is common for process variables to be set based on personal experience and preference. However, in order to insure the commercial viability of such production, a more rigorous approach to optimizing the production process is required. Many variables may potentially affect the efficiency of such a production process, and there are distinct possibilities that some of the variables investigated may interact to produce unexpected results. Since there is no theoretical model for this complex biological process, the problem must be approached by experimentation. The complexity of the physiological processes involved implies that experimental results are subject to inherent variability.

The experimentation required to solve such a problem is expensive and time-consuming. Each experiment involves many mice which must be housed and cared for during the several weeks required for the mice to begin producing antibodies. Thus, it is important to minimize the number of experimental conditions which must be investigated. The lengthy time between starting an experiment and finding out the results means that we want to conduct as few experiments as possible. Based on these constraints, we adopt the following approach:

- use a screening experiment to identify important factors,
- use an optimization experiment to identify best process performance
- minimize sample size to save time and money
- use a clear signal design to guard against interactions among factors.

Small, efficient statistical experimental designs and a statistical problem solving strategy are the key elements of this problem solving approach.

Experimental Factors   A schematic of the in vivo production of monoclonal antibodies is presented in Figure 8. After examining the production process, six process variables or factors were identified to study in a series of statistically designed experiments. Initial "best guesses" for the factors were based on previous production experience.

Some cell lines do not grow well in fully immunocompetent animals, so the mice are given an initial radiation treatment to provide partial immunosuppression. A nonlethal dose of Co^{60} gamma radiation may be used for this purpose. The dosage level of immunosuppressive radiation (Rad-Dos) is the first process variable or factor. Two dose levels were proposed; namely, 250 and 500 rads.

Additional immunosuppression is obtained by injection of a clear oil called Pristane. Both the amount (VolPrs) and timing of this injection (Prime1) may affect the monoclonal antibody yield. For the experiment, injections of 0.1 and 0.5 ml were used. The elapsed time between the Pristane injection and cell inoculation may vary from a few days to several weeks. Intervals of one and three weeks were used for the first experiment.

In order to start the production of ascites, the mice are inoculated with the antibody-secreting cells which were previously cultured in vitro. Both the number of cells injected (CelNum) and the growth state of the cells (Growth) are thought to affect yields. Investigators commonly use either 10E6 or 10E7 cells for injection. Log stage of growth and stationary, saturated cultures were proposed for study.

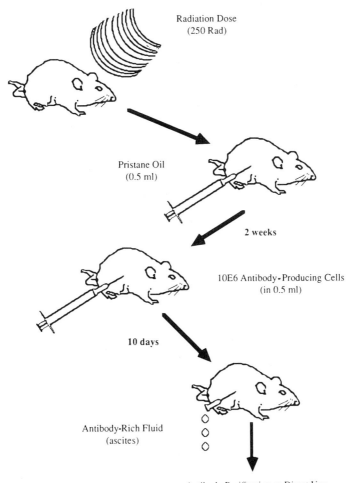

Radiation Dose
(250 Rad)

Pristane Oil
(0.5 ml)

2 weeks

10E6 Antibody-Producing Cells
(in 0.5 ml)

10 days

Antibody-Rich Fluid
(ascites)

Antibody Purification or Direct Use

Figure 8   Mice may be used to produce an antibody-rich fluid called ascites. A general production sequence for this process begins with each animal receiving an immunosuppressive treatment. In this particular case, the immunosuppressive sequence includes irradiation and injection of Pristane oil. Next, each animal is injected with antibody-secreting cells. After a growth period, the antibody-rich fluid, or ascites, can be harvested.

Finally, a second injection of Pristane may sometimes be given immediately before injection of the antibody-secreting cells (Prime2). It is unknown whether this second injection immediately before tumor inoculation is helpful, so experiments were conducted both with and without a second injection.

In order to get enough ascites to provide a reliable assay result, five mice were used for each experimental condition. At the beginning of the study, individual mice were randomly assigned to groups of five. The five mice in each set were housed together and given identical treatment. The ascites collected from the individual mice in each group of 5 were pooled to provide a sufficient volume for a reliable assay of antibody yield. The response was antibody fluid volume adjusted for titer (TtrVol), which is proportional to the number of monoclonal antibody molecules produced.

A number of other process variables were of interest but were more difficult and expensive to investigate. In particular, six-week-old, female Balb/c mice were used in this study. Six-week-old mice were used because of the prohibitively high upkeep cost while the animals age. Balb/c mice were used in the study because many strains of hybrid mice cannot be purchased in large quantities on demand. Female mice were used because of difficulties in housing males together. In addition, only one type of monoclonal antibody was considered. Since these factors were not included in the study, we don't know how changing them would have affected the outcome of the study. If they were to change at a later date, these problem solving methods could be successfully applied to reoptimize the process.

3.2   A Screening Experiment

The objective of a screening experiment is to determine which few process variables, out of many candidates, have an important affect on process performance. Designs for screening experiments are introduced in Chapter 3 of Haaland (1989). Complete descriptions of these designs and instructions for their use are included in the Design Digest in Haaland (1989) while the analysis of screening experiments is presented in Chapter 4 of Haaland (1989). In this article, we present an overview of how screening experiments are used in statistical problem solving.

Experimental Design   The factors to be included in the screening and their settings are given in Table 1. Each of the six factors are to be in-

Table 1 Variable Names and Levels for First
Experiment

	Factor Name	Units	Low Level	High Level
1.	RadDos	rads	250	500
2.	Prime1	weeks	1	3
3.	VolPrs	ml	0.1	0.5
4.	CelNum	cells	10E6	10E7
5.	Growth	state	Log	Sat
6.	Prime2	present	No	Yes

	Response Name	Units
1.	TtrVol	proportional to number of monoclonal antibody molecules produced

vestigated at two levels. Using a design from the Design Digest from Haa-land (1989), we can efficiently investigate these six factors in a sixteen run experiment; that is, by measuring antibody yields for sixteen different combinations of the settings of the process variables.

The design shown in Table 2 is called a fractional factorial design be-cause it is a specially selected subset (fraction) of the design conditions form the full matrix (factorial) design. (This design is listed in the Design Digest as FF0616—fractional factorial—6 factors—16 observations.) It is a time-saving design because it requires only a fraction of the experimental conditions of a full factorial design. In particular, a full factorial or ma-trix design would involve measuring antibody yield at all of the $2^6 = 64$ different combinations of the values of the six factors.

The use of fractional designs to reduce the sample size required by factorial designs was proposed by Finney (1945). Additional early work in this area was done by Plackett and Burman (1946), Davies and Hay (1950), Daniel (1959), and Box and Hunter (1961a,b). These and similar designs have become quite popular as tools for problem solving in industry. Some popular books which discuss fractional factorial designs are Box, Hunter, and Hunter (1978), Box and Draper (1987), Diamond (1981), Daniel (1976), and Taguchi (1986). The use of fractional factorial designs is also closely related to the use of search designs as described by Ghosh (1987), Srivastava (1975), and Patel (1987).

Table 2 Worksheet for Sixteen Run Screening Experiment

| | | | Experimental Factors | | | | Response |
Run	RadDos	Prime1	VolPrs	CelNum	Growth	Prime2	TtrVol
1.	250	1wk	0.1ml	10E6	Log	No	70
2.	250	1wk	0.1ml	10E7	Sat	No	150
3.	250	1wk	0.5ml	10E6	Sat	Yes	34
4.	250	1wk	0.5ml	10E7	Log	Yes	32
5.	250	3wk	0.1ml	10E6	Sat	Yes	137.5
6.	250	3wk	0.1ml	10E7	Log	Yes	56
7.	250	3wk	0.5ml	10E6	Log	No	123
8.	250	3wk	0.5ml	10E7	Sat	No	225
9.	500	1wk	0.1ml	10E6	Log	Yes	50
10.	500	1wk	0.1ml	10E7	Sat	Yes	2.7
11.	500	1wk	0.5ml	10E6	Sat	No	1.2
12.	500	1wk	0.5ml	10E7	Log	No	12
13.	500	3wk	0.1ml	10E6	Sat	No	90
14.	500	3wk	0.1ml	10E7	Log	No	2.1
15.	500	3wk	0.5ml	10E6	Log	Yes	4
16.	500	3wk	0.5ml	10E7	Sat	Yes	15

Note: The experiments were run in a randomized order to guard against systematic bias.

In comparison to the statistical design approach, the one-at-a-time method uses fewer conditions but would require several weeks of raising and treating mice in order to get results on each of the six factors one at a time. The statistical approach, however, consists of two small, efficient experiments; namely, a screening experiment and then an optimization experiment. This iterative approach uses small, economical experiments which are designed to reflect the objectives of different stages in the problem solving process.

The experimental design worksheet is reproduced in Table 2. The six factors are investigated in sixteen runs. Each run corresponds to a set of values for the factors at which a measurement of the response is to be made. The last column of the table lists the response values which were measured for each run. After the experimental factors and responses were specified, the run order was randomized as a guard against systematic biases. More details on this process are included in Chapter 5 of Haaland (1989), and instructions are provided in the Design Digest of Haaland (1989).

This sixteen run design is efficient because it captures the most important part of the information available from a full factorial design. That is, these sixteen runs were carefully selected to maximize the information obtained per measurement.

This fractional factorial design also has other desirable statistical properties. For example, this is a balanced design; i.e., the effect of a low-to-high change for each factor is measured at both the low and high settings of every other factor. For this design, we can independently estimate all of the factors and their interactions (in pairs) with each other. This is a "clear signal" design in that it separates the signals for each of the experimental factors.

Results of the Screening Experiment   The results of the experiment are interpreted based on our estimates of how each of the experimental factors affected the response (measure of process performance). The statistical analysis provides estimates of these effects.

In particular, the estimate of the effect of each factor is the difference in the response value associated with going from the low to the high setting of that factor. An important factor causes a large effect because the process will perform significantly better at one of its two settings. Conversely, an unimportant factor does not result in a change in process performance and so is associated with a small effect.

In a fractional factorial design, the effect of each factor is the difference between the average of measurements made at the high level of the factor and the average of the measurements made at the low level. For example, for RadDos, the high settings are runs 9 through 16. The low settings are runs 1 through 8. Thus, the estimate of the effect of going from high level of RadDos is

$$\bar{y}_{(\text{high})} - \bar{y}_{(\text{low})} = \frac{177}{8} - \frac{827.5}{8} = 22.1 - 103.4 = -81.3.$$

A Pareto chart for the main effects and all estimable interaction effects is presented in Figure 9. This Pareto chart also includes the estimated values of each of the effects. RadDos has by far the largest effect on ascites yield, and we think that it is clearly a significant factor. The next largest effects are those of Prime2, Growth, and Prime1. These factors generate smaller signals, but they still seem to be above the noise level.

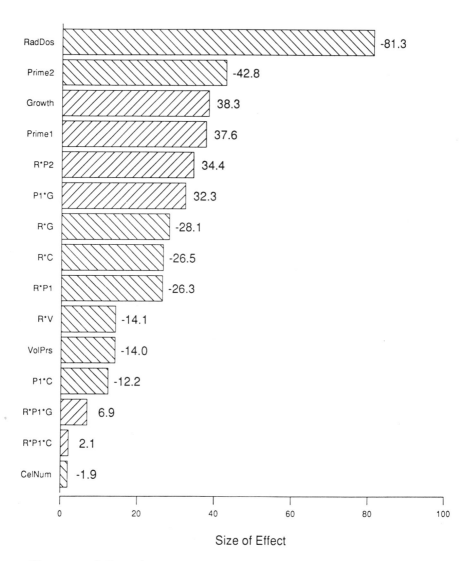

Figure 9 A Pareto chart for factor effects displays the magnitude (absolute value) of each estimated effect along the x-axis. The factor names are given on the y-axis. The Pareto principle or factor sparsity assumption suggests that only one or two factors are responsible for most of the changes in the process yield. In this case, RadDos clearly has the biggest effect on ascites yield. Three additional factors have potentially large effects. We suspect that the remaining factors represent noise.

It is not clear whether any of the two-factor interactions are important or not. Unfortunately, the Pareto chart by itself does not allow us to make this distinction.

In order to better distinguish between the signals and the noise, we use a normal plot. This plot shows the estimated effects on the x-axis and their expected values from a normal distribution on the y-axis (see Figure 10.) If there are no signals, all of the points on the normal plot will fall on a straight

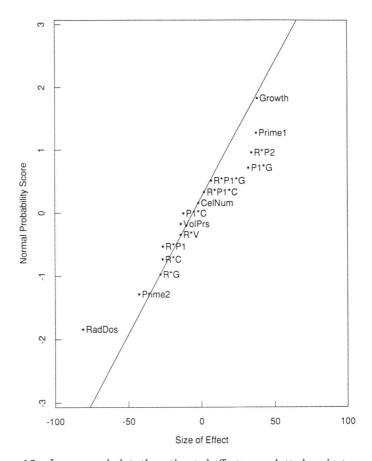

Figure 10 In a normal plot, the estimated effects are plotted against a normal probability scale. The important quality of this scale is that if there are no significant effects, all of the points fall along a straight line. Any points which fall off the line may be assumed to represent signals. In this case the likely signals are RadDos, Prime2, Growth, Prime1, RadDos $*$ Prime2 (R $*$ P2), and Prime1 $*$ Growth (P1 $*$ G).

line. Points corresponding to signals will fall away from a line going through the center of the data. In our case, Figure 10 confirms our suspicion that RadDos, Prime2, Growth, and Prime1 correspond to signals. We also think that the interactions between RadDos and Prime2 and between Prime1 and Growth are important. (Note that the design we used does not allow us to distinguish among all possible two-factor interactions. We decided on these two-factor interactions based on our knowledge of the problem. For more information on interactions, and the ability of experimental designs to distinguish among two-factor interactions, see Haaland, 1989, and Box, Hunter, and Hunter, 1978).

RadDos is by far the most important factor. We can use the "box and whisker plot" in Figure 11 to interpret its affect on ascites yield. At each level of RadDos, the box contains the middle 50% of the responses. The line inside the box is the median, which is the point above and below which lie 50% of the data. The whiskers extend out to the largest and smallest data values unless there are values which might be consider outliers in a normal population. (Box and whisker plots are discussed further in Cleveland, 1985). We see that a much better ascites yield is obtained at the low (250 rad) level than at the high (500 rad) level; that is, RadDos has a large negative effect. We should consider setting it at lower levels in our next experiment.

The other important factors are Prime1, Prime2, and Growth. Figure 12 shows a cube plot (see Haaland, 1989, and Box, Hunter, and Hunter, 1978) of these factors with the upper cube for high values of RadDos and the lower cube for low values of RadDos. Each vertex of the cube corresponds to a combination of high and low settings for the three factors Prime1, Prime2, and Growth. In this example, each vertex corresponds to a single observation. (In a larger experiment, each vertex would correspond to an average of several observations.) Because of the large effect of RadDos, we are primarily interested in the lower cube. The lower cube shows that the best results were obtained at RadDos = 250 rads, Prime1 = 3 weeks, Growth = Saturated, and Prime2 = no. In fact, the effect of Prime1 is consistently better at 3 weeks rather than 1 week (right face of the cube). The results are consistently better without the second priming (bottom face of the cube). Saturated vs. logarithmic stage cells also consistently provide better results (back face of the cube).

The factors Growth and Prime2 are qualitative in nature; that is, they take on only discrete, non-numeric values. Once a screening experiment has determined their best values, we can fix them in subsequent experiments.

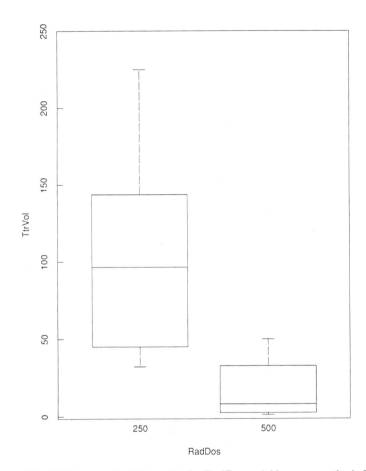

Figure 11 This box and whisker plot for RadDos quickly conveys the information that ascites yield is noticeably higher at the lower level than at the higher level; that is, RadDos has a negative effect on TtrVol. A box and whisker plot is most useful when there is only one very large factor effect because it does not control for the effect of other possibly important factors on the response. However, in this case, it is both useful and informative.

On the other hand, RadDos and Prime1 are measured or quantitative factors. Usually the results of a screening experiment suggest further changes in their values in order to increase process performance. Accordingly, we will want to make appropriate changes in the settings of the quantitative factors before we go on the next experiment.

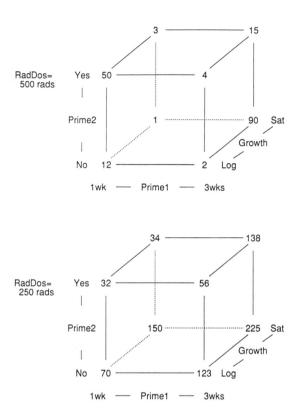

Figure 12 Each cube plot provides a means of simultaneously displaying the effects of three factors on the response. By creating a separate cube for each value of a fourth factor, we can simultaneously consider four factors. The average response value is given at each of the vertices of the cube. In this example, there is one vertex which is noticeably better than any of the others; namely, the vertex representing low RadDos, high Prime1, low Prime2, and high Growth.

Clearly, we should use saturated phase cells (Growth = Sat) without a second priming (Prime = no) in subsequent experiments. Since results seem to improve with higher values of Prime1, we should consider larger values of Prime1 in our second experiment. However, the results of the screening experiment suggest that we should consider lower values of Rad-Dos. The other two factors, CelNum and VolPrs, do not appear to be important, so they can be set at economical (low) levels of 10E6 and 0.1 ml, respectively.

Conclusions   Based on the results of the screening experiment, we identified two quantitative factors, RadDos and Prime1, for further investigation.
We can also move the ranges of these factors closer to their optimal settings.
The two qualitative factors, Growth and Prime2, were set at what appear
to be their best values. The remaining two factors, CelNum and VolPrs, do
not appear to have a significant effect on ascites yield, and we can set them
at economical levels. Thus we achieved a significant reduction in the scope
of the problem. We are now ready to conduct another small experiment.

3.3   An Optimization Experiment

In Section 1, we discussed the use of an iterative problem solving strategy.
One of the advantages of this strategy is that each experiment can be
specifically designed to meet the current objectives of the problem solver.
In particular, one or more screening experiments may be required to reduce
the scope of the problem so that an optimization experiment is feasible. In
our example, we went from six to two experimental factors as a result of
the screening experiment. Now we are ready to design an optimization
experiment.

Objectives   The factors radiation dose (RadDos) and number of days between first priming and inoculation (Prime1) survived the initial screening
for important variables. Now we want to develop a mathematical model
which can be used to predict the ascites yield (TtrVol) as a function of
RadDos and Prime1. This mathematical model provides a prediction of
the "best" settings of RadDos and Prime1 which should produce the optimum value of TtrVol.

   Optimization experiments fall in the class of response surface methods.
In the simple case of two factors and one response, the mathematical model
for the response (z) describes a surface over the x-y plane. (This can be
generalized to higher dimensions.) This is usually shown in a 3-dimensional
surface plot as in Figure 13. Note that if there is an optimum response value
somewhere within the experiment region, the surface will be higher in that
region. Thus, a curved surface must be considered. Our experimental
design includes several levels (five) of each of the two remaining factors in
order to estimate the curvature in the response surface.

Experimental Design   The design used for the optimization experiment
is given in Table 3. This is a five-level space-filling design known as a central
composite design. (This design is listed in the Design Digest of Haaland

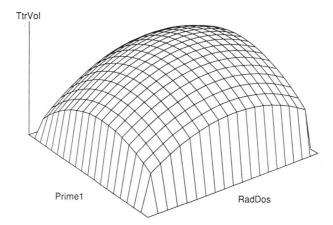

Figure 13 This response surface plot is a three-dimensional, graphical representation of the relationship between the response TtrVol and the experimental factors RadDos and Prime1. In this plot, the response is the z-value (vertical axis), and the factors are the x- and y-axes (horizontal plane). The response surface is generated by a mathematical model which approximates the true functional relationship between the response and the experimental factors.

(1989) as CC0211—central composite—2 factors—11 observations.) The center point of the design is repeated three times in order to allow a better estimate of the experimental error and to provide extra information about the yields in the interior of the experimental region.

This design efficiently fills the experimental space (Figure 14) and provides good estimates for curvature in the response surface. It is also a clear signal design. Complete information on all of the experimental designs which we commonly use for optimization experiments is provided in the Design Digest of Haaland (1989). For further information on the use of these designs, see also Box, Hunter, and Hunter (1978), Box and Draper (1987), Khuri and Cornell (1987), and Myers (1976).

The Mathematical Model The model we use to characterize TtrVol as a function of RadDos and Prime1 is a second order (quadratic) polynomial model. It is of the form

$$z = A + B * x + C * y + D * x^2 + E * y^2 + F * x * y + \text{error}.$$

The squared terms in x and y represent the curvature in the surface. The linear and two-factor interaction terms are the same as in a fractional

Table 3   Worksheet for Eleven
Run Optimization Experiment

Run	Factors RadDos	Primel	Response TtrVol
1.	100	7	207
2.	100	21	257
3.	300	7	306
4.	300	21	570
5.	200	4	315
6.	200	24	154
7.	59	14	100
8.	341	14	513
9.	200	14	630
10.	200	14	528
11.	200	14	609

Note: Experiments are run in a randomized order to prevent possible systematic bias.

factorial experiment. When we conduct an optimization experiment, we hope that the response surface will be curved (i.e., hump-shaped as in Figure 13) so that the optimum response will be found in the interior of the experimental region.

The central composite design provides good estimates for the curvature (quadratic) terms in the model. The estimated response surface is most useful if the true optimal settings fall within the experimental region. Since we successfully conducted a screening experiment, we have a good chance of satisfying this assumption. (One common error in the use of response surface methods is to do an optimization experiment without first verifying that the experimental region covers a fairly tight region around the optimum response.)

The statistical analysis of the experimental results provides the following estimated response surface model:

$$\text{predicted TtrVol} = -608.4 + 5.236 * \text{RadDos} + 77.0 * \text{Primel}$$
$$- 0.01265 * \text{RadDos}^2 - 3.243 * \text{Primel}^2$$
$$+ .07643 * \text{RadDos} * \text{Primel}.$$

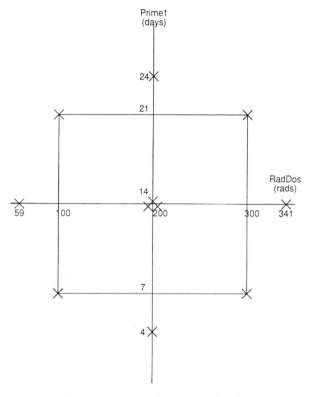

Figure 14 A central composite experimental design is a special space-filling design which we use for response surface experiments. By carefully distributing points throughout the experimental region, this design efficiently provides data which can be used to build a mathematical model. In particular, a central composite design provides an excellent estimate for the curvature of the response surface which is a key element in optimization.

This model generates a response surface plot for TtrVol and is used for numerical optimization and to generate the response surface shown in Figure 13. The dome shape of this surface indicates that within the experimental region there are settings for RadDos and Prime1 which produce an optimal response for TtrVol.

Optimization Results Using a numerical optimization program such as XSTAT, we determine that the predicted best settings are RadDos = 252 rads and Prime1 = 14.8 days. The predicted value of TtrVol at these settings is 622 units. This agrees well with the response surface in Figure 13.

A numerical optimization program is used because it is difficult to read exact values from a graph. With more factors or more than one response, such programs become essential. Two commercially available programs for this purpose are COED/RSM, which is available through the time-sharing service CompuServe, and XSTAT, which is a commercial PC program.

Further evaluation of the optimization results can be made using a contour plot (Figure 15). The contour plot shows contours of equal values of the response. It is very similar to a topographical map which shows contours of equal surface elevation. The "bulls-eye" pattern in Figure 15 is typical of the dome-shaped response surface in Figure 13.

One interesting and important feature that can be seen in the contour plot is that the values of the TtrVol do not fall off steeply if RadDos and Prime1 change slightly from their best values. This is good because the optimal solution will be robust to slight errors or variability in the experimental factors. This is very desirable property for an optimized process.

3.4   A Verification Experiment

The final step in the statistical problem solving process is to verify that the predicted solution works in practice. In order to accomplish this, a production run at the indicated settings was made. The production run produced 22 mg of monoclonal antibody per animal corresponding to a TtrVol of 602 (622 had been predicted). Previous production yields were 11 mg of monoclonal antibody per animal. Thus, a 100% increase in production yield was documented.

3.5   Summary

A screening experiment was used to evaluate six process variables in sixteen different experimental conditions. Each experimental condition corresponds to a particular set of values of the six factors. This screening experiment identified important factors and provided more information about their possible optimal settings. The data suggested that two experimental factors should be studied further in an optimization study. Two of the other six factors were qualitative factors whose best settings were determined in the screening experiment. The remaining two factors were

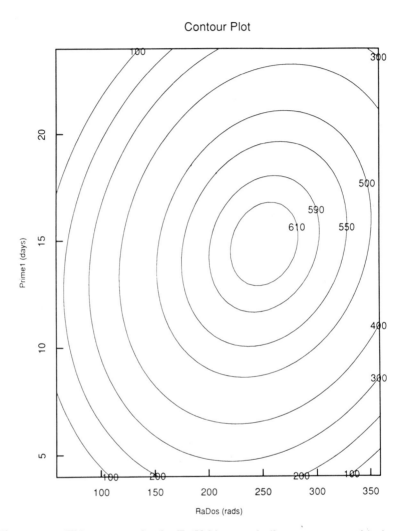

Figure 15   This contour plot for TtrVol is very similar to a topographical map used by hikers. The display consists of lines of equal value of the response. The patterns of these contours show how the response changes as a function of the two factors. If we are trying to maximize the response, we are looking for contours which represent a "hilltop."

found to have no significant effect on antibody yield and so were set at economical levels.

The second experiment used eleven experimental conditions as the basis for an optimization study. This data set was used to build a mathematical model which approximated ascites yield. Settings which were predicted to produce the optimum yield were determined for the remaining two experimental factors. Finally, the predicted yield at the best settings was verified in a production run.

Thus, a series of small, carefully designed experiments was used to solve a complex biological problem. Each experiment resulted in a small data set which provided information about several process variables. The information gained at each step was used to design the next experiment.

This problem was approached in a way characteristic of good statistical problem solving. At each step,

- well-defined questions were asked
- clear signal designs were used
- information-rich data were collected
- statistical graphics and analysis were used to provide answers.

After each experiment the results were evaluated and used to plan the next stage. Finally, a well-developed strategy was used to both facilitate and monitor progress toward the solution of the overall problem. The end result was a 100% increase in process yield.

4   CONCLUSION

Problem solving is a learning experiment. This learning occurs as we identify the important factors and learn how their settings affect process performance. The appropriate use of screening, optimization, and verification experiments provides the additional information necessary to continue progress toward the problem solution. An illustration of how knowledge about the problem increases as we approach the solution is displayed in Figure 16.

The successful problem solving strategy which we use has the following characteristics:

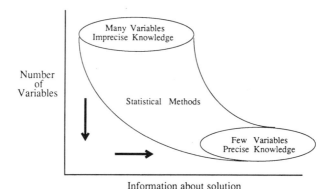

Figure 16 Two factors which affect our uncertainty about the solution to an empirical problem are which factors are important (y-axis) and how much we know about the best values for the important factors (x-axis). When we first approach an empirical problem we are in the upper left-hand corner of the x-y graph which expresses this uncertainty. The solution to the problem lies in the lower right-hand corner. Statistical problem solving is an efficient and effective method for taking us there.

- Data are collected with clear objectives in mind in order to insure that the data will determine the important facts about the process being studied.
- Information-rich data are insured by the use of statistical experimental designs for data collection.
- Answers to questions are provided by appropriate statistical analysis and graphical displays.
- A series of small experiments is used in an iterative approach to solving complex biological problems.

Although these problem solving methods have a long and successful history in many areas of science, resistance to change is universal and presents an obstacle to the adoption of statistical methods. We hear people express their resistance to changing their problem solving approach in some of the following sayings:

- "Let's just vary one thing at a time so that we don't get confused."
- "I'll investigate that factor next."
- "There aren't any interactions."

- "It's too early to use statistical methods."
- "A statistical experiment would be too large."
- "My data are too variable to use statistics."
- "We'll worry about the statistics after we've run the experiment."

These reasons are precisely why statistical problem solving methods should be used.

REFERENCES

Adrion, R. F., Siebert, G. R., C. J. Weck, D. Yen, and A. R. Manson (1984). Optimization of in vivo monoclonal antibody production using computer-assisted experimental design. *Proceedings of the First Carolina Biomedical Engineering Conference*, North Carolina Biotechnology Center, Research Triangle Park, North Carolina, pp. 125–144.

Box, G. E. P., and N. R. Draper (1987). *Empirical Model-Building and Response Surfaces*. Wiley, New York.

Box, G. E. P., and J. S. Hunter (1961a). The $2k$-p fractional factorial designs, Part I. *Technometrics*, 3, 311–352.

Box, G. E. P., and J. S. Hunter (1961b). The $2k$-p fractional factorial designs, Part II. *Technometrics*, 3, 449–458.

Box, G. E. P., W. G. Hunter, and J. S. Hunter (1978). *Statistics for Experimenters: An Introduction to Design, Data Analysis, and Model Building*. Wiley, New York.

Box, G. E. P., and R. D. Meyer (1986). An analysis for unreplicated fractional factorials. *Technometrics*, *28*: 11–18.

Cleveland, W. S. (1985). *The Elements of Graphing Data*. Wadsworth, Monterey, CA.

COED/RSM, CompuServe, Inc., ATTN: James P. Smith, Columbus, Ohio.

Daniel, C. (1959). Use of half-normal plots in interpreting factorial two-level experiments. *Technometrics*, *1*: 311–342.

Daniel, C. (1976). *Applications of Statistics to Industrial Experimentation*. Wiley, New York.

Davies, O. L., and W. A. Hay (1950). The construction and uses of fractional factorial designs in industrial research. *Biometrics*, *6*: 233–249.

Diamond, W. J. (1981). *Practical Experimental Designs*. Lifetime Learning Publications, Belmont, CA.

Finney, D. J. (1945). The fractional replication of factorial arrangements. *Annals of Eugenics*, *12*: 291–301.

Ghosh S. (1987). Non-orthogonal designs for measuring dispersion effects in sequential factor screening experiments using search linear models. *Communications in Statistics—Theory and Methods*, *10*: 2839–2850.

Haaland, P. D. (1989). *Biotechnology Experimental Design.* Marcel Dekker, New York.

Hunter, J. S. (1987). Experimental design: A winning strategy for industry, seminar series sponsored by Bolt, Beranek, and Newman.

Ishikawa, K. (1976). *Guide to Quality Control.* Nordica International Limited for The Asian Productivity Organization, Hong Kong.

Juran, J. M. (1964). *Managerial Breakthrough.* McGraw-Hill, New York.

Khuri, A. I., and J. A. Cornell (1987). *Response Surfaces: Designs and Analyses.* Marcel Dekker, New York.

Myers, R. H. (1986). *Response Surface Methodology.* Edwards Brothers, Ann Arbor, Mich.

Patel, M. S. (editor) (1987). Experiments in Factor Screening. A special issue of *Communications in Statistics—Theory and Methods, 10.*

Plackett, R. L., and J. P. Burman (1946). The design of optimum multifactorial experiments. *Biometrika, 33*: 305–325.

Srivastava, J. N. (1975). Designs for searching nonnegligible effects. *A Survey of Statistical Designs and Linear Models,* J. Srivastava, editor. North-Holland, Amsterdam, 507–519.

Taguchi, G. (1986). *Introduction to Quality Engineering: Designing Quality into Products and Processes.* Asian Productivity Organization, Tokyo, Japan.

XSTAT, Wiley Professional Software, Wiley, New York.

4

Expert Systems for the Design of Experiments

Christopher J. Nachtsheim, Paul E. Johnson, and Imran A. Zualkernan
University of Minnesota, Minneapolis, Minnesota

Kenneth D. Kotnour   3M Company, St. Paul, Minnesota

Ruth K. Meyer   St. Cloud State University, St. Cloud, Minnesota

1   INTRODUCTION AND SUMMARY

In recent years, interest in application of artificial intelligence in the practice of statistics has mushroomed. The collection of papers in Gale (1986a) and Haux (1986) and the extensive bibliographies contained therein attest to a corresponding surge in research. Much of the interest has focused on the construction of intelligent software for the analysis of data. The inevitability of such software has been discussed by Hahn (1985). Much of the recent research then has concerned the feasibility of building expert systems that would enable the relative novice to carry out appropriate analyses.

A number of expert systems for data analysis have been reported. Perhaps the best-known and most successful to data is REX (Regression Expert). REX was developed by Gale and Pregibon (1982) in an attempt to enable nonexperts to use regression analysis safely. Portier and Lai (1983) described the STATPATH system, which used production rules to help a novice select an appropriate method of analysis. Smith et al. (1983) developed BUMP, a program that would determine the user's objectives, develop formal representations, and interface to a multivariate analysis of variance package. Indeed, the RS/Explore software system (now commercially available through BBN Software Products Corporation) guides the user through the steps necessary for appropriate analyses of variance.

Considerably less emphasis has been placed on the development of expert systems for *planning* experiments. Lee et al. (1987) developed a system that automates the selection of orthogonal arrays in Taguchi experiments. A system for designing screening experiments, currently in use at Becton Dickinson Research Laboratories, has been described by Haaland et al. (1985). That very little work has been reported is not particularly surprising, since it is only recently that the value of the computer in the experimental design phase has even been recognized. In contrast to packages such as SPSS or SAS for the analysis of data, computer programs for the design of experiments are just beginning to appear. (See Nachtsheim, 1987, for review and discussion.) On the other hand, the need for intelligent systems for the design of experiments is at least as important as the need for data analytic expert systems. As is well known, even the most sophisticated analyses cannot salvage poorly designed studies. Unfortunately, few who are involved in the collection of data possess expertise in the planning of experiments; even fewer have access to statisticians. This situation has been aggravated, in some sense, by the quality movement in industry. In particular, increased interest in the application of experimental design in product and process development, generated by Genichi Taguchi, the Japanese quality expert, has been nothing short of phenomenal. There has, of course, been a concomitant increase in the demand for expertise. Automating that expertise in the form of an expert system is one potential solution.

In this paper, we summarize the results of a collaborative effort by the University of Minnesota and the Statistical Consulting group at 3M Company to assess the applicability of expert systems technology to the design of experiments in industry. We begin, in Section 2, with a discussion of strategies in experimental design. A Taguchi experiment previously reported in the literature is used to illustrate various strategic aspects. In Section 3, we describe an expert systems-building methodology that was brought to bear on the problem. This led to a model of the problem-solving process which served as a basis for the system. Details of the implementation are summarized in Section 4, and aspects of the system are demonstrated in connection with the aforementioned Taguchi example in Section 5. Finally, a discussion of directions for future work is provided in Section 6.

2 STRATEGIES IN EXPERIMENTAL DESIGN

Gale (1986) noted that a key issue in the development of statistical expert systems will be the determination and representation of strategies. Hand (1986, p. 356) gave the following definition of strategy: "a formal description of the choices, actions, and decisions to be made while using statistical methods in the course of a study." As noted by Thisted (1986), to date, most efforts in statistical computing have been directed toward the implementation of methods rather than strategies.

Various authors have identified two levels of strategy. Hand (1986) implicitly classifies strategies for choosing an appropriate method of analysis as high-level. Strategies having to do with the execution of a particular method, such as those embodied in REX, were considered low-level. Again in the context of data analysis, Thisted (1986) identified high-level strategies as those that distinguish among different classes of action. He used the term "tactics" in connection with lower-level strategies for choosing actions within a class. According to Tukey (1985, reported in Gale, 1986b, p. 2), statisticians "have been unwilling to think hard about what our strategies (not just individualized tactics) really have been or should be," and "one just cannot build an expert system without thinking through a strategy."

In experimental design, strategies are similarly partitioned. High-level strategies are concerned with identifying appropriate classes of experiments. For example, the consultant may recommend that a three-factor central composite response surface design is appropriate. Lower-level strategies would be invoked for later determination of the number of center points, locations of star points, and so forth. Work thus far in the expert systems/experimental design arena has dealt almost exclusively with the computerization of lower-level strategies.

To illustrate strategic aspects of a typical industrial experimental design problem, we consider a recent study (see Rickel and Griffith, 1987, for details) that concerned the manufacture of rubber weatherstrips in automotive production. Briefly, the rubber seals are extruded and then coated, or flocked, with material that prevents the seals from sticking when in use. Quality problems arose because the material was not adhering sufficiently to the rubber. The Taguchi method was used to investigate the problem.

Typically, the Taguchi method involves at least six steps (e.g., Kackar, 1985):

1. Brainstorming to identify control factors, noise factors, quality attributes (responses), effects to be estimated, and performance statistics.
2. Selecting an orthogonal array design.
3. Running the experiment.
4. Analyzing the results and identifying promising new settings of the control factors.
5. Testing the new settings.
6. Calculating the savings (i.e., the increase in quality) that came about as result of the experiment.

In this example, brainstorming centered around the diagram presented in Figure 1. Ishakawa, or cause-and-effect, diagrams such as this are often used to help identify factors that can potentially impact the quality attribute of interest. In this case, the four circled factors were selected as control factors in the experiment, and their associated levels are summarized in Table 1. Since each factor was to be run at three levels, an L_{27} orthogonal array was chosen. Results indicated that oven temperature, line speed, adhesive type, and the oven temperature by line speed interaction were statistically significant. The analysis indicated that the best levels of the four factors were those summarized in Table 1, and a test run corrob-

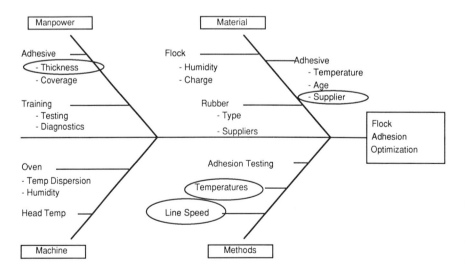

Figure 1   Cause-and-effect diagram for the rubber weatherstrip experiment.

Table 1 Factors and Associated Levels for the
Rubber Extrusion Experiment

Factor	Levels	Best level
A: Oven temperature	Low, medium, high	Medium
B: Line speed	30, 45, 60	60
C: Glue type	2, 23, 44	44
D: Wet film Thickness	4, 6, 8	4

orated the analysis. Use of the Taguchi loss function indicated that about
$500,000 would be saved on a yearly basis as a result of the experiment.

While the study was an unqualified success, there were some interesting
aspects that might have led an expert to a different design. We discuss
three of these in turn.

2.1 Selection of Control Factors

As a result of the experiment, line speed was set at its upper value, 60.
Actually, a preliminary analysis indicated that, for maximum quality, line
speed should be set at its slowest setting, 30. The authors noted, however,
that, in practice, the line must be run at its fastest setting to maximize
production. As a result, they settled on the best settings of the factors
A, C, and D, for high line speed. This raises the question: Why was line
speed included as a factor in the first place? Two-thirds of the resources
allocated to the experiment were not used efficiently. This underscores
an important characteristic of an expert system: it must be able to elicit
the objectives of the experiment effectively, objectives that may not be
clear even to the experimenter. Strategies are then necessary for selecting
number and levels of factors, sample sizes, operating ranges, etc., during
the problem formulation phase, to meet those objectives most effectively.

2.2 Choice of Design Type

An L_{27} orthogonal array design was chosen that allowed estimation of the
$A \times B$, $A \times C$, and $A \times D$ interaction. An advantage of orthogonal array
designs is that their use leads to simple graphical analyses. Since experimenters often must perform the analyses themselves, this is an important

aspect. On the other hand, different advantages accrue from the use of response surface experiments. If sample size were fixed at 27, an alternative design might consist of the 27-point central composite design with its the 2^4 corner points, 8 star points, and 3 center points. For this design, all two-, three- and four-factor interactions are estimable, and there are 2 degrees of freedom for estimation of pure error. Alternatively, a five-factor 27-point central composite design consisting of the 2^{5-1} fractional factorial design plus 10 star points and 1 center point would allow estimation of all main effects and two-factor interactions involving five factors. In this way, one additional factor from the cause-and-effect diagram of Figure 1 could have been studied. While either of these alternatives would result in increased complexity at the analysis stage, the additional information might justify selection. Our point here is not to suggest that any one of these three plans is necessarily superior; each of the three alternatives mentioned has its own advantages and disadvantages. The point is that a successful expert system must be able to generate these feasible alternatives, and it must embody a strategy for choosing among them.

2.3   The "Sliding Scale" Used for Factor Oven Temperature

The authors noted that it was not possible to run the manufacturing process at the low–oven temperature/high–line speed combination. This led to the use of what the authors termed the "sliding scale" for oven temperature summarized in Table 2. The resulting $A \times B$ design space is pictured in Figure 2. Clearly, factors A and B are no longer orthogonal: the correlation is .833. The implication of Figure 2 is that the experiment is not, in fact, being run as an orthogonal array experiment as the experimenters believed. As a

Table 2   "Sliding Scale" for Oven Temperature (A) as a Function of Line Speed (B)

	Oven Temperature		
Line speed	Low	Medium	High
30	390	400	410
45	420	435	450
50	430	450	470

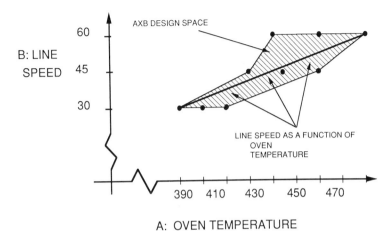

Figure 2   Line speed × oven temperature design space for the rubber weatherstrip experiment.

result, the analysis is incorrect and, in this case, misleading. A nonorthogonal (regression) analysis indicated that while A, B, and C were indeed significant as thought, the $A \times B$ interaction was not present. This leads to a different "best" setting, generating savings of about \$580,000/year. Moreover, the setting used by the authors will lead to about \$400,000/year in savings, as opposed to the indicated \$500,000/year. These results imply the potential for an additional savings of about \$180,000/year.

The problem brought about by the irregular design space could be handled by an expert system in a number of ways. First, the system could recommend use of the "modified" L_{27} array design, as employed by the authors, and simply remind the experimenter that a nonorthogonal analysis will be required. Alternatively, the system could elicit the constraints implied by Figure 2 directly and construct an optimal design for the irregular region. Regression analysis would again be required. In a somewhat different approach, the system might suggest that oven temperature and line speed may simply not be independently controllable. One of the factors, say, oven temperature, might simply be best thought of as a response to the other (line speed). This situation is summarized by the diagonal line of Figure 2. In this case, oven temperature could be dropped as a factor, leading to a potential reduction in the size of the experiment.

Clearly, the construction of expert systems for experimental design forces the identification and implementation of specific high-level strategies—something few statisticians have been willing to do. We turn now to details of an expert systems-building process that attempts to address a number of these issues.

3 BUILDING EXPERT SYSTEMS: MODELING EXPERTISE

In the preceding section, we argued that successful construction of an expert system requires the identification and modeling of expert strategies. In this section, we briefly present a methodology for construction of an expert system. (Over the last few years, several alternative methodologies for constructing expert systems have emerged. For example, see Hayes-Roth, 1983; Swartout, 1983; Freiling et al., 1986; Chandrasekaran, 1986; Clancey, 1988; Neches et al., 1985; and Gaines and Boose, 1988 for a collection of approaches to knowledge acquisition. Also, see Bobrow, 1986.) This methodology has evolved over several years of research in modeling expertise in domains such as auditing, medicine, computer hardware diagnosis, and experimental design.

3.1 Methodology

The methodology consists of the following steps.

Gain the Commitment of an Articulate Expert   By "articulate," we mean that the expert should be able to motivate and discuss the techniques, strategies, and heuristics used during a problem-solving session.

Understand and Specify the Expertise   A specification of expertise is a model of *what* is required to do a task (in this case experimental design). Marr (1982) referred to this type of specification as the *computational theory*. The data for this model-building activity are traces of behavior of the expert on selected test cases. These cases are constructed with the help of an accomplice (who is also an expert in the domain) and are, in turn, presented to the expert in simulated consulting sessions. During these sessions, the expert is required to comment (think aloud) on details of his or her approach to solving the problem. The "think-aloud" protocol, a transcription of these comments, is analyzed to understand the problem-solving process. An understanding of the problem-solving process is used to construct a specification or model of the expertise (Johnson et al., 1987a). The validity of think-aloud traces as data is discussed in Ericsson and Simon (1984) and Newell and Simon (1972).

Implement the Model of Expertise This stage is concerned with implementing the specification derived in the previous stage. The first step in this stage is the building of a theory of knowledge representation (Marr, 1982) that is best suited to the specification. However, because knowledge representation is itself an active area of research, this stage is concerned with selecting an appropriate knowledge representation scheme. (See Woods, 1987, for an overview of issues in knowledge representation.) Once an appropriate representational scheme is selected, a model of the specification is implemented using this scheme.

For the development of an expert system, it is important to distinguish between the specification and the knowledge representation scheme used for its implementation. As Chandrasekaran (1986) points out, most knowledge representation schemes such as rules-, frames-, or logic-based schemes are more like assembly languages of the field, as opposed to programming languages. These schemes should be used as modeling techniques to implement the specification model and should not be interpreted themselves as the most abstract model. For example, Clancey (1983, 1985) has shown that such an interpretation (e.g., using rules as the most abstract model of problem solving) obscures the developer's understanding of what is being modeled. This loss of understanding can lead to severe developmental problems (Zualkernan et al., 1988).

System Evolution: Testing, Validation, Debugging, and Revision
Both the specification and its implementation are typically incomplete. An exploratory style of development in which a model is continuously simulated and enhanced, is used (see Partridge, 1986; Sheil, 1984) to refine the specification and the implementation models of expert problem solving. The system's ability to evolve is critically dependent on the understanding of the problem-solving process gained at the specification stage and the extent to which that process was successfully modeled in the implementation phase. Typically, this is a cyclic process, in which the development goes through the cycles of specification, implementation, and evolution.

3.2   Application of the Methodology

In the work reported here, the implementation of the four steps just described proceeded as follows. We first obtained the commitment of a senior statistician in the Statistical Consulting Group at 3M. This expert had over 25 years of statistical consulting experience in an industrial chemical engineering environment. For comparative purposes (see below), we identified two additional statistical experts with varying degrees of familiarity with industrial problems. While both had received statistical training at the Ph.D. level, one, an academic with research interests in experimental design, had over five years of experience in designing industrial experiments. The other statistician was about to complete his Ph.D. program in statistics. We note that the academic's experience had been in food product development and other areas not directly related to chemical engineering. For this reason, refer to the academic as the "academic expert" For obvious reasons, our third statistician is referred to as the "novice expert."

With the cooperation of other members of the Statistical Consulting Group at 3M, we constructed six prototypical experimental design problems. These problems were based on consulting experiences previously encountered by members of the group (three of the problems were used to develop the model, and the other three were used to test it). Our industrial expert was not familiar with these problems. The types of problems included mixture formulation, process optimization, laboratory bench tests, a market test design, and one "fire-fighting" application in which a manufacturing line had suddenly begun producing unacceptable product. These problems were presented to the expert during a sequence of two-hour sessions in which thinking-aloud (protocol) data were collected.

Early analysis of the protocol obtained from experts working on three of the problem cases led to a model of the consultation process pictured in Figure 3. The form of this model is motivated by the heuristic classification model proposed by Clancey (1985) and the generic task characterizations of Chandrasekaran (1986). The model identifies three subprocesses or phases involved in the consultation process:

Interpretation   During this phase, the statistical consultant determines the nature of the client's problem. For example, objectives must be identified. Relevant factors and associated operating ranges are determined. The nature of the experimental unit, the need for blocking, and the ability to randomize must be assessed during this phase. Constraints on the experiment with regard to the total affordable sample size, available ma-

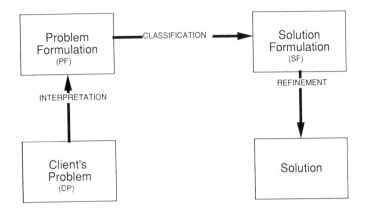

Figure 3   A problem-solving process model for experimental design.

terials, time, and personnel availability must be quantified. Sample size determination, if relevant, takes place during this phase. In essence, the client's domain-based problem is translated into a generic experimental design problem with associated constraints.

Classification   The set of feasible alternative experimental designs is determined during this phase. Here, the set of objectives, interesting factors, constraints, and so forth, summarized in the problem formulation (PF) are matched to sets of experimental designs that fit the characteristics of the problem. In the rubber extrusion example of Section 2, the three alternatives (L_{27} and two alternative 27-point response surface designs) would be identified as elements of this class.

Refinement   The term "refinement" refers to the problem of selecting a solution from the set of feasible designs and tailoring that design, if necessary, to match the particular needs of the client. A feasible design might, for instance, be a 2^{5-1} in four blocks of size 4. If the sample size can be as large as 20, the consultant must recommend whether to use the extra four observations. One strategy would place the four observations at the center point in one of the blocks (if possible), yielding an estimate of curvature and 3 degrees of freedom for pure error. If block size cannot exceed 5, an expert might recommend placing one additional center point in each block. In the context of the rubber extrusion experiment, trade-offs among the numbers of factors, numbers of replicates and orthogonality must be assessed in choosing between the orthogonal array design and the two alternative central composite designs.

To test the applicability of the process model shown in Figure 3, an experiment was conducted in which each of the three experts solved the other three prototype problems (i.e., the problems not used in the derivation of the process model). While details are given elsewhere (Johnson et al., 1988), the most interesting result is conveyed by Figure 4. To construct Figure 4, each paragraph in each protocol was analyzed to determine to which subprocess or phase it belonged. Phase was then plotted against paragraph number for each expert. The paragraph number (i.e., x-axis) scales have been adjusted to facilitate comparisons. Note the striking difference between the expert and the novice expert. The novice approach is essentially monotonic: the three phases are traversed sequentially, as represented by Figure 3. In contrast, the behavior exhibited by both the industrial expert and the academic expert was highly nonmonotonic. The industrial expert, in particular, moved frequently from interpretation to classification and back to interpretation.

There is clearly a "recognition" behavior that is not exhibited by the novice. That is, the expert, with very little exposure to the problem, is able to jump to feasible solutions and begin a process of negotiation with the client (see Johnson et al., 1987b, for details). This suggests that the industrial expert has seen similar problems previously and moves immediately to testing the applicability of his prior solutions. The tendency on the part of the novice expert is not to negotiate. The novice expert gathers all information possible during the interpretation phase and then tries to match this interpretation of the client's needs to a design type. Interestingly, the academic expert's behavior seems to be a compromise between the industrial expert and the novice. The solutions given by the expert and the academic expert were very similar, perhaps reflecting the shared industrial experience. For two of the three problems, the novice's recommended designs were substantially different.

The results summarized above suggest that powerful "recognition-based" heuristics are employed by domain experts. The problem-solving process is, we suspect, more accurately portrayed by the model of Figure 5. As shown in Figure 5, the expert moves directly form a very brief introduction to an initial solution formulation (SF). If necessary, the expert negotiates to modify the client's problem (CP). For example, the client might initially want to estimate curvature for many factors in order to optimize or fine-tune the response. This would restrict attention to designs

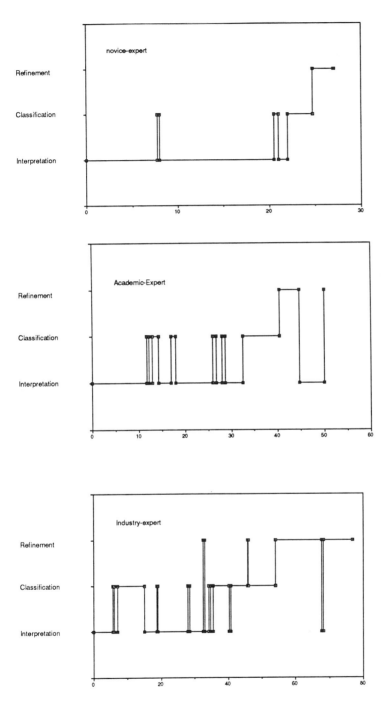

Figure 4 Problem-solving phase vs. protocol paragraph number for each of the three experts.

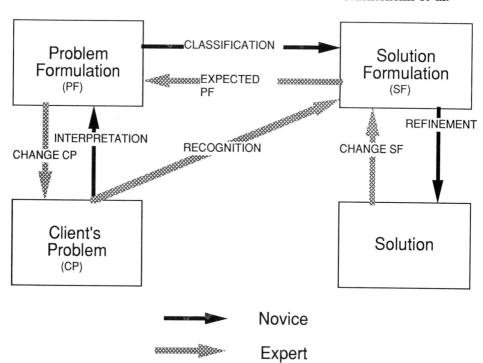

Figure 5   Revised process model for experimental design.

having at least three levels in each factor, leading to large numbers of runs.
Since linear effects can be estimated with much smaller designs, the expert
might negotiate for this change in objectives.

In the next section, we summarize our efforts to implement the
problem-solving model suggested by Figure 5 in an expert system named
MELAMPUS.[1]

4   BUILDING MELAMPUS: SOME SPECIFICS

MELAMPUS is a rule-based prototype expert system. At present, it is
based on the problem-solving model summarized in Figure 3 as a first step

---

[1]MELAMPUS is named after a Greek mythological character who was the first
mortal endowed with the powers of prediction.

toward implementation of the model suggested by Figure 5. Its capabilities are currently limited to two-level factorial (and fractional factorial) experiments with or without blocking. Taguchi-type (i.e., dispersion assessment) experiments can be constructed, as long as control and noise factors are limited to two levels. The system was developed using the Nexpert Object©️ v1.0 (Neuron Data, Inc.) expert system shell (see Shafer 1988) on a Macintosh II platform. An important aspect of our success in this exploratory phase has been the selection of this sophisticated, user-oriented expert systems shell. Choosing the appropriate shell significantly reduced the time required for prototype development.

Programming effort thus far is summarized by Figure 6, which traces the expansion in rules through accumulated development time. As the figure indicates, the current prototype comprises approximately 110 rules, a small system by most standards. The number of rules did not increase monotonically because, at times, sets of rules were consolidated, abstracted, or more economically represented. As indicated, total development time has been 140 person-hours. An additional 16 hours of interaction was required of the academic expert for the development of specific rules. Currently 50% of the rules are relevant to the interpretation phase. Because

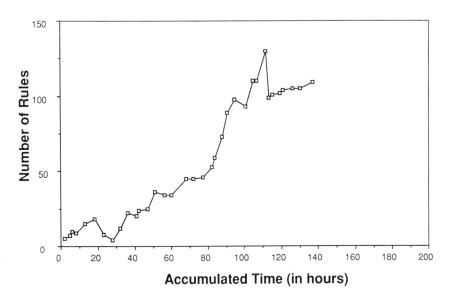

Figure 6   MELAMPUS development graph.

the breadth of available designs is limited (see below), only 20% of the rules are used during the classification phase.

As indicated, the system is still quite small. The objective has not been the development of a comprehensive system but rather development of a research prototype that could demonstrate significant expertise in specific situations. We thus decided to limit the scope of designs to two-level factorials; however, we hoped to go into some depth with regard to blocking and designing in the presence of fixed covariates. For example, to determine whether blocking is needed, the system must determine the nature of the experimental unit: blocks are constructed form collections of similar experimental units. Of course, this information must be elicited indirectly by the system—the nonexpert will rarely possess this kind of knowledge.

According to the model, the need for blocking is assessed after categorizing all factors of interest during the interpretation stage. The basis for

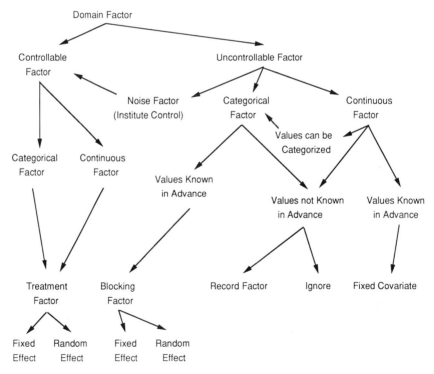

Figure 7   Factor classification scheme used by MELAMPUS.

factor categorization is given in Figure 7. If a domain variable is controllable, it must also be classified as either categorical or continuous. Obviously, certain classes of designs, such as central composite, cannot be used in the presence of categorical variables. If the domain variable is uncontrollable, it still has four potential uses. The user may wish to use the variable as a noise factor in a Taguchi experiment. In this case, the experimenter will institute control for the duration of the experiment. If the values of the uncontrollable factor are known in advance, this information can be used to help design a more efficient experiment. For example, a fixed number of experimental units, in the form of batches of raw material, may be available, having varying amounts of impurities. In this case, the fixed covariate, impurity level, can be used to determine an optimal assignment of treatments to batches (Harville, 1974). If the covariate values are categorical and there are sufficient numbers of experimental units in each class, the factor is a potential blocking variable. For example, if humidity changes from day to day and it is possible to complete 6 runs per day, maximum block size is 6. If the values of the uncontrollable factor will not be realized until the experiment is performed, the user can either record the observed values for later use in an analysis of covariance or simply ignore the factor.

Key aspects of the user interface of MELAMPUS are presented in Figures 8–11. In Figure 8, the system displays names of "application do-

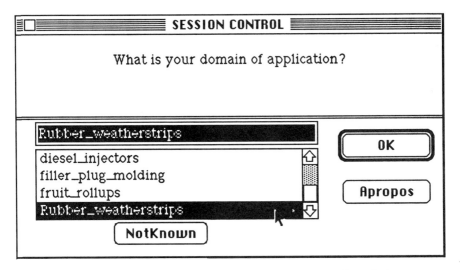

Figure 8   User interface application of domain selection window.

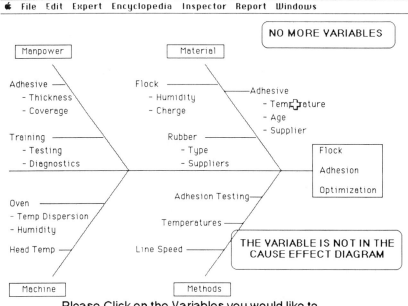

Figure 9   User interface cause-and-effect diagram for factor selection.

mains" currently stored. For example, if the user selects rubber weather-strips, the cause-and-effect display of Figure 9 appears. Potential factors are identified by cursor placement. If a cause-and-effect diagram does not exist, the user can either construct one or simply enter factors of inter-est through input windows. Since flock humidity has been identified, the user is asked, in Figure 10, whether it is controllable. The "apropos" box can be clicked to view associated help screens. Domain specific knowledge can enter into the interpretation stage. For example, the question put to the user in Figure 11 directly assesses the need for blocking on production shift.

MELAMPUS proceeds sequentially through the three phases identified in Figure 3. Thus, its behavior most closely emulates the novice expert. MELAMPUS has, nonetheless, identified appropriate designs in blocking situations.

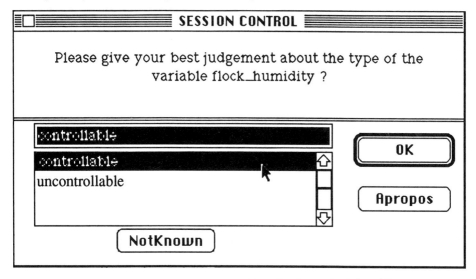

Figure 10   User interface classifying the factor flock humidity.

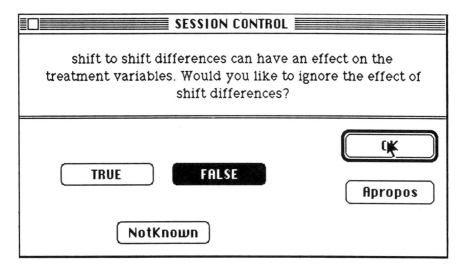

Figure 11   User interface assessing the need for blocking on production shift.

5 FUTURE WORK: THE POTENTIAL FOR EXPERT SYSTEMS IN THE PLANNING OF EXPERIMENTS

MELAMPUS has only begun to exploit the structure of the experimental design problem. In our view, the potential for application of an expert systems approach to the development of experimental design software is enormous. In what follows, some directions for additional research are discussed.

One difficulty with the development of data analytic expert systems has been the need for direct links to sophisticated graphical and statistical software. This is because the task performed during a particular step in a data analysis is typically determined by the results of an analysis performed in a preceding step. For instance, the decision to transform a variable may be the direct result of a p value computed or a residual plot observed during a previous step.

The situation is similar in the experimental design arena. Sophisticated numerical software for fractionating symmetric factorials, for determining if suitable fractions of mixed factorials exist, for determining if a symmetric fraction exists that allows estimation of particular effects (see, e.g., Turiel, 1987), and for constructing mixture, optimal, and optimal block designs for regular and irregular design spaces (Snee, 1987; Cook and Nachtsheim, 1989) is essential in the classification and planning phases. For instance, if certain factors must be run at three levels, others at two levels, and a full factorial is not possible, the system might consider the following alternatives:

- Find a suitable mixed orthogonal fraction, if one exists.
- Specify a model and construct an optimal design.
- Run all factors at two levels.

Feasible designs might exist in all three categories; in deciding which alternative is preferred, the system must evaluate the specific designs themselves. Thus, a direct link between the expert system and software for constructing designs is needed. A schematic is pictured in Figure 12.

As stated previously, "best" strategies are typically domain-specific. The expert system developed by Haaland et al. (1990) was intended for use in biological experiments. The system is devoted entirely to the construction of screening experiments that involve from 3 to 11 factors, accurately reflecting the needs of the experimenters in that domain. In contrast, our industrial expert recommended against running large, complicated exper-

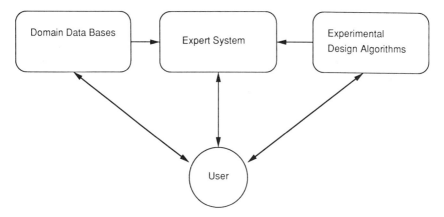

Figure 12   Schematic for future expert systems for experimental design.

iments in production environments. The probability of something going wrong, he felt, was simply too large. This leads to the need for a "generic" expert system for experimental design, which can be tailored to specific problem domains.

ACKNOWLEDGMENTS

We would like to thank Kip Smith, Rob Weiss, and the anonymous referees for useful comments.

REFERENCES

Chandrasekaran, B. (1986). Generic tasks in knowledge-based reasoning: High-level building blocks for expert systems, *IEEE Expert 1*: 23–30.

Clancey, W. J. (1983). The epistemology of a rule-based expert system—A framework for explanation, *Artificial Intelligence*, *20*: 215–251.

Clancey, W. J. (1985). Heuristic classification, *Artificial Intelligence*, *7*: 289–350.

Clancey, W. J. (1988). Acquiring, representing, and evaluating a competence model of diagnostic strategy, *The Nature of Expertise* edited by M. T. Chi, R. Glaser and M. Farr, Lawrence Earlbaum Associates, Hillsdale, New Jersey, pp. 343–418.

Cook, R. D. and Nachtsheim, C. J. (1989). Computer-aided blocking of factorial and response surface designs, *Technometrics*, in press.

Ericsson, A. K., and Simon, H. A. (1984). *Protocol Analysis*, MIT Press, Cambridge, Mass.

Freiling, M., Alexander, J., Messick, S., Rehfuss, D., and Shulman, S. (1985). Starting a knowledge engineering project: a step-by-step approach, *AI Magazine*, Fall 1985: 150–164.

Gaines, B. R. and Boose, J. H. (eds.) (1988). *Knowledge Acquisition for Knowledge-Based Systems*, Academic Press, London.

Gale, W. A. (1986a). *Artificial Intelligence and Statistics*, edited by W. A. Gale, Addison-Wesley, Reading, Mass.

Gale, W. A. (1986b). Overview of artificial intelligence and statistics, *Artificial Intelligence and Statistics*, edited by W. A. Gale, Addison-Wesley, Reading, Mass. pp. 1–16.

Gale, W. A. and Pregibon, D. (1982). An expert system for regression analysis, *Proceedings of the 14th Symposium on the Interface*, edited by Heiner, Scher, and Wilkinson, Springer-Verlag, New York, pp. 110–117.

Haaland, P. D., Lusth, J. C., Liddle, R. F., and Curry, J. W. DEXTER: a guide to selecting the best design for an industrial screening experiment, *Journal of Artificial Intelligence and Mathematics* (in press).

Hahn, G. J. (1985). More intelligent software and statistical expert systems: future directions, *American Statistician*, *39*: 1–8.

Hand, D. J. (1986). Patterns in statistical strategy, *Artificial Intelligence in Statistics*, edited by W. A. Gale, Addison-Wesley, Reading, Mass. pp. 335–387.

Harville, D. A. (1974). Nearly optimal allocation of experimental units using observed covariate values, *Technometrics*, *16*: 589–599.

Haux, R. (1986). *Expert Systems in Statistics*, edited by R. Haux, Fischer, New York.

Hayes-Roth, F., Waterman, D. A., and Lenata, D. B. (1983). *Building Expert Systems*, Addison-Wesley, Reading, Mass.

Johnson, P. E., Zualkernan I. A., Garber, S. (1987a). Specification of expertise, *International Journal of Man-Machine Studies*, *26*: 161–181.

Johnson, P. E., Nachtsheim C. J., and Zualkernan, I. A. (1987b). Consultant expertise, *Expert Systems*, *4*: 180–188.

Johnson, P. E., Meyer, R. K., Nachtsheim, C. J., and Zualkernan, I. A. (1988). Assessing a problem solving model for the design of experiments, Technical Report, Operation and Management Science Department, University of Minnesota, Minneapolis.

Kackar, R. N. (1985). Off-line quality control, parameter design, and the Taguchi method, *Journal of Quality Technology*, *17*: 176–188.

Lee, N. S., Phadke, M. S., and Keny, R. (1987). An expert system for experimental design: automating the design of orthogonal array experiments, presented at the 1987 Quality Congress, Minneapolis, Minn.

Marr D. (1982). *Vision*, W. H. Freeman, New York.

Nachtsheim, C. J. (1987). Tools for computer-aided design of experiments. *Journal of Quality Technology*, *19*: 132–160.

Neches, R., Swartout, W. R., and Moore, J. (1985). Enhanced maintenance and explanation of expert systems through explicit models of their development, *IEEE Transactions on Software Engineering, SE-11*: 1337–1351.

Newell A. and Simon, H. (1972). *Human Problem Solving*, Prentice-Hall, Englewood Cliffs. N. J.

Partridge, D. (1986). *Artificial Intelligence: Applications in the Future of Software Engineering*, Addison-Wesley, Reading, Mass.

Portier, K. M. and Lai, P. (1983). A statistical expert system for analysis determination. *Proceedings of the ASA Statistical Computing Section*, pp. 309–311.

Rickel, C. and Griffith, N. (1987). Optimization of flock adhesion on EPDM rubber weatherstrip, *Proceedings of the 5th Symposium on Taguchi methods*, ASI Press, Dearborn, Mich. pp. 727–741.

Shafer, D. (1988). Getting smart, *MacWorld, 5*: 132–139.

Sheil, B. (1984). Power tools for programmers, in *Interactive Programming Environments*, edited by D. Barstow, H. Shrobe, and E. Sandwell, New York, McGraw-Hill.

Smith, A. M. R., Lee, L. S., and Hand, D. J. (1983). Interactive user-friendly interfaces to statistical packages, *The Computer Journal: 26*, 199–204.

Snee, R. D. (1985). Computer-aided design of experiments, *Journal of Quality Technology, 17*: 222–236.

Swartout, W. J. (1983) XPLAIN: a system for creating and explaining expert consulting systems, *Artificial Intelligence, 21*: 285–325.

Thisted, R. A. (1986). Representing statistical knowledge for expert data analysis systems, *Artificial Intelligence in Statistics*, edited by W. A. Gale, Addison-Wesley, Reading, Mass., pp. 267–284.

Turiel, T. P. (1988). A FORTRAN program to generate fractional factorial experiments, *Journal of Quality Technology, 20*: 63–72.

Woods, W. A. (1987). Knowledge representation: what's important about it?, *The Knowledge Frontier*, edited by N. Cercone and G. McCalla, Springer-Verlag, New York.

Zualkernan, I. A., Tsai, W. T., Johnson, P. E., and Moller, J. H. (1988). Utility of knowledge level specifications, *Proceedings of the Fourth Annual Intelligence and Advanced Computer Technology Conference*, Long Beach, Calif.

5

The Effect of Ozone on Asthmatics and Normals: An Unbalanced ANOVA Example

Thomas J. Lorenzen   General Motors Research Laboratories, Warren, Michigan

1   INTRODUCTION

Folinsbee et al. (1984) noted that exposure to ozone has several short-term effects on lung functions of normal humans. Other studies, for example, Linn et al. (1980), have shown no quantitative differences between the responses of normal and asthmatic patients at low levels of ozone. The purpose of the experiment described in this paper was to determine whether asthmatic and normal patients differed in response to a higher (0.4 ppm) ozone concentration.

The experiment was brought to the author prior to the collection of any data and it was recognized that there were four factors of interest: health (asthmatic and normal), patients (nested within health), concentration of ozone exposure (0 and 0.4 ppm), and time (pre-exposure and every half hour up to the 2 hour exposure period). A balanced designed experiment was set up with patients as a random factor and all of the other factors fixed. The pre-analysis, indicating all tests and available information, is given in Section 2.

Nearing the completion of the experiment, the experimenters noted that there appeared to be a difference between male and female subjects (a sex effect) and, unfortunately, the number of males and females in each health category were not equal. Since it was difficult to find subjects and the experiment required specialized equipment and personnel, the experimental size was not increased in order to balance the experiment with respect to this additional factor. The result was an unbalanced five-factor design. The purpose of the experiment remained the same, the sex factor being a nuisance whose effects must be removed from the conclusions.

Previous work on unbalanced designs, for example, Speed, Hocking, and Hackney (1978), have focused on the various tests of hypotheses, assuming all factors are fixed and the effects are tested by the mean square error term. In this experiment, patients are considered random so we have a mixed rather than a fixed model. For this important difference, a premium is placed on computational formulae for expected mean squares and the determination of appropriate test statistics. Computation of expected mean squares for unbalanced designs has been given very little attention in the literature.

Section 3 lays out the unbalanced design for this five-factor experiment. New formulae for sums of squares, motivated by the desire for unbiased estimates of the usual model parameters, and the corresponding expected mean squares are given. These formulae lead to appropriate tests of hypotheses and estimation of contrasts.

Section 4 illustrates the formulae with some data collected in the ozone study. Section 5 briefly compares the estimates used in this paper with those calculated by the various methods in the SAS system. A short summary is given in Section 6.

2 ORIGINAL SETUP AND PRE-ANALYSIS

The experiment, Kreit et al. (1989), as originally brought to the author, was designed to measure the response of various lung functions caused by four factors: health (H_i, $i = 1$, 2 for asthmatics and normals), patients ($P_{j(i)}$, $j = 1, \ldots, 9$ random and nested within health since the 9 asthmatic patients differed from the 9 normal patients), ozone (O_k, $k = 1$, 2 for filtered air and 0.4 ppm ozone), and time (T_l, $l = 1, \ldots, 5$ for pre-exposure and 4 consecutive half-hour exposures). The experiment was completely randomized with the exception of the time factor, which could not be randomized (since time 2 must follow time 1, time 3 must follow time 2, and so on). As shown in Anderson and McLean (1974), this introduces a restriction error into the model. However, since the restriction error was also present in the control (0 ppm exposure) and comparisons were to be made at specific time periods, this restriction error was dropped from the model. The mathematical model describing this experiment is given by

$$y_{ijkl} = \mu + H_i + P_{j(i)} + O_k + HO_{ik} + PO_{jk(i)} + T_l + HT_{il}$$
$$+ PT_{jl(i)} + OT_{kl} + HOT_{ikl} + POT_{jkl(i)} + \epsilon_{(ijkl)}.$$

where y_{ijkl} is the measured response at time l.

The analysis of the experiment, as given in Peng (1967) for example, is indicated in Table 1. Of particular interest to the experimenter is the difference in response to ozone for asthmatics compared to normals. In terms of a hypothesis test, this can be written as

$$H_0 : \ HOT_{12l} - HOT_{11l} = HOT_{22l} - HOT_{21l}$$

vs.

$$H_1 : \ HOT_{12l} - HOT_{11l} \neq HOT_{22l} - HOT_{21l},$$

where HOT_{12l} $(= \mu + H_1 + O_2 + HO_{12} + T_l + HT_{1l} + OT_{2l} + HOT_{12l})$ is the average response for asthmatics (H_1) in ozone (O_2) at the lth time period (T_l). The appropriate test statistic for this contrast is written as

$$F_{4,64} = \frac{(\widehat{HOT}_{12l} - \widehat{HOT}_{11l} - \widehat{HOT}_{22l} + \widehat{HOT}_{21l})^2}{MSE(POT)(1/n_{12} + 1/n_{11} + 1/n_{22} + 1/n_{21})},$$

where $\widehat{HOT}_{12l}$ is the least squares estimate of the mean response for asthmatics in 0.4 ppm ozone for time period l based on $n_{12} = 9$ observations. Since the design is balanced, $\widehat{HOT}_{12l} = \bar{y}_{12.l}$, the arithmetic average of the observations for the 9 patients. The other terms are defined similarly with $n_{11} = n_{22} = n_{21} = 9$ observations.

For a balanced design, this F ratio is easily hand calculated. Alternatively, most major statistical packages will calculate the F statistic for you. In the SAS© system, this contrast is calculated using PROC GLM. Assuming the model in Table 1 is correct and the variables are labeled H, P, O, and T, respectively, the following SAS statements are used:

PROC GLM;

Table 1 ANOVA for Four Factor Experiment on Ozone

Source	df	SS	EMS
H_i	1	$90\sum_{i=1}^{2}(\bar y_{i\cdots} - \bar y_{\cdots})^2$	$\sigma_\epsilon^2 + 10\sigma_P^2 + 90\Phi(\mathrm{H})$
$P_{j(i)}$	16	$10\sum_{i=1}^{2}\sum_{j=1}^{9}(\bar y_{ij\cdots} - \bar y_{i\cdots})^2$	$\sigma_\epsilon^2 + 10\sigma_P^2$
O_k	1	$90\sum_{k=1}^{2}(\bar y_{\cdot\cdot k\cdot} - \bar y_{\cdots})^2$	$\sigma_\epsilon^2 + 5\sigma_{PO}^2 + 90\Phi(\mathrm{O})$
HO_{ik}	1	$45\sum_{i=1}^{2}\sum_{k=1}^{2}(\bar y_{i\cdot k\cdot} - \bar y_{i\cdots} - \bar y_{\cdot\cdot k\cdot} + \bar y_{\cdots})^2$	$\sigma_\epsilon^2 + 5\sigma_{PO}^2 + 45\Phi(\mathrm{HO})$
$PO_{jk(i)}$	16	$5\sum_{i=1}^{2}\sum_{j=1}^{9}\sum_{k=1}^{2}(\bar y_{ijk\cdot} - \bar y_{ij\cdots} - \bar y_{i\cdot k\cdot} + \bar y_{i\cdots})^2$	$\sigma_\epsilon^2 + 5\sigma_{PO}^2$
T_l	4	$36\sum_{l=1}^{5}(\bar y_{\cdots l} - \bar y_{\cdots})^2$	$\sigma_\epsilon^2 + 2\sigma_{PT}^2 + 36\Phi(\mathrm{T})$
HT_{il}	4	$18\sum_{i=1}^{2}\sum_{l=1}^{5}(\bar y_{i\cdot\cdot l} - \bar y_{i\cdots} - \bar y_{\cdots l} + \bar y_{\cdots})^2$	$\sigma_\epsilon^2 + 2\sigma_{PT}^2 + 18\Phi(\mathrm{HT})$
$PT_{jl(i)}$	64	$2\sum_{i=1}^{2}\sum_{j=1}^{9}\sum_{l=1}^{5}(\bar y_{ij\cdot l} - \bar y_{ij\cdots} - \bar y_{i\cdot\cdot l} + \bar y_{i\cdots})^2$	$\sigma_\epsilon^2 + 2\sigma_{PT}^2$
OT_{kl}	4	$18\sum_{k=1}^{2}\sum_{l=1}^{5}(\bar y_{\cdot\cdot kl} - \bar y_{\cdot\cdot k\cdot} - \bar y_{\cdots l} + \bar y_{\cdots})^2$	$\sigma_\epsilon^2 + \sigma_{POT}^2 + 18\Phi(\mathrm{OT})$
HOT_{ikl}	4	$9\sum_{i=1}^{2}\sum_{k=1}^{2}\sum_{l=1}^{5}(\bar y_{i\cdot kl} - \bar y_{i\cdot k\cdot} - \bar y_{i\cdot\cdot l} - \bar y_{\cdot\cdot kl} + \bar y_{i\cdots} + \bar y_{\cdot\cdot k\cdot} + \bar y_{\cdots l} - \bar y_{\cdots})^2$	$\sigma_\epsilon^2 + \sigma_{POT}^2 + 9\Phi(\mathrm{HOT})$
$POT_{jkl(i)}$	64	$\sum_{i=1}^{2}\sum_{j=1}^{9}\sum_{k=1}^{2}\sum_{l=1}^{5}(\bar y_{ijkl} - \bar y_{ijk\cdot} - \bar y_{ij\cdot l} - \bar y_{i\cdot kl} + \bar y_{ij\cdots} + \bar y_{i\cdot k\cdot} + \bar y_{i\cdot\cdot l} - \bar y_{i\cdots})^2$	$\sigma_\epsilon^2 + \sigma_{POT}^2$
$\epsilon_{(ijkl)}$	0	—	—

```
CLASSES H P O T;
MODEL Y = H | P(H) | O | T;
CONTRAST 'TIME 1' H * O +1 −1 −1 +1
                  H * O * T +1 −1 −1 +1   0 0 0 0   0 0 0 0   0 0 0 0   0 0 0 0/
                  E = P * O * T(H);
CONTRAST 'TIME 2' H * O +1 −1 −1 +1
                  H * O * T 0 0 0 0   +1 −1 −1 +1   0 0 0 0   0 0 0 0   0 0 0 0/
                  E = P * O * T(H);
        ⋮
CONTRAST 'TIME 5' H * O +1 −1 −1 +1
                  H * O * T 0 0 0 0   0 0 0 0   0 0 0 0   0 0 0 0   +1 −1 −1 +1/
                  E = P * O * T(H);
RUN;
```

In the ANOVA table printed by SAS, all factors are assumed fixed and the printed F values are not appropriate. Using the arrows in Table 1, you must either hand calculate the proper F values or give additional commands to SAS to have the package calculate the proper F values. Either way, the printed ANOVA table must be altered. You can tell SAS to calculate the proper F test on H_i by adding the command "TEST H = H E = P;." In this test statement, the first H stands for hypothesis, the second H stands for the health effect, the E for error, and the P for patient. The proper F tests on O_k and HO_{ik} are obtained with the command "TEST H = O H * O E = P * O;." The rest of the tests follow a similar pattern.

3   UNBALANCED ANALYSIS

Nearing the completion of the experiment, it was noted that the response variable was related to the sex of the patient and, unfortunately, there was not the same number of females and males within each health category. Since it was very expensive to find and include the additional patients necessary to make the experiment balanced with respect to sex, the data had to be analyzed using unbalanced methods. The remainder of this section discusses the resulting unbalanced design and presents a new method for analyzing the data.

Denote the new factor S_m, where $m = 1$ stands for a female patient and $m = 2$ stands for a male patient. The experiment had 5 asthmatic females, 5 normal females, 4 asthmatic males, and 4 normal males, resulting in an unbalanced design. The mathematical model describing this experiment now becomes

$$y_{imjkl} = \mu + H_i + S_m + HS_{im} + P_{j(im)} + O_k + HO_{ik} + SO_{mk} + HSO_{imk}$$
$$+ PO_{jk(im)} + T_l + HT_{il} + ST_{ml} + HST_{iml} + PT_{jl(im)} + OT_{kl}$$
$$+ HOT_{ikl} + SOT_{mkl} + HSOT_{imkl} + POT_{jkl(im)} + \epsilon_{(imjkl)}$$

where $j = 1, \ldots, 5$ for $m = 1$ and $j = 1, \ldots, 4$ for $m = 2$. (Note that we were lucky to have 5 female patients for both asthmatics and normals although this does not appreciably complicate the analysis. The only difference would be to have j depend on both i and m instead of depending on m alone.)

In the model given above, the usual ANOVA notation is used. That is, μ stands for the overall mean of the data, H_i for the differential effect of the ith health of the patient, HS_{im} for the interaction (added effect above or below the additive effect of each individual term) of health and sex, and so on for the other factors. The usual ANOVA assumptions on the terms in the model will be made: $\sum_{i=1}^{2} H_i = 0$, $\sum_{m=1}^{2} S_m = 0$, $\sum_{i=1}^{2} HS_{im} = \sum_{m=1}^{2} HS_{im} = 0$, $P_{j(im)}$ are Normal$(0, \sigma_P^2)$, $PO_{jk(im)}$ are Normal$(0, \sigma_{PO}^2)$ with $\sum_{k=1}^{2} PO_{jk(im)} = 0$, and so on for the remainder of the terms.

Because the design is unbalanced, the usual estimates of the ANOVA terms, that is, the estimates formed by applying the formulae for balanced designs, are biased. A few of the expected values using the formula for balanced designs are given in Table 2. These are tediously calculated by expanding the means, substituting the mathematical model for the y_{imjkl} terms, and algebraically simplifying the result. Note that the usual dot notation is employed; for example, $\bar{y}_{i \cdot \cdot k \cdot}$ is the average response obtained by summing over the m, j, and l indices and dividing by the number of responses summed. That is, $\bar{y}_{i \cdot \cdot k \cdot} = (\sum_{j=1}^{5} \sum_{l=1}^{5} y_{i1jkl} + \sum_{j=1}^{4} \sum_{l=1}^{5} y_{i2jkl})/(45)$. The bias introduced by the lack of balance in the experiment seems to follow a pattern that is heuristically described in the next paragraph.

The experiment was not balanced with respect to sex and it is the sex term and interactions with the sex term that bias the usual estimates. If the ANOVA term under consideration does not contain S_m, then the bias consists of the interaction of the term under consideration with S_m weighted according to the amount of data collected for each value of S_m. If the term under consideration contains S_m, then subtract the same term under consideration, again weighting by the amount of data collected for each value of S_m. Note that, had the design been balanced, the weights

Table 2 Some Expectations and Biases for Fixed Terms in an Unbalanced Design

ANOVA Term	Balanced Design Estimate	Expectation	Bias
μ	$\bar{y}_{\ldots\ldots}$	$\mu + (\frac{5}{9}S_1 + \frac{4}{9}S_2)$	$(\frac{5}{9}S_1 + \frac{4}{9}S_2)$
H_i	$\bar{y}_{i\cdots} - \bar{y}_{\ldots}$	$H_i + (\frac{5}{9}HS_{i1} + \frac{4}{9}HS_{i2})$	$(\frac{5}{9}HS_{i1} + \frac{4}{9}HS_{i2})$
S_m	$\bar{y}_{\cdot m\cdots} - \bar{y}_{\ldots}$	$S_m - (\frac{5}{9}S_1 + \frac{4}{9}S_2)$	$-(\frac{5}{9}S_1 + \frac{4}{9}S_2)$
HS_{im}	$\bar{y}_{im\cdots} - \bar{y}_{i\cdots} - \bar{y}_{\cdot m\cdots} + \bar{y}_{\ldots}$	$HS_{im} - (\frac{5}{9}HS_{i1} + \frac{4}{9}HS_{i2})$	$-(\frac{5}{9}HS_{i1} + \frac{4}{9}HS_{i2})$
O_k	$\bar{y}_{\cdots k\cdot} - \bar{y}_{\ldots}$	$O_k + (\frac{5}{9}SO_{1k} + \frac{4}{9}SO_{2k})$	$(\frac{5}{9}SO_{1k} + \frac{4}{9}SO_{2k})$
HO_{ik}	$\bar{y}_{i\cdots k\cdot} - \bar{y}_{i\cdots} - \bar{y}_{\cdots k\cdot} + \bar{y}_{\ldots}$	$HO_{ik} + (\frac{5}{9}HSO_{i1k} + \frac{4}{9}HSO_{i2k})$	$(\frac{5}{9}HSO_{i1k} + \frac{4}{9}HSO_{i2k})$
SO_{mk}	$\bar{y}_{\cdot m\cdot k\cdot} - \bar{y}_{\cdot m\cdots} - \bar{y}_{\cdots k\cdot} + \bar{y}_{\ldots}$	$SO_{mk} - (\frac{5}{9}SO_{1k} + \frac{4}{9}SO_{2k})$	$-(\frac{5}{9}SO_{1k} + \frac{4}{9}SO_{2k})$
HSO_{imk}	$\bar{y}_{im\cdot k\cdot} - \bar{y}_{im\cdots} - \bar{y}_{i\cdot\cdot k\cdot} - \bar{y}_{\cdot m\cdot k\cdot} + \bar{y}_{i\cdots} + \bar{y}_{\cdot m\cdots} + \bar{y}_{\cdots k\cdot} - \bar{y}_{\ldots}$	$HSO_{imk} - (\frac{5}{9}HSO_{i1k} + \frac{4}{9}HSO_{i2k})$	$-(\frac{5}{9}HSO_{i1k} + \frac{4}{9}HSO_{i2k})$
$\cdots$	$\cdots$	$\cdots$	$\cdots$

would be .5 and, by model assumptions, the bias would sum to 0. General rules for the bias of the usual estimates in the presence of unbalanced designs seem feasible but will not be offered here.

In order to write the proper sums of squares formulae, the balanced estimates must be altered in order to eliminate the bias. A few of these terms are written in Table 3. Again, general rules will not be offered although the expressions are written in such a fashion that a pattern should be ascertainable from the table. A few minutes examining the table will determine the pattern summarized here. Each ANOVA term containing the letter S_m must be adjusted to correct for the imbalance in the design. This adjustment is made by multiplying first by either $(\frac{9}{4})$ or $(\frac{9}{5})$, and then balancing the effects by multiplying by .5. Each ANOVA term not containing S_m is adjusted by subtracting the estimate of the bias found in Table 2, i.e., the interaction of that term with S_m. These terms look more complicated than the usual balanced estimates.

To construct the ANOVA table for the unbalanced experiment, we start by counting degrees of freedom (df). Instead of following the rules of thumb

Table 3   Unbiased Estimates of Some Fixed Terms in an Unbalanced ANOVA

ANOVA Term	Unbiased Estimate
μ	$\bar{y}..... - (\frac{5}{9})\hat{S}_1 - (\frac{4}{9})\hat{S}_2 \equiv \hat{\mu}$
H_i	$(\bar{y}_{i}.... - \bar{y}.....) - (\frac{5}{9})\widehat{HS}_{i1} - (\frac{4}{9})\widehat{HS}_{i2} \equiv \hat{H}_i$
S_1	$.5(\frac{9}{4})(\bar{y}._1... - \bar{y}.....) \equiv \hat{S}_1$
S_2	$.5(\frac{9}{5})(\bar{y}._2... - \bar{y}.....) \equiv \hat{S}_2$
HS_{i1}	$.5(\frac{9}{4})(\bar{y}_{i1}... - \bar{y}_{i}.... - \bar{y}._1... + \bar{y}.....) \equiv \widehat{HS}_{i1}$
HS_{i2}	$.5(\frac{9}{5})(\bar{y}_{i2}... - \bar{y}_{i}.... - \bar{y}._2... + \bar{y}.....) \equiv \widehat{HS}_{i2}$
O_k	$(\bar{y}..._k. - \bar{y}.....) - (\frac{5}{9})\widehat{SO}_{1k} - (\frac{4}{9})\widehat{SO}_{2k} \equiv \hat{O}_k$
HO_{ik}	$(\bar{y}_{i}.._k. - \bar{y}_{i}.... - \bar{y}..._k. + \bar{y}.....) - (\frac{5}{9})\widehat{HSO}_{i1k} - (\frac{4}{9})\widehat{HSO}_{i2k} \equiv \widehat{HO}_{ik}$
SO_{1k}	$.5(\frac{9}{4})(\bar{y}._1._k. - \bar{y}._1. - \bar{y}..._k. + \bar{y}.....) \equiv \widehat{SO}_{1k}$
SO_{2k}	$.5(\frac{9}{5})(\bar{y}._2._k. - \bar{y}._2. - \bar{y}..._k. + \bar{y}.....) \equiv \widehat{SO}_{2k}$
HSO_{i1k}	$.5(\frac{9}{4})(\bar{y}_{i1}._k. - \bar{y}_{i1} - \bar{y}_{i}.._k. - \bar{y}._1._k. + \bar{y}_{i}.... + \bar{y}._1... + \bar{y}..._k. - \bar{y}.....) \equiv \widehat{HSO}_{i1k}$
HSO_{i2k}	$.5(\frac{9}{5})(\bar{y}_{i2}._k. - \bar{y}_{i2} - \bar{y}_{i}.._k. - \bar{y}._2._k. + \bar{y}_{i}.... + \bar{y}._2. + \bar{y}..._k. - \bar{y}.....) \equiv \widehat{HSO}_{i2k}$
$\vdots$	$\vdots$

for the balanced case as given in Peng (1967) for example, we go back to first principles. The df are given by the number of parameters estimated minus the number of independent linear constraints. For this example, only the df for terms involving $P_{j(im)}$ differ from those given for a balanced design. For $P_{j(im)}$ itself, there are 5 asthmatic females yielding 4 df, 4 asthmatic males yielding 3 df, 5 normal females yielding 4 df, and 4 normal males yielding 3 df; for a total of 14 df. For $PT_{jl(im)}$, there are 5 time periods and one constraint, namely $\sum_{l=1}^{5} PT_{jl(im)} = 0$, for each health-sex combination. Therefore, there are $(4 \times 4) + (3 \times 4) + (4 \times 4) + (3 \times 4) = 56$ degrees of freedom. Other df terms are similarly calculated and are given in Table 4.

To calculate the sums of squares (SS) we will emulate the development for the balanced case. A test on the significance of a factor, say O_k, is given by the joint null hypothesis H_0: $O_1 = 0$, $O_2 = 0$. This is equivalent to the null hypothesis H_0: $\sum_{k=1}^{2} O_k^2 = 0$, which is equivalent to $[1/df(O)] \sum_{k=1}^{2} O_k^2 = 0$. A test statistic is created by taking the unbiased estimate of O_k, $\hat{O}_k = (\bar{y}_{...k.} - \bar{y}_{.....}) - (5/9)\widehat{SO}_{1k} - (4/9)\widehat{SO}_{2k}$, squaring each $\hat{O}_k$, and summing the squares over all of the data. The algebraically simplified results for some of the factors and interactions are given in Table 4.

Mean squares (MS) are obtained by dividing each sum of squares quantity by its appropriate degrees of freedom. These are not shown in Table 4.

While these unbalanced mean squares seem like reasonable statistics to work with, their properties can only be known by calculating their expected mean squares (EMS). Computation of the EMS is extremely tedious. Start with the formula for the mean squares, expand all of the means so the expression is written in terms of y_{imjkl}, substitute the mathematical model for y_{imjkl}, take the expectation of the complicated expression, and simplify the results using properties of the ANOVA terms. Note, in particular, that $P_{j(im)}$, $PO_{jk(im)}$, $PT_{jl(im)}$, $POT_{jkl(im)}$, and $\epsilon_{(imjkl)}$ are all normal random variables whose expectations are 0 and whose variances are given by σ_P^2, σ_{PO}^2, σ_{PT}^2, σ_{POT}^2, and σ_ϵ^2 respectively. The final results are given in Table 4. Due to the straightforward but lengthy nature of the calculations, none of the EMS calculations are illustrated here.

As can be seen in Table 4, H_i, S_m, and HS_{im} are tested by dividing their unbalanced mean squares by $(81/80)$ times the mean squares for $P_{j(im)}$; O_k, HO_{ik}, and SO_{mk} are tested by dividing their mean squares by $(81/80)$ times the mean squares for $PO_{jk(im)}$, and so forth. With the exception of the constant $(81/80)$, these are the same tests that would have been made in the balanced case! The constant $(81/80)$ is calculated as $[.5(9/4) + .5(9/5)]/2$ and can be thought of as the amount of unbalancing in the experiment. If

Table 4   ANOVA for the Five Factor Unbalanced Ozone Experiment

Source	df	SS	EMS
H_i	1	$90\sum_{i=1}^{2}\left[(\bar{y}_{i\cdots} - \bar{y}_{\cdots\cdots}) - (\frac{5}{9})\widehat{HS}_{i1} - (\frac{4}{9})\widehat{HS}_{i2}\right]^2$	$(\frac{81}{80})(\sigma_\epsilon^2 + 10\sigma_P^2) + 90\Phi(H)$
S_m	1	$100\left[.5(\frac{9}{4})(\bar{y}_{\cdot1\cdots} - \bar{y}_{\cdots\cdots})\right]^2 + 80\left[.5(\frac{9}{5})(\bar{y}_{\cdot2\cdots} - \bar{y}_{\cdots\cdots})\right]^2$	$(\frac{81}{80})(\sigma_\epsilon^2 + 10\sigma_P^2) + \Phi^*(S)$
HS_{im}	1	$\sum_{i=1}^{2}\{50[.5(\frac{9}{4})(\bar{y}_{i1\cdots} - \bar{y}_{i\cdots} - \bar{y}_{\cdot1\cdots} + \bar{y}_{\cdots\cdots})]^2$ $+ 40[.5(\frac{9}{5})(\bar{y}_{i2\cdots} - \bar{y}_{i\cdots} - \bar{y}_{\cdot2\cdots} + \bar{y}_{\cdots\cdots})]^2\}$	$(\frac{81}{80})(\sigma_\epsilon^2 + 10\sigma_P^2) + \Phi^*(HS)$
$P_{j(im)}$	14	$10\sum_{i=1}^{2}\left[\sum_{j=1}^{5}(\bar{y}_{i1j\cdots} - \bar{y}_{i1\cdots})^2 + \sum_{j=1}^{4}(\bar{y}_{i2j\cdots} - \bar{y}_{i1\cdots})^2\right]$	$\sigma_\epsilon^2 + 10\sigma_P^2$
O_k	1	$90\sum_{k=1}^{2}\left[(\bar{y}_{\cdots k\cdot} - \bar{y}_{\cdots\cdots}) - (\frac{5}{9})\widehat{SO}_{1k} - (\frac{4}{9})\widehat{SO}_{2k}\right]^2$	$(\frac{81}{80})(\sigma_\epsilon^2 + 10\sigma_{PO}^2) + 90\Phi(O)$
HO_{ik}	1	$45\sum_{i=1}^{2}\sum_{k=1}^{2}\left[(\bar{y}_{i\cdot k\cdot} - \bar{y}_{i\cdots} - \bar{y}_{\cdots k\cdot} + \bar{y}_{\cdots\cdots}) - (\frac{5}{9})\widehat{HSO}_{i1k} - (\frac{4}{9})\widehat{HSO}_{i2k}\right]^2$	$(\frac{81}{80})(\sigma_\epsilon^2 + 10\sigma_{PO}^2) + 90\Phi(HO)$
SO_{mk}	1	$\sum_{k=1}^{2}\{50[.5(\frac{9}{4})(\bar{y}_{\cdot1\cdot k\cdot} - \bar{y}_{\cdot1\cdots} - \bar{y}_{\cdots k\cdot} + \bar{y}_{\cdots\cdots})]^2$ $+ 40[.5(\frac{9}{5})(\bar{y}_{\cdot2\cdot k\cdot} - \bar{y}_{\cdot2\cdots} - \bar{y}_{\cdots k\cdot} + \bar{y}_{\cdots\cdots})]\}$	$(\frac{81}{80})(\sigma_\epsilon^2 + 10\sigma_{PO}^2) + \Phi^*(SO)$
$PO_{jk(im)}$	14	$5\sum_{i=1}^{2}\sum_{k=1}^{2}\left[\sum_{j=1}^{5}(\bar{y}_{i1jk\cdot} - \bar{y}_{i1\cdot k\cdot})^2 + \sum_{j=1}^{4}(\bar{y}_{i2jk\cdot} - \bar{y}_{i2\cdot k\cdot})^2\right]$	$\sigma_\epsilon^2 + 10\sigma_{PO}^2$
...	...	...	...

Note: $\Phi^*(S) = 100S_1^2 + 80S_2^2; \Phi^*(HS) = 50HS_{11}^2 + 40HS_{12}^2 + 50HS_{21}^2 + 40HS_{22}^2; \Phi^*(SO) = 50SO_{11}^2 + 50SO_{12}^2 + 40SO_{21}^2 + 40SO_{22}^2$

this result generalizes, and all indications are that it does generalize, then we know that all tests available in the balanced case will also be available in the unbalanced case. That is, the loss of one or two data points from a well-designed balanced experiment will not eliminate any of the tests that could be made prior to the loss of the data points. This is indeed an important result.

In a similar fashion, the test on the difference in response to ozone for asthmatics compared to normals at a specified time point is a contrast on the parameters $\mathbf{HOT}_{ikl}$ given by

$$F_{4,56} = \left(\frac{80}{81}\right) \left(\frac{(\widehat{\mathbf{HOT}}_{12l} - \widehat{\mathbf{HOT}}_{11l} - \widehat{\mathbf{HOT}}_{22l} + \widehat{\mathbf{HOT}}_{21l})^2}{\text{MSE(POT)}(1/n_{12} + 1/n_{11} + 1/n_{22} + 1/n_{21})}\right),$$

where $\widehat{\mathbf{HOT}}_{12l}$ is the unbiased estimate of the average response for asthmatics in 0.4 ppm ozone for time period l based on $n_{12} = 9$ observations. Note that the unbiased estimate $\widehat{\mathbf{HOT}}_{12l}$ contains the unbiased estimates of μ, H_1, O_2, T_l, HO_{12}, HT_{1l}, OT_{2l}, and HOT_{12l}, namely $\widehat{\mathbf{HOT}}_{12l} = \mu + \hat{H}_1 + \hat{O}_2 + \widehat{HO}_{12} + \hat{T}_l + \widehat{HT}_{1l} + \widehat{OT}_{2l} + \widehat{HOT}_{12l}$. The other terms in the above contrast are written similarly with $n_{11} = n_{22} = n_{21} = 9$ observations.

4 SOME DATA AND CALCULATIONS

For the experiment summarized in the previous section, several different lung function measurements were recorded. We illustrate the calculations using one of the lung function measurements, percent change in forced expiratory volume. Forced expiratory volume is the amount of air that can be expelled from the lungs. Since this volume is highly dependent on the size of the patient, percent change from a baseline is calculated. The baseline is the first reading taken the day of the test, prior to any of the experimental protocol. Readings are taken just before entering the air chamber and every half hour up to 2 hours. The readings for all of the patients are given in Table 5. The imbalance with respect to the sex variable is indicated by the lesser number of males than females.

The first step in the analysis is the calculation of unbiased estimates of the ANOVA terms. These are found by adjusting the cell means according to the formulae given in Table 3. Some of these estimates are given in

Table 5   Percent Change in Forced Expiratory Volume

HEALTH	SEX	PATIENT	OZONE EXPOSURE									
			0 ppm (O_1) TIME (Minutes)					0.4 ppm (O_2) TIME (Minutes)				
			(T_1) Pre	(T_2) 30	(T_3) 60	(T_4) 90	(T_5) 120	(T_1) Pre	(T_2) 30	(T_3) 60	(T_4) 90	(T_5) 120
Asthmatic (H_1)	Female (S_1)	AK ($P_{1(11)}$)	-5.6	1.0	-33.9	-26.6	-39.2	1.6	5.8	-21.1	-30.8	-26.3
		JS ($P_{2(11)}$)	-12.9	-14.0	-20.3	-25.1	-22.9	-5.1	-10.9	-33.0	-35.9	-49.3
		LA ($P_{3(11)}$)	-1.4	3.8	3.5	2.8	4.2	-1.8	-0.4	-0.4	1.1	-3.6
		LD ($P_{4(11)}$)	2.3	-4.3	-14.7	-3.1	-8.1	-4.1	1.4	-18.2	-15.1	-21.3
		SD ($P_{5(11)}$)	-1.0	0.7	4.3	4.7	5.0	2.9	1.8	-9.1	-12.0	-24.3
	Male (S_2)	BT ($P_{1(12)}$)	6.1	5.5	8.3	16.2	27.2	2.4	8.2	-18.2	-35.8	-36.1
		DW ($P_{2(12)}$)	2.7	-5.1	-31.4	-21.7	-12.0	-15.5	-13.2	-50.0	-39.7	-22.6
		JR ($P_{3(12)}$)	4.6	6.2	4.3	8.9	6.2	-4.4	-2.9	-9.3	-11.1	-17.4
		TP ($P_{4(12)}$)	-2.3	-0.8	1.9	4.9	2.4	-1.0	-2.2	1.2	1.0	-0.2
Normal (H_2)	Female (S_1)	GS ($P_{1(21)}$)	-3.8	-4.1	-1.4	0.2	-2.4	-2.2	-1.4	-1.0	-1.9	-9.9
		KG ($P_{2(21)}$)	-3.9	-1.8	-0.9	3.0	1.5	-6.6	-6.0	-18.4	-17.0	-23.3
		RF ($P_{3(21)}$)	-6.7	-8.0	-1.6	-1.3	-3.2	-8.8	-5.4	0.0	-3.0	-4.0
		RS ($P_{4(21)}$)	-3.1	-2.4	-1.5	-0.9	-1.7	-2.6	-3.2	-9.2	-11.9	-13.4
		SW ($P_{5(21)}$)	-4.7	-1.3	0.3	2.6	1.8	-6.9	-6.2	-5.2	-6.4	-7.9
	Male (S_2)	DT ($P_{1(22)}$)	-7.6	-2.5	-2.2	-1.0	0.5	-6.0	-4.8	-7.6	-5.6	-8.7
		JK ($P_{2(22)}$)	-3.4	-1.2	0.6	1.4	0.6	-3.9	-1.6	-1.8	-3.3	-6.8
		MB ($P_{3(22)}$)	-4.3	-5.6	-5.8	-2.5	-2.7	-0.9	2.6	-1.3	-6.2	-2.1
		RB ($P_{4(22)}$)	-4.3	-3.3	0.0	-1.7	0.7	-3.2	-4.3	-7.5	-13.3	-18.6

Table 6. The particular estimates given in Table 6 are those that will be needed to form the contrasts of interest.

Using the unbiased estimates of the ANOVA terms, the next step is to calculate the sums of squares. This is accomplished by squaring the unbiased estimator of the appropriate ANOVA term and summing over all of the data. Some specific formulae are given in Table 4 of the previous section, with the numerical results given in Table 7. In this table, tests are indicated by arrows with the base of the arrow indicating the denominator and the tip of the arrow indicating the numerator. Note that a multiplicative constant must be used to account for the unbalanced nature of the design.

Table 7 also contains the calculated mean squares and the F statistics. The mean squares column is simply the sum of squares column divided by the degrees of freedom column. Note, however, that the F column is formed by taking (80/81) times the appropriate ratio, the constant (80/81) entering into the picture because the data are unbalanced. The critical F is unchanged by the unbalanced nature of the experiment.

The interpretation of Table 7 is the same for the balanced and the unbalanced case. We see that there is a significant change in the normalized

Table 6   Unbiased Estimates of Percent Change in Forced Expiratory Volume

ANOVA Term	Unbiased Estimate	ANOVA Term	Unbiased Estimate	ANOVA Term	Unbiased Estimate	ANOVA Term	Unbiased Estimate
μ	-6.079375	HT_{11}	3.367125	OT_{15}	3.722875	HOT_{124}	-0.862250
H_1	-1.943375	HT_{12}	3.073375	OT_{21}	2.895875	HOT_{125}	-0.914125
H_2	1.943375	HT_{13}	-2.769125	OT_{22}	3.257125	HOT_{211}	0.668375
O_1	3.465875	HT_{14}	-2.051000	OT_{23}	0.037125	HOT_{212}	1.083375
O_2	-3.465875	HT_{15}	-1.620375	OT_{24}	-2.467250	HOT_{213}	0.024625
HO_{11}	1.343375	HT_{21}	-3.367125	OT_{25}	-3.722875	HOT_{214}	-0.862250
HO_{12}	-1.343375	HT_{22}	-3.073375	HOT_{111}	-0.668375	HOT_{215}	-0.914125
HO_{21}	-1.343375	HT_{23}	2.769125	HOT_{112}	-1.083375	HOT_{221}	-0.668375
HO_{22}	1.343375	HT_{24}	2.051000	HOT_{113}	-0.024625	HOT_{222}	-1.083375
T_1	2.938125	HT_{25}	1.620375	HOT_{114}	0.862250	HOT_{223}	-0.024625
T_2	3.925625	OT_{11}	-2.895875	HOT_{115}	0.914125	HOT_{224}	0.862250
T_3	-2.178125	OT_{12}	-3.257125	HOT_{121}	0.668375	HOT_{225}	0.914125
T_4	-1.757500	OT_{13}	-0.037125	HOT_{122}	1.083375		
T_5	-2.928125	OT_{14}	2.467250	HOT_{123}	0.024625		

Table 7 ANOVA for Percent Change in Forced Expiratory Volume

Source	df	SS[a]	MS	F[b]	F(.05)
⟶ H_i	1	679.81	679.81	1.33	4.60
⟶ S_m	1	295.20	295.20	0.58	4.60
⟶ HS_{im}	1	137.69	137.69	0.27	4.60
⟶ $P_{j(im)}$	14	7179.25	512.80	—	—
⟶ O_k	1	2162.21	2162.21	16.79	4.60
⟶ HO_{ik}	1	324.84	324.84	2.52	4.60
⟶ SO_{mk}	1	190.17	190.17	1.48	4.60
⟶ HSO_{imk}	1	481.40	481.40	3.74	4.60
⟶ $PO_{jk(im)}$	14	1803.00	1803.00	—	—
⟶ T_l	4	1456.20	364.05	6.34	2.54
⟶ HT_{il}	4	1270.20	317.55	5.53	2.54
⟶ ST_{ml}	4	207.90	51.97	0.90	2.54
⟶ HST_{iml}	4	166.25	41.56	0.72	2.54
⟶ $PT_{jl(im)}$	56	3216.61	57.44	—	—
⟶ OT_{kl}	4	1401.96	350.49	10.79	2.54
⟶ HOT_{ikl}	4	115.20	28.80	0.89	2.54
⟶ SOT_{mkl}	4	19.72	4.93	0.15	2.54
⟶ $HSOT_{imkl}$	4	24.09	6.02	0.19	2.54
⟶ $POT_{jkl(im)}$	56	1819.38	32.49	—	—
Total	179	22882.93			

[a]SS are calculated using unbiased estimates of ANOVA terms.
[b]$F = (81/80) \times$ (indicated ratio) since the data are unbalanced.

forced expiratory volume due to ozone exposure, due to time, and due to the health-time and ozone-time interactions. Since time is really exposure time, the ozone-time interaction is to be expected. The real question of interest to the investigators is whether ozone affects asthmatics differently than it affects normals. The significant health-time interaction indicates that there is a differential effect.

Table 8   Difference in Response to Ozone
for Asthmatics and Normals

Exposure Time	Estimated Difference	Calculated Contrast	Critical F(.05)
Pre	−2.7000	0.50	2.54
30	−1.0400	0.07	2.54
60	−5.2750	1.90	2.54
90	−8.8225	5.32	2.54
120	−9.0300	5.58	2.54

To better understand the differential effect of ozone on asthmatics and normals, we wish to directly compare differences at each time point l. At a given time point l, the difference in response under ozone minus the response under air gives the effect of ozone. By comparing the ozone difference for asthmatics to the ozone difference for normals, we can determine whether asthmatics differ in response to ozone compared to normals. Let $\widehat{HOT}_{ikl} = \hat{\mu} + \hat{H}_i + \hat{O}_k + \widehat{HO}_{ik} + \hat{T}_l + \widehat{HT}_{il} + \widehat{OT}_{kl} + \widehat{HOT}_{ikl}$ be the unbiased estimate of average response for health i and ozone exposure j at time l. Then, at time l, the difference in response for asthmatics compared to normals is given by $\widehat{HOT}_{12l} - \widehat{HOT}_{11l} - \widehat{HOT}_{22l} + \widehat{HOT}_{21l}$ with the test statistic as indicated in the previous section. The numerical values of the differences and the test contrasts are given in Table 8. We see that there is a significant degradation in lung function for asthmatics relative to normals after 90-min exposure to 0.4 ppm ozone. Note also that all sex effects have been removed from this comparison.

5   COMPARISON WITH OTHER UNBALANCED METHODS

The calculation of sums of squares as indicated in Table 4 differs from all of the SAS types I through IV calculations. SAS type I sums of squares are sequential in nature and are not meant to be applied to analysis of this nature. Type IV sums of squares have the drawback that their calculations depend on the order in which the ANOVA terms are entered into the model statement (Freund and Littell, 1981).

SAS type II sums of squares correspond to the usual sums of squares not adjusted for the bias caused by the unbalanced experiment. That is, type II sums of squares correspond to the usual estimates given in the second column of Table 2. This results in a confounding of the term of interest and the interaction of that term with the unbalanced sex term. For example, the type II sums of squares for H_i is given by $90 \sum_{i=1}^{2} (\bar{y}_{i\dots} - \bar{y}_{\dots})^2$ which has an expected mean square of $\sigma_\epsilon^2 + 10\sigma_P^2 + 90 \sum_{i=1}^{2} [H_i + (5/9)HS_{i1} + (4/9)HS_{i2}]^2$, confounding H_i with the HS_{im} interaction term.

It is difficult to express the SAS type III sums of squares in terms of mean quantities for these data since their descriptions are given in terms of the $R(\cdot)$ notation described in Speed and Hocking (1976). However, the SAS PROC GLM command can print out expected mean squares through the use of the RANDOM statement. While SAS assumes each main effect or interaction is either fixed or random [e.g., $PO_{jk(im)}$ are normal $(0, \sigma_{PO}^2)$ but $\sum_{k=1}^{2} PO_{jk(im)} \neq 0$] and thus computes a different EMS than it would if it used the assumptions of this paper, the printed quadratic term will indicate the confounding for each fixed term. The type III EMS for H_i has a quadratic form involving H_i and all interactions involving H_i. Therefore, type III sums of squares is not desirable for analyzing this experiment.

The type IV sums of squares agree with the type III sums of squares as long as there are no missing cells. However, SAS treats all factors as fixed and therefore thinks there are missing cells associated with the patient term $P_{j(im)}$. As a result, the type IV sums of squares differ from the type III sums of squares. Unfortunately, the type IV sums of squares depend on the labeling used when entering the data. (See Freund 1980 for a more detailed explanation.) For the particular labeling used when entering these data, the type IV sums of squares for H_i agreed with the type II sums of squares for H_i while the type IV sums of squares for S_j agreed with the type III sums of squares for S_j, neither of which is acceptable for analyzing this experiment.

A final method of analyzing unbalanced data is the method of weighted squares of means described in Yates (1934). Extension from the simple fixed model given in the reference to the present ozone example is far from straightforward. However, SAS can be "fooled" into calculating the weighted squares of means for every term not containing $P_{j(im)}$ by simply deleting the patient term and its interactions from the MODEL statement. It will still be necessary to calculate the patient SS's by hand. For this model, the types III and IV SS are the same and equal the weighted squares of means. The type II SS are as described three paragraphs above.

The expected mean squares cannot be calculated by SAS since there are no random terms left in the model. However, they can be calculated by hand. The expected mean squares for the method of weighted squares of means do lead to legitimate tests. For example, the EMS for the health factor is $\sigma_\epsilon^2 + 10\sigma_P^2 + 88.889(H_1^2 + H_2^2)$ while the EMS for the sex factor is $\sigma_\epsilon^2 + 10\sigma_P^2 + 79.01S_1^2 + 98.77S_2^2$, both terms being tested by the patient $P_{j(im)}$ term. Thus, besides the new method presented here, the method of weighted squares of means represents another viable technique for analyzing this particular data set. It can be calculated by appropriately "fooling" SAS but the tests on contrasts must be calculated by hand. It is difficult to see how the weighted squares of means method extends to models having more than one random factor.

6 SUMMARY

Unbalanced designs commonly occur and create special problems in terms of analysis and interpretation. When all of the factors are fixed, dummy variable regression techniques such as those given in Dodge (1985) and used in computer packages such as SAS may give good analysis and reasonable interpretation. However, when some factors are fixed and others are random, both the analysis and the interpretation become extremely difficult.

In this paper, we considered one unbalanced design, arising because an important variable was discovered only after the experiment was completed. A step-by-step approach was used to model, analyze, and interpret the design. Using the usual ANOVA notation and assumptions, it was shown that the usual mean estimates were biased and therefore unacceptable. The pattern of the biases was derived and it appears that some general rules can be specified.

To analyze the data, sum of squares formulae were developed along the same lines as those used for the balanced case, i.e., by summing over all of the data the square of the unbiased estimate of the parameter under consideration. The same thought process can be used for any unbalanced design, although no automated method for directly writing the SS's was presented. It turns out that this method of computing sums of squares disagrees with common methods for computing sums of squares of unbalanced data. All of the type I through type IV SS's calculated by SAS resulted in unacceptable tests for this example. However, SAS could be "fooled" into calculating the weighted squares of means which did result in valid test statistics. It is not known whether such a trick can always be performed.

To form test statistics using this new method, an extremely tedious algebraic method of computing EMS was used. Such a method is obviously unacceptable for general use. However, the ratios forming the test statistics agreed with the ratios used in the balanced case (for which simple algorithms exist) up to a multiplicative constant. General procedures may be possible if this relation holds for other cases as well.

General methods for the analysis and interpretation of balanced designs exist, are well understood, and are commonly used in practice. The same cannot be said for unbalanced designs, even though unbalanced designs arise quite often and for a variety of reasons. This paper considered one complicated unbalanced design, exhaustively analyzed and interpreted the design, and laid out the results so generalizations should be possible. Hopefully, some day, unbalanced designs will be as well understood and used as balanced designs.

ACKNOWLEDGMENTS

This problem was brought to my attention by Dr. Ken Gross of the Biomedical Science Department of General Motors Research Laboratories and Drs. John Kreit and William Eschenbacher of the University of Michigan Medical Center. They ran the original four-factor experiment exactly as it was designed and recognized the importance of a fifth factor. Thanks also go to the attendees of the 1988 ASA Winter Conference whose discussion led to alternate methods of analyzing this unbalanced data.

REFERENCES

Anderson, V. L. and McLean, R. A. (1974). *Design of Experiments A Realistic Approach*, Marcel Dekker, New York.

Dodge, Y. (1985). *Analysis of Experiments with Missing Data*, Wiley, New York.

Folinsbee, L. J., Bedi, J. F., and Horvath, S. M. (1984). "Pulmonary function changes after 1 h continuous heavy exercise in 0.21 ppm ozone," *Journal of Applied Physiology*, *57*: 984–988.

Freund, R. J. (1980). "The case of the missing cell," *The American Statistician*, *34*: 94–98.

Freund, R. J. and Littell, R. C. (1981). *SAS for Linear Models: A Guide to the ANOVA and GLM Procedures*, SAS Institute Inc., Cary, N. C.

Kreit, J. W., Gross, K. B., Moore, T. B., Lorenzen, T. J., D'Arcy, J. B., and Eschenbacher, W. L. (1989). "Ozone-induced changes in pulmonary function and bronchial responsiveness in asthmatics," *Journal of Applied Physiology*, *66*: 217–222.

Linn, W. S., Jones, M. P., Backmayer, C. E., Spier, S., Fasciano, M., Avol, E. L., and Hackney, J. D. (1980). "Health effects of ozone exposure in asthmatics," *American Review of Respiratory Disorders*, *121*: 243–252.

Peng, K. C. (1967). *The Design and Analysis of Scientific Experiments An Introduction with Some Emphasis on Computation*, Addison-Wesley, Reading, Mass.

Speed, F. M. and Hocking, R. R. (1976). "The Use of R( ) Notation with Unbalanced Data," *The American Statistician*, *30*: 30–33.

Speed, F. M., Hocking, R. R., and Hackney, O. P. (1978). "Methods of Analysis of Linear Models with Unbalanced Data," *Journal of the American Statistical Association*, *73*: 105–112.

Yates, F. (1934). "The Analysis of Multiple Classifications with Unequal Numbers in the Different Classes," *Journal of the American Statistical Association*, *29*: 52–66.

6

Sensitivity of an Air Pollution and Health Study to the Choice of a Mortality Index

Diane I. Gibbons and Gary C. McDonald   General Motors Research Laboratories, Warren, Michigan

1   INTRODUCTION

Over the last fifteen years Lave and Seskin have published a series of articles assessing the quantitative effects of air pollution on human mortality, and much of this work is summarized in their book *Air Pollution and Human Health* (1977). Due to the policy implications of these findings, these analyses have become the target of rather intense statistical and scientific review and reanalysis. It is our intent here to also engage in such retrospection. The initial approach of Lave and Seskin is very direct and straightforward. They assemble for 117 Standard Metropolitan Statistical Areas (SMSAs) a data base which includes a mortality rate and indices of air pollution, population density, age distribution, poverty, racial mix, and other quantities thought to possibly affect the mortality pattern of a community. The mortality rate is then regressed on the explanatory indices using a "standard" least squares multiple linear regression model. Based on this regression quantification, they note that a 50% reduction in air pollution corresponds to a 4.7% reduction in the mortality rate (Lave and Seskin, 1977, p. 218). A substantial portion of the Lave and Seskin book

(1977) is subsequently devoted to examining the sensitivity of this conclusion to a maze of potential technical pitfalls underlying this direct regression approach to the question. Each of these sensitivity checks appears to roughly validate the initial regression findings and empirically attest to the robustness of the usual regression approach to a practical problem. The findings as originally reported appear statistically significant and possess substantial policy implications.

The Lave and Seskin analyses have not gone without review and critique. Thibodeau et al. (1980) have undertaken a lengthy critique of the data and methodology used in the Lave and Seskin studies. After correcting errant data, removing outliers, and reanalyzing the resultant data, Thibodeau et al. did not dispute the existence of an association between air pollution and mortality but did express some reservations about its reported magnitude. Their reanalysis of the data using eleven explanatory variables yields air pollution regression coefficient estimates as given in Table 1. Their reanalysis indicates a shift in the sulfate impact from the minimum biweekly reading to the arithmetic mean of the biweekly readings and for the particulate impact from the mean to the minimum reading.

The Thibodeau study, which examines the accuracy of the data and the sensitivity of the regression estimates to aberrant SMSAs, represents one critique of the Lave and Seskin analyses. Another approach is described by Cooper and Hamilton (1979) and Gibbons and McDonald (1980). These studies explore the model building aspects of the regression problem; i.e., which of a large number (61) of possible explanatory variables should be included in "the" regression model and in what form should these variables be entered (log transforms, inverses, etc.). If we set aside the question of form, there are $2^{61} = 2.3 \times 10^{18}$ linear regression models which can be con-

Table 1 1960 Total Mortality Rates: Lave and Seskin Air Pollution Regression Estimates for 117 SMSAs and the Thibodeau et al. Reanalysis for 108 SMSAs

Biweekly Pollution Reading	Sulfates		Particulates	
	Lave & Seskin	Reanalysis	Lave & Seskin	Reanalysis
Minimum	.47	.08	.20	.57
Mean	.17	.26	.30	.01
Maximum	.03	.08	−.02	.01

structed from this set of 61 variables. (Cooper and Hamilton augmented the Lave and Seskin data base of 60 explanatory variables with an additional variable—median age). How can we be assured that the 11 variable model and the 7 variable model proposed by Lave and Seskin are the "best" or at least acceptably good models—all things considered? There are computer routines to answer these questions if "best" is defined as a function of the residual sum of squares corresponding to a model fit (Furnival and Wilson, 1974). Our experience, however, indicates that these routines are not able to search all possible regressions using all 61 explanatory variables. The "best" equations obtained, however, when these routines were applied to a subset of 46 possible explanatory variables did not include the air pollution variables (sulfates and particulates). Moreover, when these "best" equations were augmented by the air pollution variables, the resulting fit was not significantly improved (see Gibbons and McDonald, 1980). Other references on this topic can be found in Gibbons, Gunst, and McDonald (1987), who use this problem area to illustrate the interplay between regression diagnostics and robust regression estimation procedures.

This article addresses yet another important aspect of this problem— the sensitivity of the conclusions to the choice of mortality index. An unadjusted mortality index reflects the age-sex-race structure of a particular SMSA. An adjusted mortality index standardizes the mortality rate to compensate for differing age-sex-race structures. It is hoped that by adjusting for age, sex, and race in the mortality index, differences in death rates due to other causes will be more clearly discernible.

There are many ways to adjust the total mortality rate for age, sex, and race. Duffy and Carroll (1967) and Bishop, Fienberg, and Holland (1975) discuss direct and indirect standardization and the properties of each. The ordering of the SMSAs with respect to mortality index depends on the method of standardization. No one method, however, is considered practically or theoretically superior to another. Since there is no universally preferred method of standardization, it is important to examine empirically (or numerically) the sensitivity of the conclusions in a study to the form of mortality index used.

In this study, four indices will be considered as the regression-dependent variable using the same explanatory variables employed in Chapter 3 of Lave and Seskin (1977). The four mortality indices include an unadjusted and adjusted rate studied by Lave and Seskin, an adjusted rate examined by McDonald and Schwing (1973), and a new adjusted rate constructed within this study. The definitions and characteristics of these mortality rates will be described in Section 2 along with the other variables used in the subsequent analyses. The regression summaries are presented in Section 3, and the final section of this article summarizes the conclusions drawn from these analyses.

It should be noted that the topic of this article involves indices constructed to "adjust" for influential factors in a deterministic fashion. Data-dependent transforms for the regression variables are not discussed here. Methods such as those described by Snee (1986) could be used in conjunction with any one of the mortality indices.

2   DATA DESCRIPTION

This study is based on an analysis of 1960 cross-sectional data for 114 SMSAs[1] (geographical areas designated by the U.S. Census Bureau which include urban centers and their suburbs). The eleven independent variables included in the model developed by Lave and Seskin (1977 and 1979) are used in these analyses. These variables are:

SMIN   Smallest biweekly sulfate reading (μg/m^3 × 10)

SMEAN   Arithmetic mean of biweekly sulfate readings (μg/m^3 × 10)

SMAX   Largest biweekly sulfate reading (μg/m^3 × 10)

PMIN   Smallest biweekly suspended particulate reading (μg/m^3)

PMEAN   Arithmetic mean of biweekly suspended particulate readings (μg/m^3)

PMAX   Largest biweekly suspended particulate reading (μg/m^3)

PM2   SMSA population density (per square mile × .1)

GE65   Percent of SMSA population at least 65 years old (×10)

PNOW   Percent of non-whites in SMSA population (×10)

POOR   Percent of SMSA families with income below the poverty level (×10)

LPOP   The logarithm (base 10) of SMSA population (×100)

---

[1] Actually, standard economic areas (SEAs) were used as the unit of observation in the New England region.

Measurements on these variables for 117 SMSAs were provided by Michael J. Chappie who assisted Lave and Seskin in their studies and subsequently documented the data. The adjusted mortality rates considered in this study cannot be computed for three of these SMSAs: Manchester, New Hampshire; Scranton, Pennsylvania; and Wilkes-Barre, Pennsylvania.[2] Hence these SMSAs are deleted in this study. The data set was modified further by correcting the three sulfate measurements for Bridgeport, Connecticut. These measurements are incorrect in the Lave and Seskin data base as noted by Thibodeau et al. (1980)

The three-digit codes associated with the SMSAs are identified in Appendix A, followed by a complete listing of the data in Appendix B. The summary statistics (means, standard deviations, minimum and maximum values, and correlations) for these variables are included in Table 2.

Four different indices of 1960 mortality are considered as a dependent variable. The first is unadjusted total mortality rate (deaths per 100,000), designated TMR. The second is an age-sex-race adjusted mortality rate, denoted ASR. Lave and Seskin calculate detailed age-sex-race specific mortality rates, and combine them into an overall adjusted mortality rate by ASR $= \sum_i MR_i P_i$, where MR_i is the SMSA mortality rate for age-sex-race group i and P_i is the proportion of the total 1960 U.S. population in age-sex-race group i. These calculations are described in Appendix E of the Lave-Seskin book (1977).

In order to examine the stability of the conclusions to the form of mortality index, we consider two additional adjusted mortality rate indices in the regression models. These indices utilize the SMSA direct age adjusted rates tabulated by Duffy and Carroll (1967, Table 5)[3] for the four sex-race categories: male white, female white, male non-white, and female non-white (say, categories 1, 2, 3, 4, respectively). The two additional indices differ in the manner of combining the four sex-race rates to obtain an overall SMSA mortality index.

The third mortality index considered in this study corresponds to the index defined by McDonald and Schwing (1973). This SMSA mortality

---

[2] Duffy and Carroll (1967) note that the reported deaths in certain age categories for these SMSAs exceed the number of people in the category.

[3] The age-sex-race specific death rates tabulated by Duffy and Carroll differ from the Lave and Seskin rates in that Duffy and Carroll represent average rates over the three years 1959–1961 in order to avoid extremely small numbers of deaths in an individual age-sex-race group.

Table 2 Summary Statistics for Air Pollution/Mortality Data ($n = 114$)

Variable	Mean	Standard Deviation	Minimum Value	Maximum Value
SMIN	47.123	31.990	1.000	189.000
SMEAN	100.526	54.175	26.000	283.000
SMAX	227.877	124.679	58.000	940.000
PMIN	45.395	18.556	10.000	99.000
PMEAN	117.728	40.167	54.000	247.000
PMAX	267.061	130.972	94.000	958.000
PM2	70.836	137.104	1.600	1357.200
GE65	83.105	20.790	45.000	171.000
PNOW	128.009	103.548	5.000	400.000
POOR	180.649	65.545	87.000	327.000
LPOP	566.429	40.911	493.739	702.917
TMR	903.132	142.834	618.000	1338.000
ASR	1011.509	69.264	789.000	1180.000
MSR	950.715	78.957	774.248	1207.477
ADJR	959.057	63.698	786.865	1135.650

Correlation Coefficients

	SMIN	SMEAN	SMAX	PMIN	PMEAN	PMAX	PM2	GE65	PNOW	POOR	LPOP	TMR	ASR	MSR	ADJR
SMIN	1.000														
SMEAN	.622	1.000													
SMAX	.374	.844	1.000												
PMIN	.285	.558	.414	1.000											
PMEAN	.220	.561	.535	.747	1.000										
PMAX	.022	.372	.520	.371	.778	1.000									
PM2	.360	.408	.239	.268	.177	.004	1.000								
GE65	.295	.292	.218	.031	-.069	-.177	.139	1.000							
PNOW	-.146	-.202	-.253	-.090	-.186	-.101	-.046	-.509	1.000						
POOR	-.172	-.348	-.314	-.253	-.219	-.067	-.235	-.227	.642	1.000					
LPOP	.038	.395	.323	.345	.258	.069	.353	.117	-.070	-.417	1.000				
TMR	.418	.424	.292	.145	.008	-.135	.272	.857	-.220	-.098	.125	1.000			
ASR	.373	.414	.301	.278	.221	.106	.284	-.009	.207	.044	.099	.455	1.000		
MSR	.188	.189	.109	.166	.107	.116	.214	.318	.625	.372	.036	.183	.819	1.000	
ADJR	.369	.430	.324	.290	.251	.168	.333	-.026	.227	.072	.082	.444	.955	.873	1.000

index, denoted MSR, is defined by

$$\text{MSR} = \left(\sum_{i=1}^{4} D_i \right) \left[\sum_{i=1}^{4} (D_i/R_i) \right]^{-1}$$

where D_i and R_i are the number of deaths and direct age adjusted death rate of the ith sex-race category. The units of this index are deaths per 100,000 "standard" population, and "standard" refers to a fixed age distribution as given in Duffy and Carroll (1967, p. xvi). The denominator of the MSR expression is composed of four terms—each being the size of a "standard" population (expressed in units of 100,000) required so that the expected number of deaths in a specific sex-race category matches that actually observed in the SMSA.

Utilizing the 1960 U.S. population distribution as given in Table 3, the fourth mortality index considered is constructed by taking a weighted average of the four sex-race direct age adjusted rates. This index, denoted

Table 3 1960 United States Population

Age Group	White		Non-White	
	Male	Female	Male	Female
0–4	8,857,828	8,507,730	1,481,643	1,474,663
5–14	15,679,720	15,065,005	2,371,351	2,359,039
15–24	10,551,416	10,594,798	1,430,653	1,513,741
25–34	9,941,904	10,219,622	1,231,665	1,428,904
35–44	10,556,349	11,007,175	1,182,842	1,329,166
45–54	9,165,462	9,450,641	974,209	1,035,068
55–64	6,874,668	7,406,414	694,485	732,277
65–74	4,609,991	5,371,192	412,419	454,297
75–84	1,805,828	2,375,114	147,364	167,247
85+	304,603	492,219	28,780	38,183
Total	78,347,769	80,489,910	9,955,411	10,532,585

Source: Table 189. Nativity by Age, Race, and Sex: 1970 and 1960, *Census of Population: 1970*, U.S. Government Printing Office, Washington, DC.

ADJR, is given by

$$\text{ADJR} = \sum_{i=1}^{4} f_i R_i,$$

where R_i is as defined previously and f_i is the fraction of the 1960 U.S.
population which belongs to the ith sex-race category.

These four mortality indices are listed in Appendix B, and the summary
statistics and correlations for these measurements are reported in Table 2.
The unadjusted total mortality rate, TMR, has a relatively low correlation
with all of the adjusted indices—most notably with the MSR index for
which the correlation is 0.183. Figure 1 provides a scatter plot of these
two indices and notes a few SMSAs for which there is a relatively large

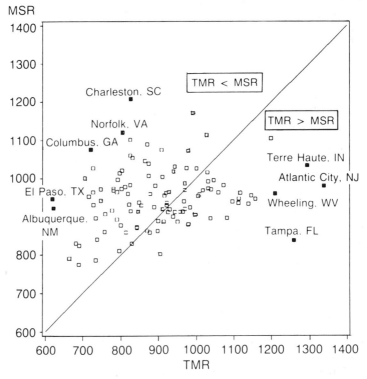

Figure 1   Unadjusted total mortality rate vs. the MSR Adjusted Mortality
Rate. (All units are deaths per 100,000 population.)

difference in the values. The four SMSAs noted in Figure 1 for which TMR is substantially greater than MSR have relatively large values for the GE65 variable; conversely, the SMSAs for which TMR is much smaller than MSR have relatively small values for this variable. The Lave and Seskin adjusted rate ASR is best correlated (0.955) with the ADJR index. This is not surprising since the adjustment made for each of these indices is of the same direct type. The correlation between the MSR index and the ADJR index is .873.

3 REGRESSION SUMMARIES

Each mortality index (TMR, ASR, MSR, and ADJR) is regressed on the eleven explanatory variables introduced in Section 2. The R^2 and F statistics are presented in Table 4 as are the regression coefficient estimates and the associated t statistics.[4] As indicated, the R^2 is much lower when the dependent variable is adjusted mortality (ASR, MSR, or ADJR) than when the dependent variable is unadjusted total mortality rate (TMR). However a test of the hypothesis that the variables under consideration have no explanatory power results in rejecting the hypothesis (the observed F statistic in each case exceeds the upper 1% value of an F distribution with 11 and 102 degrees of freedom).

In examining the individual regression estimates one observes that the following coefficients are statistically insignificant ($t < 2$) for all the mortality indices: SMIN, SMEAN, SMAX, PMIN, PMEAN, PMAX, POOR, and LPOP. The insignificance of the air pollution variables examined *individually* does not imply that taken collectively they are insignificant. Assuming variables 1 through 6 correspond to the air pollution variables, the hypothesis $H : \beta_1 = \beta_2 = \beta_3 = \beta_4 = \beta_5 = \beta_6 = 0$ is tested with the statistic

$$F = \frac{(R_p^2 - R_q^2)/(p - q)}{(1 - R_p^2)/(n - p - 1)}$$

where R_p^2 and R_q^2 are the sample multiple correlation coefficients when the model is fitted with p variables and $q(< p)$ variables respectively ($p = 11$

---

[4]The regression estimates for TMR and ASR differ from those reported by Lave and Seskin because of the three deleted SMSAs and the modified sulfate measurements for Bridgeport, Conn., as reported in Section 2.

Table 4   The R^2, F, and Regression Estimates for 114 SMSAs. (The t-statistics are below the regression estimates in parentheses.)

	Eleven Explanatory Variables				Seven Explanatory Variables			
	TMR	ASR	MSR	ADJR	TMR	ASR	MSR	ADJR
R^2	.852	.303	.555	.347	.846	.266	.528	.306
F	53.40	4.03	11.55	4.94	83.01	5.49	16.97	6.68
Constant	408.184	988.441	979.156	954.620	341.250	912.038	929.850	891.112
	(3.84)	(8.84)	(9.61)	(9.59)	(3.40)	(8.58)	(9.58)	(9.38)
Variables								
SMIN	.232	.366	.296	.288	.488	.657	.482	.539
	(.91)	(1.36)	(1.21)	(1.20)	(2.45)	(3.12)	(2.50)	(2.86)
SMEAN	.256	.283	.191	.275				
	(.88)	(.92)	(.69)	(1.01)				
SMAX	.044	.059	.057	.049				
	(.44)	(.56)	(.60)	(.53)				
PMIN	.191	.201	.479	.278				
	(.35)	(.35)	(.93)	(.55)				
PMEAN	.222	.157	−.143	.021	.369	.335	.353	.373
	(.60)	(.40)	(−.40)	(.06)	(2.42)	(2.07)	(2.39)	(2.58)
PMAX	−.045	−.048	.055	.014				
	(−.51)	(−.52)	(.66)	(.17)				
PM2	.094	.062	.103	.094	.103	.071	.107	.101
	(2.03)	(1.26)	(2.30)	(2.15)	(2.24)	(1.46)	(2.40)	(2.31)
GE65	6.624	−.019	−.327	−.007	6.847	.235	−.147	.210
	(18.22)	(−.05)	(−.94)	(−.02)	(19.83)	(.64)	(−.44)	(.64)
PNOW	.474	.208	.470	.192	.506	.242	.502	.225
	(5.16)	(2.15)	(5.34)	(2.24)	(5.77)	(2.61)	(5.92)	(2.72)
POOR	−.094	.004	.067	.030	−.147	−.055	.019	−.022
	(−.70)	(.03)	(.52)	(.24)	(−1.12)	(−.40)	(.15)	(−.18)
LPOP	−.310	−.143	−.259	−.181	−.210	−.029	−.188	−.090
	(−1.76)	(−.77)	(−1.53)	(−1.10)	(−1.24)	(−.16)	(−1.15)	(−.56)

and $q = 5$). This statistic has an F distribution with $p - q$ and $n - p - 1$ degrees of freedom. For this study, the following results are observed:

	TMR	ASR	MSR	ADJR
		Dependent Variable		
R_p^2	.852	.303	.555	.347
R_q^2	.823	.138	.457	.175
F	3.33	4.02	3.74	4.48

Since the upper 1% point of an F distribution with 6 and 102 degrees of freedom is approximately 3.0, the hypothesis that all the pollution variables taken together do not add significantly to the explanatory power of the regression can be rejected.

The variables which are statistically significant in one or more of the four regression equations are: PM2, GE65, and PNOW. The population density variable (PM2) is significant for three mortality indices (TMR, MSR, and ADJR). The percent of the population 65 years of age and older variable (GE65) is significant only when the dependent variable is unadjusted total mortality (TMR), and here it is highly significant ($t = 18.22$). The percent non-white variable is significant in the equations involving all four mortality indices and, as Lave and Seskin note (1977, p. 56), may be "exerting an independent effect on the mortality rate in addition to its role in controlling for the population at risk; that is, a concentration of non-whites could indicate a geographically concentrated group with poor housing or high crime rates."

A second model for these data is investigated, corresponding to the model studied by Lave and Seskin (1977, pp. 31 and 58–59; 1979, p. 181). This model has seven explanatory variables: two pollutants (SMIN and PMEAN) and the five socioeconomic variables included in the previous study (PM2, GE65, PNOW, POOR, and LPOP). The regression estimates, t values, and F and R^2 statistics for this study are included in Table 4. Again, despite the low R^2 values, one can reject the hypothesis that the variables under consideration have no explanatory power. The significance or non-significance of the five socioeconomic variables is identical to that observed in the eleven-variable model. The major difference between the eleven-variable model and the seven-variable model is that the two air pollution variables remaining in the seven-variable model are statistically significant.

It should be emphasized, however, that the significance or non-significance of the results as measured by the t value is dependent on the model being correctly specified. The consequences of model misspecification in regression analysis have been studied by Rao (1971), Rosenberg and Levy (1972), Hocking (1976), and Deegan (1976). These studies show that if the true model is the model with eleven explanatory variables and if one estimates the coefficients based on the seven-variable model, then the least squares estimator in the reduced model is biased and its variance is decreased. Some coefficients may then test to be significant because of the reduced variance-covariance matrix. The apparent significance of the results, therefore, can be misleading since they may occur as a consequence of model misspecification.

4   SUMMARY AND CONCLUSIONS

We have constructed eight regression equations, examining four mortality indices as the dependent variable in each of two models. The effect of these four indices of mortality on R^2, on the regression estimates (both in magnitude and significance), and on the overall conclusions has been examined. This section summarizes the major findings of this study.

The R^2 is much lower when the dependent variable is adjusted mortality than when it is total unadjusted mortality. However, it is not so low that the overall regression equation tests to be insignificant.

The significance or non-significance of the explanatory variables does not appear to depend on the choice of mortality index with a couple of exceptions; namely, the percent of population 65 and older is a significant variable only when the dependent variable is total unadjusted mortality, and the population density variable is not a significant variable only when the dependent variable is the age-sex-race adjusted mortality rate studied by Lave and Seskin.

The estimated elasticities are less affected by the choice of mortality index than by the model specification. In this study, the estimated elasticity is the model implication of a 50% decrease in the air pollution indices from the mean values (everything else remaining constant in the model). For an individual pollution index, this elasticity is computed by $50\hat{\beta}_i(\bar{x}_i/\bar{y})$ where

$\hat{\beta}_i$ is the estimated least squares coefficient, $\bar{x}_i$ is the mean of the pollution variable, and $\bar{y}$ is the mean of the dependent mortality rate variable. The estimated effect on mortality of a 50% decrease in the three sulfate indices and the three particulate indices (obtained by summing the individual estimates) are presented in Table 5. Estimates for each mortality index in both the full (eleven-variable) and the reduced (seven-variable) model are given. There are small observed differences between the elasticities for the four mortality indices in a given model. However, the observed differences going from the full model to the reduced model are striking; the sulfates appear to have a greater effect on mortality than the particulates when estimating from the full model whereas the opposite is observed when estimating from the reduced model. This further highlights the importance of model specification for this study.

The sensitivity of the estimates to the three deleted SMSAs and the corrected sulfate measurements for Bridgeport, Connecticut, is examined by comparing the Lave and Seskin elasticities with their counterparts in Table 5. The results are presented in Table 6. The elasticities of the particulates in the full model are particularly sensitive to the deletion of these cities. This particular aspect is examined systematically within a regression diagnostic context by Gibbons and McDonald (1983).

In reducing the regression model from eleven variables to seven, Lave and Seskin retain all five socioeconomic variables in addition to one sulfate and one particulate variable. The retained sulfate variable (SMIN) is that one which has the largest t value among the three sulfate variables in the full model (and likewise for the retained particulate variable). Using this criteria for pollution variable selection with the four mortality indices yields Table 6 (p. 166).

Table 5 Estimated Percentage Change in Mortality (Standard deviation of combined elasticity is in parentheses.)

50% Change:	Full Model				Reduced Model			
	TMR	ASR	MSR	ADJR	TMR	ASR	MSR	ADJR
All sulfates	2.59	2.92	2.43	2.73	1.27	1.53	1.19	1.32
All particulates	1.26	.73	1.03	.98	2.41	1.95	2.19	2.29
Combined	3.85	3.65	3.46	3.71	3.68	3.48	3.38	3.61
	(1.03)	(.97)	(.94)	(.91)	(1.00)	(.94)	(.92)	(.89)

Table 6   Estimated Percentage Change in Mortality—Effect of SMSA Deletion. (Standard deviation of combined elasticity is in parentheses.)

50% Change:	Full Model				Reduced Model			
	117 SMSAs		114 SMSAs		117 SMSAs		114 SMSAs	
	TMR	ASR	TMR	ASR	TMR	ASR	TMR	ASR
All sulfates	2.52	2.85	2.59	2.92	1.63	1.92	1.27	1.53
All particulates	2.19	1.98	1.26	.73	2.93	2.71	2.41	1.95
Combined	4.71	4.83	3.85	3.65	4.56	4.63	3.68	3.48
	(1.12)	(1.06)	(1.03)	(.97)	(1.07)	(1.01)	(1.00)	(.94)

	Mortality Index			
Variables Retained	TMR	ASR	MSR	ADJR
Sulfates	SMIN	SMIN	SMIN	SMIN
Particulates	PMEAN	PMAX	PMIN	PMIN

The minimum sulfate reading is uniformly retained whereas the particulate variable retained depends on the mortality index. The results of applying stepwise procedures for determining subsets of variables maximizing R^2 to these data are described in a report devoted to the model building aspects of this problem (Gibbons and McDonald, 1980).

ACKNOWLEDGMENTS

The authors gratefully acknowledge the cooperation of Professor Lester B. Lave and Mr. Michael J. Chappie in providing the data and documentation, the comments of Dr. L. A. Thibodeau related to data modification, and the suggestions and comments of Dr. Richard C. Schwing.

Appendix A   SMSA CODE NUMBERS

401	Birmingham	AL	475	Springfield	MA*
402	Mobile	AL	476	Worcester	MA*
403	Montgomery	AL	478	Detroit	MI
405	Phoenix	AZ	479	Flint	MI
407	Little Rock	AR	480	Jackson	MI
410	Fresno	CA	482	Lansing	MI
413	Los Angeles	CA	483	Saginaw	MI
417	Sacramento	CA	484	Duluth	MN
419	San Diego	CA	485	Minneapolis	MN
420	San Francisco	CA	487	Jackson	MS
421	San Jose	CA	488	Kansas City	MO
422	Denver	CO	489	St. Louis	MO
423	Bridgeport	CT*	492	Omaha	NE
424	Hartford	CT*	493	Las Vegas	NV
426	New Haven	CT*	494	Manchester	NH**
429	Wilmington	DE	495	Atlantic City	NJ
430	Washington	DC	499	Jersey City	NJ
431	Jacksonville	FL	500	Newark	NJ
432	Miami	FL	501	Albuquerque	NM
433	Orlando	FL	510	New York	NY
434	Tampa	FL	518	Charlotte	NC
436	Atlanta	GA	519	Greensboro	NC
437	Augusta	GA	520	Raleigh	NC
138	Columbus	GA	524	Canton	OH
439	Macon	GA	525	Cincinnati	OH
440	Savannah	GA	526	Cleveland	OH
443	Chicago	IL	527	Columbus	OH
445	Rockford	IL	528	Dayton	OH
449	Gary	IN	529	Hamilton	OH
451	Indianapolis	IN	530	Lorain	OH
452	South Bend	IN	531	Springfield	OH
453	Terre Haute	IN	532	Toledo	OH
455	Des Moines	IA	533	Youngstown	OH
457	Topeka	KS	534	Oklahoma City	OK
458	Wichita	KS	538	Portland	OR
460	Baton Rouge	LA	539	Allentown	PA
461	New Orleans	LA	542	Harrisburg	PA
462	Shreveport	LA	543	Johnstown	PA
463	Portland	ME*	544	Philadelphia	PA
464	Baltimore	MD	545	Pittsburgh	PA
465	Boston	MA*	546	Reading	PA
466	Brockton	MA*	547	Scranton	PA**
468	Fall River	MA*	548	Wilkes Barre	PA**

549	York	PA	567	Houston	TX
551	Providence	RI	568	San Antonio	TX
552	Charleston	SC	570	Waco	TX
553	Columbia	SC	571	Salt Lake City	UT
554	Greenville	SC	575	Norfolk	VA
555	Sioux Falls	SD	577	Richmond	VA
556	Chattanooga	TN	578	Roanoke	VA
557	Knoxville	TN	579	Seattle	WA
558	Memphis	TN	581	Tacoma	WA
559	Nashville	TN	582	Charleston	WV
560	Austin	TX	583	Huntington	WV
561	Beaumont	TX	584	Wheeling	WV
563	Dallas	TX	585	Madison	WI
564	El Paso	TX	586	Milwaukee	WI
565	Forth Worth	TX	894	Winston-Salem	NC
566	Galveston	TX			

*State Economic Area (SEA)
**SMSAs deleted in this study

Appendix B   DATA LISTING

OBS	SMSA	SMIN	SMEAN	SMAX	PMIN	PMEAN	PMAX	PM2	GE65	PNOU	POOR	LPOP	TMR	ASR	MSR	ADJR
1	401	55	145	341	38	146	400	56.8	77	345	258	580.268	943	1011	1030.38	960.31
2	402	47	67	248	29	129	284	25.3	57	323	254	549.735	823	1072	1059.40	1011.06
3	403	30	42	70	26	72	157	21.4	71	383	311	522.843	908	1014	1074.32	982.93
4	405	15	86	266	98	247	573	7.2	72	55	191	582.185	758	953	906.53	922.84
5	407	19	61	175	29	72	147	31.7	86	215	260	538.557	910	954	918.57	895.19
6	410	18	34	198	45	119	304	6.1	81	75	215	556.342	845	983	912.92	922.14
7	413	25	123	280	50	156	344	139.3	89	88	124	682.883	869	943	861.83	890.46
8	417	22	41	147	22	81	149	51.1	69	75	103	570.138	810	1016	976.99	980.06
9	419	28	77	149	34	81	166	24.3	73	55	151	601.410	737	928	839.71	862.82
10	420	16	62	202	32	70	183	84.0	90	125	119	644.457	925	975	911.70	929.45
11	421	21	37	105	50	108	302	49.3	70	32	102	580.775	664	848	790.73	801.34
12	422	2	61	188	54	126	229	25.4	82	42	131	596.819	841	970	871.77	903.00
13	423	137	205	308	32	91	182	103.3	94	53	93	581.530	938	1018	899.53	945.28
14	424	28	128	344	53	114	241	93.2	91	75	87	583.857	899	1018	887.47	932.95
15	426	39	124	288	42	88	248	108.3	99	53	101	581.975	983	1038	923.23	963.22
16	429	57	228	445	99	221	403	46.5	77	124	121	556.367	910	1077	1003.50	1019.93
17	430	51	124	210	76	135	242	134.8	62	249	105	630.144	780	986	967.80	949.61
18	431	46	66	133	23	106	193	58.6	62	234	224	565.840	863	1076	1045.31	1011.29
19	432	44	57	68	33	54	124	45.5	100	149	228	597.083	897	883	861.44	867.13
20	433	21	47	106	23	76	164	25.7	92	166	239	550.309	828	877	828.27	832.59
21	434	60	105	197	48	94	233	59.2	171	115	300	528.787	1259	914	836.25	868.51
22	436	46	91	139	46	112	236	59.0	65	228	208	600.740	823	1019	982.29	963.36
23	437	31	46	158	28	66	142	15.2	60	298	322	533.574	823	1099	1100.94	1026.66
24	438	34	60	145	28	99	160	19.6	49	293	310	533.843	721	1109	1075.34	1033.79
25	439	18	27	128	22	122	754	28.6	62	310	263	525.624	869	1050	1088.90	1027.48
26	440	49	71	120	46	82	192	42.7	65	341	280	527.485	990	1152	1169.82	1071.59
27	443	75	166	328	88	182	296	167.5	86	148	106	679.385	1000	1085	1024.88	1038.55
28	445	33	66	210	36	86	143	40.3	85	42	118	532.173	842	959	869.66	909.23

Appendix B   DATA LISTING (cont.)

OBS	SMSA	SMIN	SMEAN	SMAX	PMIN	PMEAN	PMAX	PM2	GE65	PNOW	POOR	LPOP	THR	ASR	MSR	ADJR
29	449	113	190	290	56	170	420	61.1	62	153	109	575.857	796	1060	1013.02	1002.27
30	451	50	139	269	92	178	275	173.5	85	144	128	584.359	969	1058	968.66	990.82
31	452	73	77	261	28	90	164	51.1	84	60	116	537.770	888	1024	924.03	954.17
32	453	63	135	214	53	118	203	26.1	123	45	219	503.526	1294	1119	1032.82	1074.23
33	455	31	88	188	61	183	329	44.8	97	41	129	542.540	950	998	885.63	941.58
34	457	19	37	91	52	101	158	25.9	99	73	159	515.010	904	916	801.15	838.12
35	458	23	54	139	22	102	174	34.4	67	64	131	553.559	690	901	823.76	853.48
36	460	27	58	113	57	125	352	49.8	52	318	219	536.184	706	1024	1000.18	931.36
37	461	49	96	187	62	87	117	77.7	73	310	243	593.876	1027	1180	1113.16	1087.29
38	462	84	88	272	47	104	197	16.3	74	341	304	544.945	891	985	979.95	925.83
39	463	42	140	287	50	82	147	21.4	114	5	169	526.186	1146	1058	953.88	1026.46
40	464	24	165	414	48	148	495	95.6	76	222	145	623.730	978	1119	1071.29	1068.64
41	465	42	163	337	55	141	252	174.5	109	27	115	649.275	1112	1072	934.70	981.01
42	466	46	142	332	30	79	165	37.4	113	20	123	539.524	1139	1114	932.19	983.44
43	468	62	79	136	18	102	254	71.7	116	13	171	560.042	1157	1107	944.96	1021.70
44	475	37	162	396	24	77	182	46.4	105	30	123	572.639	996	1018	904.15	958.25
45	476	71	90	204	32	99	229	38.5	116	7	128	576.584	1082	981	895.70	908.44
46	478	52	128	260	59	146	235	191.5	72	151	135	657.546	817	1009	959.22	961.30
47	479	64	80	229	49	124	468	58.2	62	99	122	557.323	747	1005	941.18	959.04
48	480	41	52	138	39	77	124	18.7	90	57	135	512.055	928	980	963.87	944.69
49	482	46	57	226	34	76	160	17.6	78	28	144	547.560	799	946	876.53	899.57
50	483	33	51	107	42	101	202	23.5	79	100	143	528.047	854	999	929.00	952.31
51	484	46	72	251	28	74	135	3.6	114	8	181	544.185	1117	1038	961.11	984.06
52	485	23	69	202	43	100	231	70.2	92	18	101	617.086	876	958	857.62	906.02
53	487	29	70	161	27	74	124	21.3	64	400	309	527.195	789	945	963.87	894.94
54	488	38	141	350	70	142	343	63.3	93	114	141	601.682	969	1015	919.73	948.07
55	489	60	182	299	63	168	295	64.6	93	145	151	631.389	1004	1037	953.56	975.78
56	492	20	74	148	39	107	198	29.9	90	60	136	566.075	922	1026	935.48	983.06
57	493	31	79	201	50	145	389	1.6	45	95	113	510.386	727	1039	1026.45	1021.63

OBS	SMSA	SMIN	SMEAN	SMAX	PMIN	PMEAN	PMAX	PM2	GE65	PNOW	POOR	LPOP	TMR	ASR	MSR	ADJR
58	495	54	75	110	25	71	118	28.0	140	177	230	520.650	1338	1045	979.43	989.19
59	499	155	229	340	63	147	253	1357.2	103	69	127	578.585	1199	1162	1103.78	1135.65
60	500	54	131	297	42	113	232	242.0	94	134	104	622.774	1017	1040	967.82	993.35
61	501	2	50	91	61	244	646	22.5	47	33	153	541.863	621	1003	922.28	954.88
62	510	60	228	531	79	162	270	497.7	97	120	132	702.917	1046	1055	994.65	1015.08
63	518	43	81	147	62	124	234	50.2	57	246	205	543.475	791	1029	996.14	957.76
64	519	38	60	71	28	60	94	37.9	61	209	206	539.185	747	980	971.12	954.62
65	520	34	49	97	39	80	132	19.6	65	261	291	522.810	801	982	971.87	921.96
66	524	189	273	399	81	175	323	59.4	95	54	132	553.192	964	1021	912.35	950.06
67	525	45	125	194	63	145	316	146.8	96	121	149	603.004	1039	1055	970.47	1004.93
68	526	69	160	282	86	174	336	261.1	89	145	114	625.445	969	1048	985.95	1000.17
69	527	94	161	276	74	119	190	127.2	74	119	137	583.440	877	1033	958.84	984.96
70	528	18	106	241	50	132	327	53.9	74	102	129	584.175	847	1047	936.23	962.02
71	529	82	100	225	42	86	163	42.3	71	53	130	529.902	776	996	915.62	959.82
72	530	59	81	351	37	144	417	43.9	72	57	120	533.746	821	1052	931.38	957.09
73	531	58	73	212	39	111	255	32.7	101	94	160	511.873	978	972	878.21	909.59
74	532	67	86	309	52	104	193	133.2	98	95	139	565.985	1031	1041	972.46	998.34
75	533	75	145	263	58	148	371	49.0	89	92	129	570.672	915	1012	954.44	967.91
76	534	18	26	63	42	84	173	24.0	79	94	186	570.913	812	942	857.95	888.60
77	538	28	75	212	21	70	185	22.5	113	30	148	591.482	1037	956	893.99	926.72
78	539	129	146	305	60	135	261	45.5	104	8	136	569.211	1059	1073	962.35	1040.52
79	542	73	119	220	50	92	189	32.1	98	67	140	553.791	1008	1041	956.77	1007.66
80	543	88	123	245	70	166	452	15.8	103	13	267	544.829	1072	1096	981.02	1010.26
81	544	87	229	620	70	160	342	122.4	91	157	130	663.778	1029	1081	1015.02	1032.79
82	545	55	150	345	43	166	475	78.8	95	68	147	638.119	1031	1074	991.29	1017.22
83	546	50	94	186	34	120	242	31.9	112	18	139	543.999	1113	1030	946.18	992.41
84	549	120	162	488	28	147	408	26.2	97	23	152	537.719	957	1004	911.82	963.04
85	551	30	163	349	56	119	223	116.1	109	21	161	585.645	1096	1063	938.50	992.96
86	552	37	56	152	35	78	233	22.9	48	365	327	533.522	825	1108	1207.48	1078.93
87	553	11	84	167	55	112	274	17.9	59	290	308	541.635	727	939	963.39	914.35
88	554	51	69	212	39	125	285	26.6	62	176	264	532.176	829	1095	1053.04	1045.70

Appendix B   DATA LISTING (cont.)

OBS	SMSA	SMIN	SMEAN	SMAX	PMIN	PMEAN	PMAX	PM2	GE65	PMOU	POOR	LPOP	TMR	ASR	MSR	ADJR
89	555	18	55	121	25	108	358	10.6	92	7	176	493.739	795	789	809.71	807.61
90	556	10	105	191	69	186	361	27.7	77	176	260	545.205	940	1046	1017.61	1004.19
91	557	56	77	157	28	135	302	25.8	74	75	275	556.594	825	1015	914.05	947.70
92	558	35	48	69	46	102	201	83.5	73	364	275	579.728	873	989	1006.49	951.52
93	559	54	160	362	45	130	310	75.1	79	192	235	560.178	919	1014	961.01	951.77
94	560	40	46	58	10	78	157	20.9	76	128	248	532.661	689	825	774.25	786.87
95	561	27	71	144	32	76	190	23.5	58	207	201	548.574	728	963	930.39	907.73
96	563	31	69	148	22	96	230	29.7	71	146	186	603.487	757	928	860.10	877.08
97	564	47	87	207	49	150	373	29.8	45	33	221	549.703	618	1027	946.62	954.67
98	565	9	32	73	28	79	152	35.8	73	107	193	575.832	789	973	891.71	916.12
99	566	62	72	86	23	55	125	32.7	64	214	232	514.726	873	1084	1037.66	1018.38
100	567	18	65	171	26	117	385	72.7	54	201	181	609.453	716	1014	952.53	944.91
101	568	16	68	233	50	124	296	55.1	68	69	272	583.705	734	945	895.56	927.02
102	570	20	28	88	10	76	156	14.5	98	161	298	517.635	914	908	885.78	892.08
103	571	27	79	260	47	121	309	50.1	70	14	118	558.324	682	896	830.21	855.28
104	575	49	112	198	39	89	242	86.7	53	264	269	576.231	803	1129	1119.81	1061.47
105	577	49	115	214	26	82	206	56.3	82	264	168	561.119	974	1050	1025.50	1014.14
106	578	29	80	313	42	112	343	52.4	89	127	220	520.086	929	999	918.01	937.15
107	579	1	47	179	32	69	141	26.2	96	48	112	604.423	938	961	899.26	914.53
108	581	36	126	264	46	143	347	19.2	94	51	167	550.730	943	992	911.48	933.54
109	582	15	283	940	55	225	958	27.9	70	58	214	540.299	780	992	972.70	993.38
110	583	60	70	137	56	122	219	18.1	93	30	268	540.617	967	1040	939.28	985.27
111	584	152	194	437	74	198	444	20.1	122	24	222	527.953	1210	1061	959.03	1004.45
112	585	39	65	166	30	83	215	18.6	82	12	128	534.654	734	858	785.68	810.59
113	586	65	134	236	49	150	299	150.2	88	56	96	607.711	921	1005	929.15	958.20
114	894	28	58	128	72	147	306	44.7	62	242	201	527.744	802	976	1019.65	948.06

REFERENCES

Bishop, Y. M. M., Fienberg, S. E., and Holland, P. W. (1975), *Discrete Multivariate Analysis: Theory and Practice*, The MIT Press, Cambridge, Mass.

Chappie, Michael J. (1978), *Documentation for Data Used in Air Pollution and Human Health*, School of Urban and Public Affairs, Carnegie-Mellon University, Pittsburgh.

Cooper, D. E. and Hamilton, W. C. (1979), Atmospheric Sulfates and Mortality— The Phantom Connection, *Mining Congress Journal*, Vol. 65, pp. 1–16.

Deegan, John, Jr. (1976), The Consequences of Model Misspecification in Regression Analysis, *Multivariate Behavioral Research*, Vol. 11, pp. 237-248.

Duffy, E. A. and Carroll, R. E. (1967), *United States Metropolitan Mortality, 1959–1961*, PHS Publication No. 999-AP-39, U.S. Public Health Service, National Center for Air Pollution Control, Cincinnati, Ohio.

Furnival, G. M. and Wilson, R. W. (1974), Regressions by Leaps and Bounds, *Technometrics*, Vol. 16, pp. 499–511.

Gibbons, D. I. and McDonald G. C. (1980), Examining Regression Relationships Between Air Pollution and Mortality, General Motors Research Publication GMR-3278, General Motors Research Laboratories, Warren, Michigan.

Gibbons, D. I. and McDonald, G. C. (1983), Illustrating Regression Diagnostics with an Air Pollution and Mortality Model, *Computational Statistics & Data Analysis*, Vol. 1, pp. 201–220.

Gibbons, D. I., Gunst, R. F., and McDonald, G. C. (1987), The Complementary Use of Regression Diagnostics and Robust Estimators, *Naval Research Logistics*, Vol. 34, pp. 109–131.

Hocking, R. R. (1976), The Analysis and Selection of Variables in Linear Regression, *Biometrics*, Vol. 32, pp. 1–49.

Lave, L. B. and Seskin, E. P. (1977), *Air Pollution and Human Health*, The John Hopkins University Press, Baltimore.

Lave, L. B. and Seskin, E. P. (1979), Epidemiology, Causality, and Public Policy, *American Scientist*, Vol. 67, pp. 178–186.

McDonald, G. C. and Schwing, R. C. (1973), Instabilities of Regression Estimates Relating Air Pollution to Mortality, *Technometrics*, Vol. 15, pp. 463–481.

Rao, R. (1971), Some Notes on Misspecification in Multiple Regressions, *The American Statistician*, Vol. 25, pp. 37–39.

Rosenberg, S. H. and Levy P. S. (1972), A Characterization on Misspecification in the General Linear Regression Model, *Biometrics*, Vol. 28, pp. 1129-1133.

Snee, R. D. (1986), An Alternative Approach to Fitting Models When Re-Expression of the Response is Useful, *Journal of Quality Technology*, Vol. 18, pp. 211–225.

Thibodeau, L. A., Reed, R. B., Bishop, Y. M. M., and Kammerman, L. A. (1980), Air Pollution and Human Health: A Review and Reanalysis, *Environmental Health Perspectives*, Vol. 34, pp. 165–183.

U. S. Bureau of the Census (1973), *Census of Population: 1970*, Vol. 1, Characteristics of the Population, Part 1, United States Summary—Section 2, U.S. Government Printing Office, Washington, D.C.

II

METHODS

7

Mixture Experiments

*John A. Cornell** Colorado State University, Fort Collins, Colorado

1   INTRODUCTION

In a mixture experiment, two or more ingredients are mixed or blended together to form an end product. Measurements of the physical properties of the end product are taken on several blends of the ingredients in an attempt to find the blend that produces the "best" result. Characteristics of the end products, such as the quality of the product, depend only on the relative proportions of the ingredients (components) present in the mixture and not on the total amount of mixture. For example, the ingredients of a liquid detergent used for cleaning oil stains are water, urea, ethyl alcohol, and sodium xylene sulphonate. The viscosity of the liquid detergent is an important characteristic, and a suitable blend of water, urea, ethyl alcohol, and sodium xylene sulphonate is the proportionate amounts 0.30, 0.40, 0.25, and 0.05, respectively, of the total mixture.

A feature of a mixture experiment is that the controllable variables represent proportionate amounts of the mixture where the proportions are

---

*Current affiliation: University of Florida, Gainesville, Florida

by volume, by weight, or by mole fraction. When expressed as fractions of the mixture, the proportions sum to unity. Thus, if the number of ingredients (or components) in the system under investigation is denoted by q and if the proportion of the ith component in the mixture is represented by x_i, then,

$$x_i \geq 0, \qquad i = 1, 2, \ldots, q \tag{1}$$

and

$$\sum_{i=1}^{q} x_i = x_1 + x_2 + \cdots + x_q = 1.0 \tag{2}$$

Note that the x_i could represent nonnegative percentages of the mixtures but, when divided by 100%, they would be fractions as in Eq. (2).

The following is an example of a simple mixture experiment. Two chemicals, Vendex and Kelthane, are to be mixed with three other ingredients in a 30% : 70% ratio of chemical to other ingredients in forming a liquid pesticide used for killing mites on strawberry plants. Each chemical can be mixed alone with the other ingredients, as well as combinations of the two chemicals mixed with the other ingredients. Tolerance to each chemical by the mites after a period of time is of concern to the experimenter. It is suspected that if the two chemicals are blended together along with the other ingredients, the efficacy of the pesticide will be longer-lasting than if each chemical appears by itself in the pesticide and the single-chemical pesticides are applied sequentially.

An experiment is set up consisting of five different combinations of the two chemicals mixed with the other ingredients to form five pesticide blends. Each of the pesticide blends is to be sprayed on six groups of 10 strawberry plants each, where the 30 groups (5×6) of 10 plants each were uniformly infested with mites before the pesticides were applied. The same amount of pesticide is sprayed on each group so as to remove any effect the amount may have. The objective of the experiment is to find the blend of Vendex and Kelthane that produces the highest percent mortality of the mites after a specified period of time.

The five different blends consist of the following ingredient proportions. One blend contains Vendex (V) only with the other ingredients in a 30% : 70% ratio, while another blend contains only Kelthane (K) with the other ingredients in a 30% : 70% ratio. The single chemical blends are called *single-component mixtures*. The remaining three pesticide blends consist of 22.5% : 7.5% : 70% of V : K : other ingredients, 15% : 15% : 70% of V : K : other ingredients and 7.5% : 22.5% : 70% of V : K : other ingredients. In the 70% of each blend that is contributed by the other three ingredients, the relative proportions of the three ingredients remain fixed. When expressed in terms of the proportion of chemical in each of the five pesticide blends, the ratios of V : K are 1.0 : 0, 0 : 1.0, 0.75 : 0.25, 0.50 : 0.50, and 0.25 : 0.75, respectively. The average percent mortality, taken over 60 strawberry plants, of each of the five blends and the chemical proportion of Vendex and Kelthane in each blend are presented in Table 1, while a plot of the percent mortalities is drawn in Figure 1.

Additivity in the efficacy of the two chemicals, Vendex and Kelthane, would result in the average mortality of any blend consisting of both V and K being the volumetric average of the average mortalities of the single-chemical blends. For example, the $(0.50, 0.50)$ blend, if additive blending is present, would have an average mortality that is calculated to be $(67\% \times 0.50) + (35\% \times 0.50) = 51\%$ since 67% is the average percent mortality of Vendex alone and 35% is the average percent for Kelthane alone. However, since the average percent mortality of the $(0.50, 0.50)$ blend is 79% and this value exceeds the volumetric average of 51%, the chemicals Vendex and Kelthane are said to be *synergistic* (beneficial) in their blending with each other. [Had the average percent mortality of the $(0.50, 0.50)$ blend been less than 51%, then the chemicals would be said to be *antagonistic* in their blending with each other.] The average mortalities of the remaining blends, $(0.75, 0.25)$ and $(0.25, 0.75)$, exceed their respective volumetric averages, 59% and 44.5%, respectively, further supporting the synergistic blending properties of the two chemicals. We shall illustrate how to test

Table 1 Pesticide Experimental Data

Vendex		Kelthane		Average percent mortality
%	(x_1)	%	(x_2)	
100	1.00	0	0	67
75	.75	25	.25	75
50	.50	50	.50	79
25	.25	75	.75	58
0	0	100	1.00	35

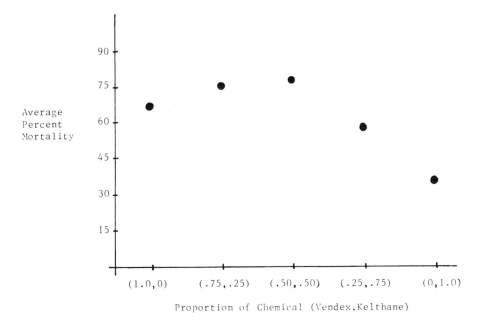

Figure 1   Percent mortality of mites for each of the five chemical blends.

for synergism or antagonism between a pair of components later in the example presented in the section on models for mixture experiments.

Let us return to Eqs. (1) and (2), which represent the principal restrictions on the component proportions in mixture experiments. As a result of the restrictions (1) and (2) on the values of x_i, the experimental region or factor space of interest is a regular $(q - 1)$-dimensional simplex. For $q = 2$ components, the factor space is a straight line, as shown in Figure 1, where the horizontal axis lists the proportions associated with the two chemicals that make up the five blends. For $q = 3$, the factor space is an equilateral triangle, and for $q = 4$, the factor space is a tetrahedron (see Figure 2). Note that since the proportions sum to unity as shown in Eq. (2), the x_i are constrained variables, and altering the proportion of one component in a mixture will cause a change in the proportions of at least one other component in the experimental region.

The coordinate system for the mixture proportions is a simplex coordinate system. With three components, for example, the vertices of the triangle represent single-component mixtures $x_i = 1$, $x_j = x_k = 0$ for

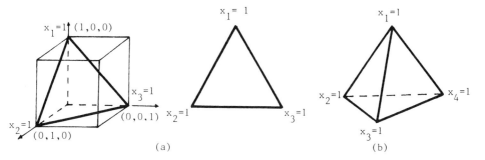

Figure 2   The three- and four-component simplex factor spaces: (a) with three components, the simplex is a triangle; (b) with four components, the simplex is a tetrahedron.

$i, j, k = 1$, 2, and 3, $i \neq j \neq k$, and are denoted by $(1,0,0)$, $(0,1,0)$, and $(0,0,1)$. The interior points of the triangle represent mixtures in which none of the components are absent, that is, $x_1 > 0$, $x_2 > 0$, and $x_3 > 0$. The centroid of the triangle corresponds to the mixture with equal proportions $(1/3, 1/3, 1/3)$ from each of the components.

In mixture experiments, the experimental data are defined on a quantitative scale and are said to be the yield or some other physical property of a product formed from the blend. The purpose of the experimental program will be to *model the blending surface with some form of mathematical equation* so that either or both of the following will result:

1. Predictions of the response for any mixture or combination of the ingredients can be made empirically.
2. Some measure of the influence on the response of each component singly and in combination with the other components (joint blending of the components) can be obtained.

We shall refer to these experimental objectives throughout this chapter.

2   MODELING A RESPONSE SURFACE

A convenient way to evaluate the performance of a mathematical equation in representing the mixture system is through the concept of a response surface. Initially, it is assumed that there exists some functional relationship

$$\eta = \phi(x_1, x_2, \ldots, x_q)$$

which defines the dependence of the response η on the proportions x_1, $x_2, \ldots, x_q$ of the components. The function ϕ is a continuous function in the x_i, and ϕ is represented usually by a first- or second-degree polynomial. On some occasions, a third-degree polynomial (a cubic or reduced form of a cubic equation with certain terms omitted from the complete cubic equation) may be necessary to represent the surface.

In an experimental program consisting of N trials, the observed value of the response in the uth trial, denoted by y_u, is assumed to vary about a mean of η, with common variance σ^2 for all $u = 1, 2, \ldots, N$. The observed value contains additive experimental error ϵ_u

$$y_u = \eta + \epsilon_u, \qquad 1 \leq u \leq N \tag{3}$$

where the errors ϵ_u are assumed to be uncorrelated and identically distributed with zero mean and common variance σ^2.

After the N observations are collected, the unknown parameters or coefficients in the model are estimated by the *method of least squares*. The coefficient estimates are substituted into the model for use in predicting response values. To illustrate, let us assume that there are only two components in the system, whose proportions are denoted by x_1 and x_2, and that Eq. (3) is written as a first-degree polynomial

$$y_u = \beta_1 x_1 + \beta_2 x_2 + \epsilon_u \tag{4}$$

[The absence of the parameter β_0 in Eq. (4) is owing to the restriction that $x_1 + x_2 = 1$. We shall discuss the derivation of the form of Eq. (4) in Section 4.] With $N \geq 2$ observations collected on y_u, we can obtain the estimates b_1 and b_2 of the parameters β_1 and β_2, respectively. If it is decided that the estimates are not zero, then the unknown parameters in Eq. (4) are replaced by their respective estimates to give the approximating equation

$$\hat{y}(\mathbf{x}) = b_1 x_1 + b_2 x_2$$

where $\hat{y}(\mathbf{x})$ (read "y hat") denotes the predicted or estimated value of η for given values of x_1 and x_2.

Statistical methods are used to measure product characteristics and improve product performance. These methods include (1) selecting the type of model equation to be fitted to the experimental data, (2) choosing the design program that defines which blends to run for collecting the data, and (3) using the appropriate techniques in the analysis of such data. We briefly discuss these methods for experiments in which all combinations of the ingredients are possible, as well as for experiments in which only certain combinations are feasible. Some of the expository papers written on methods for analyzing data from mixture experiments are by Cornell (1973, 1979), Gorman and Hinman (1962), Hare (1974), and Snee (1971, 1973, 1979). Mixture designs, models, and techniques used in the analysis of data are discussed in considerable detail in Cornell (1981).

3 SIMPLEX-LATTICE DESIGNS FOR EXPLORING THE WHOLE SIMPLEX REGION

For investigating the response surface over the entire simplex region, a natural choice for a design would be one with points that are positioned uniformly over the simplex factor space. Such a design is the $\{q, m\}$ simplex lattice introduced by Scheffé (1958), where the points are defined by the following coordinate settings: the proportions assumed by each component take the $m + 1$ equally spaced values from 0 to 1,

$$x_i = 0, \frac{1}{m}, \frac{2}{m}, \ldots, 1 \tag{5}$$

and all possible combinations (mixtures) of the components are considered, using the proportions in Eq. (5) for each component.

For a $q = 3$ component system, suppose each component is to take the proportions $x_i = 0, 1/2$, and 1, for $i = 1$, 2, and 3, which is the same as setting $m = 2$ in Eq. (5). The $\{3, 2\}$ simplex lattice consists of the six points on the boundary of the triangular factor space,

$$(x_1, x_2, x_3) = (1, 0, 0), (0, 1, 0), (0, 0, 1), (1/2, 1/2, 0),$$
$$(1/2, 0, 1/2), (0, 1/2, 1/2)$$

The three vertices $(1, 0, 0)$, $(0, 1, 0)$, and $(0, 0, 1)$ represent the individual components, while the points $(1/2, 1/2, 0)$, $(1/2, 0, 1/2)$, and $(0, 1/2, 1/2)$ represent the binary blends or two-component mixtures and are located at the midpoints of the three sides of the triangle. The $\{3, 2\}$, $\{3, 3\}$, and $\{4, 2\}$ simplex lattices are shown in Figure 3.

An alternative arrangement to the $\{q, m\}$ simplex lattice is the simplex-centroid design introduced by Scheffé (1963). In a q-component simplex-centroid design, the number of points is $2^q - 1$. The design points correspond to the q permutations of $(1, 0, 0, \ldots, 0)$, the $\binom{q}{2}$ permutations of

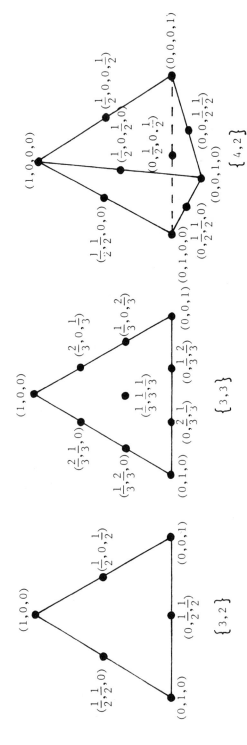

Figure 3 The {3,2}, {3,3}, and {4,2} simplex-lattice arrangements and the coordinate settings of the design points.

$(1/2, 1/2, 0, 0, \ldots, 0)$, the $\binom{q}{3}$ permutations of $(1/3, 1/3, 1/3, 0, \ldots, 0)$, $\ldots$, and the centroid point $(1/q, 1/q, \ldots, 1/q)$. A four-component simplex-centroid design consists of $2^4 - 1 = 15$ points.

Besides experimental regions, mixture experiments also differ from the ordinary regression problems in the form of the polynomial model to be fitted. Scheffé (1958, 1963) introduced the canonical polynomials for use with the simplex-lattice and simplex-centroid designs.

4 CANONICAL FORM OF MIXTURE POLYNOMIALS

The canonical form of the mixture polynomial is derived by applying the restrictions $x_1 + x_2 + \cdots + x_q = 1$ to the terms in the standard polynomial and then simplifying. For example, with two components, x_1 and x_2, the standard first-degree polynomial is written as

$$\eta = \beta_0 + \beta_1 x_1 + \beta_2 x_2$$

However, since $x_1 + x_2 = 1$, we can replace β_0 by $\beta_0(x_1 + x_2 = 1)$ in η to get

$$\eta = (\beta_0 + \beta_1)x_1 + (\beta_0 + \beta_2)x_2$$
$$= \beta_1' x_1 + \beta_2' x_2$$

so that the constant term β_0 is removed from the model. For the second-degree polynomial, the quadratic terms $\beta_{11}x_1^2$ and $\beta_{22}x_2^2$ are also removed from the model, along with the constant term β_0. Thus, the mixture models have fewer terms than the standard polynomials. Hereafter, we shall periodically refer to the canonical form of the polynomial equation as "Scheffé-type" models.

In general, the canonical forms of the mixture models (with the primes removed from the β_i) are:

Linear:

$$\eta = \sum_{i=1}^{q} \beta_i x_i \tag{6}$$

Quadratic:

$$\eta = \sum_{i=1}^{q} \beta_i x_i + \sum \sum_{i<j}^{q} \beta_{ij} x_i x_j \tag{7}$$

Full cubic:

$$\eta = \sum_{i=1}^{q} \beta_i x_i + \sum_{i<j}^{q} \sum \beta_{ij} x_i x_j + \sum_{i<j}^{q} \sum \gamma_{ij} x_i x_j (x_i - x_j)$$

$$+ \sum_{i<j<k}^{q} \sum \sum \beta_{ijk} x_i x_j x_k \quad (8)$$

Special cubic:

$$\eta = \sum_{i=1}^{q} \beta_i x_i + \sum_{i<j}^{q} \sum \beta_{ij} x_i x_j + \sum_{i<j<k}^{q} \sum \sum \beta_{ijk} x_i x_j x_k \quad (9)$$

With $q = 3$ components, for example, the quadratic and special-cubic models, respectively, are

$$\eta = \beta_1 x_1 + \beta_2 x_2 + \beta_3 x_3 + \beta_{12} x_1 x_2 + \beta_{13} x_1 x_3 + \beta_{23} x_2 x_3$$
$$\eta = \beta_1 x_1 + \beta_2 x_2 + \beta_3 x_3 + \beta_{12} x_1 x_2 + \beta_{13} x_1 x_3 + \beta_{23} x_2 x_3 + \beta_{123} x_1 x_2 x_3$$

The linear (6), quadratic (7), and full-cubic models (8) are generally associated with the $\{q, 1\}$, $\{q, 2\}$, and $\{q, 3\}$ simplex lattices, respectively. The special-cubic equation is a reduced form of third-degree polynomial that provides measures of the ternary blends of the three components i, j, and k. It represents the lowest form of polynomial of degree higher than 2 and contains $q(q^2 + 5)/6$ terms while the full-cubic model contains $q(q + 1)(q + 2)/6$ terms.

The canonical form of the polynomial in q components that is to be fitted to data collected at the points of the simplex-centroid design is

$$\eta = \sum_{i=1}^{q} \beta_i x_i + \sum_{i<j}^{q} \sum \beta_{ij} x_i x_j + \sum_{i<j<k}^{q} \sum \sum \beta_{ijk} x_i x_j x_k + \cdots$$

$$+ \beta_{12\cdots q} x_1 x_2 \ldots x_q \quad (10)$$

The model of Eq. (10) contains $2^q - 1$ terms, and this number corresponds exactly to the number of points in the simplex-centroid design.

The terms in the canonical polynomial models have simple interpretations. In Eq. (7), for example, if $x_i = 1$ and therefore $x_j = 0$ for $j \neq i$, then $\eta = \beta_i$, that is, β_i is the expected response to the pure component i. When the blending is strictly additive, the model is $\eta = \sum_{i=1}^{q} \beta_i x_i$, which is the equation of a planar surface. With a quadratic model, the second-degree terms describe quadratic departure of the response surface from a plane. Higher-degree terms such as $\beta_{123}x_1x_2x_3$ and $\gamma_{ij}x_ix_j(x_i-x_j)$ describe additional departures in the shape of the response surface from that of a plane beyond those described by the second-degree terms.

When data are collected at the points of the $\{q, m\}$ simplex lattice, the formulas for the estimates of the coefficients in the canonical polynomials are expressed as simple functions of the observed values of the response collected at the points of the design. For instance, suppose that n_i observations are taken on the single component i ($x_i = 1$, $x_j = 0$, $j \neq i$) and that the average of the n_i observations is denoted by $\bar{y}_i$. Further suppose that n_{ij} observations are taken on the 50% : 50% binary mixture ($x_i = 1/2$, $x_j = 1/2$, $x_k = 0$ for all $i < j < k$) of components i and j and that the average of the n_{ij} observations is denoted by $\bar{y}_{ij}$. Then the least-squares formulas for calculating the coefficient estimates b_i and b_{ij} in the second-degree model of Eq. (7) are

$$b_i = \bar{y}_i, \qquad i = 1, 2, \ldots, q$$

$$b_{ij} = 4\bar{y}_{ij} - 2(\bar{y}_i + \bar{y}_j), \qquad i, j = 1, 2, \ldots, q, \quad i < j \qquad (11)$$

Moreover, if, with the simplex-centroid design, n_{ijk} observations are taken on the $x_i = x_j = x_k = 1/3$ blend and the average is denoted by $\bar{y}_{ijk}$, then the formula for calculating the estimate of β_{ijk} in Eq. (9) is

$$b_{ijk} = 27\bar{y}_{ijk} - 12(\bar{y}_{ij} + \bar{y}_{ik} + \bar{y}_{jk}) + 3(\bar{y}_i + \bar{y}_j + \bar{y}_k) \qquad (12)$$

Note that the scalar quantities 4 and 2 in the formula for b_{ij} and the quantities 27, 12, and 3 in the formula for b_{ijk} do not depend on the values of n_i, n_{ij}, and n_{ijk} but rather come from the values of x_i, x_j, and x_k. As a reminder, the simple formulas in Eqs. (11) and (12) occur only when fitting the models (6), (7), and (9) to their respective simplex-lattice and simplex-centroid arrangements. In cases in which additional blends are added to these designs, the estimation formulas for the coefficients are more complicated. Formulas for estimating the parameters in models up to degree 4 are presented in Cornell (1981), Chapter 2.

4.1 Numerical Example

In producing vinyl for automobile seat covers, blends of three plasticizers (P_1, P_2, P_3), individually and in combination with one another, along with several other ingredients (stabilizers, lubricants, fillers, drying agents, and resins) are chosen for study. The percentage of the total formulation contributed by the plasticizers is 42%, while the remaining ingredients contribute 58% of the total.

To study the blending properties of the three plasticizers individually, two at a time, and all three together, a 7-point simplex-centroid design was set up in the three plasticizers. In each of the blends, the proportion of plasticizer was held fixed at 42% of the total batch weight. The seven blends differed only in the proportions of the three plasticizers that make up the total plasticizer proportion.

Three batches of each blend were prepared. The response of interest taken on each batch is a vinyl thickness measurement (in millimeters). High thickness values are considered to be more desirable than low thickness values. Table 2 lists the individual thickness values for each of the three batches of each blend along with the plasticizer proportions that make up each blend. The vinyl thickness values have been scaled to facilitate the computations, and the plasticizer percentages are scaled to represent proportions, $x_i = P_i/42\%$, $i = 1, 2, 3$.

The fitted special-cubic (9) model representing vinyl thickness as a function of the plasticizer proportions is

$$\hat{y}(\mathbf{x}) = 11.67x_1 + 5.33x_2 + 8.67x_3 + 23.32x_1x_2 - 4.68x_1x_3$$
$$\;\;\;\;\;(1.01)\;\;\;\;\;(1.01)\;\;\;\;\;(1.01)\;\;\;\;\;\;\;(4.94)\;\;\;\;\;\;\;\;(4.94)$$
$$+ 12.00x_2x_3 + 1.05x_1x_2x_3 \quad (13)$$
$$(4.94)\;\;\;\;\;\;\;\;(34.74)$$

Table 2   Vinyl Thickness Experimental Data

Plasticizer(Proportion)			Scaled vinyl	Average
P_1 (x_1)	P_2 (x_2)	P_3 (x_3)	thickness (y_u)	$(\bar{y})$
1	0	0	10, 12, 13	11.67
0	1	0	5, 4, 7	5.33
0	0	1	8, 8, 10	8.67
1/2	1/2	0	12, 16, 15	14.33
1/2	0	1/2	10, 8, 9	9.00
0	1/2	1/2	9, 9, 12	10.00
1/3	1/3	1/3	13, 9, 14	12.00

The coefficient estimates are calculated using Eqs. (11) and (12),

$$b_1 = 11.67, \qquad b_2 = 5.33, \qquad b_3 = 8.67$$
$$b_{12} = 4(14.33) - 2(11.67 + 5.33) = 57.32 - 34.00 = 23.32$$
$$b_{13} = 4(9.00) - 2(11.67 + 8.67) = -4.68,$$
$$b_{23} = 4(10.00) - (5.33 + 8.67) = 12.00$$
$$b_{123} = 27(12.00) - 12(14.33 + 9.00 + 10.00) + 3(11.67 + 5.33 + 8.67)$$
$$= 1.05$$

The quantities in parenthesis below the coefficient estimates in formula (13) are the estimated standard errors of the coefficient estimates. They are calculated as

$$\text{s.e.}(b_i) = (s^2/3)^{1/2} = 1.01, \qquad \text{s.e.}(b_{ij}) = (24s^2/3)^{1/2} = 4.94,$$
$$\text{s.e.}(b_{123}) = (1188s^2/3)^{1/2} = 34.74$$

where an estimate s^2 of the variance of each vinyl thickness value is calculated using the pooled within-blend variance of $s^2 = 3.05$ with 14 degrees of freedom. As in formulas (11) and (12) for the coefficient estimates, the formulas for calculating the values of the standard errors of the coefficient estimates are simplified by the fitting of the canonical polynomial to data collected only at the points of the associated simplex-lattice or simplex-centroid design. This is because there is a one-to-one correspondence between the number of terms in the canonical polynomial and the number of points in the associated simplex design.

The interpretations of the coefficient estimates in the special-cubic model (13) are as follows. The linear blending estimates (the b_i) are the average vinyl thickness values for the single-plasticizer blends and represent the heights of the estimated vinyl thickness surface at the vertices of the triangle; see Figure 4. Since $b_1 > b_3 > b_2$, plasticizer 1 produced the highest average vinyl thickness value of the three plasticizers when used alone, while plasticizer 2 produced the lowest average vinyl thickness value. The planar surface (or linear blending) portion of the model (13) consisting only of the first-degree terms is $b_1 x_1 + b_2 x_2 + b_3 x_3 = 11.67x_1 + 5.33x_2 + 8.67x_3$ and the surface, $\hat{y}(\mathbf{x}) = \sum_{i=1}^{3} b_i x_i$, is shown in Figure 4.

The binary cross-product coefficient estimates (the b_{ij}) measure curvilinear departure of the vinyl thickness surface from the planar surface of Figure 4, arising from the two-plasticizer blends. In particular, the syn-

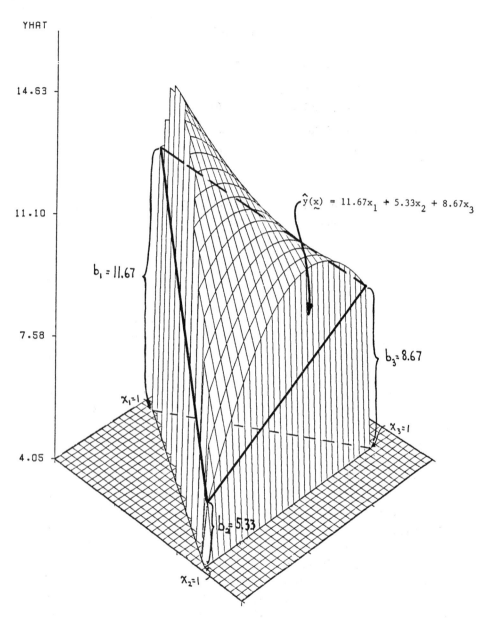

Figure 4   The estimated special-cubic surface modeled by (13) and the planar
only portion of the surface modeled by $\hat{y}(\mathbf{x}) = 11.67x_1 + 5.33x_2 + 8.67x_3$.

ergistic blending of plasticizers 1 and 2 appears to be real because of the magnitude of the coefficient estimate $b_{12} = 23.32$ relative to the magnitude of its standard error.

An approximate test for the synergistic blending of two components, or for testing the hypothesis $H_0 : \beta_{ij} = 0$ versus $H_A : \beta_{ij} \neq 0$, consists of computing the value of the test statistic

$$t = \frac{b_{ij}}{\text{s.e.}(b_{ij})}$$

and comparing the calculated value of t against a value of t taken from the table of "Student t" values. In the case of testing the synergistic blending of plasticizers 1 and 2, the calculated value of t is $t = 23.32/4.94 = 4.72$. This value greatly exceeds the table value, $t_{.005,14} = 2.977$, which prompts us to infer, at the 0.01 level of significance, that plasticizers 1 and 2, when blended together, produce a higher vinyl thickness value (blend synergistically) than we would have expected from simply averaging the vinyl thickness values of the single-component blends. The tests of the coefficient estimates reflecting the joint blending of plasticizers 1 and 3 and of 2 and 3 produced calculated t values of -0.99 and 2.43, respectively. Neither t value is greater in absolute value than 2.977, and so we remark that there is not sufficient evidence, at the 0.01 level of significance, to indicate that plasticizer 3 blends nonlinearly with either plasticizer 1 or 2. At the 0.05 level of significance, the table t value is $t_{.025,14} = 2.145$, which leads us to suspect that components 2 and 3 also blend nonlinearly. The shape of the estimated surface along the x_2-x_3 edge of the triangle in Figure 4 supports this latter conclusion.

The average vinyl thickness value (12.00) owing to the ternary blending of the three plasticizers does not depart greatly from the average of the vinyl thickness values obtained from the two-plasticizer blends, $(14.33 + 9.00 + 10.00)/3 = 11.11$. This is reflected in the coefficient estimate $b_{123} = 1.05$, not being significantly different from zero at the 0.05 level of significance. Hence, because of the nonsignificance of the estimates $b_{13} = b_{123}$, an alternative model form for expressing the vinyl thickness as a function of the plasticizer proportions is the reduced fitted model form

$$\hat{y}(\mathbf{x}) = 11.25x_1 + 5.36x_2 + 8.26x_3 + 23.60x_1x_2 + 12.26x_2x_3 \quad (14)$$
$$(0.89) \quad (0.97) \quad (0.89) \quad (4.46) \quad (4.46)$$

The fitted model (14) was obtained using a least-squares regression program, and the estimate of σ^2 obtained with the model (14) is $s^2 = 2.86$.

Many times, in order to form a valid mixture from which an acceptable product can be made, component i must be present in some minimum amount, $L_i > 0$. The quantity L_i is called a *lower bound* for x_i, that is, $x_i \geq L_i > 0$. Moreover, the placing of a lower bound on x_i creates upper bounds for the remaining x_j, i.e., $x_j \leq U_j = 1 - L_i$ and $\sum_{j\neq i}^q x_j = 1 - L_i$. When upper and lower bound constraints are placed on the x_i, we have what is called *a constrained mixture problem*.

5   MULTIPLE CONSTRAINTS ON THE COMPONENT PROPORTIONS

In the formulation of a rocket propellant consisting of the three components, fuel (x_1), oxidizer (x_2), and binder (x_3), each blend must contain at least 20% fuel, at least 40% oxidizer, and at least 20% binder. This forces the following constraints on the component proportions

$$0.20 \leq x_1, \qquad 0.40 \leq x_2, \qquad 0.20 \leq x_3 \tag{15}$$

The placing of lower bounds on the x_i of the form in (15) creates implied upper bounds for the x_i. The upper bound 0.40 for x_3 is forced by the presence of lower bounds 0.20 and 0.40 for x_1 and x_2, respectively. Similarly, x_1 and x_2 have implied upper bounds of 0.40 and 0.60, respectively. Note that the effect of forcing bounds on the x_i of the form in (15) produces combinations of the components or mixtures that are limited to the subregion of the simplex defined by the additional constraints; see Figure 5.

For constructing designs in which to collect data for modeling the surface over the smaller subregion, generally *pseudocomponents* are used. The pseudocomponents proportions are denoted by x_i' and are defined as follows. Let $L_i > 0$ denote the lower bound for component i. Then the ith pseudocomponent is

$$x_i' = \frac{x_i - L_i}{1 - L} \tag{16}$$

where $L = \sum_{i=1}^q L_i < 1$. For example, with the lower bounds defined in (15), where $L = 0.20 + 0.40 + 0.20 = 0.80$, then $1 - L = 0.20$ and the

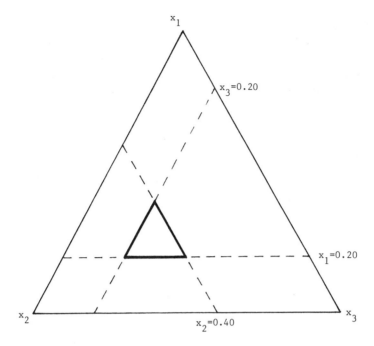

Figure 5 The subregion defined by the constraints $0.20 \leq x_1$, $0.40 \leq x_2$, $0.20 \leq x_3$.

pseudocomponents are

$$x_1' = \frac{x_1 - 0.20}{0.20}, \qquad x_2' = \frac{x_2 - 0.40}{0.20}, \qquad x_3' = \frac{x_3 - 0.20}{0.20} \qquad (17)$$

The factor space shown in Figure 5 is a regular two-dimensional simplex in the pseudocomponents x_i', $i = 1$, 2, 3. The placing of lower bounds on some or all of the x_i will always produce a simplex-shaped subregion.

Suppose it is desired to fit a quadratic mixture model over the subregion defined by the constraints (15). Since the subregion is a simplex, we could set up a $\{3,2\}$ simplex lattice in the pseudocomponent proportions and augment the design with one or more interior points. The interior points enable a check to be made on the adequacy of the fitted quadratic model if an independent estimate of the experimental error is known. For illustrative purposes, we shall assume that 10 different blends are desired (6 from the $\{3,2\}$ simplex lattice plus 4 additional interior blends) and that an external estimate of σ^2 is available.

    The pseudocomponent settings of the 6 blends of the $\{3,2\}$ simplex lattice plus the 4 interior blends are listed in Table 3. Also listed are the component proportions for fuel (x_1), oxidizer (x_2), and binder (x_3). The proportions of the original components, fuel, oxidizer, and binder, are obtained by reversing Eq. (16) to obtain

$$x_i = L_i + (1 - L)x_i'$$

For example, corresponding to the blend ($x_1' = 1/2$, $x_2' = 1/2$, $x_3' = 0$), the proportions of fuel, oxidizer, and binder, respectively, are

$$x_1 = 0.20 + (0.20)1/2, \qquad x_2 = 0.40 + (0.20)1/2, \qquad x_3 = 0.20 + (0.20)0$$
$$= 0.30 \qquad\qquad\qquad = 0.50 \qquad\qquad\qquad = 0.20$$

    Once data values are collected from the 10 blends, the quadratic model can be fitted in the pseudocomponent proportions or in the original component (fuel, oxidizer, binder) proportions. For this particular example, Khuri and Cornell (1987) fitted special-cubic models in the pseudocomponent proportions as well as in the original component proportions and discussed the difference in the interpretations of the estimated coefficients associated with the two separate models. Briefly, while the two models are

Table 3   Rocket Propellant Design Blends

Pseudocomponents			Original components		
x_1'	x_2'	x_3'	x_1	x_2	x_3
1	0	0	.40	.40	.20
0	1	0	.20	.60	.20
0	0	1	.20	.40	.20
1/2	1/2	0	.30	.50	.20
1/2	0	1/2	.30	.40	.30
0	1/2	1/2	.20	.50	.30
1/3	1/3	1/3	.27	.46	.27
2/3	1/6	1/6	.33	.43	.23
1/6	2/3	1/6	.23	.53	.23
1/6	1/6	2/3	.23	.43	.33

equivalent surface representations over the subregion defined by $0.20 \leq x_1$, $0.40 \leq x_2$, and $0.20 \leq x_3$, the coefficient estimates in the pseudocomponent model describe the shape of the surface directly above the pseudocomponent subregion. The estimates of the coefficients in the original component model are extrapolations of the shape of the surface above the subregion back to the boundaries of the original triangle, whose vertices are defined as $x_1 = 1$, $x_2 = x_3 = 0$; $x_2 = 1$, $x_1 = x_3 = 0$; and $x_3 = 1$, $x_1 = x_2 = 0$. Generally speaking, the fitted model in the pseudocomponents is the easier model to use for describing the shape of the surface (or interpreting the type of blending that is present) over the subregion. This is because the coefficient estimates in the original component model are extrapolations of the type of blending that is present among the components, and the making of inferences based on extrapolations obtained from fitting polynomial models involves a risk that should not be taken.

When both upper (U_i) and lower (L_i) bounds are placed on the x_i of the form

$$0 < L_i \leq x_i \leq U_i < 1, \qquad i = 1, 2, \ldots, q \tag{18}$$

the resulting factor space is a convex polyhedron. The shape of the polyhedron is generally more complicated than the simplex-shaped subregion, which is defined by placing only lower bounds (L_i) on the x_i. As an example, in Figure 6 is drawn the subregion defined by the constraints

$$0.20 \leq x_1 \leq 0.60, \qquad 0.10 \leq x_2 \leq 0.60, \qquad 0.10 \leq x_3 \leq 0.50 \tag{19}$$

The region (19) has six vertices and six edges joining the extreme vertices. The extreme vertices have the coordinate settings $(x_1, x_2, x_3) = (0.6, 0.3, 0.1)$, $(0.3, 0.6, 0.1)$, $(0.2, 0.6, 0.2)$, $(0.2, 0.3, 0.5)$, $(0.4, 0.1, 0.5)$, and $(0.6, 0.1, 0.3)$.

With constrained regions of the type shown in Figure 6, the extreme vertices and convex combinations (or midpoints of the edges connecting the extreme vertices) of some of the vertices are used as design points for fitting the Scheffé-type model forms. For example, to fit the first-degree model, $y = \beta_1 x_1 + \beta_2 x_2 + \beta_3 x_3 + \epsilon$, at least three distinct design points are required, and one suggestion is to collect data at three or four of the vertices (points 1., 3., 4., and 6., say) and also at the centroid of the subregion. The vertices 1., 3., 4., and 6. are recommended as a four-point design because these four points provide a greater coverage of the subregion than would any other four-point design; and the greater the spread of the design points,

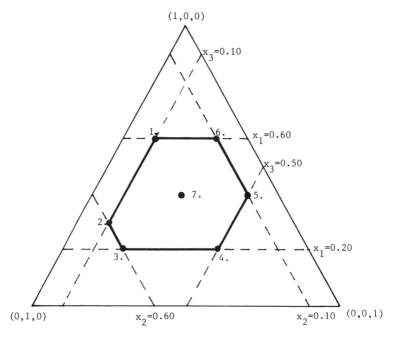

Figure 6   The subregion defined by the constraints $0.20 \leq x_1 \leq 0.60$, $0.10 \leq x_2 \leq 0.60$, and $0.10 \leq x_3 \leq 0.50$. The extreme vertices are numbered 1. to 6., and point 7 is the centroid of the subregion.

the smaller will be the variances and covariances of the coefficient estimates in the fitted first-degree model. The centroid point is an excellent location for detecting the presence of curvature in the shape of the surface through a test for lack of fit of the first-degree model. Such a test for lack of fit requires an independent estimate of the experimental error variance to be obtained; see Shelton et al. (1983). An independent estimate of the experimental error variance is obtained by collecting replicate observations at one or more of the design points and by pooling the variation among the replicates at each point.

   To fit the second-degree model, $y = \beta_1 x_1 + \beta_2 x_2 + \beta_3 x_3 + \beta_{12} x_1 x_2 + \beta_{13} x_1 x_3 + \beta_{23} x_2 x_3 + \epsilon$, at least six distinct design points are required. One suggestion of a design for this example would be to use all six vertices, the midpoints of the three longest edges, and the centroid of the region. The three longest edges connect vertices 1 and 2, vertices 3 and 4, and vertices 5 and 6. Snee (1975) discusses the computer-aided construction of designs for fitting quadratic models in constrained mixture regions. Some of the algorithms that have been suggested for locating the extreme vertices of constrained regions and that have appeared in the literature are the extreme vertices algorithm by McLean and Anderson (1966), the XVERT algorithm by Snee and Marquardt (1974), CONSIM by Snee (1979), and XVERT1 by Nigam et al. (1983). Algorithms used for selecting subsets of points from a candidate list are DETMAX by Mitchell (1974), CADEX by Kennard and Stone (1969), and ACED by Welch (1982). A software program on diskette for personal computers used for generating the extreme vertices of constrained region is XSTAT, and another program on diskette for generating the extreme vertices, as well as selecting a subset of the vertices, is ECHIP. These programs and others are discussed in detail by Nachtsheim (1987).

In the analysis of data collected from constrained mixture regions, generally one begins by fitting the first- or second-degree Scheffé-type mixture model, using a least-squares regression program. Summary statistics, such as the coefficient of determination and the overall test for regression, measure how well the model fits the data; Snee and Marquardt (1974). Approximate tests can be performed on the estimated coefficients to determine which of the coefficients are significantly different from zero. An alternative strategy to testing the coefficients is to use a stepwise regression procedure to arrive at a final model form. When the final form of the model is decided on, a technique that can be used to measure the change in the estimated response brought about by varying the proportions of each of the components is known as *plotting the response trace.*

The response trace is a plot of the predicted values of the response taken at a reference point and at points along certain directions in the constrained region. The reference point or reference blend is usually the centroid of the constrained region. The directions from the reference point are toward the vertices of the original simplex itself or toward the vertices of a pseudocomponent simplex inside the original simplex; Piepel (1982). Since moving from the reference blend toward the vertices of the simplex in the original constrained components is akin to adding more of the particular component, whose vertex is $x_i = 1$, into the mixture, this direction has some intuitive appeal for one who is studying the change in the predicted response to a change made in the proportion of a specific component.

To illustrate the technique of plotting the response trace, let us refer to the constrained region in Figure 6 and consider fitting the Scheffé second-degree model to the data at the 10 different blends listed in Table 4. We

Table 4　Data from the Constrained Region

Design point	Original components			Pseudocomponents			Data values
	x_1	x_2	x_3	x_1'	x_2'	x_3'	y
1.	.600	.300	.100	.667	.333	0	8
2.	.300	.600	.100	.167	.833	0	6
3.	.200	.600	.200	0	.833	.167	7
4.	.200	.300	.500	0	.333	.667	14
5.	.400	.100	.500	.333	0	.667	12
6.	.600	.100	.300	.667	0	.333	4
7.	.383	.334	.283	.305	.390	.305	12
8.	.450	.450	.100	.417	.583	0	11
9.	.200	.450	.350	0	.583	.417	11
10.	.500	.100	.400	.5	0	.5	8

shall fit both the second-degree model in the original components, the x_i, as well as in the pseudocomponents, the x_i', defined in (16) simply to illustrate the differences in the magnitudes of the coefficient estimates with the two equivalent models. The response trace will be generated from the fitted model in the original components, however. The 10 blends, which are listed as points 1. to 10. in Table 4, represent the six extreme vertices of the region (points 1. to 6.), the centroid of the region (point 7.), and the midpoints of the longest three edges (points 8., 9., and 10.). Also listed in Table 4 are values of the response taken at each of the 10 blends, along with the pseudocomponent proportions corresponding to each of the 10 blends.

The fitted model in the original components, whose proportions are constrained in (19), is

$$\hat{y}(\mathbf{x}) = -21.2x_1 - 24.6x_2 + 20.6x_3 + 129.3x_1x_2 + 25.2x_1x_3 + 36.5x_2x_3 \quad (20)$$
$$\phantom{\hat{y}(\mathbf{x}) = }(6.6)\quad\;\;(5.5)\quad\;\;(9.3)\quad\;\;(21.2)\quad\quad(26.7)\quad\;\;(21.2)$$

The quantities in parentheses directly below the coefficient estimates are the estimated standard errors of the coefficient estimates, where an estimate of σ^2 is the residual mean square (MSE = 0.86) taken from the analysis of variance table, Table 5. The value of the coefficient of determination for the model (20), as well as for the model (21), which is fitted in the pseudocomponent proportions, is $R^2 = 0.9618$.

Table 5 ANOVA Table for the Fitted Models (20) and (21)

Source	d.f.	Sum of squares	Mean square
Regression (fitted model)	5	86.66	17.33
Residual	4	3.44	MSE = 0.86
Total	9	90.10	

The fitted model in the pseudocomponent proportions x_1', x_2', and x_3' is

$$\hat{y}(\mathbf{x'}) = -4.7x_1' + 1.8x_2' + 16.4x_3' + 46.5x_1'x_2' + 9.1x_1'x_3' + 13.2x_2'x_3' \quad (21)$$
$$\;\;(2.8)\quad(1.3)\quad(2.8)\quad\;\;(7.6)\quad\;\;(9.6)\quad\;\;(7.6)$$

With both fitted models (20) and (21), it is questionable whether component 3 blends nonlinearly with components 1 and 2, while the latter components (1 and 2) do appear to blend nonlinearly in the presence of one another.

Although a subset regression program would probably omit the two terms, $b_{13}x_1x_3$ and $b_{23}x_2x_3$, from the model (20), we shall include all six terms in the fitted model (20) in producing the plots of the response trace. Let us designate design point 7., which is located at the center of the subregion in Figure 6, to be the reference blend. Let us subscript the coordinate settings of the reference blend by using the letter r so that, at point 7., the coordinate settings are ($x_{1r} = 0.383$, $x_{2r} = 0.334$, $x_{3r} = 0.283$). Next, let us envision three lines drawn from the three vertices of the triangle and extending to the three sides of the triangle opposite the vertices in Figure 6, where each line passes through the reference blend, point 7. These lines are called rays, and it is along these rays that we will define blends for which to predict the response in plotting the response trace.

Consider a blend along the x_1 ray in which the proportion of component 1 is increased by an amount equal to 0.02 over that of the reference blend. Along the x_1 ray, when x_1 is increased from 0.383 to 0.403, the proportion $x_2 + x_3$ is decreased by 0.02, and the relative proportions x_2/x_3 remain constant and equal to the value at the reference blend, $x_2/x_3 = 1.180 = x_{2r}/x_{3r}$. At the first blend defined on the x_1 ray where $x_1 = 0.403$, the

Table 6 Coordinate Values and
Predicted Response Values Along x_1
Ray

Coordinates			Predicted response
x_1	x_2	x_3	$(\hat{y})$
.203	.431	.366	11.58
.223	.421	.356	11.86
.243	.410	.347	12.11
.263	.399	.338	12.30
.283	.388	.329	12.44
.303	.377	.320	12.51
.323	.366	.311	12.52
.343	.356	.301	12.47
.363	.345	.292	12.37
.383	.334	.283[a]	12.21
.403	.323	.274	11.99
.423	.313	.264	11.71
.443	.302	.255	11.38
.463	.291	.246	10.99
.483	.280	.237	10.54
.503	.269	.228	10.03
.523	.258	.219	9.46
.543	.247	.210	8.83
.563	.237	.200	8.16
.583	.226	.191	7.42

[a] Reference blend

values of the remaining component proportions are

$$x_2 = x_{2r} - \frac{0.02x_{2r}}{1 - x_{1r}},$$

$$= 0.334 - \frac{0.02(0.334)}{1 - 0.383}$$

$$= 0.323$$

$$x_3 = x_{3r} - \frac{0.02x_{3r}}{1 - x_{1r}}$$

$$= 0.283 - \frac{0.02(0.283)}{1 - 0.383}$$

$$= 2.74$$

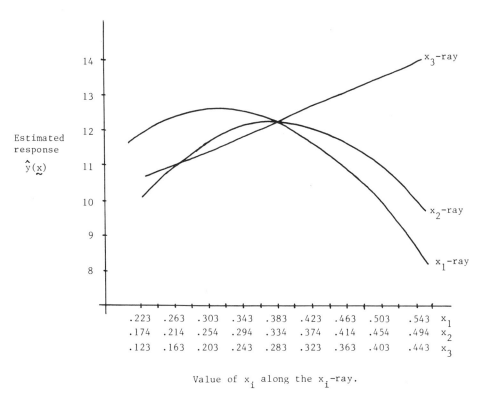

Value of x_i along the x_i-ray.

Figure 7 Plots of the response trace taken along the three x_1 rays, $i = 1, 2, 3$.

The predicted value of the response at the blend whose coordinate are (0.403, 0.323, 0.274) is calculated using Eq. (20) to be

$$\hat{y}(\mathbf{x}) = -21.2(0.403) - 24.6(0.323) + 20.6(0.274) + 129.3(0.403)(0.323)$$
$$+ 25.2(0.403)(0.274) + 36.5(0.323)(0.274)$$
$$= 11.99.$$

At a second point along the x_1 ray where x_1 is equal to 0.423, the values of the proportions for components 2 and 3 are $x_2 = 0.313$ and $x_3 = 0.264$, and the predicted value of the response for this blend is $\hat{y}(\mathbf{x}) = 11.71$.

Table 6 lists the coordinate values and predicted response values for 20 blends taken along the x_1 ray. Only points along the x_1 ray falling inside of the constrained subregion in Figure 6 are of interest, so that the first blend that is listed in Table 6 and that is farthest away from the vertex $x_1 = 1$ has the coordinate setting ($x_1 = 0.203$, $x_2 = 0.431$, $x_3 = 0.366$), while the last blend that is listed and that is the closest blend to the vertex $x_1 = 1$ has the coordinates ($x_1 = 0.583$, $x_2 = 0.226$, $x_3 = 0.191$). In Figure 7, predicted values of the response using the fitted model (20) are plotted (or traced) for blends defined on each of the three rays. The blends along the x_i ray are listed in increments of 0.02 for x_i, $i = 1$, 2, 3. Only for the reference blend (0.383, 0.334, 0.283) do the values of x_1, x_2, and x_3 along the ordinate in Figure 7 sum to unity. As a reminder, only the blends falling inside the constrained subregion of Figure 6 are used to generate the plots in Figure 7. For additional discussion on plotting the response trace, see Hare (1985) and Sahrman et al. (1987).

The plots in Figure 7 illustrate curvature in the shape of the estimated response surface along the x_1 and x_2 rays. Along the x_3 ray, the predicted response is linear. These plots are consistent with our inferences made earlier concerning the nonlinear blending terms in the fitted model (20), where we said that only components 1 and 2 appear to blend nonlinearly with one another.

6   ALTERNATIVE MODELS

Much of the success of response surface methodology is based on the assumption that the response behavior can be modeled within the range of the data by a low-order (first- or second-order) polynomial. Quenouille (1959) criticized the Scheffé-type polynomials for their inability to account correctly for components that are inert or that have additive effects. To model additive effects, Becker (1968) introduced the following homogeneous models of degree 1.

$$H1: \quad E(y) = \sum_{i=1}^{q} \alpha_i x_i + \sum_{i<j}^{q}\sum \alpha_{ij} \min(x_i, x_j)$$

$$+ \sum_{i<j<k}\sum\sum^{q} \alpha_{ijk} \min(x_i, x_j, x_k)$$

$$H2: \quad E(y) = \sum_{i=1}^{q} \alpha_i x_i + \sum_{i<j}^{q}\sum \alpha_{ij} \frac{x_i x_j}{(x_i + x_j)}$$

$$+ \sum_{i<j<k} \sum \sum^{q} \alpha_{ijk} \frac{x_i x_j x_k}{(x_i + x_j + x_k)^2} \tag{22}$$

$$H3: \quad E(y) = \sum_{i=1}^{q} \alpha_i x_i + \sum_{i<j} \sum^{q} \alpha_{ij}(x_i x_j)^{1/2}$$

$$+ \sum_{i<j<k} \sum \sum^{q} \alpha_{ijk}(x_i x_j x_k)^{1/3}$$

Snee (1973) discusses the types of curvature modeled by $H1$, $H2$, and $H3$ and provides plots of the curvature of each model over the range $0 \leq x_i \leq 1$. Contour plots of estimated surfaces of mite population numbers modeled by $H1$, $H2$, and $H3$ over a three-component triangle are presented in Cornell (1981, p. 231).

It is sometimes meaningful to study ratios of the component proportions. For example, in the development of a particular type of porcelain glass, considering the ratio of silica to soda and of silica to lime is a more meaningful approach to studying the three ingredients rather than looking at the specific proportions of each in a blend. Since the sum of the component proportions equals unity in every blend, then only $q - 1$ ratios or ratio variables are required in studying the q component system.

Ratio variables can be defined in a variety of ways. With three components whose proportions we denote x_1, x_2, and x_3, there are several ways to define the two ratio variables r_1 and r_2. For example, three arbitrary sets of ratio variables are

$$r_1 = \frac{x_1}{x_2} \quad \text{or} \quad r_1 = \frac{x_1}{x_3}, \quad r_2 = \frac{x_2}{x_3}$$

$$r_1 = \frac{x_1}{x_2 + x_3}, \quad r_2 = \frac{x_2}{x_3}$$

$$r_1 = x_1, \quad r_2 = \frac{x_2}{x_3}$$

Standard factorial designs can be set up, and ordinary polynomial models can be fitted in the ratio variables. To show this, suppose we have three components and are interested in looking at the proportions of each of components 1 and 2 relative to the proportion of component 3. The ratio variables are $r_1 = x_1/x_3$ and $r_2 = x_2/x_3$. Values of the ratio variable $r_1 = x_1/x_3$ define rays drawn from the vertex $x_2 = 1$, while values of the ratio variable $r_2 = x_2/x_3$ define rays drawn from the $x_1 = 1$ vertex. In

Figure 8, three rays are drawn corresponding to the values $r_1 = 1$, 2, and 3, and three rays are drawn corresponding to the values $r_2 = 2$, 3, and 4.

The points of intersection of the rays for $r_1 = 1$, 2, 3 with the rays for $r_2 = 2$, 3, 4 in Figure 8 define the nine points of a 3^2 factorial arrangement in the variables r_1 and r_2. Data collected at the nine points can be used to fit the second-degree model

$$y(\mathbf{r}) = \alpha_0 + \alpha_1 r_1 + \alpha_2 r_2 + \alpha_{12} r_1 r_2 + \alpha_{11} r_1^2 + \alpha_{22} r_2^2 + \epsilon$$

The settings in the mixture components corresponding to the nine combinations of the values of r_1 and r_2 are found by substituting the values of r_1 and r_2 into $x_1 = r_1/(1 + r_1 + r_2)$, $x_2 = r_2/(1 + r_1 + r_2)$, and $x_3 = 1/(1 + r_1 + r_2)$. For the nine combinations of $r_1 = 1$, 2, 3 and $r_2 = 2$, 3, 4, the settings

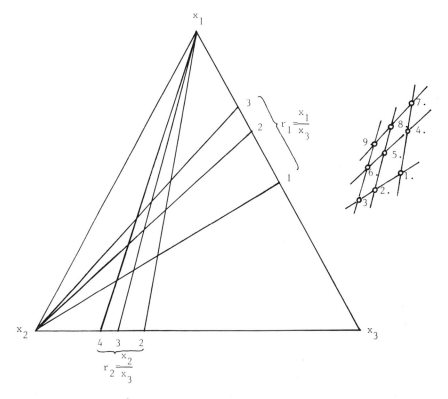

Figure 8 The 3^2 factorial arrangement in the ratio variables r_1 and r_2.

in x_1, x_2, and x_3 are

$$
\begin{array}{cccccccccc}
r_1 = & 1 & 1 & 1 & 2 & 2 & 2 & 3 & 3 & 3 \\
r_2 = & 2 & 3 & 4 & 2 & 3 & 4 & 2 & 3 & 4 \\
x_1 = & \frac{1}{4} & \frac{1}{5} & \frac{1}{6} & \frac{2}{5} & \frac{2}{6} & \frac{2}{7} & \frac{3}{6} & \frac{3}{7} & \frac{3}{8} \\
x_2 = & \frac{2}{4} & \frac{3}{5} & \frac{4}{6} & \frac{2}{5} & \frac{3}{6} & \frac{4}{7} & \frac{2}{6} & \frac{3}{7} & \frac{4}{8} \\
x_3 = & \frac{1}{4} & \frac{1}{5} & \frac{1}{6} & \frac{1}{5} & \frac{1}{6} & \frac{1}{7} & \frac{1}{6} & \frac{1}{7} & \frac{1}{8}
\end{array}
$$

For modeling an extreme change in the response as the value of x_i approaches zero, Draper and St. John (1977) suggest that inverse terms $1/x_i$ be added to the Scheffé polynomials. The first- and second-degree Scheffé models, with the inverse terms added, are

$$
E(y) = \sum_{i=1}^{q} \beta_1 x_i + \sum_{i=1}^{q} \beta_{-1} x_i^{-1}
$$

$$
E(y) = \sum_{i=1}^{q} \beta_1 x_i + \sum_{i<j}^{q} \sum \beta_{ij} x_i x_j + \sum_{i=1}^{q} \beta_{-1} x_i^{-1}
$$

7 INCLUSION OF PROCESS VARIABLES

In many types of mixture experiments, there are factors in addition to the component proportions that may influence the blending properties of the mixture components. An example is in coating photographic film where, in addition to the ingredients in the coating solution being studied, the pH of the dip solution is varied along with the length of time the film is dipped. Such factors as pH and dip time are known as *process variables*.

When process variables are included in mixture experiments, the simplest design strategy is to set up a lattice arrangement in the mixture component proportions (a $\{q, m\}$ simplex-lattice or a simplex-centroid arrangement) at each point of a factorial arrangement in the levels of the process variables. For example, suppose we have three components, x_1, x_2, and x_3, in which we want to fit the quadratic mixture model. Further, suppose we also have two process variables and denote the coded variables z_1 and z_2 for process variable 1 and 2, respectively. Then the combined design might be a $\{3, 2\}$ simplex lattice $\times 2^2$ factorial arrangement as shown in Figure 9. The $\{3, 2\}$ simplex lattice is used for fitting the second-degree

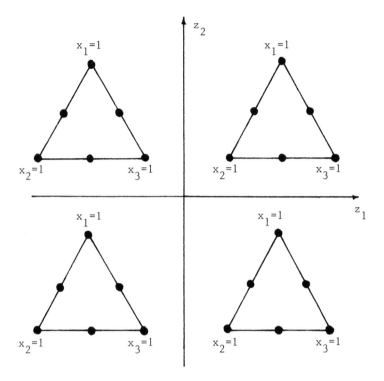

Figure 9   The 24 points of the $\{3,2\}$ simplex lattice $\times$ 2^2 factorial arrangement.

model

$$y(\mathbf{x}) = \beta_1 x_1 + \beta_2 x_2 + \beta_3 x_3 + \beta_{12} x_1 x_2 + \beta_{13} x_1 x_3 + \beta_{23} x_2 x_3 + \epsilon$$

while the 2^2 factorial arrangement is used for fitting the model

$$y(\mathbf{z}) = \alpha_0 + \alpha_1 z_1 + \alpha_2 z_2 + \alpha_{12} z_1 z_2 + \epsilon$$

in the coded process variables.

The combined $6 \times 4 = 24$-term model in the mixture components and the process variables is

$$y(\mathbf{x}, \mathbf{z}) = [\beta_1 x_1 + \beta_2 x_2 + \beta_3 x_3 + \beta_{12} x_1 x_2 + \beta_{13} x_1 x_3 + \beta_{23} x_2 x_3]$$
$$\times [\alpha_0 + \alpha_1 z_1 + \alpha_2 z_2 + \alpha_{12} z_1 z_2] + \epsilon$$

$$= \sum_{i=1}^{3} \gamma_i^0 x_i + \sum_{i<j}^{3} \gamma_{ij}^0 x_i x_j + \sum_{l=1}^{2} \left[\sum_{i=1}^{3} \gamma_i^l x_i + \sum_{i<j}^{3} \gamma_{ij}^l x_i x_j \right] z_l$$

$$+ \left[\sum_{i=1}^{3} \gamma_i^{12} x_i + \sum_{i<j}^{3} \gamma_{ij}^{12} x_i x_j \right] z_1 z_2 + \epsilon \qquad (23)$$

where $\gamma_i^l = \beta_i \alpha_l$, $\gamma_{ij}^l = \beta_{ij} \alpha_l$, $l = 0,1,2$, $\gamma_i^{12} = \beta_i \alpha_{12}$ and $\gamma_{ij}^{12} = \beta_{ij} \alpha_{12}$, $i,j = 1,2,3$, and $i < j$. The coefficients γ_i^0, $i = 1,2,3$ and γ_{ij}^0, $i < j$, in (23) are measures of the linear blending of component i and nonlinear blending of components i and j, respectively, when averaged over all combinations of the levels of the process variables. The coefficients γ_i^l and γ_{ij}^l, $l = 1, 2$, represent the effect of changing the level of process variable l on the linear blending of component i and the nonlinear blending of components i and j, respectively. Furthermore, γ_i^{12} and γ_{ij}^{12} are measures of the interaction effect of process variables 1 and 2 on the linear and nonlinear blending properties of the mixture components.

The method of testing the blending properties of the mixture components and the effects of the process variables on the component blending properties depends on the way in which the data are collected during the experimental program. Cornell (1981) discusses the testing of the combined model coefficient estimates when the experimentation is completely random and, later (1988), Cornell presents the testing formulas when the design is of the split-plot type. For additional reading on the inclusion of process variables in mixture experiments, see Cornell and Gorman (1984), Gorman and Cornell (1982), and Hare (1979).

7.1 Example

Three plasticizers ($P1$, $P2$, and $P3$) were blended in the formation of vinyl for flooring. Six combinations of the plasticizers were prepared at each of two molding temperatures (50 and 85°C). The response recorded on the vinyl blends was a toughness measurement. The plasticizer proportions, mold temperatures, and toughness values are shown in Table 7.

Fitted to the toughness values at the 50°C temperature, the quadratic model is

$$\hat{y}_{50^\circ} = 6x_1 + 4.5x_2 + 12x_3 + 15x_1 x_2 - 8x_1 x_3 + 7x_2 x_3$$

Table 7   Example Values

P1	P2	P3	Temp., °C 50	85
1	0	0	6	8.5
1/2	1/2	0	9	12
0	1	0	4.5	5
0	1/2	1/2	10	14
0	0	1	12	16
1/2	0	1/2	7	10

At the 85°C temperature, the fitted quadratic model is

$$\hat{y}_{85°} = 8.5x_1 + 5x_2 + 16x_3 + 21x_1x_2 - 9x_1x_3 + 14x_2x_3$$

If we combine the data from both temperatures and define the coded variable, $z = (\text{temp.} - 67.5°)/17.5°$, the fitted combined equation is

$$\begin{aligned}
\hat{y}(\mathbf{x}, \mathbf{z}) = {}& 7.25x_1 + 4.75x_2 + 14.0x_3 + 18.0x_1x_2 \\
& - 8.5x_1x_3 + 10.5x_2x_3 + 1.25x_1z + 0.25x_2z \qquad (24) \\
& + 2.0x_3z + 3.0x_1x_2z - 0.5x_1x_3z + 3.5x_2x_3z
\end{aligned}$$

The first six terms (in the x_i only) in (24) represent the blending properties of the three plasticizers averaged over both temperatures (or at the mean temperature, 67.5°), and the last six terms represent the effect of raising the temperature 17.5°C on the blending properties of the three plasticizers. The coefficient estimates in the combined model (24) are simple functions of the coefficient estimates in the separate models at the two temperatures.

8   SUMMARY

In this chapter, we have presented some of the more popular methods used in the design, modeling, and analysis of data for mixture experiments. Starting with the most general description of a mixture experiment, where all combinations of the ingredients (or components) are possible and the whole simplex region is to be explored, the designs presented were of the simplex-lattice and simplex-centroid types. Models fitted to data collected from the simplex designs are the Scheffé canonical polynomials.

In many mixture experiments, *all* the ingredients must be present in each of the blends, which forces all the component proportions to be nonzero. When lower bounds are placed on some or all of the component proportions, the factor space is a smaller simplex located inside the original simplex region. Pseudocomponents were introduced for constructing designs in the subregion and for fitting models for studying the shape of the response surface over the subregion. The placing of both lower and upper bounds on the component proportions was also discussed, and several computer programs were mentioned for locating the extreme vertices of the constrained region to be used as design points. A numerical example was given, showing how the plotting of the response trace is helpful in learning how changes in the component proportions affect the value of the predicted response over the constrained region.

Alternative model forms were mentioned briefly. Some of these were homogeneous models of degree 1, polynomials with terms expressed in ratios of the component proportions, and polynomials augmented with inverse terms. The chapter concluded with a short discussion on the inclusion of process variables in mixture experiments.

REFERENCES

Becker, N. G. (1968). Models for the response of a mixture. *J. Royal Statist. Soc., B, 30*: 349–358.

Cornell, J. A. (1973). Experiments with mixtures: A review. *Technometrics, 15*: 437–455.

Cornell, J. A. (1979). Experiments with mixtures: An update and bibliography. *Technometrics, 21*: 95–106.

Cornell, J. A. (1981). *Experiments With Mixtures: Designs, Models, and the Analysis of Mixture Data.* Wiley, New York.

Cornell, J. A. (1988). Analyzing data from mixture experiments containing process variables: A split-plot approach. *J. Qual. Technology, 20*: 2–23.

Cornell, J. A. and Gorman, J. W. (1984). Fractional design plans for process variables in mixture experiments. *J. Qual. Technology, 16*: 20–38.

Draper, N. R. and St. John, R. C. (1977). A mixtures model with inverse terms. *Technometrics, 19*: 37–46.

Gorman, J. W. and Hinman, J. E. (1962). Simplex-lattice designs for multicomponent systems. *Technometrics, 4*: 463–487.

Gorman, J. W. and Cornell, J. A. (1982). A note on model reduction for experiments with both mixture components and process variables. *Technometrics*, *24*: 243–247.

Hare, L. B. (1974). Mixture designs applied to food formulation. *Food Technology*, *28*: 50–62.

Hare, L. B. (1979). Designs for mixture experiments involving process variables. *Technometrics*, *21*: 159–173.

Hare, L. B. (1985). Graphical display of the results of mixture experiments. *Experiments in Industry: Design, Analysis and Interpretation of Results*, C&PI Division Technical Supplement, ASQC, Milwaukee, Wisc., 99–109.

Kennard, R. W. and Stone, L. A. (1969). Computer aided design of experiments. *Technometrics*, *11*: 137–148.

Khuri, A. I. and Cornell, J. A. (1987). *Response Surfaces: Designs and Analysis*, Marcel Dekker, New York.

McLean, R. A. and Anderson, V. L. (1966). Extreme vertices design of mixture experiments. *Technometrics*, *8*: 447–454.

Mitchell, T. J. (1974). An algorithm for the construction of D-optimal experimental designs. *Technometrics*, *16*: 203–210. Computer construction of D-optimal first-order designs. *Technometrics*, *16*: 211–220.

Nachtsheim, C. J. (1987). Tools for computer-aided design of experiments. *J. Qual. Technology*, *19*: 132–160.

Nigam, A. K., Gupta, S. C., and Gupta, S. (1983). A new algorithm for extreme vertices designs for linear mixture models. *Technometrics*, *25*: 367–371.

Piepel, G. F. (1982). Measuring component effects in constrained mixture experiments. *Technometrics*, *24*: 29–39.

Quenouille, M. H. (1959). Experiments with mixtures. *J. Royal Statist. Soc.*, *B*, *21*: 201–202.

Sahrman, H. F., Piepel, G. F., and Cornell, J. A. (1987). In search of the optimum Harvey Wallbanger recipe via mixture experiment techniques. *The Amer. Statistician*, *41*: 190–194.

Scheffé, H. (1958). Experiments with mixtures. *J. Royal Statist. Soc.*, *B*, *20*: 344–360.

Scheffé, H. (1963). The simplex-centroid design for experiments with mixtures. *J. Roy. Statist. Soc.*, B, 25: 235–263.

Shelton, J. T., Khuri, A. I., and Cornell, J. A. (1983). Selecting check points for testing lack of fit in response surface models. *Technometrics*, *25*: 20–38.

Snee, R. D. (1971). Design and analysis of mixture experiments. *J. Qual. Technology*, *3*: 159–169.

Snee, R. D. (1973). Techniques for the analysis of mixture data. *Technometrics*, *15*: 517–528.

Snee, R. D. (1975). Experimental designs for quadratic models in constrained mixture spaces. *Technometrics*, *17*: 149–159.

Snee, R. D. (1979). Experimental designs for mixture systems with multicomponent constraints. *Comm. in Statistics, Theory and Methods*, *A8*: 303–326.

Snee, R. D. and Marquardt, D. W. (1974). Extreme vertices designs for linear mixture models. *Technometrics*, *16*: 399–408.

Welch, W. J. (1982). Branch-and-bound search for experimental designs based on D-optimality and other criteria. *Technometrics*, *24*: 41–48.

8

Response Surface Designs and the Prediction Variance Function

Raymond H. Myers   Virginia Polytechnic Institute and State University, Blacksburg, Virginia

1   INTRODUCTION

In recent years there has been considerable interest in statistical methods to enhance quality. Engineers and physical scientists have begun upgrading their training in areas of quality control, process control, and design of experiments. An area in which considerable interest has resurfaced is that of response surface methodology (RSM) and, as a result, response surface design. RSM allows for the design of an experiment and analysis of data for building empirical models that are used for either process improvement or determination of optimum operating conditions. Box and Wilson (1951) developed the notion of response surface exploration, and Myers, Khuri, and Carter (1989) provide a broad review of the subject. In addition, textbooks by Myers (1976), Khuri and Cornell (1987), and Box and Draper (1987) are dedicated to the subject.

2 THE RESPONSE SURFACE PROBLEM—DESIGN AND ANALYSIS

The scientist or engineer is generally interested in a region in the controllable variables in which a model should be fit and the relationship between these variables and some response (often multivariate) y is to be explored. It is convenient to re-express the controllable variables, the ξ in *design units*. Namely,

$$x_j = \frac{\xi_j - c}{d}, \qquad j = 1, 2, \ldots, k$$

The x_j represent a centered and scaled version of the natural units. For example, one is allowed, then, to center and scale so that the design region is the unit cube or perhaps the unit sphere.

The RSM models that are usually fit include the *first order model*

$$E(y) = \beta_0 + \sum_{i=1}^{k} \beta_i x_i \qquad (1)$$

the first order model with two-factor interactions,

$$E(y) = \beta_0 + \sum_{i=1}^{k} \beta_i x_i + \sum \sum_{i<j} \beta_{ij} x_i x_j \qquad (2)$$

and the second order model

$$E(y) = \beta_0 + \sum_{i=1}^{k} \beta_i x_i + \sum_{i=1}^{k} \beta_{ii} x_i^2 + \sum \sum_{i<j} \beta_{ij} x_i x_j \qquad (3)$$

Obviously, less empirical models that are nonlinear in the parameters can be used but we will confine ourselves to the above models in this paper.

The response surface analysis that follows the experimental strategy revolves around (1) prediction of response values, (2) exploring the response surface in the region of the designed experiment, and (3) possibly finding conditions on the design variables that give rise to optimum response. Generally the model is fit by the method of least squares. Often a steepest ascent procedure (Myers 1976) is used in the case of the first order model to allow the user to move in the direction of improved response. This, of course, may very well be followed by additional experimental runs as confirmatory trials or for the purpose of fitting a higher order model. In the case of a second order model, "exploration of a response surface" may very well involve canonical analysis (Box and Draper 1987) of the fitted function or ridge analysis (Hoerl 1959). For both of these analysis, the user gains insight into areas in the design space in which an improvement in the response may be experienced. In the case of ridge analysis, an analyst essentially computes a locus of points that produce maximum (or minimum) response at a fixed distance from the center of the design $[(0, 0, \ldots, 0)$ in design units]. Hence a "ridge" is produced that gives conditionally optimum response. In a later section we shall use ridge analysis in an illustration to show the importance of taking into account the properties of the experimental design.

2.1   The Response Surface Design

The proper choice of an experimental design can often have a profound effect on the success of the response surface exploration. The properties of the predicted response are governed in large part by the general linear model

$$\mathbf{y} = X\boldsymbol{\beta} + \epsilon$$

where $\mathbf{y}$ is a vector of observed responses, X is a data/model matrix, $\boldsymbol{\beta}$ is a vector of regression coefficients, and ϵ is a vector of random disturbances around $N(\mathbf{0}, \sigma^2 I)$. The method of least squares is used to estimate $\boldsymbol{\beta}$. The impact of experimental design is best seen through the variance covariance matrix of the coefficients, namely,

$$\mathrm{Var}(\hat{\boldsymbol{\beta}}) = \sigma^2 (X'X)^{-1}$$

The $(X'X)^{-1}$ matrix is influenced by the model and by the design. A very important and popular criterion for choice of design is to choose the design so that $|(X'X)^{-1}|$ is minimized, or so that $|X'X|$ is maximized. This is the so-called *D-optimality* criterion discussed elsewhere in this text. There are other optimality criteria. Some are designed to emphasize estimation of coefficients. However, others are attentive to prediction of response. Here, of course, we refer to the value

$$\hat{y}(\mathbf{x}) = \mathbf{x}'\hat{\boldsymbol{\beta}}$$

is a vector which is dependent on the model and the point in ___ ₁ space at which the prediction is to be made. The variance of a predicted value is, then, given by

$$\frac{N \operatorname{Var} \hat{y}(\mathbf{x})}{\sigma^2} = \mathbf{x}'(X'X)^{-1}\mathbf{x} \tag{4}$$

Here we divide by σ^2 in order to consider a criterion that is scale free. Of course, N is the number of experimental runs.

2.2 Behavior of Prediction Variance

The choice of an experimental design for response surface analysis does not involve simple concepts, though many different classes of designs have been proposed by statisticians. The "exploration of a response surface" is done through the use of the fitted function $\hat{y}(\mathbf{x})$, which of course allows for prediction of response at a location (quantified by $\mathbf{x}$) inside or on the perimeter of the design *region*. Unfortunately, the variance function in Eq. (4) is not constant in the region of the design. As a result, it is important that the user take into account the *distribution* of $(N \operatorname{Var} \hat{y}(\mathbf{x}))/\sigma^2$ in the region of the design. Since it is so difficult to visualize the characterization of this "variance function," many users ignore it. Much of what we do in this paper revolves around the behavior of this function. Clearly, its appearance, and thus the performance of the fitted response function, depends on the experimental plan, geometry of the design, etc. In a later section we shall illustrate how the user who does not take the variance into account may very will experience difficulty in interpretation of the analysis.

As early as 1957 Box and Hunter urged consideration of the behavior of $(N \operatorname{Var} \hat{y}(x))/\sigma^2$ by introducing the notion of a *rotatable design*. A rotatable design is one in which $(N \operatorname{Var} \hat{y}(\mathbf{x}))/\sigma^2$ is constant on spheres (in the metric of the design units). This allows for a certain stability in prediction variance that seems desirable. Box and Hunter noticed that for the case of a second order model typically a plot of the variance for a rotatable design against the radius of a sphere centered at the center of the design region resembled that in Figure 1.

The appearance of the plot can be altered substantially by the use of *center runs* in the design. Box and Hunter introduced a *uniform precision design* that results in the choice of center runs of a rotatable design to be that which produces stability in the variance function for a portion of the

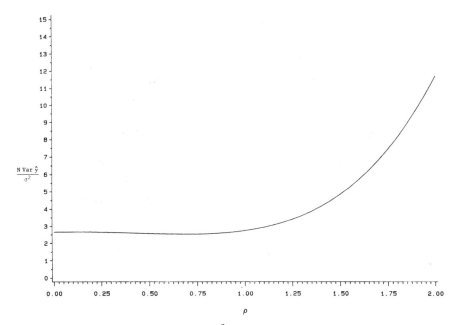

Figure 1   Plot of $N\operatorname{Var}\hat{y}(\mathbf{x})/\sigma^2$ for a second order rotatable design.

design region. This is discussed at length in Box and Draper (1987), Khuri and Cornell (1987), and Myers (1976).

2.3   Robustness in RSM Designs

Since the early 1970s much attention has been devoted by statisticians to robust methods. The origin of robust methods resides with hypothesis testing and estimation. However, by the mid 1970s attention was deflected toward robust designs, particularly in an RSM setting. While the definition of robustness is altered a bit, the basic principle remains intact. A robust design is one whose impact on the results is resistant to conditions that are nonideal. See Box and Draper (1975), Herzberg and Andrews (1979), and Myers, Khuri, and Carter (1989). Certainly, when one does a response surface analysis, many conditions may arise to create nonideal conditions. To cite only a few, we have

1. Missing data points
2. Outliers
3. Non-normal errors
4. Model misspecification

In addition to the above and other sources of nonideality, the notion of robustness in response surface designs center around the fact that in almost every practical RSM situation the ideal analysis is sequential. By this we mean that the scientist or engineer begins analysis of the data with a sizable amount of uncertainty. A robust design, for example, is one that perhaps will be appropriate for more than one model, for surely a model may very well need to be changed as one proceeds in the analysis. This implies, of course, that the design should require ample degrees of freedom for "lack of fit." In addition, replication of design points may be necessary in order to estimate experimental error variance. The above represent some requirements of a robust response surface design. There are others. It is this notion of robustness that has to some extent deflected interest from use of *optimal design* to that of designs that have balanced performance in several areas. Often optimal designs, conceived to be best in one area, do not possess characteristics of robustness.

3   DISCUSSION OF SPECIFIC RESPONSE SURFACE DESIGNS

In this section we discuss specific families of RSM designs. It is not our intention to give an exhaustive review. However, we indicate a few of the more useful designs.

3.1   First Order Designs

The appropriate type of first order design is the 2-level orthogonal array (see Box, Hunter, and Hunter 1978). The Resolution III design allows for orthogonal estimation of coefficients of the first order response model. The Resolution IV design allows for orthogonal estimation of first order, even in the presence of 2-factor interactions, though all 2-factor interactions cannot be estimated.

The 2-level orthogonal design enjoys advantages that are very important in an RSM problem. The design is D-optimal for first order models and also results in minimum variance of all regression coefficients on a per observation basis. As an illustration, consider the following 2^3 factorial array

$$
D = \begin{array}{c} \begin{array}{ccc} x_1 & x_2 & x_3 \end{array} \\ \begin{bmatrix} -1 & -1 & -1 \\ 1 & -1 & -1 \\ -1 & 1 & -1 \\ -1 & -1 & 1 \\ 1 & 1 & -1 \\ 1 & -1 & 1 \\ -1 & 1 & 1 \\ 1 & 1 & 1 \end{bmatrix} \end{array}
$$

Of course, each row of the *design matrix* represents a *design point* in which the factor levels are at the assigned high or low values. The ±1 representation is in *design levels*. They imply a centering and scaling of the variables from their natural metric to *design units*.

The orthogonal arrays (the 2^k factorial being an important special case) has received considerable attention in recent years. They are in the class of *D*-optimal designs for first order RSM models. In addition, they are rotatable designs for the first order case. Any 2-level design that is Resolution III (main effects mutually orthogonal) will possess these properties in the case of a first order model. There are many Resolution III 2-level designs. Many are discussed in Box, Hunter, and Hunter (1978) and Box and Draper (1987). When one of these designs is used, the engineer or scientist can be confident that the variance properties associated with the estimated regression coefficients are optimal. In addition, one may be assured that the predicted variance function behaves extremely well inside the region of the experiment. The variance is constant on spheres in the design metric and the increase in $(N \operatorname{Var} \hat{y}(\mathbf{x}))/\sigma^2$ from a radius $\rho = 0$ to a radius $p = \sqrt{k}$ is as small as one would expect for a design in which N runs are used.

3.2 Prediction Variance Function for the Orthogonal Array

Many of the points that we attempt to make in this paper deal with the behavior of the prediction variance. In the case of the orthogonal design we have

$$
\frac{N \operatorname{Var} \hat{y}(\mathbf{x})}{\sigma^2} = \frac{N}{\sigma^2} \left\{ \operatorname{Var}[\hat{\beta}_0 + \hat{\beta}_1 x_1 + \hat{\beta}_2 x_2 + \cdots + \hat{\beta}_k x_k] \right\}
$$

The $(X'X)$ matrix is merely NI_N and thus all coefficients are mutually independent and

$$\mathrm{Var}(\hat{\beta}_i) = \sigma^2/N, \qquad (i = 0, 1, 2, \ldots, k)$$

As a result,

$$\frac{N \, \mathrm{Var} \, \hat{y}(\mathbf{x})}{\sigma^2} = 1 + \rho^2$$

Thus Figure 2 depicts the complete prediction variance for the first order orthogonal array for $k = 4$.

It is natural that prediction of response will not be as good at the exterior of the design region. The experimenter must be aware of this as he or she carries out the analysis. However, the stability of prediction variance in the orthogonal array, depicted in Figure 2 for the case of $k = 4$, represents the most desirable distribution of prediction variance among all first order designs. In other words, the orthogonal array not only allows for constant prediction variance on spheres but it also results in *minimum rate of increase of* $(N \, \mathrm{Var} \, \hat{y}(\mathbf{x}))/\sigma^2$ as one approaches the design perimeter. See Giovannitti-Jensen and Myers (1988). We will subsequently show examples

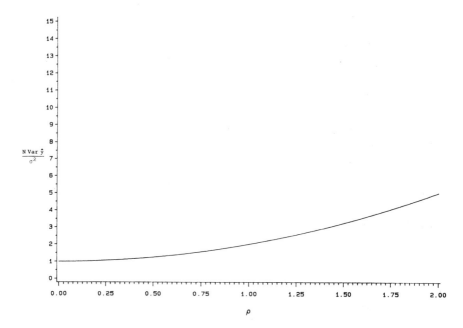

Figure 2 $N \, \mathrm{Var} \, \hat{y}/\sigma^2$ for a first order orthogonal design for $k = 4$.

that illustrate the value in RSM problems of being aware of the behavior of $(N \operatorname{Var} \hat{y}(\mathbf{x}))/\sigma^2$.

3.3 Second Order Designs

Statisticians have developed many different classes of second order designs for doing RSM. Box and Wilson first proposed the set of *composite* designs, the rationale being that the second order model in Eq. (3) is quite easily accommodated by adding certain design points to the 2^k factorial array. The 2-level factorial nicely estimates linear and 2-factor interaction coefficients. The *central composite design* features a 2-level factorial with additional *star* or axial points given by the following $2k$ points.

$$
\begin{array}{cccc}
x_1 & x_2 & \cdots & x_k \\
\begin{bmatrix}
-\alpha & 0 & . & \cdots & 0 \\
\alpha & 0 & . & \cdots & 0 \\
0 & -\alpha & 0 & \cdots & 0 \\
0 & \alpha & 0 & \cdots & 0 \\
\vdots & \vdots & \vdots & & \vdots \\
0 & 0 & 0 & \cdots & -\alpha \\
0 & 0 & 0 & \cdots & \alpha
\end{bmatrix}
\end{array}
$$

Here, of course, $\alpha \geq 0$.

The design consists of $2^k + 2k$ points. Often additional points are placed *in the center of the design* $(0, 0, \ldots, 0)$. Of course, the number of center runs has an impact on the degrees of freedom for replication error and on the behavior of the prediction variance function. It is recommended that at $k \geq 5$, the user should consider replacing the 2^k factorial with a 1/2 fraction. For optimum efficiency in estimation of 2-factor interaction coefficients, the fraction should be of Resolution V or higher.

The axial points in the central composite design allow for 5 levels of each factor (3 levels in cases where $\alpha = 1.0$) and reasonably efficient estimation of first and second order coefficients. In addition, the choice of n_0 (the number of center runs) and α provides for flexibility and a certain amount of control over the behavior of the prediction variance function. For example, if $\alpha = \sqrt[4]{F}$, where F is the number of factorial points, the resulting design is rotatable. Choice of the number of center runs will affect the stability of prediction variance.

A popular class of 3-level second order designs is the Box–Behnken designs developed in 1960. See Box and Behnken (1960). An example of a Box–Behnken ($k = 3$) is given by

$$
D = \begin{bmatrix}
-1 & -1 & 0 \\
-1 & 1 & 0 \\
1 & -1 & 0 \\
1 & 1 & 0 \\
-1 & 0 & -1 \\
-1 & 0 & 1 \\
1 & 0 & -1 \\
1 & 0 & 1 \\
0 & -1 & -1 \\
0 & -1 & 1 \\
0 & 1 & -1 \\
0 & 1 & 1
\end{bmatrix}
$$

A through discussion of Box–Behnken designs is given in Box and Draper (1987). The Box–Behnken designs compete favorably with the central composite designs.

Both the central composite design and the Box–Behnken design represent reasonable design plans that allow for efficient estimation of coefficients. In addition, both allow for *orthogonal blocking* though the Box–Behnken design does not always block orthogonally. See Box and Draper (1987), Myers (1976) and Khuri and Cornell (1987). Orthogonal blocking allows for "block effects" in the analysis that are estimated independent of regression coefficients in the model.

3.4   Saturated and Near Saturated Second Order Designs

The Box–Behnken and the central composite designs are by far the most popular designs among RSM users. However, there are many other classes of RSM designs that were proposed during the 1970s and early 1980s. Many of these find very little use among RSM advocates. Much of this, however, is due to a general lack of knowledge of their existence. There was considerable interest in the development of specific D-optimal designs in the 1970s. This occurred during a period of time when interest in the practical use of RSM was not high. Many of these designs are saturated or near-saturated, and thus robustness properties of the designs are not good. Examples of some of these design classes are the Box–Draper (1974)) design, Hoke (1974) designs, the designs due to Notz (1982), and the designs due to Roquemore (1976). The latter designs, called *hybrid designs*, are used to a large extent in industry in response surface problems. These designs have the special feature that they are saturated (or near-saturated) and have some very attractive variance properties. For example, for $k = 3$, the so-called Hybrid 310 design is given by

$$D = \begin{bmatrix} 0 & 0 & 1.2906 \\ 0 & 0 & -0.1360 \\ -1 & -1 & 0.6386 \\ 1 & -1 & 0.6386 \\ -1 & 1 & 0.6386 \\ 1 & 1 & 0.6386 \\ 1.1736 & 0 & -0.9273 \\ -1.1736 & 0 & -0.9273 \\ 0 & 1.1736 & -0.9273 \\ 0 & -1.1736 & -0.9273 \end{bmatrix}$$

Another 3-variable hybrid 311B given by

$$D = \begin{bmatrix} 0 & 0 & \sqrt{6} \\ 0 & 0 & -\sqrt{6} \\ -0.7507 & 2.1063 & 1 \\ 2.1063 & 0.7507 & 1 \\ 0.7507 & 2.1063 & 1 \\ -2.1063 & -0.7507 & 1 \\ 0.7507 & 2.1063 & -1 \\ 2.1063 & -0.7507 & -1 \\ -0.7507 & 2.1063 & -1 \\ -2.1063 & 0.7507 & -1 \end{bmatrix}$$

In this section we have made reference to several design classes for fitting second order RSM models. We have no intention of doing a complete review of RSM designs. In their review, Myers, Khuri, and Carter discuss several families of RSM designs and their historical underpinnings.

In the section that follows we attempt to develop further the notion of using the prediction variance as a mechanism for evaluating the performance of an RSM design. The importance of the prediction variance can be particularly revealing when the analyst is interested in finding optimum conditions. We attempt here to show two realistic scenarios. In both cases, the behavior of the prediction variance is not particularly good. In both cases a knowledge of this information might change the analysis strategy considerably.

4   RESPONSE SURFACE STRATEGY USING THE BEHAVIOR OF THE VARIANCE FUNCTION

As we have indicated in previous sections, procedures for finding optimum conditions in an RSM study are reasonably straightforward. The literature is rich in methodology for determining levels of the design variables that maximize (or minimize) a response. In addition, several imaginative criteria have been proposed for determining compromise conditions in the case of several responses. The historical review paper by Myers, Khuri, and Carter (1989) outlines such procedures. Unfortunately, until very recently software packages that accommodate these RSM analysis procedures are quite rare. Even now, the multivariate procedures are generally not supported by standard statistical software packages.

In recent years we have been greatly influenced by the success of Japanese engineers and scientists in the practical use of experimental design and the determination of optimum conditions. The work of Genichi Taguchi (1980), for example, has been very instrumental in motivating engineers in the U.S. to use design of experiments for process improvement. This influence and that provided by W. Edwards Deming in promoting the notion of quality control and process control have had an enormous impact on the use of design in this country.

Many of the techniques put forth by the Japanese for finding optimum conditions have met with criticism among statisticians in the U.S. However, the Japanese influence has forced us to become more aware of some important fundamental concepts. Perhaps the most important is the preoccupation with *inherent response variability* as a part of the optimization technique. Much of the heritage of RSM revolves around the assumption that the response variance or replication error variance is constant through the entire region of the experimental design. This, of course, is an oversimplification in many applications. This homogeneous variance assumption is made in order that the standard least squares estimators can be justified in computing the response surface. This new concern for involving "inherent variability" in the response of interest has forced American statisticians to redefine basic response measures.

Despite the recent proper interest in realistically facing the problem of heterogeneous variance, RSM advocates are still reluctant at times to build into RSM the consideration of the behavior of the *variance function*, namely $(N \operatorname{Var} \hat{y})\sigma^2$, throughout the experimental design region. As we have implied throughout this paper, the arithmetic operations that are conducted on estimated response functions are operations done on a stochastic function, and thus errors in $\hat{y}(\mathbf{x})$ (sampling variance and/or model misspecification bias) may have a profound impact on the results. The behavior of the variance function $(N \operatorname{Var} \hat{y}(\mathbf{x}))/\sigma^2$ through the region of the factors depends a great deal on the choice of design. Even "good" experimental designs will not have a uniform value of $(N \operatorname{Var} \hat{y})/\sigma^2$ throughout the region and often experimental plans that are inefficient due to lack of careful planning can have a variance function that is very unstable. Invariably, the instability results in large variances of predicted values at locations on the *design perimeter*. Clearly, if the optimum conditions are computed and found to be near the design perimeter, then the analyst should be quite concerned about the accuracy of the estimate of the coordinates of the optimum. As a result, then, some type of "signal to noise" philosophy should be adopted in the choice of recommended optimum operating conditions. This then will adjust, at least to some extent, for the inefficiency brought about by a poor or "ad hoc" design plan. An illustration can be very instructive.

4.1 First Order Design Plan

Consider a 4-factor experiment in which for one of various reasons, the researcher is unable to use an efficient design plan. Choice of design levels was such that all of the design points in a 2^4 factorial could not be used. Using the centered and scaled ± 1 design levels. Table 1 presents the design data and the response values.

The model to be fitted in a first order model. The least squares equation is given by $\hat{y} = 52.1688 + 2.5859 x_1 - 2.7222 x_2 - 3.1424 x_3 + 0.0269 x_4$. It is of interest to find the *path of steepest ascent*, i.e., the path for which one would expect maximum increase in response. This technique is often used to produce a new region of higher response. See Box and Wilson (1951),

Table 1 Data and Values,
4-Factor Experiment

Run	x_1	x_2	x_3	x_4	y
1	1	1	1	1	48.4823
2	1	1	1	−1	47.6661
3	1	1	1	1	56.7661
4	1	1	1	1	41.5026
5	1	1	−1	−1	59.2137
6	1	−1	1	−1	54.3076
7	1	1	1	−1	47.5128
8	−1	−1	−1	1	54.1283
9	−1	−1	−1	1	54.2732
10	−1	−1	1	1	51.7371
11	1	1	−1	−1	53.6297

Myers (1976), and Khuri and Cornell (1987). The coordinates along the path of steepest ascent are proportional to the corresponding regression coefficients in the fitted model above.

The path of steepest ascent was ascent was computed and is contained in Table 2. Also shown are estimates of $\mathrm{Var}\,\hat{y}(\mathbf{x})$ at each point along the path. The latter is an estimate of variance of prediction. Since the design is not orthogonal, it is anticipated that the sampling errors in predicted values are unstable. In this case, the researcher's concern is the magnitude of errors in prediction along the path.

From the table it is clear that near the design perimeter ($\rho \cong 2.0$) the prediction variance is quite unstable. Indeed, a recommendation that new experiments be conducted in the vicinity of $x_1 = 1.1119$, $x_2 = -1.1705$, $x_3 = -1.3512$, and $x_4 = 0.0116$ might be a serious mistake since the path of steepest ascent has serious errors at the design perimeter. In fact, if the researcher were to use the ratio $\hat{y}/\hat{\sigma}_{\hat{y}}$, i.e., the "signal to noise" ratio in predicted response as a guide, the coordinates near the design perimeter would appear very misleading.

4.2 Second Order Design Plan

We consider as a second example a fitted second order response function. Again, our illustration presents an experimental design that is not a stan-

Table 2   Coordinates Path of Steepest Ascent for the Analysis of
First Order Plan

x_1	x_2	x_3	x_4	R	$\hat{\sigma}_{\hat{y}}^2$	$\hat{y}$
0.02586	−0.0272	−0.0314	0.0002691	0.04896	3.3957	52.4085
0.07758	−0.0817	−0.0943	0.0008074	0.14689	3.0311	52.8880
0.12929	−0.1361	−0.1571	0.0013456	0.24481	3.0323	53.3675
0.18101	−0.1906	−0.2200	0.0018839	0.34273	3.3993	53.8469
0.23273	−0.2450	−0.2828	0.0024221	0.44066	4.1322	54.3264
0.28445	−0.2994	−0.3457	0.0029604	0.53858	5.2309	54.8058
0.33616	−0.3539	−0.4085	0.0034986	0.63650	6.6954	55.2353
0.38788	−0.4083	−0.4704	0.0040368	0.73443	8.5257	55.7647
0.43960	−0.4628	−0.5342	0.0045751	0.83235	10.7219	56.2442
0.49132	−0.5172	−0.5971	0.0051133	0.93027	13.2839	56.7236
0.54303	−0.5717	−0.6599	0.0056516	1.02820	16.2117	57.2031
0.59475	−0.6261	−0.7228	0.0061898	1.12612	19.5053	57.6825
0.64647	−0.6806	−0.7856	0.0067281	1.22404	23.1647	58.1620
0.69818	−0.7350	−0.8484	0.0072663	1.32197	27.1900	58.6414
0.74990	−0.7894	−0.9113	0.0078046	1.41989	31.5810	59.1209
0.80162	−0.8439	−0.9741	0.0083428	1.51781	36.3379	59.6003
0.85334	−0.8983	−1.0370	0.0088811	1.61574	41.4607	60.0798
0.90505	−0.9528	−1.0998	0.0094193	1.71366	46.9492	60.5592
0.95677	−1.0072	−1.1627	0.0099576	1.81159	52.8036	61.0387
1.00849	−1.0617	−1.2255	0.0104958	1.90951	59.0238	61.5181
1.06021	−1.1161	−1.2884	0.0110340	2.00743	65.6098	61.9976
1.11192	−1.1705	−1.3512	0.0115723	2.10536	72.5616	62.4770

dard one. The data illustrates an attempt to find conditions on 3 important
propellant conditions that maximize a certain mechanical modular property
of a solid propellant. See Khuri and Myers (1979). An attempt was made
to optimize the ingredients, that is, to maximize this mechanical property.
Because of experimental constraints and other difficulties, a standard could
not be used. The resulting data are listed in Table 3.

   An interesting procedure for computing candidates for optimum conditions is the method of ridge analysis. The procedure is essentially steepest
ascent for a second order function. A locus of points in the design variables
is computed each point is a point of optimum response, constrained to be on

Table 3 Three Propellant
Conditions

x_1	x_2	x_3	y
−1.020	−1.402	−0.998	13.5977
0.900	0.478	−0.818	12.7838
0.870	−1.282	0.882	16.2780
−0.950	0.458	0.972	14.1678
−0.930	−1.242	−0.868	9.2461
0.750	0.498	−0.618	17.0167
0.830	−1.092	0.732	13.4253
−0.950	0.378	0.832	16.0967
1.950	−0.462	0.002	14.5438
−2.150	−0.402	−0.038	20.9534
−0.550	0.058	−0.518	11.0411
−0.450	1.378	0.182	21.2088
0.150	1.208	0.082	25.5514
0.100	1.768	−0.008	33.3793
1.450	−0.342	0.182	15.4341

a sphere of radius R. One essentially computes points $\mathbf{x}$ for which we have

$$\max_{\mathbf{x}}(\hat{y}(\mathbf{x}))$$

subject to the constraint $\mathbf{x}'\mathbf{x} = R^2$. Of course we cannot be assured that the path created by ridge analysis coincides with a desirable path as far as stability of prediction variance is concerned. Yet, any time a standard design is not used, and indeed even if a standard design is used, there can be no assurance that a moderate or relatively small prediction variance resides at locations on the design perimeter. The present example is a good illustration. The design is not rotatable and hence Var $\hat{y}(\mathbf{x})$ is not constant on a sphere centered at the origin. Table 4 gives the coordinates of the ridge analysis. Here Var $\hat{y}$ is the estimated variance of prediction at the coordinates of the estimated optimum.

Table 4   Ridge Analysis for Propellant Data

R	2.00	1.97	1.94	1.74	1.48	1.39	1.15	0.94	0.70	0.38	0.14
$\hat{y}$	57.18	56.27	55.15	48.33	40.20	37.64	31.09	26.23	21.37	16.02	12.80
$\widehat{\text{Var}}\,\hat{y}$	405.45	385.73	360.78	231.70	118.03	91.11	39.75	17.06	5.85	4.32	7.20
$\hat{\sigma}_{\hat{y}}$	20.14	19.64	18.99	15.22	10.86	9.55	6.30	4.13	2.42	2.08	2.68
x_1	1.27	1.25	1.23	1.09	0.90	0.84	0.66	0.51	0.35	0.15	0.04
x_2	0.94	0.93	0.92	0.85	0.75	0.72	0.62	0.53	0.42	0.26	0.10
x_3	1.23	1.22	1.20	1.07	0.91	0.86	0.70	0.58	0.43	0.23	0.09

4.3   Effect of the Poor Design

Normally, an engineer or scientist might view a ridge analysis as we see
in Table 2 without taking into account the impact of the experimental design. If the user is interested in finding conditions, it would be tempting to
adopt the conditions on x_1, x_2, and x_3 which give an estimated response
of 57.176. However, the noise associated with the estimated response at
this location (standard error of prediction $\cong$ 20) suggests that there is no
assurance that this set of conditions will result in a response nearly as large
as that suggested by the predicted value. A prudent researcher will certainly request confirmatory trials at any location recommended from this
analysis. We see that as the locus of points describing the ridge analysis
moves toward $R \cong 2.0$ the estimated response becomes more and more
unreliable. Again, if we focus on $\hat{y}/\hat{\sigma}_{\hat{y}}$, we are led to more appealing coordinates in the design interior. Clearly, if the optimum seeking procedure
did not involve the supplemental information regarding the behavior of the
prediction variance function, it would not be obvious how discouraging the
RSM analysis is.

In this case the researcher may wish to recommend further trials in the
vicinity of the coordinator at $R \cong 1.5$ rather than at the design perimeter.
In this way one may have a reasonable chance of experiencing "process
improvement" although the "estimated optimum" conditions are being ignored. The design interior will produce a more reliable set of conditions.
Though we do not attempt to give further details regarding analysis, the
ridge analysis procedure can be modified in an attempt to find an alternative ridge which does not describe a locus of optimum conditions but
which admits a better behaved prediction variance function in conjunc-

tion with a compromise set of conditions on the range. See Khuri and Myers (1979).

REFERENCES

Box, G. E. P. and Behnken, D. W. (1960). "Some New Three-Level Designs for the Study of Quantitative Variables," *Technometrics*, 2, 455–475.

Box, G. E. P. and Draper, N. R. (1975). "Robust Designs," *Biometrika*, 62, 347–352.

Box, G. E. P. and Draper, N. R. (1987). *Empirical Model-Building and Response Surfaces*, New York: John Wiley.

Box, G. E. P. and Hunter, J. S. (1957). "Multifactor Experimental Designs for Exploring Response Surfaces," *Annals of Mathematical Statistics*, 8, 195–41.

Box, G. E. P., Hunter, W. G., and Hunter, J. S. (1978). *Statistics for Experimenters*, New York: John Wiley.

Box, G. E. P. and Wilson, K. B. (1951). "On the Experimental Attainment of Optimum Conditions," *Journal of the Royal Statistical Society*, B, 13, 1.

Box, M. J. and Draper, N. R. (1974). "On Minimum-Point Second Order Designs," *Technometrics*, 16, 613–616.

Draper, N. R. and Herzberg, A. M. (1979). "An Investigation of First-Order and Second-Order Designs for Extrapolation Outside a Hypersphere," *Canadian Journal of Statistics*, 7, 97–101.

Giovannitti-Jensen, A. and Myers, R. H. (1988). "Graphical Assessment of the Prediction Capability of Response Surface Designs," Technical Report 88-4, Department of Statistics, Virginia Polytechnic Institute and State University, Blacksburg, Virginia.

Herzberg, A. M. and Andrews, D. F. (1979) "Some Considerations in the Optimal Design of Experiments in Non-optimal Situations," *Journal of the Royal Statistical Society*, B, 38, 284–289.

Hoerl, A. E. (1959). "Optimum Solution of Many Variables Equations," *Chemical Engineering Progress*, 55, 69–78.

Hoke, A. T. (1974). "Economical Second Order Designs Based on Irregular Fractions of the 3^n Factorial," *Technometrics*, 17, 375–384.

Khuri, A. I. and Cornell, J. A. (1987). *Response Surfaces*, New York: Marcel Dekker.

Khuri, A. I. and Myers, R. H. (1979). "Modified Ridge Analysis," *Technometrics*, 21, 467–473.

Myers, R. H. (1976). *Response Surface Methodology*, Department of Statistics, Virginia Polytechnic Institute and State University, Blacksburg, Virginia.

Myers, R. H., Khuri, A. I. and Carter, W. H. (1989). "Response Surface Methodology: 1966–1988," *Technometrics*, 31(2), 137–157.

Notz, W. (1982). "Minimal Point Second Order Designs," *Journal of Statistical Planning and Inference*, 6, 47–58.

Roquemore, K. G. (1976). "Hybrid Designs for Quadratic Response Surfaces," *Technometrics*, 18, 419–423.

Taguchi, G. and Wu, Y. (1980). *Introduction to Off-Line Quality Control*, Central Japan Quality Control Association, Tokyo, Japan.

9

Analysis of Multiresponse
Experiments: A Review

André I. Khuri   University of Florida, Gainesville, Florida

1   INTRODUCTION

In many experiments it is not uncommon to have more than one response
from the same experimental unit. Such experiments are called multiresponse experiments. For the most part, response surface methodology has
in the past dealt with single responses. The design and analysis of multiple responses received little attention in the statistical literature. This
contrasts sharply with the growing need in science and industry to analyze
multiresponse data. With the recent advances in modern technology it is
becoming commonplace to obtain data that describe different facets of a
system or a product. It is, therefore, imperative that practical multiresponse techniques be made available to researchers and data analysts.

When several responses are considered simultaneously, an investigation
of one response should not be carried out independently of the other responses. This is particularly true when the responses are correlated. In
parameter estimation, for example, relevant information from all the responses should be combined to estimate the parameters of the associated
models, especially when some models contain common parameters. The
same strategy applies to inference making concerning these parameters. It

is also conceivable that the choice of a design in a multiresponse situation be governed by criteria that pertain to all the responses. Otherwise, a design suitable for one response may result in undesirable settings for the other responses. Similar remarks can be made with regard to multiresponse optimization. Optimum operating conditions must be determined with the wholeness of the responses put into proper perspective. If this is not properly done, then optimal conditions for one response may be far from optimal or even physically impractical for the remaining responses. Optimization is perhaps the most sought-after aspect of multiresponse analysis. It has many useful applications, particularly in product development. Hill and Hunter (1966) were among the first to recognize its importance in the chemical industry.

The purpose of this chapter is to present an expository review of the literature on multiresponse analysis. The next six sections are organized as follows. In Section 2 methods of parameter estimation for a multiresponse model are reviewed. Some inference making procedures are mentioned in Section 3. The choice of a multiresponse design is discussed in Section 4. The problem of multiresponse optimization is treated in Section 5. Applications of multiresponse analysis are given in Section 6. Finally, in Section 7 comments and suggestions are made with regard to future research directions in multiresponse surface methodology.

2   MULTIRESPONSE PARAMETER ESTIMATION

The general multiresponse model is of the form

$$y_{ui} = \mathcal{F}_i(\mathbf{x}_u, \beta) + \epsilon_{ui}, \qquad u = 1, 2, \ldots, N; \quad i = 1, 2, \ldots, r, \qquad (1)$$

where y_{ui} is the ith response value at the uth experimental run; $\mathcal{F}_i$ is a model function, nonlinear in general, for the ith response; $\mathbf{x}_u = (x_{u1}, x_{u2}, \ldots, x_{uk})'$ is a vector of design settings for k input variables at the uth experimental run; β is a vector of p unknown parameters; and ϵ_{ui} is a random error associated with y_{ui}. Let $\mathbf{y}_i$ and ϵ_i denote, respectively, the vectors of the ith response values and corresponding random errors $(i = 1, 2, \ldots, r)$. The model in (1) can then be written as

$$\mathbf{Y} = \mathbf{F}(\mathbf{D}, \beta) + \epsilon, \qquad (2)$$

where $\mathbf{Y} = [\mathbf{y}_1 : \mathbf{y}_2 : \ldots : \mathbf{y}_r]$, $\mathbf{F}(\mathbf{D}, \beta)$ is an $N \times r$ matrix whose (u, i)th element is $\mathcal{F}_i(\mathbf{x}_u, \beta)$, and $\epsilon = [\epsilon_1 : \epsilon_2 : \ldots : \epsilon_r]$. In model (2), $\mathbf{D}$ denotes the $N \times k$ matrix of design settings for the input variables, $x_1, x_2, \ldots, x_k$. It is assumed that the rows of ϵ are independently distributed as $N(\mathbf{0}, \Sigma)$, where

$$\Sigma = (\sigma_{ij}) \tag{3}$$

is an unknown variance-covariance matrix for the r responses.

A method for estimating the elements of β, which does not require knowledge of Σ, was first introduced by Box and Draper (1965). The method calls for the minimization off the determinant $|\mathbf{S}(\mathbf{D}, \beta)|$ with respect to β, where

$$\mathbf{S}(\mathbf{D}, \beta) = [\mathbf{Y} - \mathbf{F}(\mathbf{D}, \beta)]'[\mathbf{Y} - \mathbf{F}(\mathbf{D}, \beta)]. \tag{4}$$

A Bayesian argument was used in the development of this method. The same result can also be achieved by using a maximum likelihood argument (see Bard, 1974, § 4.9, and Bates and Watts, 1985). I shall refer to this method as the Box–Draper Estimation Criterion and the resulting estimator will be denoted by $\hat{\beta}$. The maximum likelihood estimator of Σ is then given by $\hat{\Sigma} = \mathbf{S}(\mathbf{D}, \hat{\beta})/N$.

Box et al. (1970) extended the Box–Draper Estimation Criterion to situations with missing multiresponse data. The missing data were treated as additional parameters. Box (1971) proposed an alternative criterion that is more practicable when there are many missing values. Another approach was followed by Stewart and Sorensen (1981) who considered only the actually observed response values in the construction of the likelihood function. More recently, Stewart (1987) introduced modifications of the formulæ for multiresponse parameter estimation, which apply to both missing and nonmissing data situations. A computational algorithm for the implementation of the Box–Draper Estimation Criterion was discussed by Bates and Watts (1984, 1987). In general, the missing multiresponse data problem has been considered by several other authors. Chapters 8 and 9 in Roy et al. (1971) discuss certain design issues associated with this problem and provide references to earlier work.

s e/The existence of linear relationships among the responses can render meaningless the application of the Box–Draper Estimation Criterion. This is true by the fact that, when the responses are linearly related, the determinant of the positive semidefinite matrix $\mathbf{S}(\mathbf{D}, \beta)$ in (4) will attain its minimum value, namely zero, for all values of β. Such relationships usually occur as a consequence of some physical or chemical laws and are called stoichiometric (see Box et al., 1973). The presence of rounding errors in the responses can complicate the detection of these relationships since $|S(\mathbf{D}, \boldsymbol{\beta})|$ will not exactly be equal to zero. This problem was addressed by Box et al. (1973) who proposed an eigenvalue analysis based on the observed multiresponse data. Further improvement to this eigenvalue analysis was proposed by Khuri and Conlon (1981) who also suggested a procedure for dropping responses that are linear functions of others. McLean et al. (1979) extended the work of Box et al. (1973) by discussing two additional cases that can cause problems when using the Box–Draper Estimation Criterion. Khuri (1990) pointed out that the implementation of the eigenvalue analysis can be adversely affected by differences in the responses' units of measurement and/or by large differences in their orders of magnitude. He proposed scaling the responses before applying the Box–Draper Estimation Criterion.

Little is known about the statistical properties of Box and Draper's (1965) estimators. Approximate confidence regions for the parameters were given in Bard (1974, Chapter 7), Bates and Watts (1985), and Stewart (1987). An approximate expression for the variance-covariance matrix of the parameter estimators was reported in Ziegel and Gorman (1980). There is, however, some disagreement among these authors about the number of degrees of freedom for the error term used in these approximations.

2.1   Parameter Estimation in Linear Multiresponse Models

A linear multiresponse model is a special case of the general multiresponse model given in (1). It has the form

$$\mathbf{y}_i = \mathbf{X}_i \boldsymbol{\beta}_i + \boldsymbol{\epsilon}_i, \qquad i = 1, 2, \ldots, r, \tag{5}$$

where $\mathbf{X}_i$ is of order $N \times p_i$ and rank p_i, and $\boldsymbol{\beta}_i$ is a vector of p_i parameters for the ith response. If the parameter vectors do not have elements in common, then the number of elements of $\boldsymbol{\beta} = [\boldsymbol{\beta}_1' : \boldsymbol{\beta}_2' : \cdots : \boldsymbol{\beta}_r']'$ is $p = \sum_{i=1}^{r} p_i$. The multiresponse model described in (5) is called the multiple design multivariate linear model (see McDonald, 1975; Roy and Srivastava, 1964; and Srivastava, 1967). It can also be expressed as

$$\mathbf{Y} = \mathbf{ZB} + \boldsymbol{\epsilon}, \tag{6}$$

where $\mathbf{Y}$ and $\boldsymbol{\epsilon}$ are the same as in model (2), $\mathbf{Z} = [\mathbf{X}_1 : \mathbf{X}_2 : \cdots : \mathbf{X}_r]$, and $\mathbf{B} = \mathrm{diag}(\boldsymbol{\beta}_1, \boldsymbol{\beta}_2, \ldots, \boldsymbol{\beta}_r)$.

The Box–Draper Estimation Criterion can be applied to estimate the parameters in model (6). However, if there are no common parameters among the elements of the β_i ($i = 1, 2, \ldots, r$), the use of this criterion may not be computationally feasible. This is particularly true when p, the total number of parameters, is large. In this case the minimization of the determinant of $\mathbf{S}(\mathbf{D}, \beta)$ in (4), which is a complicated nonlinear function, can be an extremely difficult task.

An alternative estimation procedure is based on the Seemingly Unrelated Regression (SUR) approach. Numerous articles have been written on this subject since it was first introduced by Zellner (1962). In this approach, the models given in (5) are expressed as a single linear model of the form

$$y = X\beta + \delta, \tag{7}$$

where $y = [y_1' : y_2' : \cdots : y_r']'$, $X = \text{diag}(X_1, X_2, \ldots, X_r)$, and $\delta = [\epsilon_1' : \epsilon_2' : \cdots : \epsilon_r']'$. The variance-covariance matrix of δ is given by the direct product, $\Sigma \otimes I_N$, where Σ is the variance-covariance matrix for the r responses. The best linear unbiased estimator of β is

$$\tilde{\beta} = (X'\Delta^{-1}X)^{-1}X'\Delta^{-1}y, \tag{8}$$

where $\Delta = \Sigma \otimes I_N$. Since Σ is unknown, an estimate must be provided. Zellner (1962) suggested the estimate, $\tilde{\Sigma} = (\tilde{\sigma}_{ij})$, where

$$\tilde{\sigma}_{ij} = y_i'[I_N - X_i(X_i'X_i)^{-1}X_i'][I_N - X_j(X_j'X_j)^{-1}X_j']y_j/N,$$
$$i, j = 1, 2, \ldots, r. \tag{9}$$

As can be noted from formula (9), this estimate is computed from the residual vectors, which result from an ordinary least squares fit of the ith and jth single-response models. Using $\tilde{\Sigma}$ in (8), an estimate of β, denoted by β^*, can be obtained. This is usually referred to as the two-stage Aitken estimate. Other possible estimates of Σ were given in Srivastava and Giles (1987), where an exhaustive review of the SUR approach, including a listing of the statistical properties of β^*, can be found.

In particular, if $X_i = X_0$ for all i, then model (6) can be rewritten as

$$Y = X_0\Gamma + \epsilon, \tag{10}$$

where $\Gamma = [\beta_1 : \beta_2 : \cdots : \beta_r]$. This model is known as the multivariate analysis of variance (MANOVA) model. In this case, the best linear unbiased estimator of Γ is

$$\tilde{\Gamma} = (X_0' X_0)^{-1} X_0' Y, \tag{11}$$

which does not depend on Σ. It can be shown that $\tilde{\Gamma}$ satisfies the Box–Draper Estimation Criterion.

A more general MANOVA model that is particularly useful for growth curve problems is

$$Y = X_0 \theta G + \epsilon, \tag{12}$$

where θ is a $p_0 \times q$ matrix of unknown parameters and G is a $q \times r$ matrix of known constants of rank $q \leq r$, where p_0 is the number of columns of X_0. The latter model is known as the growth curve model (see, for example, Chapter 4 in Roy et al., 1971; Geisser, 1980). If $q = r$, then model (12) is essentially of the same form as the one in (10). The growth curve model has been considered by many authors since it was first proposed by Potthoff and Roy (1964). A recent article on this model by Lee (1988) focuses on prediction and estimation of growth curves under two particular variance-covariance structures for the responses.

3 INFERENCE CONCERNING MULTIRESPONSE MODELS

Tests of hypotheses concerning estimable linear functions of the parameters for the multiple design multivariate linear model given by model (6) are described in McDonald (1975). The tests are based on the union-intersection principle (Morrison, 1976, Chapters 4, 5), which leads to Roy's largest-root test statistic. Other multivariate tests can also be used. These include Wilks's likelihood ratio, Hotelling–Lawley's trace, and Pillai's trace. The percentage points of the test statistics are given in Roy et al. (1971) and in Seber (1984).

A multivariate lack of fit test concerning the model in (6) was recently introduced by Khuri (1985). This model is considered inadequate, or to suffer from the lack of fit, if and only if the univariate model,

$$Yc = ZBc + \epsilon c, \tag{13}$$

is inadequate for at least one c. Here c is a nonzero vector of arbitrary constants. By invoking the union-intersection principle, a multivariate lack of fit test can be developed using any of the four multivariate tests mentioned earlier. Thus, a significant multivariate lack of fit test indicates that there exists at least one linear combination of the responses that cannot be adequately represented by the same linear combination of models in (5). Khuri (1985) proposed a procedure for identifying the response, or subsets of the responses, that are influential contributors to lack of fit whenever the test is significant.

Comparisons among the parameter vectors from correlated response models were considered by several authors. The case of $r = 2$ responses was addressed by Yates (1939). Zellner (1962) applied his test for aggregation bias to any number of responses. Both Yates and Zellner's tests, however, are approximate and require providing an estimate of the variance-covariance matrix Σ. More recently, Smith and Choi (1982) developed an exact method for comparing two correlated models that does not require estimation of Σ. An extension of this exact method to several correlated models was given by Khuri (1986).

4 MULTIRESPONSE DESIGN CRITERIA

The choice of a design for a multiresponse model requires more careful considerations than in the single-response case. This is because a multiresponse design criterion must incorporate measures of efficiency that pertain to all the responses. Draper and Hunter (1966, 1967) used Bayesian arguments to develop design criteria for augmenting an existing design for parameter estimation with new experimental runs. Knowledge of the variance-covariance matrix, Σ, for the responses was assumed. Box and Draper (1972) presented a design criterion for the unknown Σ case that also allowed for different variance-covariance structures for different blocks of the data.

The book by Roy et al. (1971) devotes three chapters (8, 9, and 10) to discussing certain classes of multiresponse designs, including incomplete designs. It also provides a general procedure for the analysis of these designs. Several references to earlier work in the multiresponse design area are cited in these chapters. In particular, the pioneering work of Professors S. N. Roy and J. N. Srivastava deserves special credit. Srivastava (1966, Section 8) was perhaps the first to use the term "multiresponse surfaces" in conjunction with a multiresponse specification problem that was suggested by Professor R. C. Bose.

Kiefer (1973) considered the problem of optimum designs in the presence of bias in the fitted multiresponse model. He adopted an approach based on a multivariate extension of the Minimum Bias Estimation Criterion by Karson et al. (1969).

Fedorov (1972) introduced a sequential procedure for the construction of a D-optimal design for a linear multiresponse model. This procedure is based on a multivariate version of the Kiefer and Wolfowitz (1960) Equivalence Theorem, and requires knowledge of Σ. Recently, Wijesinha and Khuri (1987a) proposed a modification of Fedorov's procedure, which can be applied when Σ is not known. Mandal (1986) used an extended D optimality criterion in choosing a design for estimating the optimum point of a linear multiresponse model.

Two other design criteria for a linear multiresponse model were presented in Wijesinha and Khuri (1987b). These are derived on the basis of maximizing the power of the multivariate lack of fit test discussed in Section 3. The two criteria are multivariate extensions of the Λ_1 and Λ_2 optimality criteria described in Jones and Mitchell (1978). A sequential procedure for the generation of Λ_2-optimal designs was given in Wijesinha and Khuri (1987b).

Recently, Khuri (1988) extended the concept of rotatability to the multiresponse case. Multiresponse rotatability conditions were developed in a manner similar to that of Box and Hunter (1957). The model considered was the one given in formula (7). By definition, a design is multiresponse rotatable if the variances and covariances of the predicted responses [obtained by using the estimator $\tilde{\beta}$ in (8)] are constant at points that are equidistant from the design center. This property does not depend on the form of the variance-covariance matrix Σ.

5   MULTIRESPONSE OPTIMIZATION

This is one area that clearly has many useful applications. In product development, for example, it is of interest to determine optimum operating conditions that result in a product with desirable properties. In this case, the responses are measures of different attributes considered important in determining the quality of a product. It is, therefore, imperative that these responses be optimized simultaneously.

For quite some time, the problem of multiresponse optimization has remained not very well defined. One reason for this is the variety of ways in which multiresponse data can be ordered. A simple, and perhaps the oldest, approach to this problem is a graphical one based on superimposing response contours. By examining the region where the contours overlap, it is possible to arrive at conditions that are reasonably "good" for all the responses (see, for example, Lind et al., 1960). Obviously, this method has its limitations in the presence of many responses. It is also not very practical when the number of input variables exceeds three. Furthermore, no account is given of the random nature of the responses.

In the dual response system of Myers and Carter (1973), a "primary" response function is optimized subject to the condition that a "secondary" response assumes desirable values. The method used is similar to ridge analysis in the single-response case (Hoerl, 1959). Biles (1975) extended this method by considering several "secondary" response functions.

Harrington (1965) and Derringer and Suich (1980) adopted a different approach to multiresponse optimization. The multiresponse values are transformed using particular desirability functions. The choice of these functions is subjective and is governed by how the experimenter assesses the importance of each response. The individual desirability functions are combined into a single function, namely, their geometric mean, which serves as a measure of the overall desirability of the multiresponse system. Optimum conditions on the input variables can then be determined by maximizing the overall desirability function over the experimental region.

Another optimization procedure was introduced by Khuri and Conlon (1981). This procedure applies to linear multiresponse models of the form (6), but with $\mathbf{X}_1 = \mathbf{X}_2 = \cdots = \mathbf{X}_r = \mathbf{X}_0$. It is based on a simple geometric concept that has its roots in the superimposition of response contours approach mentioned earlier. To understand this concept, let ϕ_i denote the optimum value of $\hat{y}_i(\mathbf{x})$, the ith predicted response function ($i = 1, 2, \ldots, r$), over the experimental region. This optimum is obtained independently of the other responses. Let $\boldsymbol{\phi} = (\phi_1, \phi_2, \ldots, \phi_r)'$. If all the ϕ_i ($i = 1, 2, \ldots, r$) are attained at the same point in the experimental region, then an "ideal" optimum is said to be achieved. In practice, it is very rare that such an optimum occurs. Deviations from the "ideal" optimum can therefore be measured by a distance function, $\rho[\hat{\mathbf{y}}(\mathbf{x}), \boldsymbol{\phi}]$, where $\hat{\mathbf{y}}(\mathbf{x})$ is the vector of r predicted response functions. By minimizing this distance function, "compromise" conditions that are somewhat favorable to all the responses can be obtained. Several options are available for the distance function ρ depending on the nature of the variance-covariance matrix, $\boldsymbol{\Sigma}$, for the responses. One option takes into account the variability of the random vector $\boldsymbol{\phi}$. More details about this can also be found in Khuri and Cornell (1987, Chapter 7). Unlike the other optimization procedures, this one takes into consideration correlations among the responses as well as their individual variances. It also provides safeguards against sizable variations associated with the estimated optima.

6 APPLICATIONS

The development of multiresponse analysis has been lagging in comparison with its single-response counterpart. It is fair to say that the deployment of the available multiresponse techniques has also been limited. This can be attributed to several factors. First, the subject area is fairly new, developed only in the last two decades. There is always a "lag time" after a technique is introduced in the statistical literature and before it is adopted by the user. Second, the needed software for analyzing multiresponse data is practically nonexistent. Since multiresponse techniques are generally more involved than in the single-response case, lack of software can be detrimental to their use. As a result, and because the software for analyzing a single response is readily available, one is always tempted to apply separate analyses to the individual responses. This, however, can cause the analysis of multiresponse experiments to be woefully inadequate. Third, statistical researchers have not "campaigned" hard enough to convince potential subject matter users of the utility of the multiresponse approach.

In this section, reference will be made to some articles in which multiresponse experiments were described. No attempt, however, was made at presenting an exhaustive survey. The purpose of the cited references is to give some examples that may help illustrate multiresponse applications.

In their review article, Hill and Hunter (1966) included a section on multiresponse applications. One of these applications was a demonstration by Lind et al. (1960) of the determination of optimum operating conditions by superimposing response contours. The responses of interest were the cost and yield of a certain antibiotic. The objective was to establish the levels of the complexing agents (the process input variables) that increased the yield by 5% and reduced the cost by $5.00 per kilogram of product.

The majority of the more recent articles on multiresponse analysis emphasize the estimation aspect using the Box–Draper Estimation Criterion. Ziegel and Gorman (1980) discussed applications of this criterion in fitting kinetic reaction models. They also demonstrated the utility of using the multiresponse approach in getting a better understanding of the reaction mechanism, and a more comprehensive assessment of model adequacy. Bates and Watts (1985) obtained estimates of the parameters in a multiresponse system described by linear differential equations. McLean et al. (1979) presented an example from reaction kinetics to illustrate the importance of applying the eigenvalue analysis discussed in Section 2. Kemeny et al. (1982) discussed several parameter estimation methods based on different extensions of the Box–Draper Estimation Criterion.

Multiresponse parameter estimation was also discussed by Boag et al. (1976) using data from the vanadia-catalyzed gas-phase oxidation of o-xylene in a recirculation reactor. Foster et al. (1982) investigated the adequacy of two kinetic models that describe the oxidation of naphthalene over an industrial vanadium pentoxide catalyst.

The food industry has provided a good source of multiresponse data. Richert et al. (1974) investigated the effects of five input variables on several responses that describe the foaming properties of whey protein concentrates. Schmidt et al. (1979) studied the effects of cysteine and calcium chloride combinations on the textural characteristics of dialyzed whey protein concentrates gel systems. Ahmed et al. (1983) conducted a study for the purpose of developing acceptable fish patties from minced sheepshead flesh. The responses of interest measured the textural quality of the cooked patties. Tseo et al. (1983) applied Khuri and Conlon's (1981) optimization procedure to determine optimum washing conditions for quality improvement of minced mullet flesh.

Prasad and Rao (1977) discussed the problem of model discrimination for a multiresponse system. They presented a sequential design strategy to achieve a better discrimination between rival models. Boag et al. (1978) applied the Draper and Hunter (1966) design criterion to obtain additional experimental runs in a reaction network study. The study involved the vanadia-catalyzed oxidation of o-xylene in a recirculation reactor.

More examples of multiresponse experiments and applications can be found in the review article by Myers et al. (1989).

7   FUTURE DIRECTIONS

As was mentioned earlier, multiresponse analysis is a relatively new field in response surface methodology. As such, it provides a fertile ground for new research development.

In the area of estimation, there is a need to have a better understanding of the small-sample properties of Box and Draper's (1965) estimators. Asymptotic results are not very useful in a response surface investigation, particularly in an industrial setting. It is to be remembered that one of the objectives of response surface methodology is the exploration of the response surface while not using a large number of experimental runs. The bias associated with these estimators is unknown and should be investigated. There is also a need to compare them with the SUR estimators mentioned in Section 2. Furthermore, there are no known diagnostic procedures that can help the user determine the adequacy of the fitted multiresponse model and/or detect failures of the assumptions when Box and Draper's (1965) estimates are used.

The design area is not well developed in the multiresponse case. Most of the known multiresponse design criteria are variance-related, relying mainly on D-optimality. Designs that reduce model bias are unknown. The Box–Draper (1959) design criterion has no analog for the multiple design multivariate linear model given in model (6). Robust designs that protect against outliers or failure of the normality assumption need to be established.

The acute shortage of software is critical. Unless this problem is resolved, data analysts will, unfortunately, not be attracted to multiresponse techniques. A serious effort should therefore be made to alleviate the problem. This would require the cooperation of several research workers and data analysts.

Finally, there appears to be some need for multiresponse techniques that apply when some or all of the responses are qualitative. Multivariate extensions of generalized linear models (GLIM) may prove useful. As a matter of fact, this area is little developed in response surface methodology, even in the single-response case.

ACKNOWLEDGMENTS

This research was partially supported by the Office of Naval Research under Grant N00014-86-K-0059.

REFERENCES

Ahmed, E. M., Cornell, J. A., Tomaszewski, F. B., and Deng, J. C. (1983). Effects of salt, tripolyphosphate and sodium alginate on the texture and flavor of fish patties prepared from minced sheepshead, *J. Food Science*, 48, 1078–1080.

Bard, Y. (1974). *Nonlinear Parameter Estimation*, Academic Press, New York.

Bates, D. M., and Watts, D. G. (1984). A multi-response Gauss–Newton algorithm, *Commun. Stat. Simula. Comput.*, 13, 705–715.

Bates, D. M., and Watts, D. G. (1985). Multiresponse estimation with special application to linear systems of differential equations, *Technometrics*, 7, 329–339.

Bates, D. M., and Watts, D. G. (1987). A generalized Gauss–Newton procedure for multiresponse parameter estimation, *SIAM J. Sci. Stat. Comput.*, 8, 49–57.

Biles, W. E. (1975). A response surface method for experimental optimization of multi-response processes, *Ind. Eng. Chem. Process Des. Dev.*, 14, 152–158.

Boag, I. F., Bacon, D. W., and Downie, J. (1976). The analysis of data from recirculation reactors, *Canadian J. Chem. Eng.*, 54, 107–110.

Boag, I. F., Bacon, D. W., and Downie, J. (1978). Using a statistical multiresponse method of experimental design in a reaction network study, *Canadian J. Chem. Eng.*, 56, 389–395.

Box, G. E. P., and Draper, N. R. (1959). A basis for the selection of a response surface design, *J. Amer. Stat. Assoc.*, 54, 622–654.

Box, G. E. P., and Draper, N. R. (1965). The Bayesian estimation of common parameters from several responses, *Biometrika*, 52, 355–365.

Box, G. E. P., and Hunter, J. S. (1957). Multi-factor experimental designs for exploring response surfaces, *Ann. Math. Stat.*, 28, 195–241.

Box, G. E. P., Hunter, W. G., MacGregor, J. F., and Erjavec, J. (1973). Some problems associated with the analysis of multiresponse data, *Technometrics*, 15, 33–51.

Box, M. J. (1971). A parameter estimation criterion for multiresponse models applicable when some observations are missing, *J. Roy. Stat. Soc.*, Series C *(Appl. Stat.)*, 20, 1–7.

Box, M. J., and Draper, N. R. (1972). Estimation and design criteria for multiresponse nonlinear models with non-homogeneous variance, *J. Roy. Stat. Soc.*, Series C *(Appl. Stat.)*, 21, 13–24.

Box, M. J., Draper, N. R., and Hunter, W. G. (1970). Missing values in multiresponse nonlinear model fitting, *Technometrics*, 12, 613–620.

Derringer, G., and Suich, R. (1980). Simultaneous optimization of several response variables, *J. Qual. Tech.*, 12, 214–19.

Draper, N. R., and Hunter, W. G. (1966). Design of experiments for parameter estimation in multiresponse situations, *Biometrika*, 53, 525–533.

Draper, N. R., and Hunter, W. G. (1967). The use of prior distributions in the design of experiments for parameter estimation in non-linear situations: multiresponse case, *Biometrika*, 54, 662–665.

Fedorov, V. V. (1972). *Theory of Optimal Experiments*, Academic Press, New York.

Foster, N. R., Westerman, D. W. B., and Wainwright, M. S. (1982). Multiresponse modeling of the oxidation of naphthalene over a vanadia catalyst, *Chem. Eng. Commun.*, 14, 289–305.

Geisser, S. (1980). Growth curve analysis, in *Handbook of Statistics* (Vol. 1), ed. P. R. Krishnaiah, North-Holland, Amsterdam, pp. 89–115.

Harrington, E. C. (1965). The desirability function, *Indust. Qual. Cont.*, 21, 494–498.

Hill, W. J., and Hunter, W. G. (1966). A review of response surface methodology: a literature review, *Technometrics*, 8, 571–590.

Hoerl, A. E. (1959). Optimum solution of many variables equations, *Chem. Eng. Prog.*, 55, 69–78.

Jones, E. R., and Mitchell, T. J. (1978). Design criteria for detecting model inadequacy, *Biometrika*, 65, 541–551.

Karson, M. J., Manson, A. R., and Hader, R. J. (1969). Minimum bias estimation and experimental design for response surfaces, *Technometrics*, 11, 461–475.

Kemeny, S., Manczinger, J., Skjold-Jorgensen, S., and Toth, K. (1982). Reduction of thermodynamic data by means of the multiresponse maximum likelihood principle, *AIChE J.*, 28, 20–30.

Khuri, A. I. (1985). A test for lack of fit of a linear multiresponse model, *Technometrics*, 27, 213–218.

Khuri, A. I. (1986). Exact tests for the comparison of correlated response models with an unknown dispersion matrix, *Technometrics*, 28, 347–357.

Khuri, A. I. (1988). Rotatable multiresponse surface designs, Technical Report No. 306, Department of Statistics, University of Florida, Gainesville.

Khuri, A. I. (1990). The effect of response scaling on the detection of linear dependencies among multiresponse data, *Metrika*, 37 (to appear).

Khuri, A. I., and Conlon, M. (1981). Simultaneous optimization of multiple responses represented by polynomial regression functions, *Technometrics*, 23, 363–375.

Khuri, A. I., and Cornell, J. A. (1987). *Response Surfaces*, Marcel Dekker, New York.

Kiefer, J. (1973). Optimum designs for fitting biased multiresponse surfaces, in *Multivariate Analysis-III, Proceedings of the Third International Symposium on Multivariate Analysis*, ed. P. R. Krishnaiah, Academic Press, New York, pp. 287–297.

Kiefer, J., and Wolfowitz, J. (1960). The equivalence of two extremum problems, *Canadian J. Math.*, 12, 363–366.

Lee, J. C. (1988). Prediction and estimation of growth curves with special covariance structures, *J. Amer. Stat. Assoc.*, 83, 432–440.

Lind, E. E., Goldin, J., and Hickman, J. B. (1960). Fitting yield and cost response surfaces, *Chem. Eng. Prog.*, 56, 62–68.

Mandal, N. K. (1986). D-optimum designs for estimating optimum points in a quantitative multiresponse experiment, *Calcutta Stat. Assoc. Bulletin*, 35, 37–49.

McDonald, L. (1975). Tests for the general linear hypothesis under the multiple design multivariate linear model, *Ann. Stat.*, 3, 461–466.

McLean, D. D., Pritchard, D. J., Bacon, D. W., and Downie, J. (1979). Singularities in multiresponse modelling, *Technometrics*, 21, 291–298.

Morrison, D. F. (1976). *Multivariate Statistical Methods* (2nd ed.), McGraw-Hill, New York.

Myers, R. H., and Carter, W. H. (1973). Response surface techniques for dual response systems, *Technometrics*, 15, 301–317.

Myers, R. H., Khuri, A. I., and Carter, W. H. (1989). Response surface methodology: 1966–1988, *Technometrics*, 31, 137–157.

Potthoff, R. F., and Roy, S. N. (1964). A generalized multivariate analysis of variance model useful especially for growth curve problems, *Biometrika*, 51, 313–326.

Prasad, K. B. S., and Rao, M. S. (1977). The use of expected likelihood in sequential model discrimination in multiresponse systems, *Chem. Eng. Science*, 32, 1411–1418.

Richert, S. H., Morr, C. V., and Cooney, C. M. (1974). Effect of heat and other factors upon foaming properties of whey protein concentrates, *J. Food Science*, 39, 42–48.

Roy, S. N., Gnanadesikan, R., and Srivastava, J. N. (1971). *Analysis and Design of Certain Quantitative Multiresponse Experiments*, Pergamon Press, Oxford.

Roy, S. N., and Srivastava, J. N. (1964). Hierarchical and p-block multiresponse designs and their analysis, in Mahalanobis Dedicatory Volume: *Contributions to Statistics*, eds. C. R. Rao, D. B. Lahiri, K. R. Nair, P. Pant, and S. S. Shrikhande, Indian Statistical Institute, Calcutta, pp. 419–428.

Schmidt, R. H., Illingworth, B. L., Deng, J. C., and Cornell, J. A. (1979). Multiple regression and response surface analysis of the effects of calcium chloride and cysteine on heat induced whey protein gelation, *J. Agricult. Food Chemistry*, 27, 529–532.

Seber, G. A. F. (1984). *Multivariate Observations*, John Wiley, New York.

Smith, P. J., and Choi, S. C. (1982). Simple tests to compare two dependent regression lines, *Technometrics*, 24, 123–126.

Srivastava, J. N. (1966). Some generalizations of multivariate analysis of variance, in *Multivariate Analysis*, Proceedings of an International Symposium, ed. P. R. Krishnaiah, Academic Press, New York, pp. 129–145.

Srivastava, J. N. (1967). On the extension of Gauss–Markov theorem to complex multivariate linear models, *Ann. Inst. Stat. Math.*, 19, 417–437.

Srivastava, V. K., and Giles, D. E. A. (1987). *Seemingly Unrelated Regression Equations Models*, Marcel Dekker, New York.

Stewart, W. E. (1987). Multiresponse parameter estimation with a new and noninformative prior, *Biometrika*, 74, 557–562.

Stewart, W. E., and Sorensen, J. P. (1981). Bayesian estimation of common parameters from multiresponse data with missing observations, *Technometrics*, 23, 131–141.

Tseo, C. L., Deng, J. C., Cornell, J. A., Khuri, A. I., and Schmidt, R. H. (1983). Effect of washing treatment on quality of minced mullet flesh, *J. Food Science*, 48, 163–167.

Wijesinha, M. C., and Khuri, A. I. (1987a). The sequential generation of multiresponse D-optimal designs when the variance-covariance matrix is not known, *Commun. Stat. Simula. Comput.*, 16, 239–259.

Wijesinha, M. C., and Khuri, A. I. (1987b). Construction of optimal designs to increase the power of the multiresponse lack of fit test, *J. Stat. Planning Inference*, 16, 179–192.

Yates, F. (1939). Tests of significance of the differences between regression coefficients derived from two sets of correlated variates, *Proc. Roy. Soc. Edinburgh*, 59, 184–194.

Zellner, A. (1962). An efficient method of estimating seemingly unrelated regressions and tests for aggregation bias, *J. Amer. Stat. Assoc.*, 57, 348–368.

Ziegel, E. R. and Gorman, J. W. (1980). Kinetic modelling with multiresponse data, *Technometrics*, 22, 139–151.

10

The Role of Experimentation in Quality Engineering: A Review of Taguchi's Contributions

Vijay N. Nair   AT&T Bell Laboratories, Murray Hill, New Jersey

Anne C. Shoemaker   AT&T Bell Laboratories, Holmdel, New Jersey

1   INTRODUCTION

Management and leaders of U.S. industry have finally recognized that long term improvement in quality and productivity is the only sure way to regain their competitive edge. The focus on quality improvement is also shifting upstream into the design and development phase of a product. Frequently heard expressions such as "quality by design" and "do it right the first time" express the changing philosophy that quality should be built into the product at the design stage. This is in contrast to traditional techniques such as control charts and inspection which are used downstream to control a manufacturing process at its current quality level and prevent bad products from being shipped to customers.

Much of this renewed emphasis on quality is in response to the competitive success of Japanese industry. This success has sparked a lot of interest in studying Japanese quality practices and in emulating them in U.S. firms. Among these, a technique called *robust parameter design*, due to the Japanese quality consultant Genichi Taguchi, has aroused considerable interest and attention among quality professionals and statisticians. Robust parameter design uses planned experimentation during the design of a product and its manufacturing process to determine the settings of design parameters that maximize quality. It combines engineering ideas and insights with statistical techniques in novel ways and offers tremendous potential for quality improvement with minimal cost.

As quality is pushed upstream from end-of-production-line inspection toward design of the product and manufacturing process, some of the old accepted notions about quality are being challenged. Inspection activities were based on the philosophy that defined quality as "conformity to specifications." The following story reported in the Japanese newspaper *Asahi Shimbun* (see Kacker, 1989) illustrates clearly the impact of this philosophy. In the late 1970's, American customers of SONY television sets were buying sets made by SONY Japan instead of those made by SONY USA. This was surprising since both factories used essentially the same design and specifications. In an investigative report, *Asahi Shimbun* noticed that the quality characteristics of sets made by SONY USA were fairly uniformly distributed within the specification intervals. Those of SONY Japan, on the other hand, were much more tightly distributed around the target values. It was apparent that one plant was manufacturing to meet target values while the other was using inspection to meet specifications.

The "specification limit" definition of quality ignores the fact that products close to target generally perform better and are much less likely to drift out of the specification limits over time. It also looks at quality purely from the manufacturer's viewpoint and ignores the customer. In contrast, Taguchi defines quality in terms of a loss function that increases as the product's performance deviates from its target value. If one considers a squared error loss function, as Taguchi does for most of his applications, the average (expected) loss has two components: the squared bias and the variance. It then becomes clear that minimizing variability is an integral part of maximizing quality. Robust parameter design uses experimentation to identify the effects of design parameters on both variability and mean and to set their levels so that variability is minimized and bias is small or zero. For more on the quality philosophies that lead to robust parameter design, see Kackar (1985).

Robust parameter design has been used in some major Japanese manufacturing firms, where engineers receive extensive training in quality control, design of experiments, and Taguchi's ideas (see Box et al., 1988). AT&T was one of the first companies in the United States to be introduced to robust parameter design. It was applied in 1980 at AT&T Bell Laboratories to improve the design of a window-forming process in integrated circuit (IC) fabrication (see Phadke et al., 1983). Since 1980, applications at AT&T and other U.S. firms have grown rapidly.

The success of many applications has demonstrated the power of Taguchi's overall approach, but many of the specific statistical techniques he has proposed for implementing robust parameter design have generated a great deal of controversy. There are efforts under way at various places to understand the specific techniques he has proposed and, where possible, provide better alternatives.

This paper reviews the important ideas behind robust parameter design and provides an overview of the ongoing research. In doing so, we have drawn not only on our own work but on that of various other people working actively in the area. In Section 2, we explain the fundamental concepts in an uncritical way, laying out Taguchi's basic approach. Section 3 illustrates how the approach was followed in the window forming example mentioned above. The controversies over the statistical methods employed by Taguchi are outlined in Section 4 where we also describe some of the research under way to understand and improve these methods. In Section 5 we briefly summarize the status of current applications and educational programs on robust parameter design at AT&T and other U.S. companies.

2   THE BASIC CONCEPTS OF ROBUST PARAMETER DESIGN

2.1   The Building Blocks

Figure 1 shows a block diagram representation of a simple robust parameter design problem. The block represents the system under study, which might be a product such as a power supply circuit, or part of a manufacturing process such as photolithography steps in integrated circuit (IC) fabrication. There are one or more outputs or "responses" that characterize the performance of the system. For example, responses might be the line-width and cross-section profile of an IC conductor path, or the output voltage of a power circuit. The goal is to make the system perform as close as possible to its target level. A loss is incurred whenever any response deviates from its target value.

The system's performance depends on two types of inputs: a) design parameters (θ)—parameters that can be easily controlled and manipulated by the system's designer; and b) noise variables (N)—variables which are difficult or expensive for the system's designer to control. Examples of

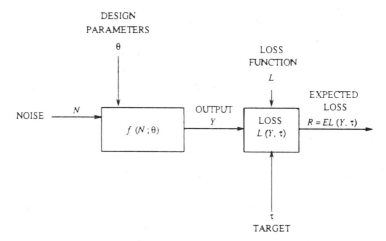

Figure 1 A block diagram representation of a simple *robust parameter design* problem. The output Y depends on the design parameters θ and the noise variables N through the transfer function f. Loss is incurred if the output deviates from the target value τ. The goal is to choose the settings of the design parameters θ to minimize average loss.

design parameters are the nominal values of the components in a power circuit and the baking temperature and baking time in an IC fabrication process. Noise variables include manufacturing variations such as process parameter drift, impurities in raw materials, variation of temperature and humidity in the manufacturing environment, dust and dirt, and operator variations, as well as deterioration and wear-out of system components over time and variations in the customer's environment. Variations in these noise variables may lead to deviation of the system's performance from its target value.

In manufacturing, response variation has traditionally been reduced by controlling the noise variations that cause it. Quality control efforts focused on identifying and removing "assignable causes" of variation. Raw material variations were controlled by specifying higher grade materials that have fewer impurities. Variation in component parts was reduced by inspecting incoming supplies. Temperature and humidity variations and dust and dirt in the factory were reduced by maintaining a controlled environment around the manufacturing process.

Although many of these measures for controlling noise variations are often necessary, they are expensive. In each case mentioned above, controlling the noise adds cost to each unit of product produced: higher grade materials cost more; inspection is labor intensive and leads to scrap and rework of defective parts; and controlled manufacturing environments are expensive to set up and maintain.

To reduce the cost of tightening response variation, Taguchi emphasizes the importance of decreasing the system's sensitivity to noise variations. Thus, robust parameter design attempts to identify the design parameter values that minimize the system's sensitivity to noise variables. More formally, suppose the quality loss when the response Y deviates from its target value τ can be quantified through a loss function $L(Y, \tau)$. Let $R(\theta) = E_N L(Y, \tau)$, the loss averaged over the possible values of the noise variables. Then, the robust parameter design problem can be stated as:

Find the value of θ that minimizes $R(\theta) = E_N L(Y, \tau)$.

To see how robust parameter design can work, consider the example of a power circuit (see Taguchi, 1986, page 77, for a similar example). We will illustrate how the nominal values of two of the circuit's components, a transistor and a resistor, can be chosen to reduce the expected loss in the response, an output voltage. Let G denote the transistor gain, R the resistance, and V the output voltage. For illustrative purposes, we assume that the relationship between the output voltage and the two components can be approximated by the additive model

$$V = f(G) + h(R)$$

where f is a monotone function and h is linear. (see Figure 2). Figures 2a and 2b show how the output voltage changes as one component varies and the other is held fixed at some value.

The target output voltage is 115 volts, but the actual voltage deviates from this target because of unit-to-unit variation of the circuit components. More specifically, $G = G_0 + \epsilon_1$ where G_0 is the nominal value of gain and ϵ_1 is random with zero mean and variance σ_1^2. Similarly, $R = R_0 + \epsilon_2$ where R_0 is the nominal value for resistance and ϵ_2 is random with zero mean and variance σ_2^2.

Suppose the goal is to minimize squared error loss, i.e., $L(Y, 115) = (Y - 115)^2$, subject to the unbiasedness constraint that the output voltage equals 115 volts on the average. Thus, minimizing mean squared error of V is equivalent to minimizing the variance of V subject to the unbiasedness

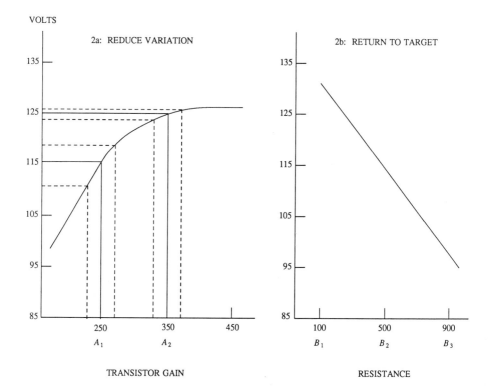

Figure 2 Power circuit design: A simple example to illustrate some basic ideas in robust parameter design. The nonlinear relationship between output voltage and transistor gain is first exploited to minimize variation. The linear relationship between gain and resistance is then used to get the mean back on target without affecting variability.

condition. We can approximate the variance of V by

$$[(f'(G_0))^2\sigma_1^2 + (h')^2\sigma_2^2].$$

The unbiasedness constraint dictates that either G_0 or R_0 should be chosen so that $f(G_0)R_0$ equals 115 volts. To minimize the variance (and hence the mean squared error), we only have to choose the value of G_0 so that $f'(G_0)$ is small. Changing the value of R_0 does not affect variability since h', the derivative of h, is constant. Figure 2a shows how variations around a nominal value of the transistor gain map into variations in the output voltage. Specifying a nominal gain of 350 will cause less variability in the output voltage compared to the nominal gain of 250 since the derivative is much flatter around 350 than around 250. If the average output voltage is now far from its target value, the nominal value for the resistance R_0 can be adjusted so that $f(G_0)R_0$ equals the target value of 115 volts.

If the goal is to minimize mean squared error without the unbiasedness constraint, the nominal value R_0 must be selected to adjust the mean to an off-target value. For details and a general discussion of this type of problems, see Section 2.2 of Leon, Shoemaker and Kacker (1987). Parameters such as this resistance, which can be used to adjust the output's mean without affecting its performance characteristic, are called adjustment parameters or scaling and leveling factors (Taguchi and Phadke, 1984; Leon, Shoemaker and Kacker, 1987).

The power circuit example illustrates how judicious choice of nominal design parameter values can reduce response variation when the source of that variation is the design parameters themselves. This exploitation of nonlinearity to reduce response variation extends to situations where variation also comes from external variables such as temperature, humidity, and impurities. Basically, robust parameter design seeks to exploit interactions between design parameters and noise variables, identifying values of design parameters where the effects of noise variables are small.

2.2 Signal-to-Noise (SN) Ratios

Although the objective is to minimize average loss, Taguchi usually maximizes a measure which he calls the "signal to noise (SN) ratio." He classifies the parameter design problems into different categories and defines an SN ratio for each class (see Taguchi and Phadke, 1984). In some cases, the SN ratio is just a monotone function of the mean squared error loss. For example, the SN ratio for problems with positive response and target value of zero (the so-called *smaller-the-better problems*) is

$$SN \text{ ratio} = -\log EY^2,$$

which is equivalent to the mean squared error. For problems where the goal is to make the response as large as possible (*larger-the-better problems*) he uses the SN ratio

$$SN \text{ ratio} = -\log E\left(\frac{1}{Y}\right)^2.$$

This is equivalent to converting the larger-the-better problem to the smaller-the-better problem by applying the reciprocal transformation and then using a squared error loss function.

In a third category of parameter design problems, the objective is to get the response as close as possible to a positive, finite target value (*on-target problems*). For this, Taguchi considers the mean response and the following SN ratio:

$$SN \text{ ratio} = -\log \frac{\text{Var}\, Y}{(EY)^2}.$$

This SN ratio is a monotone function of the coefficient of variation. The motivation and theory behind it will be discussed in more detail in Section 4. For now, we can think of this as a measure of variability which considers only the part of the variance that does not depend on the mean.

2.3   Design of Experiments

When knowledge of the functional relationship between the system's responses, design parameters, and noise variables is incomplete, experiments can be conducted to investigate this relationship and identify the parameter values that will reduce loss caused by noise. This is true for physical experiments as well as for computer experiments, although the nature of the analysis may well vary in the two situations. Taguchi uses fractional factorial experiments to estimate the effect of design parameters on both the mean response and measures of variability and loss caused by noise.

The standard experimental setup recommended by Taguchi is shown in Figure 3. The design matrix (which Taguchi calls the "inner array") is a fractional factorial design, constructed from one of a collection of standard orthogonal arrays (Taguchi, 1986, 1987) Most of the orthogonal arrays (OAs) that Taguchi uses are familiar 2^{k-p}, 3^{k-p}, or Plackett–Burman designs. Among mixed designs, he most often recommends the L_{18} and the L_{36} (Taguchi, 1986, 1987).

In some applications, the orthogonal arrays are tailored. That is, three or more columns are combined to create a single column with a larger number of levels, or the number of levels within a column is reduced by redefining one level to be equal to one of the other levels. While this approach makes it easier to study parameters at three or four levels, it makes it harder to allow for interactions between parameters. We will discuss Taguchi's philosophy on interactions in Section 4.

Design Matrix: L_{18} Orthogonal Array

Experiment Number	Column Number & Factor							
	1	2	3	4	5	6	7	8
	A	BD	C	E	F	G	H	I
1	1	1	1	1	1	1	1	1
2	1	1	2	2	2	2	2	2
3	1	1	3	3	3	3	3	3
4	1	2	1	1	2	2	3	3
5	1	2	2	2	3	3	1	1
6	1	2	3	3	1	1	2	2
7	1	3	1	2	1	3	2	3
8	1	3	2	3	2	1	3	1
9	1	3	3	1	3	2	1	2
10	2	1	1	3	3	2	2	1
11	2	1	2	1	1	3	3	2
12	2	1	3	2	2	1	1	3
13	2	2	1	2	3	1	3	2
14	2	2	2	3	1	2	1	3
15	2	2	3	1	2	3	2	1
16	2	3	1	3	2	3	1	2
17	2	3	2	1	3	1	2	3
18	2	3	3	2	1	2	3	1

Noise Plan: This is repeated for each experiment (row) in the design matrix

Replication Number	Wafer	Location of Chips
1	1	Top
2	1	Center
3	1	Bottom
4	1	Left
5	1	Right
6	2	Top
7	2	Center
8	2	Bottom
9	2	Left
10	2	Right

Figure 3 A typical robust design experiment plan: This is the plan used for the contact window experiment discussed in Section 3. (a) An eighteen-run design matrix (L_{18} orthogonal array) is used to study the effects of one two-level factor (A) and seven three-level factors (BD–I). (b) The effects of noise variables are studied indirectly through the noise plan whereby multiple measurements are taken from two wafers and five different locations within each wafer. The noise plan is repeated for each row of the design matrix.

Just as the design parameters are varied according to the design matrix, the noise variables are often also varied according to a plan which we call the "noise matrix." The noise matrix (which Taguchi calls the "outer array") may be constructed by assigning noise variables to the columns of an orthogonal array, in much the same way the design matrix is constructed. Unlike the design matrix, however, the purpose of the noise matrix is not to obtain estimates of factor effects. Instead, it is intended to provide a measure of overall response variability caused by the noise variables.

In many robust parameter design applications, however, noise variables are not investigated formally through a noise matrix. Instead, the effect of noise is studied indirectly. For example, response measurements might be taken on silicon wafers baked at different locations in a furnace in order to capture the effects of variation in temperature and gas composition inside the furnace. These replications are chosen systematically with the same sampling plan used for all the runs in the design matrix.

Whether it be a matrix or simply a systematic replication plan, in most cases the noise plan provides only a rough indication of the effects of noise on the response. Nevertheless, this measure is useful for gauging the relative effect of each design parameter setting on response variability.

3 ILLUSTRATIVE EXAMPLE

The case study reported in Phadke et al. (1983) is, to our knowledge, the first documented application of robust parameter design in the United States. The goal of this study was to improve the process of forming contact windows in complementary metal-oxide semiconductor circuits at AT&T Bell Laboratories. The experiment resulted in the reduction in the variability of the process by a factor of four and led to a substantial reduction in the processing time. We use this application to illustrate the basic concepts of robust parameter design.

Window forming is one of the critical steps in the IC fabrication process. Contact windows facilitate the interconnections between the gates, sources, and drains in circuits. It is very important that the window sizes are close to the target dimension; windows that are not open or are too small result in loss of contact to the devices, while excessively large windows lead to shorted device features. The window forming process in operation at that time resulted in high variation in window sizes and at times produced windows that failed to open and print.

The process of forming the contact windows involves photolithography. Phadke et al. (1983) identified nine parameters that were considered important in controlling window size: A—mask dimension, B—viscosity, C—spin speed, D—bake temperature, E—bake time, F—aperture, G—exposure time, H—developing time, and I—plasma etch time. The operating levels of these parameters as well as the test levels for the experiment are given in Table 1. Factors A, B, and D had two test levels while all the others had three. After collapsing B and D to form a composite three level parameter, Phadke et al. (1983) used the L_{18} orthogonal array given in Figure 3 as the design matrix. This design has eighteen runs with eight parameters, one at two levels and seven at three levels.

Table 1 Design Parameters and Test Levels for the Contact Window Experiment

		Levels		
Label	Design Parameters		Standard Levels	
A	Mask Dimension (μm)		2	2.5
B	Viscosity		204	206
C	Spin Speed (rpm)	Low	Normal	High
D	Bake temperature (°C)	90	105	
E	Bake Time (min)	20	30	40
F	Aperture	1	2	3
G	Exposure Time	20% Over	Normal	20% Under
H	Developing Time (s)	30	45	60
I	Plasma Etch Time (min)	14.5	13.2	15.8

The effect of noise variables, such as variations in process parameters from nominal values and uneven temperature distributions, was not studied explicitly through a noise matrix. Instead ten observations were made for each design configuration: two wafers to study wafer-to-wafer variation, and five observations within each wafer corresponding to five chips located in five different positions. In this way, the effects of oven temperature variations and gas composition and density nonuniformities were captured indirectly.

The relevant response or quality measure in this study was the window size with the target value of 3.5 μm. But the sizes of the windows in the functional circuits could not be measured easily. Therefore, a line-width test pattern and a window test pattern were provided on each chip. These test patterns were considered to be reasonable surrogate measures of the sizes of the windows in the functional circuits. Three measurements were made on these test patterns: pre-etch line width, post-etch line width, and post-etch window size. The complete set of data can be found in Phadke et al. (1983). There were, in fact, some missing observations in this experiment. For illustrative purposes, we will follow Phadke et al. (1983) and ignore this additional level of complexity.

Tables 2–4 give the results from the analyses of these data. Since factor I (etch time) should not affect pre-etch line width, it was ignored in

Table 2   Analysis of Pre-etch Line-width Data

		2a: ANOVA for SN ratio			
	Source	Degrees of Freedom	Sum of Squares	Mean Square	F
A	Mask Dimension	1	0.2399	0.2399	6.30
BD	Viscosity Bake Temperature	2	0.0169	0.0085	0.22
C	Spin Speed	2	0.0668	0.0334	0.88
E	Bake Time	2	0.0804	0.0402	1.06
F	Aperture	2	0.2210	0.1105	2.90
G	Exposure Time	2	0.0634	0.0317	0.83
H	Developing Time	2	0.0017	0.0009	0.02
Error		4	0.1522	0.0381	
Total		17	0.8423		

		2b: Pooled ANOVA for mean line width			
	Source	Degrees of Freedom	Sum of Squares	Mean Square	F
A	Mask Dimension	1	1.05	1.05	19.81
B	Viscosity	1	0.83	0.83	15.74
C	Spin Speed	2	0.73	0.37	6.89
G	Exposure Time	2	0.19	0.10	1.79
Error		11	0.58	0.05	
Total		17	3.37		

analyzing this data set. The pre-etch and post-etch line-width data were continuous responses with a fixed target value of 3.5 μm. The SN ratio used by Taguchi for such problems is $-\log[\mathrm{Var}\, Y/(EY)^2]$. Sample SN ratios are computed from the ten replications at each design run, and a linear model is fit to these empirical values using the L_{18} design matrix. Tables 2a and 3a give the results of the ANOVA computations for the pre-etch and post-etch line width data, respectively. Using the ANOVA table and the F statistics in a formal way, one can conclude that parameters A and F are important in controlling variability of pre-etch line width and that none of the parameters are important in controlling variability of post-etch line width. Tables 2b and 3b provide corresponding results from the analyses

Table 3 Analysis of Post-etch Line-width Data

3a: ANOVA for SN ratio

	Source	Degrees of Freedom	Sum of Squares	Mean Square	F
A	Mask Dimension	1	0.005	0.005	0.02
B	Viscosity	1	0.134	0.134	0.60
D	Bake Temperature	1	0.003	0.003	0.01
C	Spin Speed	2	0.053	0.027	0.12
E	Bake Time	2	0.057	0.028	0.13
F	Aperture	2	0.065	0.043	0.19
G	Exposure Time	2	0.312	0.156	0.70
H	Developing Time	2	0.156	0.078	0.35
I	Etch Time	2	0.008	0.004	0.02
Error		2	0.444	0.222	
Total		17	1.257		

3b: Pooled ANOVA for mean line width

	Source	Degrees of Freedom	Sum of Squares	Mean Square	F
A	Mask Dimension	1	0.677	0.677	16.92
B	Viscosity	1	2.512	2.512	63.51
C	Spin Speed	2	1.424	0.712	17.80
G	Exposure Time	2	1.558	0.779	19.48
H	Developing Time	2	0.997	0.499	12.48
Error		9	0.356	0.040	
Total		17	7.524		

of the sample means of the pre- and post-etch line-width data. This is the usual kind of analysis done in design of experiments to identify parameters that affect location. Table 2b and supplementary analysis to separate the effects of B and D suggest that parameters A, B, and C are important in affecting mean pre-etch line width. Similarly, Table 3b suggests that parameters B, G, C, A, and H are important for adjusting mean post-etch line width.

Two points are worth noting. (i) Taguchi recommends using the F statistics as informal tools to determine the relative importance of the various effects, and not as formal tests of significance. (ii) The ANOVA computations are based on the assumption of constant variance, an assumption that is unlikely to be satisfied for the response surfaces corresponding to the mean observations (Table 2b) and some of Taguchi's SN ratios. For a precise analysis, one must take into account the unequal variances and do a weighted least squares analysis.

For the post-etch window size data, many of the test windows did not open, and window sizes could only be measured from those that were open. Phadke et al. (1983) grouped this mixed categorical-continuous data into five categories and analyzed it as ordered categorical data. They used a technique called accumulation analysis proposed by Taguchi (1974) for analyzing such data. A description of accumulation analysis and a discussion of its properties are given in the next section. Table 4 gives the results from accumulation analysis of the window size data. This suggests that parameters B, C, A, H, and G affect post-etch window size.

Table 4   Accumulation Analysis of Post-etch Window Size Data

	Source	Degrees of Freedom	Sum of Squares	Mean Square	F
A	Mask Dimension	4	26.64	6.66	2.67
BD	Viscosity-Bake Temperature	8	112.31	14.04	5.64
C	Spin Speed	8	125.52	15.69	6.30
E	Bake Time	8	36.96	4.62	1.88
F	Aperture	8	27.88	3.49	1.40
G	Exposure Time	8	42.28	5.29	2.12
H	Developing Time	8	45.57	5.70	2.29
I	Etch Time	8	23.80	2.98	1.20
Lack of Fit		8	17.25	2.16	0.87
Error Between Wafers Within Experiment		60	149.33	2.49	11.69
Error Between Replicates Within Wafers Within Experiment		528	112.45	0.21	
Total		656	720.00		

For on-target problems, Taguchi proposes the following general strategy for predicting parameter settings to improve performance: (i) reduce variability as much as possible by appropriately choosing the settings of the parameters that affect the SN ratio; and (ii) adjust the settings of those parameters that affect the mean but not the SN ratio to get the mean on target. This two-step strategy will be discussed in more detail in the next section. In the present application, there are multiple measures of quality, and the results from all the analyses have to be compared and reconciled with engineering judgment before a final recommendation is made. Based on such an iterative process, Phadke et al. (1983) concluded the following: (i) change the mask dimension (A) from the operating condition of 2 μm to 2.5 μm; (ii) change spin speed (C) from 3000 rpm to 4000 rpm; and (iii) change developing time (H) from 45 sec to 60 sec. The settings of the other parameters were unchanged. Subsequent experiments suggested that exposure setting (G) should be changed from the operating condition of 90 to 140 PEP to achieve the desired target of 3.5 μm. Phadke et al. (1983) report that these new parameter settings led to the reduction in variability by a factor of four, and a threefold reduction in defect density due to unopen windows. The study also resulted in a more stable, robust process which in turn led to the elimination of several in-process checks and a substantial reduction in the processing time.

This was the first application of robust parameter design methodology at AT&T. The success has spawned a number of other applications within AT&T. A discussion of the current state of the applications is given in Section 6.

4 SOME ISSUES

There is general consensus that Taguchi's robust parameter design methodology is based on sound engineering principles. Reducing quality loss by designing the products and processes to be insensitive to variation in the noise variables is a novel concept to statisticians and quality engineers. Taguchi was also one of the first to recognize that statistically planned experiments can be used to identify parameters that affect variability as well as those that affect location. He, as much as anyone else, is responsible for the renewed interest in the use of experimentation and fractional factorial designs in industry.

However, some of the specific statistical techniques Taguchi uses to implement his robust design methodology have been the subject of controversy and debate. Taguchi himself does not provide any theoretical justification for using these techniques. Therefore, it has been left to others to investigate the properties of his techniques and, wherever possible, provide justification or better alternatives. In this section, we provide a brief overview of some of these topics and indicate the current state of research.

4.1 Experimental Design

The theory and practice of design of experiments has a long history predating the quality revolution. Most of the statistical principles underlying design of experiments were developed by Fisher during his pioneering work at Rothamsted Experimental Station in the 1920's and 1930's. Factorial designs, for studying several factors simultaneously, were developed by Fisher and Yates at Rothamsted (Fisher, 1926). Fractional factorial designs were introduced by Finney (1945) for studying the main effects and low-order interactions in a small number of runs. Many of the orthogonal arrays and their theoretical development are due to Rao (1946, 1947) and Bose and Bush (1952). Most of the early applications of experimental design techniques were in agriculture. Their use in the chemical industry was promoted by the extensive work of Box and his collaborators on response surface designs (see Box and Draper, 1987).

Taguchi emphasizes the use of highly fractional, orthogonal main-effect designs. He argues that knowledge of the underlying physical relationships should be used to choose the quality characteristics and SN ratios so that the effect of the parameters can be reasonably approximated by an additive model with main effects only (see Phadke and Taguchi, 1987). This enables one to study the effects of many parameters economically. As a protection against unanticipated interactions, Taguchi recommends the use of OAs like L_{12} and L_{18} which smear the effects of interactions among several main effects. The rationale is that if an interaction is present, then it is likely to affect all the main effects about equally.

The idea of choosing responses appropriately so that the response surface can be approximated by an additive model is a familiar one in statistics. It is a basis of data transformations (Box and Cox, 1964) and recent work on ACE (Breiman and Friedman, 1985) and generalized additive models (Hastie and Tibshirani, 1986). Taguchi, however, does not make use of any data-analytic techniques, but rather selects responses and SN ratios based on the underlying physics of the problem. He relies exclusively on the confirmation experiment to detect any violations of the additive model. When, through appropriate choice of response and SN ratio, the additive model is a good approximation, Taguchi's approach works well. But what if the response surface cannot be approximated reasonably by an additive model?

In situations where the response surface is highly nonlinear, it is unlikely that one can induce additivity by judicious choice of the response variable or by transformations. See, for example, Wu, Mao, and Ma (1987). Often the goal is to approximate the unknown response surface (or a known complicated one) empirically by a polynomial. Even for a second-order model, one has to consider both the quadratic terms and the interaction terms. Taguchi recommends taking more than two levels for continuous parameters to study quadratic effects but typically ignores interactions.

However, the study of interactions is the very essence of robust parameter design. The objective is to identify the parameter levels at which the effect of the noise variables is smallest. This assumes interaction between the design parameters and noise variables. It is unreasonable to assume a priori that the design variables interact only with the noise variables and not among themselves, especially as classification of factors into design parameters and noise variables is sometimes arbitrary.

One important argument for using orthogonal designs is that the resulting estimators of parameter effects are uncorrelated. But this holds only under variance homogeneity, an assumption that is likely to be violated for the response surfaces corresponding to some of Taguchi's SN ratios and the mean response. Hunter (1984) also argues that, from the viewpoint of approximating a response surface by a polynomial model, the important criterion is not orthogonality but minimum bias and variance. Thus, one could consider non-orthogonal designs. Hunter (1985), Box et al. (1988), and others take a sequential approach to experimentation, using both orthogonal and nonorthogonal designs. They assume the hypothesis of *factor sparsity*, i.e., only a small subset of the factors is important, and that the nature of the interactions among these factors is unknown a priori. Hunter (1985) proposes starting with two level fractional factorial designs with the center point added to get an idea of the quadratic effects. Attention can then be restricted to the subset of the factors that are identified as important (active), and the relationship among these factors can be explored by collapsing the design onto the reduced factor space and supplementing the design to estimate both quadratic and interaction effects if necessary. When the assumption of factor sparsity is not satisfied, Hunter (1985) recommends a full second-order design such as a rotatable central composite design.

This kind of iterative experimentation may be useful in some problems.

In contrast, Taguchi's objective seems to be to get satisfactory improvement in one iteration. He recommends the use of a confirmation experiment to verify that the predicted settings do in fact lead to improved performance. For a number of reasons, the confirmation experiment can fail to verify the prediction in improved performance. Taguchi does not offer any specific guidelines on what further analyses or experimentation should be done to isolate these problems. The iterative aspect of the experimentation process is also not anticipated in selecting the original design.

4.2 SN Ratios, PerMIAs and Transformations

As discussed in Section 2, Taguchi maximizes measures called SN ratios in order to maximize quality. He does not, however, provide any explanation or justification for these measures, and they have been a source of mystery and controversy. Leon, Shoemaker, and Kacker (1987) were the first to provide a systematic, mathematical framework for studying these SN ratios and the two-step optimization implicit in Taguchi's methodology. We briefly discuss the rationale behind the SN ratio $-\log(\operatorname{Var} Y/EY^2)$, the performance criterion for on-target problems.

Recall that the goal of robust parameter design can be stated formally as choosing the setting of the design parameters θ that minimizes the average loss $R(\theta) = E_N L(Y, \tau)$. Suppose we can divide the design parameters θ into two classes $\theta = (\mathbf{d}, \mathbf{a})$, and there is a short-cut method for calculating $P(\mathbf{d}) = \min_{\mathbf{a}} R(\mathbf{d}, \mathbf{a})$. Then, the expected loss can be minimized more easily using the following two-step procedure:

1. Find the setting $\mathbf{d} = \mathbf{d}^*$ that minimizes $P(\mathbf{d})$;
2. Find $\mathbf{a}^*$ that minimizes $R(\mathbf{d}^*, \mathbf{a})$.

Leon, Shoemaker, and Kacker (1987) call $\mathbf{a}$ adjustment parameters and $P(\mathbf{d})$ performance measure independent of adjustment (PerMIA). They show that Taguchi's SN ratio $-\log(\operatorname{Var} Y/(EY)^2$ is a PerMIA when the loss function is quadratic and the transfer function follows a multiplicative model. More specifically, suppose the loss function can be approximated by a quadratic loss function $L(Y, \tau) = k(Y - \tau)^2$ (where k is some proportionality constant, assumed here equal to 1). Further, assume that the transfer function (see Figure 1) is given by the multiplicative model

$$Y = \mu(\mathbf{d}, \mathbf{a})\epsilon(N, \mathbf{d}),$$

where $EY = \mu(\mathbf{d}, \mathbf{a})$ is a strictly monotone function of each component of $\mathbf{a}$ for each $\mathbf{d}$. Then, the expected loss is

$$R(\mathbf{d}, \mathbf{a}) = \mu^2(\mathbf{d}, \mathbf{a})\sigma^2(\mathbf{d}) + (\mu(\mathbf{d}, \mathbf{a}) - \tau)^2,$$

where $\sigma^2(\mathbf{d}) = \text{Var } \epsilon(\mathbf{N}, \mathbf{d})$. Then, it is easily shown that the above two-step procedure reduces to the following:

1. Find the setting of $\mathbf{d} = \mathbf{d}^*$ that minimizes $P(\mathbf{d}) = (\tau^2\sigma^2(\mathbf{d}))/(1 + \sigma^2(\mathbf{d}))$;
2. Find $\mathbf{a}^*$ such that $\mu(\mathbf{d}, \mathbf{a}^*(\mathbf{d})) = \tau/(1 + \sigma^2(\mathbf{d}))$.

Taguchi's SN ratio $-\log[\text{Var } Y/(EY)^2]$ equals $-\log \sigma^2(\mathbf{d})$, which is a decreasing function of the PerMIA $P(\mathbf{d})$. Hence, the use of Taguchi's SN ratio in Step 1 above does indeed minimize the expected loss. However, when this model fails to hold, the use of Taguchi's SN ratio would lead to non-optimal parameter settings. Leon, Shoemaker and Kacker (1987) also consider other classes of parameter design problems and develop the general notion of PerMIAs for two-step optimization.

The above discussion of SN ratios and PerMIAs assumed knowledge of the general form of the transfer function. In practice, however, even the form of the transfer function may be unknown and may have to be estimated from the data. Box (1986, 1988), Nair and Pregibon (1986), and Tsui (1990) investigate the use of empirical data-analytic tools for identifying the underlying model, and propose an analysis based on data transformations. Specifically, Nair and Pregibon (1986) consider the model $\text{Var } Y = f(\mu(\mathbf{d}, \mathbf{a}))P(\mathbf{d})$, where the function $f(\cdot)$ is unknown, whereas Box (1986) considers the approximately equivalent model $\text{Var } h(Y) = P(\mathbf{d})$. They both propose the following strategy for solving the parameter design problem:

1. Use data-analytic methods to separate the design parameters into $\mathbf{d}$ and $\mathbf{a}$ and find the transformation $h(\cdot)$ to make $\text{Var } h(Y)$ independent of $\mathbf{a}$;
2. Choose $\mathbf{d}$ to minimize variance of $h(Y)$;
3. Adjust $\mathbf{a}$ to get the mean of Y on target.

Taguchi's use of the SN ratio $-\log[\text{Var } Y/(EY)^2]$ corresponds to the model $f(\mu) = \mu^2$. In this case, Box (1986, 1988) and Nair and Pregibon (1986) recommend the use of the approximately equivalent criteria $\text{Var}(\log Y)$. There is no reason to assume that the model $f(\mu) = \mu^2$ will hold in all situations. One can use data-analytic methods to try to identify the appropriate relationship. Box (1988) proposes the use of λ plots to identify a member of the power family of transformations (see also Tsui, 1990). Nair and Pregibon (1986) propose a general data-analysis strategy based on mean-variance plots and other graphical techniques. Box (1988) and Nair and Pregibon (1986) also emphasize the use of graphical tools like normal probability plots instead of the use of formal F statistics from ANOVA tables to identify the important parameters.

We now illustrate the data-analytic approach outlined in Nair and Pregibon (1986) by reanalyzing the pre-etch line-width data discussed in the last section. Figure 4 is a mean-variance plot that displays the sample variance as a function of the sample mean. The labels denote the particular experimental run. The use of the log-log axes is intended to detect the

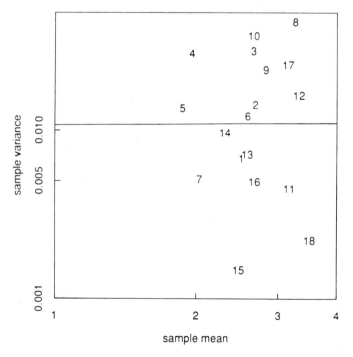

Figure 4   Mean-variance plot of the pre-etch data. The sample variance from the pre-etch experiment is plotted against the corresponding sample mean on a log-log scale. The experiment number is used as the plotting character. The line superimposed on the scatter plot was fitted by eye. There appears to be no significant relationship between variance and mean.

extent of the dependence of the form $\operatorname{Var} Y \propto \mu^k$. The slope of the straight line plotted by eye suggests that there is no strong relationship between the mean and the variance. Therefore, no transformation is necessary, and it suffices to analyze just the mean and variance of the original data. In the PerMIA terminology, the model for the transfer function is

$$Y = \mu(\mathbf{d}, \mathbf{a}) + \epsilon(N, \mathbf{d}),$$

and the corresponding PerMIA for two-step optimization is $P(\mathbf{d}) = \operatorname{Var} Y$. Figures 5 and 6 are the half-normal probability plots of location and dispersion contrasts, respectively. The latter were obtained by fitting an additive model to the log-variances. The lines on the plot were filled by eye and separate the "real" effects from noise. Figure 5 suggests that there are five important location effects—B, A, and linear effects of C, G, and H. (As before, we note that, since the response could have unequal variances, the usual assumptions underlying the use of a normal probability plot could be violated.) Figure 6 suggests that there are no really large dispersion effects, with A and the linear effects of F and G possibly being important. The conclusions from this analysis, together with a similar analysis of the post-etch data and an analysis of the window size data, can then be used to predict new settings of the design parameters to reduce quality loss. Readers are referred to Nair and Pregibon (1986) for some additional details.

For the pre-etch line-width data, the analysis of Taguchi's SN ratio leads to essentially the same conclusions as an analysis based on log-variances. This is because Taguchi's SN ratio is equivalent, approximately, to variance of $\log Y$, and over the range of the pre-etch line-width data, the log-transformation is approximately linear. This is why the mean-variance plot does not suggest the need for a transformation. In general, if the observations are all of the same order of magnitude, analyses in different metrics are likely to lead to the same qualitative conclusions concerning model identification. When the data span several orders of magnitude, it will typically be possible to identify an appropriate metric for analysis using the data-analytic tools discussed in this section.

The analyses considered thus far ignore the structure from the systematic sampling of noise variables. One can, however, do a more sensitive analysis by taking this structure into account. Specifically, we can treat the design parameters and noise variables symmetrically by combining the design and noise matrix and doing a fixed-effects model analysis. In such an analysis, dispersion effects would be captured by the interactions between

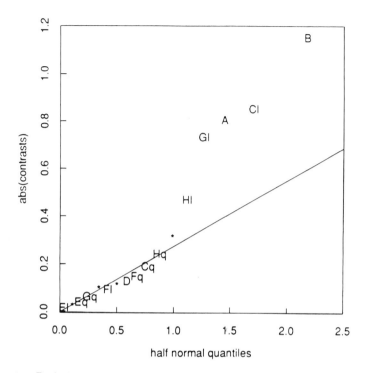

Figure 5 Probability plot of location effects for the pre-etch data. The absolute values of the single degree-of-freedom contrasts are plotted against the quantiles of the half-normal distribution. The factor level combination is used as the plotting character. The contrasts labeled "*" correspond to "error contrasts" rather than effects of interest. The effects of three-level factors were decomposed into linear (l) and quadratic (q) contrasts. The line superimposed on the plot was fitted (to the nonsignificant contrasts) by eye. The plot shows that location effects B, Cl, A, Gl, and, to a lesser extent, Hl are important.

design factors and noise variables. Easterling (1984) noted the possibility of doing such an analysis. Welch et al. (1989) and Shoemaker, Tsui, and Wu (1990) also investigate this and related approaches.

4.3   Robust Parameter Design Using Computer Models

Increasingly, products and processes are being modeled and their performances simulated using computer-aided design (CAD) systems. Computer

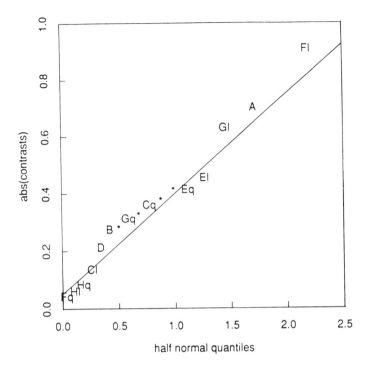

Figure 6 Probability plot of dispersion effects for the pre-etch data. The absolute values of the single degree-of-freedom contrasts are plotted against the quantiles of the half-normal distribution. The factor level combination is used as the plotting character. The contrasts labeled "*" correspond to "error contrasts" rather than effects of interest. The line superimposed on the plot was fitted by eye. The plot shows that dispersion effects Fl, A, and Gl are marginally important.

models are now routinely used in the design of most circuits and in many physical and thermal design problems. Computer modeling of manufacturing processes is also becoming more common.

The objectives of robust parameter design still apply in the CAD environment, but several key characteristics are different, among them:

- The response is generated by a computer model, rather than through a physical experiment.
- All response variation must be caused by varying the input parameters in the computer model. Differences between the nominal and

actual levels of the design parameters are a major source of this variation. There may be "external" sources of variation such as temperature or power supply variations that are part of the computer model.

- There is no measurement error or error due to unknown noise variables.
- Experimental runs may be much easier to make, but can be costly in terms of computer time.
- The validity of the results depends critically on the accuracy of the computer simulation model.

Because of these special characteristics, different techniques (both design and analysis) for implementing robust parameter design may be called for in the CAD environment.

Box and Fung (1986) view the problem of minimizing variation induced by local changes in the design parameters as a standard sensitivity analysis problem, which can be solved by a variety of numerical analysis techniques. Even if the response surface is studied systematically using an experimental design such as an OA, the minimization of an appropriate performance criteria (SN ratio, if necessary) can be done numerically rather than through fitting an additive model. Box and Fung (1988) reanalyze the Wheatstone bridge circuit example in Taguchi (1986) and show that Taguchi's additive model fit leads to a suboptimal design configuration. In situations where derivatives of the transfer functions can be computed or evaluated numerically, one can estimate the variation induced by local changes in the design parameters directly using a Taylor series approximation. This removes the need for noise arrays, thus reducing the number of runs. In any statistical analysis of data from computer experiments, it must be remembered that the only errors around are systematic errors from a lack of fit of the statistical model to the computer model. Thus, some of the usual assumptions underlying the statistical procedures may be violated.

A recent application of robust parameter design techniques to computer models is Welch et al. (1989), who study a clock skew circuit. As discussed at the end of the last section, they treat the design parameters and noise variables symmetrically and develop an overall regression model involving all the parameters. The predicted model is used to compute loss caused by variations in the noise variables.

4.4 Accumulation Analysis of Ordered Categorical Data

The available responses from some experiments are in the form of categorical data with an ordering in the categories. For example, the responses can be subjective ratings of the quality of individual units as poor, fair, and good. Taguchi (1974) recommends a procedure called accumulation analysis for analyzing such data. His original motivation for developing this procedure is to improve upon Pearson's chi-square test, which does not take into account the ordering in the categories. Takeuchi and Hirotsu (1982) and Nair (1987) have examined the relationship between accumulation analysis and other goodness-of-fit tests in the literature. The behavior of accumulation analysis in the context of robust parameter design experiments was studied by Nair (1986), Box and Jones (1986), and Hamada and Wu (1986).

Accumulation analysis is an ANOVA-like procedure that is applied to the cumulative frequencies corresponding to the ordered categories. With K ordered categories, one computes the cumulative frequencies for the first $K - 1$ categories and obtains from these $K - 1$ sums of squares for each design parameter. These $K - 1$ sums of squares are then combined in a specified way to obtain a single ANOVA table for the entire data set. Table 4 in Section 3 gives the accumulation analysis ANOVA table for the window size data.

Nair (1986) and Box and Jones (1986) show that the accumulation analysis statistic can be expressed as a weighted sum of $K - 1$ orthogonal components with decreasing weights. The Pearson's chi-square statistic can be expressed as an unweighted sum of these components. The accumulation analysis statistic gives most weight to the first component which is essentially equivalent to the Wilcoxon/Kruskal–Wallis procedure for grouped data. Since this test has good power against location alternatives, accumulation analysis performs reasonably in detecting parameters with important location effects. However, it is unnecessarily complicated for this purpose. Moreover, it is not even as powerful as Pearson's chi-square test for detecting parameters with dispersion effects. Hamada and Wu (1986) show that the accumulation analysis statistic can detect effects spuriously in fractional factorial experiments.

As an alternative, Nair (1986) and Box and Jones (1986) recommend analyses based on simple scoring schemes. Table 5 gives the results from a reanalysis of the post-etch window size data obtained by assigning a simple

Table 5 Reanalysis of
Post-etch Window Size Data
Using Linear Scores

Source	Degrees of Freedom	Mean Square
A	1	7.50
BD	2	19.15
C	2	21.65
E	2	2.52
F	2	3.04
G	2	7.13
H	2	6.14
I	2	1.82

set of scores to each category and analyzing the weighted scores (see Nair, 1986, for details). The resulting procedure is equivalent to the Kruskal–Wallis test for grouped data. For this data set, the conclusions from this analysis are qualitatively the same as those in Table 4 obtained using accumulation analysis. The analysis based on scoring schemes, however, is much simpler and more efficient. There are also other procedures, especially those based on logit and log-linear models, that have been proposed in the literature for analyzing ordered categorical data. Among these, McCullagh's (1980) regression models for ordinal data can be used to model and estimate both location and dispersion effects from ordered categorical data. See Nair (1986) and the discussions therein for further details.

4.5   Minute Analysis of Failure Time Data

Many robust parameter design experiments involve life tests where the responses are times to failure. See Taguchi and Wu (1980) for an experiment on clutch springs. Failure time data are often subject to censoring. Taguchi (1987) proposes a method called minute analysis for analyzing censored data. Briefly, this method partitions the failure time scale into several time intervals and creates an indicator variable at each interval corresponding to whether or not the unit survived that time point. The cumulative values of these indicator variables are analyzed to determine the parameter effects. A more detailed ("minute") analysis examines the interaction between the parameters and the time scale. This analysis can be interpreted as a test for differences in the shapes of the survival functions for the different runs. Interestingly enough, during the estimation phase, Taguchi applies a logit transform so that the effects are approximately additive on the transformed scale. During the testing phase, however, he ignores the binary nature of the indicator variables and also the fact that they are highly correlated across different time points. These techniques are rather ad hoc with little theoretical justification.

Fung (1986) and Hamada and Wu (1987) have looked at alternative methods for analyzing failure time data from highly fractional factorial experiments. Hamada and Wu (1987) consider a normal location model and develop an iterative algorithm for estimating the location effects of the design parameters. This technique can be extended to other parametric models. The performance of Cox's proportional hazards models in analyzing censored data from highly fractional experiments needs to be explored.

5   CONCLUDING REMARKS

The case study discussed in Section 3 was conducted during Taguchi's first visit to AT&T Bell Laboratories in 1980. Since that time, the methodology has been applied within AT&T to a variety of problems ranging from IC fabrication and router-bit life improvement to response time optimization of a UNIX system (see Kacker and Shoemaker, 1986; Phadke, 1986; Pao et al., 1985; Nazaret and Klingler, 1985; and Lin and Kackar, 1985).

In the early applications, the studies were conducted according to the techniques proposed by Taguchi. As more efficient techniques were developed, they were incorporated into the later experiments and analyses. Bell Labs statisticians and engineers have developed a methodology for robust parameter design that combines many of Taguchi's ideas and some of his statistical techniques with improved experimentation and analysis methods used at AT&T. The methodology is taught extensively to engineers in AT&T development and manufacturing organizations in a three-day workshop.

Most of the applications at AT&T thus far have been in the areas of integrated circuit fabrication and circuit pack assembly. Experimental design techniques and the objective of reducing variability are naturally applicable and clearly beneficial in these areas. In addition, exploratory applications are extending into the areas of physical design, especially thermal design, and circuit design. These are areas where computer-aided design systems are used heavily, and, as discussed above, alternative methods are being developed to be more effective in these situations.

Taguchi's ideas are also being used in many other U.S. companies such as Ford and Xerox. There are also many courses on robust parameter design offered by organizations like American Supplier Institute, Rochester Institute of Technology, and the Center for Quality and Productivity Improvement at the University of Wisconsin in Madison. The American Supplier Institute also has an annual symposium where case studies on the applications of the so-called Taguchi methods are presented.

We have described some of the research in progress to understand and, where possible, provide better alternatives to, the statistical techniques that Taguchi uses to implement robust parameter design. However, the controversy surrounding these issues should not detract from the fact that Taguchi has made original, fundamental contributions to quality engineering. The widespread attention given to his ideas by both statisticians and quality professionals in the last few years is a testimony to their importance.

REFERENCES

Bose, R. C. and Bush, K. A. (1952), "Orthogonal Arrays," *Annals of Mathematical Statistics*, 23, 508–54.

Box, G. E. P. (1986), "Studies in Quality Improvement: Signal to Noise Ratios, Performance Criteria and Statistical Analysis: Part I," Technical Report 11, Center for Quality and Productivity Improvement, University of Wisconsin–Madison.

Box, G. E. P. (1988), "Signal-to-Noise Ratios, Performance Criteria, and Transformations (with discussion)," *Technometrics*, 30, 1–40.

Box, G. E. P., Bisgaard, S. and Fung, C. (1988), "An Explanation and Critique of Taguchi's Contributions to Quality Engineering," Technical Report, Center for Quality and Productivity, University of Wisconsin–Madison.

Box, G. E. P. and Cox, D. R. (1964), "An Analysis of Transformations (with discussion)," *Journal of the Royal Statistical Society B*, 26, 211–252.

Box, G. E. P. and Draper, N. R. (1987), *Empirical Model-Building and Response Surfaces*, John Wiley, New York.

Box, G. E. P. and Fung, C. (1986), "Studies in Quality Improvement: Minimizing Transmitted Variation by Parameter Design," Technical Report, Center for Quality and Productivity, University of Wisconsin–Madison.

Box, G. E. P., and Jones, S. (1986), "Discussion of Testing in Industrial Experiments with Ordered Categorical Data by V. N. Nair," *Technometrics*, 8, 295–301.

Box, G. E. P., Kacker, R. N., Nair, V. N., Phadke, M. S., Shoemaker, A. C., and Wu, C. F. J. (1988), "Quality Practices In Japan," *Quality Progress*, 37–41.

Breiman, L. and Friedman, J. H. (1985), "Estimating Optimal Transformations for Multiple Regression and Correlation (with discussion)," *Journal of the American Statistical Association*, 30, 580–619.

Dehnad, K. (1988), *Quality Control, Robust Design and Taguchi Method*, Wadsworth, Belmont, CA.

Easterling, R. G. (1984), "Discussion of Kacker's Paper," *Journal of Quality Technology*, Vol. 17, No 4, 191–192.

Finney, D. J. (1945), "Fractional Replication of Factorial Arrangements," *Annals of Eugenics*, 12, 291–301.

Fisher, R. A. (1926), "The Arrangement of Field Experiments," *Journal of the Ministry of Agriculture*, 33, 503–513.

Fung, C. A. (1986), "Statistical Topics in Off-Line Quality Control," Ph.D. Thesis, Department of Statistics, University of Wisconsin–Madison.

Hamada, M. and Wu, C. F. J. (1986), "Discussion of Testing in Industrial Experiments with Ordered Categorical Data by V. N. Nair," *Technometrics*, 28, 302–306.

Hamada, M. and Wu, C. F. J. (1987), "Analysis of Incomplete Data from Highly Fractional Industrial Experiments," Report of the Center for Quality and Productivity, University of Wisconsin–Madison.

Hastie, T. and Tibshirani, R. (1986), "Generalized Additive Models (with discussion)," *Statistical Science*, 1, 297–318.

Hunter, J. S. (1985), "Statistical Design Applied to Product Design," *Journal of Quality Technology*, 17, 210–221.

Kackar, R. N. (1985), "Off-Line Quality Control, Parameter Design, and the Taguchi Method (with discussion)," *Journal of Quality Technology*, 17, 176–209.

Kacker, R. N. (1989), "Taguchi Methods," Chapter in *Handbook of Statistical Methods for Scientists and Engineers*, Ed. by Harrison and Wadsworth, McGraw Hill, New York.

Kacker, R. N. and Shoemaker, A. C. (1986), "Robust Design: A Cost-Effective Method for Improving Manufacturing Processes," *AT&T Technical Journal*, March–April, 65, 39–50.

Leon, R., Shoemaker, A. C. and Kacker, R. N. (1987), "Performance Measures Independent of Adjustment: An Explanation and Extension of Taguchi's Signal-to-Noise Ratios (with discussion)," *Technometrics*, 9, 253–285.

Lin, K. M. and Kackar, R. N. (1985), "Optimizing the Wave Soldering Process," *Electronic Packaging and Production*, February, 108–115.

McCullagh, P. (1980), "Regression Models for Ordinal Data" (with discussion), *Journal of the Royal Statistical Society*, Ser. B, 109–142.

Nair, V. N. (1986), "Testing in Industrial Experiments with Ordered Categorical Data (with discussion)," *Technometrics*, 28, 283–311.

Nair, V. N., and Pregibon, D. (1986), "A Data-Analysis Strategy for Quality Engineering Experiments," *AT&T Technical Journal*, 65, 73–84.

Nair, V. N. (1987), "Chi-Square Type Tests for Ordered Alternatives in Contingency Tables," *Journal of the American Statistical Association*, 82, 283–291.

Nazaret, W. A. and Klingler, W. (1985), "Tuning Computer Systems for Maximum Performance: A Statistical Approach," Unpublished manuscript.

Pao, T. W., Phadke, M. S., and Sherrerd, C. S. (1985), "Computer Response Time Optimization Using Orthogonal Array Experiments," *IEEE International Communication Conference Record*, pp. 890–895.

Phadke, M. S., Kackar, R. N., Speeney, D. V. and Greico, M. J. (1983), "Off-Line Quality Control in Integrated Circuit Fabrication Using Experimental Design," *The Bell System Technical Journal*, 62, 1273–1310.

Phadke, M. S. (1986), "Design Optimization Case Studies," *AT&T Technical Journal*, March–April, 65, 51–84.

Phadke, M. S. and Taguchi, G. (1987), "Selection of Quality Control Characteristics and *SN* Ratios for Robust Design," Unpublished manuscript.

Plackett, R. L. and Burman, J. P. (1946), "The Design of Optimum Multifactorial Experiments," *Biometrika*, 33, 305–325.

Rao, C. R. (1946), "Hypercubes of strength *d* leading to confounded designs in factorial experiments," *Bulletin of the Calcutta Mathematics Society*, 38, 67–78.

Rao, C. R. (1947), "Factorial Experiments Derivable from Combinatorial Arrangements of Arrays," *Journal of the Royal Statistical Society, Supplement*, 9, 128–139.

Shoemaker and Kacker (1988), "Methodology for Planning Experiments in Robust Product and Process Design," *Quality and Reliability Engineering International*, 4, 95–103.

Shoemaker, A. C., Tsui, K. L., and Wu, C. F. J. (1990), "Economical Modeling and Experimentation Methods for Robust Parameter Design," *Technometrics* (submitted).

Taguchi, G. (1974), "A New Statistical Analysis for Clinical Data, the Accumulating Analysis, in Contrast with the Chi-Square Test," *Shaishin Igaku (The Newest Medicine)*, 9, 806–813.

Taguchi, G. and Wu, Y. (1980), *Introduction to Off-Line Quality Control*, Central Japan Quality Control Association, Nagoya, Japan.

Taguchi, G. (1986), *Introduction to Quality Engineering*, Asian Productivity Organization, Tokyo.

Taguchi, G. (1987), *System of Experimental Design*, Vol. I and II, UNIPUB, New York.

Taguchi, G. and Phadke, M. S. (1984), "Quality Engineering Through Design Optimization," *Proceedings of GLOBECOM 84 Meeting*, IEEE Communications Society, pp. 1106–1113.

Takeuchi, K. and Hirotsu, C. (1982), "The Cumulative Chi-Squares Method Against Ordered Alternatives in Two-Way Contingency Tables," *Reports of Statistical Application Research, Japanese Union of Scientists and Engineers*, 29, 1–13.

Tsui, K. L. (1990), "Empirical Two-Step Procedure for Parameter Design," Unpublished manuscript.

Welch, W. J., Yu, T. K., Kang, S. M., and Sacks, J. (1990), "Computer Experiments for Quality Control by Parameter Design," *Journal of Quality Technology* (to appear).

Wu, C. F. J., Mao, S. S., and Ma, F. S. (1987), "An Investigation of OA-Based Methods for Parameter Design Optimization," Report of the Center for Quality and Productivity, University of Wisconsin–Madison.

11

SEL: A Search Method Based on Orthogonal Arrays

C. F. Jeff Wu   University of Waterloo, Waterloo, Ontario, Canada

S. S. Mao   East China Normal University, Shanghai, People's Republic of China

F. S. Ma   Tianjin University, Tianjin, People's Republic of China

1   INTRODUCTION

Experimental design usually consists of the following steps: (1) understanding the problem, (2) identifying important factors and choosing factor levels, (3) choosing an experimental plan and implementing the experiment according to the plan, (4) analyzing data and, based on it, choosing new factor levels for confirmation. These steps may be repeated until satisfactory results are obtained. Orthogonal arrays, including fractional factorial designs, are often used for the experimental plan in step (3). In many industrial experiments, including those advocated by G. Taguchi for quality improvement, arrays of economic run size are used to accommodate a large number of factors. For these arrays, few degrees of freedom are left for estimating interactions. Confounding patterns among interactions further complicate the task. Without adding a substantial number of additional runs, which can be costly, these interactions can not be disentangled. Methods for analysis of data from such arrays are therefore limited.

A method for step (4), advocated by Taguchi (1986) and commonly used in certain sectors, is to determine the "optimum" level for each factor based on the marginal means (details in Section 2). We call this method *analysis of marginal means* (ANOMM). Its main justification is the orthogonality among main effects of different factors. If the model for the response is not separable in the factors, which is usually manifested by the presence of interactions or more complex quantities, this method can fail miserably in predicting optimum level combinations. As a reference for comparison, we consider the simple method of choosing the level combination with the best response among the runs in the experiment. This is called the *pick-the-winner* (PW) rule. This rule has a rationale quite different from the previous method. In addition to orthogonality, points in an orthogonal array usually provide a uniform coverage of the factor space. Maximum among the responses at these points is a good guide to the maximum (global or local) response over the factor space. PW is a method of search. In contrast, ANOMM is a method of estimation based on a particular model for the response. Comparisons of these two methods are given in Section 2. For more details, see Wu et al. (1987).

In Section 3, we propose a class of methods called SEL (sequential elimination of levels) for searching over the orthogonal array to get better settings. It includes ANOMM and PW as special cases. Orthogonal arrays have been used in the search for maximum in both physical and computer experiments. Many successful cases are reported in the monograph *Three-Stage Design of Experiments with Known Transfer Functions* (see Chen et al. (1983) or Cui et al. (1983). However, if the factor space is large, single-stage search requires a large orthogonal array to cover the space properly. The proposed search method is multistage. It starts with an array of economic size. By eliminating levels that give poor results, search is concentrated on a smaller array on the remaining levels. A detailed description and an illustrative example are given in Section 3. The relative merits of two versions of SEL, SEL(mean) and SEL(mini), are compared in Section 4. When searching over a large space of factors, these methods can be more effective if applied iteratively to different regions of the factor space as determined by the results of the previous experiments. In Section 5, several such iterative schemes are compared. Some concluding remarks are given in Section 6.

Sequential elimination of levels can also be used to select good starting points for nonlinear optimization. If the function (or the response in a physical experiment) to be optimized has many local maxima, standard optimization methods do not guarantee that the global maximum be found since the local maximum reached by the algorithm depends on its starting point. Choice of starting points is crucial to the performance of a global optimization algorithm. SEL is an iterative method for selecting starting points based on orthogonal arrays. Its performance in this context will be studied later.

Three examples of robust design, of a heat exchanger, an OTL circuit, and a voltage stabilizer circuit, are used to compare and illustrate different methods under consideration. The objective of *robust product design*, an original idea due to Taguchi (1986), is to select the levels of the control factors so that the product at the chosen setting performs satisfactorily and stably over a wide range of conditions of the noise factors. See Section 2 for a detailed description of robust design.

The proposed methods are not tailored for robust design problems, for which steps such as the minimization of a performance measure (including SN ratios) and the adjustment of mean can be taken to advantage (see Leon et al., 1987; Leon and Wu, 1989). Some methods considered here can be useful supplements to existing methods for robust design. This will be reported elsewhere. It should be emphasized that the proposed methods and our study are also applicable to design problems without noise factors, e.g., traditional factorial design of experiment. In each example, there is an explicit mathematical model. These models are chosen so that simulations can be conducted to compare different strategies. Our findings are also applicable to problems without mathematical models. In the simulation studies, the mathematical models are treated as a "black box." No attempt is made to exploit the knowledge of the model, such as solving the equations or fine-tuning an adjustment factor. Such measures would definitely improve the performance, but the resulting study would lose its generality.

2 ANALYSIS OF MARGINAL MEANS OR PICK-THE-WINNER: AN EXAMPLE FROM PARAMETER DESIGN OPTIMIZATION

For robust design, Taguchi (1986) advocates the use of an orthogonal array (OA) for the control factors and an orthogonal array for the noise factors. The former array is called the *control array* (or inner array), the latter *noise array* (or outer array). For each control factor setting, a performance measure is defined as a function of the observations over the noise array. In parameter design optimization, the control factor setting that gives the best performance measure is selected.

We will use the design of a heat exchanger from Cui et al. (1983) to illustrate the use of orthogonal arrays for obtaining better control factor settings.

Example A

In the heat exchanger under consideration, the inlet temperature T_1 fluctuates in the range 670°C ±30°C, and the flow rate V fluctuates in the range $(42000 \pm 2000) \times 1/21 \times [1 + (T_1/273)]$. The three control factors are d (outside diameter of pipe), D (diameter of heat exchanger), and L/D, where L is the length of pipe. The objective of design is to select the levels of control factors so that the outlet temperature T_2 does not deviate from the target temperature 360°C by more than 15°C for values of T_1 and V in their respective ranges.

According to the derivations in Cui et al. (1983), the outlet temperature T_2 is related to the other factors by the following equation:

$$T_2 = (T_1 - T_g)e^{-A} + T_g, \qquad T_g = 222.7°C$$

and

$$A = \frac{57.1(L/D)D^3\lambda_t}{Vd^2\rho_t C_{pm}} \left[1.53 \times 10^{-3} \frac{d_i \rho_t}{\mu_t} \frac{V}{D^2} \left(\frac{d}{d_i}\right)^2 \right]^{0.8} \left[\frac{C_{pm}\mu_t}{\lambda_t} \right]^{0.4} \qquad (1)$$

where d_i = inside diameter of pipe (m), ρ_t = gaseous density (Kg/m³), μ_t = gaseous viscosity, λ_t = thermal conductivity (KCal/s · m · °C) = 3.335×10^{-5}, C_{pm} = specific heat (KCal/Kg · °C).

Since T_1 and V cannot be fixed at the design stage but may vary during operation of the exchanger, they are the noise factors. Unlike other examples in the paper, the levels of noise factors here do not form a rectangular grid, i.e., the level of V increases as T_1 increases. The layouts of control factors and noise factors are given in Tables 1 and 2. The parameter d_i in Eq. (1) is a deterministic function of d (Table 3); ρ_t, C_{pm}, and μ_t in Eq. (1) are deterministic functions of T_1 (Table 4).

The level combinations of the control factors are chosen using the orthogonal array $L_9(3^3)$ given in Table 5. This orthogonal array is called the *control array*. For each setting (i.e., combination of levels) of the control factors, the T_2 value given in formula (1) is computed for the nine combinations of the T_1 and V values in Table 2. These nine combinations form the *noise array*. Let $\Delta_i = 360 - T_2$ denote the difference between the target 360°C and the outlet temperature T_2 for the ith combination of T_1 and

Table 1 Levels of
Control Factors

	Factor		
Level	d	D	L/D
1	.025	.8	3
2	.032	1.0	4
3	.038	1.2	5

Table 2 Levels of Noise Factors

	Factor	
Level	T_1	V
1	640	$\frac{40000}{21} \times \left(1 + \frac{T_1}{273}\right)$
2	670	$\frac{42000}{21} \times \left(1 + \frac{T_1}{273}\right)$
3	700	$\frac{44000}{21} \times \left(1 + \frac{T_1}{273}\right)$

V. One such set of Δ_i values is given in Table 6. Then the performance measure for the given combination of control factor levels is defined as

$$\Delta = \max_{1 \le i \le 9} |\Delta_i|$$

**Table
3** d_i as
Function
of d

d	d_i
.025	.019
.032	.025
.038	.031

Table 4 ρ_t, C_{pm}, and μ_t as Functions of T_1

T_1	ρ_t	C_{pm}	μ_t
640	5.286	1.024	2.83×10^{-5}
670	5.185	1.029	2.89×10^{-5}
700	5.089	1.031	2.93×10^{-5}

Table 5 Layout of Control Factors and Δ Values

Run	d	D	L/D	Δ
1	1	1	2	54.90
2	1	2	1	58.18
3	1	3	3	125.64
4	2	1	1	67.03
5	2	2	3	81.71
6	2	3	2	85.25
7	3	1	3	19.78
8	3	2	2	19.82
9	3	3	1	14.97

Level	Mean Δ		
1	79.57	47.23	46.72
2	78.00	53.24	53.32
3	18.19	75.28	75.71

Level	Minimum Δ		
1	54.90	19.78	14.97
2	67.03	19.82	19.82
3	14.97	14.97	19.78

Table 6 Layout of Noise Factors and Δ_i Values (for Run No. 1 in Table 5)

Run	T_1	V	Δ_i
1	640	6370.14	54.90
2	670	6579.45	46.97
3	700	6788.77	39.49
4	640	6688.65	53.59
5	670	6908.42	45.56
6	700	7128.21	37.97
7	640	7007.15	52.34
8	670	7237.40	44.20
9	700	7467.64	36.52

Note that Δ is not a smooth function of the control factors. The objective of design is to make Δ less than 15. From the Δ values given in Table 5, only the ninth run satisfies the requirement $\Delta \leq 15$.

Based on the Δ values of the nine runs in Table 5, is it possible to find a better setting of the control factors? One method favored by Taguchi (1986) is described as follows. Compute the average of the Δ values, denoted by $\bar{\Delta}$, at each level of a control factor. For each control factor, select the level with the smallest $\bar{\Delta}$ value. Taguchi calls the resulting combination of levels "optimum." One more experiment is then conducted at the "optimum" setting of the control factors. We call this method the *analysis of marginal means* (ANOMM). The word "optimum" is misleading; as shown later, it often gives results far from optimal.

Let us now examine its performance on the heat exchanger example. The $\bar{\Delta}$ value for the three levels of factors d, D, L/D are given in their respective columns in Table 5. For example, the value 79.57 is the average of Δ values for $d = 1$. The levels selected by the ANOMM method are $d = 3$, $D = 1$, and $L/D = 1$, which gives $\Delta = 107.71$ (see Table 7) and is much worse than most runs in the L_9 array (Table 5). On the other hand, the best Δ value from the nine runs is 14.97. This selection rule is called the *pick-the-winner* (PW) rule since it picks the winner among the existing runs. The PW method is a convenient benchmark for comparing methods for parameter design optimization.

Since the Δ surface is rugged, ANOMM is inferior to PW. A more thorough study of the two methods can be found in Section 3 of Wu et al. (1987).

3   SEQUENTIAL ELIMINATION OF LEVELS (SEL)

A general method, which includes the analysis of marginal means (ANOMM) and the pick-the-winner (PW) rule as special cases, is proposed as follows:

1. For each factor, *eliminate* those level(s) with the *worst* mean value(s) of the performance measure computed from the current array.
2. Choose an orthogonal array (typically of a smaller size) for the remaining levels, and replace the array in step 1 with the new array.
3. Conduct another experiment on the new array.
4. Repeat steps 1–3 if necessary.

This method is called the *sequential elimination of levels* (SEL). In Step 1, if the mean is replaced by another descriptive statistic x, we call the method SEL(x). We consider the mean and the minimum (mini) for x in this paper. Another choice of x is the mean of the few smallest values. Comparison of SEL(mean) and SEL(mini) will be taken up in the next section. Note that more than one level may be eliminated each time.

SEL is a very general method as it includes both ANOMM and PW as extreme cases. If, in step 1 all levels but one are eliminated, it is easy to see that SEL(mean) reduces to ANOMM and SEL(mini) reduces to PW. Note that SEL exploits both the search aspect and the modeling aspect of an orthogonal array (see Section 1).

Typically, there is no need to retain more than three levels. If one level is distinctly better than the rest, only one is retained. If the best two or three levels are not significantly different, they are retained. The new array in step 2 is of the type $2^m 3^n$ and typically does not exceed 18 runs, since $L_{18}(2^1 \times 3^7)$ can accommodate a fair number of two- and three-level factors. SEL can also handle interactions. If it is desired to investigate the interaction of A and B, factors A and B will be replaced by the new "factor" $A \times B$ in the procedure.

We now illustrate the SEL method on the heat exchanger example. The mean and minimum Δ values for each factor level are given in the bottom part of Table 5. According to SEL(mean), $d = 1$, $D = 3$, and

Table 7 Second-Round
Design Using SEL(mean)

Run	d	D	L/D	Δ
1[a]	2	1	1	67.03
2	2	1	2	16.69
3	2	2	1	12.80
4	2	2	2	54.11
5	3	1	1	107.71
6	3	1	2	58.88
7	3	2	1	54.56
8[a]	3	2	2	19.82

[a] Runs already in the first-
round design (see Table 5).

$L/D = 3$ are eliminated (step 1). Each factor now has two levels. Select
the full factorial $L_8(2^3)$ for the remaining levels (step 2). The second
array has two combinations already in the first array (compare Tables 5
and 7). Conduct six additional runs (step 3) with the results given in
Table 7, which include the smallest Δ value 12.8. Similar results from

Table 8 Second-Round
Design Using SEL(mini)

Run	d	D	L/D	Δ
1	1	1	1	13.69
2	1	1	3	82.37
3	1	3	1	85.51
4[a]	1	3	3	125.64
5	3	1	1	107.71
6[a]	3	1	3	19.78
7[a]	3	3	1	14.97
8	3	3	3	83.32

[a] Runs already in the first-
round design

using SEL(mini) are given in Table 8. A better Δ value 13.69 is found, although the best value 12.8 is missed.

The performance of SEL is tested on all the 12 distinct L_9 orthogonal arrays (obtained by changing the L/D column in Table 5). In all the 8 cases in which the first-round design does not include the smallest value 12.8, SEL(mean) captures it in the second-round design. SEL(mini) does not do as well. In 5 cases, 12.8 is not captured. But in 2 of these 5 cases, it results in improvement. However, for more complicated problems studied later, SEL(mini) outperforms SEL(mean).

SEL can be applied to more complex situations such as the following example, which has more factors and levels than Example A.

Example B   (from Chen et al., 1983)

In designing an output transformerless (OTL) pull-push circuit (see Figure 1), an objective is to make the midpoint voltage V_m stable around 6 V in the presence of variations of the components of the circuit.

The midpoint voltage V_m is related to the other factors by the equation

$$V_m = (V_{b1} + V_{be1})\frac{\beta R_0}{\beta R_0 + R_f} + (E_c - V_{be3})\frac{R_f}{\beta R_0 + R_f} + \frac{V_{be2}R_f\beta R_0}{(\beta R_0 + R_f)R_{c1}} \quad (2)$$

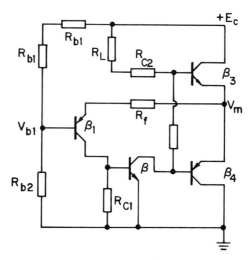

Figure 1   OTL pull-push circuit.

where

$$V_{b1} = E_c \frac{R_{b2}}{R_{b1} + R_{b2}}, \qquad R_0 = R_{c2} + R_L, \qquad R_L = 9\Omega, \qquad E_c = 12 \text{ V},$$

$$V_{be1} = V_{be3} = 0.65 \text{ V}, \qquad V_{be2} = 0.74 \text{ V}$$

R_{b1}, R_{b2}, R_f, R_{c1}, and R_{c2} are resistances, and β is the current gain. The ranges of these six factors are

$$R_{b2} : 25 \text{ K} \sim 70 \text{ K} \qquad R_{b1} : 50 \text{ K} \sim 150 \text{ K} \qquad R_f : 0.5 \text{ K} \sim 3 \text{ K}$$

$$R_{c2} : 0.25 \text{ K} \sim 1.2 \text{ K} \qquad R_{c1} : 1.2 \text{ K} \sim 2.5 \text{ K} \qquad \beta : 50 \sim 300$$

In the original treatment by Chen et al. (1983), these six factors are used as control factors and also as noise factors. By examining Eq. (2) more closely, it is found that V_m depends on R_{b1} and R_{b2} through R_{b2}/R_{b1}. Therefore, in our approach, we will use the following five factors:

$$A = R_{b2}/R_{b1}, \qquad B = R_f, \qquad C = R_{c2}, \qquad D = R_{c1}, \qquad E = \beta$$

The levels of A–E as control factors are given in Table 9. The levels of B–E are the same as in Chen et al. (1983). The levels of A are chosen to be close to the levels of R_{b1} and R_{b2} in Chen et al. (1983, Table 1). The resistors, being of first grade, have 5% variations. The transistor β, being of third grade, has a 50% variation. This determines the choice of the levels of the noise factors A–E (Table 10).

The purpose of parameter design is to choose the nominal values of A–E such that the variation of V_m (as A–E vary from their nominal values) is small. Since the total number of combinations of control factor levels is too large ($5^5 = 3125$), we select 25 combinations according to the array

Table 9   Levels of Control Factors

			Factor		
Level	A	B	C	D	E
1	.215	649.38	237.14	1271.1	73
2	.316	865.96	316.23	1467.8	102
3	.464	1154.8	421.70	1695.0	143
4	.681	1539.9	562.34	1957.3	200
5	1.00	2053.5	749.89	2260.3	280

Table 10   Levels of Noise Factors

	Factor	
Level	A–D	E
1	0.95 of level 2	0.5 of level 2
2	From control array	From control array
3	1.05 of level 2	1.5 of level 2

$L_{25}(5^6)$ given in Table 13. This is the control array. For each combination in the control array, the levels of the factors A–E can vary by 5% or 50% of the nominal values (Table 10). The total number of possibilities $3^5 = 243$ is too large. Instead, we use the first five columns of the array $L_{18}(3^6)$ (see Appendix A) to represent the variations of factors around the nominal value chosen in the control array. Here $L_{18}(3^5)$ is the noise array. For each combination of levels in the noise array, V_m can be determined from Eq. (2). The mean squared error of the 18 V_m values

$$v = \frac{1}{18} \sum_1^{18} (V_m - 6)^2$$

is used as a stability measure of the setting of the control factors. One such set of v values is given in Table 13. Both arrays L_{25} and L_{18} are chosen in accordance with Chen et al. (1983).

Since each factor has five levels, SEL may be applied to eliminate one level at a time or, for reducing the number of iterations, to eliminate several levels at a time. Both SEL(mean) and SEL(mini) are considered. For each version, three sequential schemes are considered. See Table 11.

The six schemes are compared in a simulation study. The array $L_{25}(5^5)$ is obtained from randomly selecting 5 columns from the array $L_{25}(5^6)$ in Table 13; $L_{16}(4^5)$ from randomly selecting (i.e., permuting) 5 columns from the array in Table 14; $L_{18}(3^5)$ from randomly selecting 5 columns from the array $L_{18}(3^6)$ in Appendix A; and $L_{16}(2^5)$ from randomly permuting columns 1, 2, 4, 8, 15 of the array $L_{16}(2^{15})$ in Taguchi and Wu (1982, Appendix). In each round, the best (smallest) v value among the runs in the control array is recorded. The frequencies of the best v values (based on 500 simulations) are summarized in Table 12, the main points of which are listed below.

Table 11   Sequential Schemes Using SEL

Round	Scheme 1	Cumulative sample size	Scheme 2	Cumulative Sample Size	Scheme 3	Cumulative sample size
1	$L_{25}(5^5)$	25	$L_{25}(5^5)$	25	$L_{25}(5^5)$	25
2	$L_{16}(4^5)$	41	$L_{18}(3^5)$	43	$L_{16}(2^5)$	41
3	$L_{18}(3^5)$	59	$L_{16}(2^5)$	59		
4	$L_{16}(2^5)$	75				

1. Except for $L_{16}(4^5)$ (second round), SEL(mini) outperforms SEL (mean). The difference is more significant when there are fewer levels (i.e., in later rounds). Comparison of SEL(mean) and SEL(mini) will be addressed in the next section.
2. When SEL(mean) is used, for approximately the same cumulative sample size, scheme 1 is better than scheme 2, which is better than scheme 3. That is, schemes that eliminate fewer levels in each round perform better.
3. When SEL(mini) is used, the situation in 2 is reversed.
4. Design schemes with larger cumulative sample sizes do better.
5. From 1 to 3, sequential schemes that use SEL(mini) and eliminate more than one level in the early rounds are recommended. In addition to producing good results for the same sample size, they require fewer iterations. In practice, the latter may be as important a consideration as sample size reduction.

A limited study (omitted here) on the sensitivity of v to the choice of columns in the noise array $L_{18}(3^5)$ was conducted. It appears that there is little change in the v values and their relative order in the control array.

So far we have treated the mathematical model (2) as a black box for reasons given in Section 1. If this knowledge is exploited, we can use an $L_{25}(5^4)$ for the levels of four factors, say, A–D, and, for each level combination of A–D, we can choose the level of E so that $V_m(A, B, C, D, E)$ in Eq. (2) is close to the target 6. The resulting v values would be much smaller than those in Table 13.

In this section, we only consider the use of SEL on a *fixed* grid determined by a small number of levels. Better results can be obtained by

Table 12 Frequencies[a] of Best v Values from Designs Using SEL (Based on 500 Simulations)

Design		Scheme 1							Scheme 2							Scheme 3						
		I	II	III	IV	V	VI	VII	I	II	III	IV	V	VI	VII	I	II	III	IV	V	VI	VII
$L_{25}(5^5)$		46	48	130	151	95	20	10														
$L_{16}(4^5)$	SEL(mean)	97	87	163	107	35	5	6														
	SEL(mini)	67	74	167	155	33	4	0														
$L_{18}(3^5)$	SEL(mean)	142	114	169	59	15	1	0	80	84	171	114	41	7	3							
	SEL(mini)	137	138	171	51	3	0	0	90	125	195	79	11	0	0							
$L_{16}(2^5)$	SEL(mean)	185	164	130	17	3	1	0	114	173	159	39	14	1	0	59	78	154	135	59	15	0
	SEL(mini)	210	188	94	8	0	0	0	148	177	152	21	2	0	0	110	165	142	75	8	0	0

[a]Frequencies in the seven intervals: I = (.0148,.015], II = (.015,.016], III = (.016,.018], IV = (.018,.022], V = (.022,.03], VI = (.03,.04], VII = (.04,.0664].

Table 13   Layout of Control Factors and v Values

Run	A	B	C	D	E		v
1	4	2	1	3	1	2	.066
2	5	4	2	5	1	5	2.153
3	1	1	3	4	1	4	7.704
4	3	3	4	1	1	1	.471
5	2	5	5	2	1	3	1.326
6	3	4	1	4	2	3	.326
7	2	1	2	1	2	2	3.685
8	4	3	3	2	2	5	.085
9	5	5	4	3	2	4	2.917
10	1	2	5	5	2	1	8.005
11	5	1	1	2	3	1	1.164
12	1	3	2	3	3	3	6.177
13	3	5	3	5	3	2	.423
14	2	2	4	4	3	5	4.189
15	4	4	5	1	3	4	.260
16	2	3	1	5	4	4	3.560
17	4	5	2	4	4	1	.248
18	5	2	3	1	4	3	1.464
19	1	4	4	2	4	2	5.391
20	3	1	5	3	4	5	1.522
21	1	5	1	1	5	5	3.150
22	3	2	2	2	5	4	1.085
23	2	4	3	3	5	1	2.850
24	4	1	4	5	5	3	.076
25	5	3	5	4	5	2	1.254

Level			Mean v		
1	6.085	2.830	1.654	1.806	2.344
2	3.122	2.962	2.670	1.811	3.003
3	.765	2.310	2.505	2.706	2.443
4	.147	2.196	2.610	2.744	2.437
5	1.791	1.613	2.473	2.843	1.683

Level			Minimum v		
1	3.150	.076	.066	.260	.066
2	1.326	.066	.248	.085	.085
3	.326	.085	.085	.066	.260
4	.066	.260	.076	.248	.248
5	1.165	.248	.260	.076	.076

combining the use of SEL, adaptive choice of grid, and search over a broader range of factor levels. This will be addressed in Section 5.

4   SEL(MEAN) OR SEL(MINI)?

In order to understand why SEL(mini) does substantially better than SEL(mean) in the simulation study for Example B, we will study one of the simulations more closely. Its control array is given by the A–E columns of Table 13. The mean and minimum v values for the five levels of each factor are given at the bottom of the table. Using the analysis of marginal means (ANOMM), the levels (4,5,1,1,5) or (4,5,1,2,5) are chosen for factors A–E with their respective v values 0.7805 and 0.5341, both much bigger than the smallest v value 0.066 from the array L_{25} in Table 13. Again, ANOMM is worse than PW.

The application of SEL(mini) results in the elimination of the levels (1,4,5,1,3). The second-round design using an $L_{16}(4^5)$ is given in Table 14.

Table 14   Second-Round
Design Using SEL(mini)

Run	A	B	C	D	E	v
1	2	1	1	2	1	3.359
2	3	1	2	3	2	1.246
3	4	1	3	4	4	.051
4	5	1	4	5	5	.794
5	3	2	1	4	5	1.268
6	2	2	2	5	4	4.269
7	5	2	3	2	2	1.421
8	4	2	4	3	1	.018
9	4	3	1	5	2	.060
10	5	3	2	4	1	1.772
11	2	3	3	3	5	3.538
12	3	3	4	2	4	.800
13	5	5	1	3	4	3.011
14	4	5	2	2	5	.477
15	3	5	3	5	1	.226
16	2	5	4	4	2	1.990

Its best v value is .018. On the other hand, SEL(mean) eliminates the levels (1,2,2,5,2). The second-round design using the same $L_{16}(4^5)$ is given in Table 15 and has a bigger best v value .042. Applying SEL(mean) to the second-round design results in the elimination of level 3 of factor B. It turns out from a more extensive study that all combinations with low v values have level 2 or 3 for factor B. Therefore, SEL(mean) cannot give good results in this case.

So, where does SEL(mean) go wrong? Let us find out how level 2 of factor B is eliminated. The five values for $B = 2$ are

$$.066, 1.085, 1.464, 4.189, 8.00$$

including the smallest value, 0.066, of the first-round design. The mean 2.96 of the five values is influenced by the two large values 4.189 and 8.00. The small values play little role in pulling down the mean. The same reason explains why level 3 of B is eliminated after the second round. The

Table 15 Second-Round
Design Using SEL(mean)

Run	A	B	C	D	E	v
1	2	1	1	1	1	3.184
2	3	1	3	2	3	1.303
3	4	1	4	3	4	.042
4	5	1	5	4	5	.843
5	3	3	1	3	5	.851
6	2	3	3	4	4	3.666
7	5	3	4	1	3	1.947
8	4	3	5	2	1	.068
9	4	4	1	4	3	.156
10	5	4	3	3	1	2.416
11	2	4	4	2	5	2.598
12	3	4	5	1	4	.346
13	5	5	1	2	4	3.486
14	4	5	3	1	5	.664
15	3	5	4	4	1	.216
16	2	5	5	3	3	2.015

Table 16   SEL(mean) and SEL(mini)
Equally Good

Level	Low (unobserved)	Medium value	High
1	×	×	×
2	×	×	×

behavior of SEL(mini) is quite different. Level 2 of B is kept because its minimum value 0.066 is small.

Example B should not lead the readers to believe that SEL(mini) is always better than SEL(mean). Comparison of these methods will be discussed below using typical but hypothetical situations.

The sequential design problem may be viewed as a problem of predicting factor levels with low v values from the observed data. If a low (observed) value is a clue to where lower values can be found, SEL(mini) is a good method since the levels of this low value will be retained. Similarly, if a low mean (observed) value is a clue to where lower values can be found, SEL(mean) is good. One such situation, in which both SEL(mean) and SEL(mini) correctly select level 2, is described in Table 16. Here the values in the "low" category are not observed.

Table 17 typifies situations in which SEL(mini) outperforms SEL (mean). Based on the observed values in the "medium" and "high" categories, the level containing the lower low value is to be predicted. (For Table 17, it is level 2.) SEL(mini) correctly selects level 2 because the observed minimum value (in the "medium" category) at level 1 is higher. On the other hand, SEL(mean) incorrectly selects level 1 because the mean value at level 2, being influenced by the value in the "high" category, is

Table 17   SEL(mini) Better than
SEL(mean)

Level	Low (unobserved)	Medium value	High
1	×	×	×
2	×	×	×

Table 18 SEL(mean) Better than
SEL(mini)

Level	Low (unobserved)	Medium value	High	
1	×	×		×
2	×		×	×

larger. In Table 17, the data pattern in the "low" category is similar to that in the "medium" category but not to that in the "high" category. SEL(mini) bases its prediction on values in "medium", while SEL(mean) bases its prediction on mean values.

Let us now go back in Example A to see why SEL(mean) outperforms SEL(mini) there. The data patterns for Example A may be described as in Table 18. In this situation, SEL(mean) correctly selects level 2, while SEL(mini) incorrectly eliminates level 2. The reason is that the data pattern in "low" is similar to that in "high" and to the average of "medium" and "high" but not to that in "medium."

For the comparison of SEL(mean) and SEL(mini), one central question is which of the scenarios in Table 16–18 is more likely to occur. This needs further investigation. From our limited experience, including a simulation study of a more complex problem reported in Section 5, it appears that SEL(mini) is quite effective for situations in which substantially better values are yet to be found.

5   ITERATIVE SEARCH USING ORTHOGONAL ARRAYS

So far, we have considered only the use of orthogonal array and SEL over a *fixed* grid of factor levels. The technique can be made more powerful if applied iteratively to different regions of the factor space as determined by the results of the previous experiments. This will be illustrated with the following example taken from Zhang and Zheng (1983). It is closely related to the TV power circuit example of Taguchi and Wu (1982, Section 5.3) We do not consider the latter example since some of the figures reported there are not reproducible.

Example C

The function of a stabilizer circuit in a color TV is to convert a 100-V AC input into a 115-V DC output. The output voltage E_0 is determined by Eq. (3). The layout of the circuit can be found in Zhang and Zheng (1983) and closely resembles the one in Taguchi and Wu (1982, Figure 5.5). There are 13 factors that influence E_0,

$$z_1 \text{ to } z_3 \text{ and } z_5 \text{ to } z_{10} = \text{resistances}$$

$$z_{11} = \text{Zener voltage}, \qquad z_{13} = h_{\text{FE}} \text{ of } TR_3$$

$$z_{15} = h_{\text{FE}} \text{ of } TR_1, \qquad z_{17} = h_{\text{FE}} \text{ of } TR_2$$

The equation is

$$E_0 = \frac{136.67(a + b/z_9) + d(c + e)g/f - 1.2}{1 + (de)/f + b(0.006 + 1.08202/z_9) + 0.08202 \times a} \tag{3}$$

where

$$a = \frac{z_2}{z_1 + z_2}, \qquad b = \frac{1}{z_{15}z_{17}}\left(\frac{z_1 z_2}{z_1 + z_2} + z_3\right) + z_{10}$$

$$c = z_5 + \frac{1}{2}z_7, \qquad d = \frac{z_1 z_2}{z_1 + z_2}z_{13}, \qquad e = z_6 + \frac{1}{2}z_7$$

$$f = (c + e)(1 + z_{13})z_8 + ce, \qquad g = z_{11} + 0.6$$

(In Zhang and Zheng, 1983, the formula for d has a typographical error.)

    The objective of design is to choose the nominal values of z_i (treated as control factors) so that the output E_0 is stable around the target 115 V in the presence of the variations of z_i (treated as noise factors) around their nominal values. It is known that the nine resistors (z_1 to z_3, z_5 to z_{10}) have 10% variations and the four rectifier tubes (z_{11}, z_{13}, z_{15}, z_{17}) have 50% variations. For the noise array, we choose three levels for each noise factor, with the midlevel coming from the control array, and the high and low levels equal to 1.1 and 0.9 of the midlevel for z_1 to z_{10}, and equal to 1.5 and 0.5 of the midlevel for z_{11}, z_{13}, z_{15}, z_{17}. We use $L_{27}(3^{13})$ (in Appendix B) for the noise array, which is the same as in Zhang and Zheng (1983). For each setting of the control factors, 27 combinations of the noise factor levels are determined by $L_{27}(3^{13})$. For each of the 27 combinations, the E_0 value is computed by Eq. (3). We then use the mean squared error

$$v = \frac{1}{27} \sum_1^{27} (E_0 - 115)^2$$

as a measure of stability of the given setting of the control factors.
We rewrite the factors as

$$A = z_1, \quad B = z_2, \quad C = z_3, \quad D = z_5, \quad E = z_6,$$
$$F = z_7, \quad G = z_8, \quad H = z_9, \quad I = z_{10},$$
$$J = z_{11}, \quad K = z_{13}, \quad L = z_{15}, \quad M = z_{17}$$

The layout of the control factor levels is given in Table 19. We choose the three levels given in Table 20 for the initial design in any iterative scheme. Before describing the simulation study, we study two iterative schemes more closely. In scheme I of Table 21, the $L_{27}(3^{13})$ array (Appendix B) is used

Table 19 Layout of Factor Levels

Factor	51	52	53	54	55	56	Incremental[a] constant c
			Level number				
A	162	237	348	511	750	1100	10
B	147	215	316	464	681	1000	10
C				Same as B			10
D	562	825	1210	1780	2610	3830	10
E				Same as D			10
F				Same as B			10
G				Same as D			10
H				Same as D			10
I	1	1.1	1.21	1.33	1.47	1.62	1.78
J	12	12.455	12.927	13.416	13.925	14.452	1.25
K	110	114.168	118.494	122.984	127.644	132.480	1.25
L	45	46.705	48.475	50.312	52.218	54.196	1.25
M				Same as K			1.25

[a]The next six higher (lower) levels are obtained from multiplying (dividing) the six levels displayed in the table by c. Other levels are obtained by using the constant c^2, c^3, and so forth.

Table 20   Factor Levels of
Initial Design

Factor	Level		
	Low	Medium	High
A	57	63	69
B	64	70	76
C	50	56	62
D	57	63	69
E	51	57	63
F	53	59	65
G	39	45	51
H	41	47	53
I	45	51	57
J–M		Same as I	

for the control array at the initial stage. For the second and subsequent stages, the two combinations with the best v values among *all* the previous runs are selected. Two control arrays both using $L_{27}(3^{13})$ are chosen for the above two combinations. The midlevels of each array, denoted by m in Table 21, are equal to the levels of the corresponding combination. The low and high levels are chosen to be $m - i$ and $m + i$ for the $(7 - i)$th stage (see Table 21). The gap between levels is smaller for the later stages. The only change for scheme V is that, at each stage, after the experiment with $L_{27}(3^{13})$, SEL(mean) is used to eliminate one level of each factor and then, for the remaining two levels, 16 more runs are conducted according to the array $L_{16}(2^{13})$ (the first 13 columns of the $L_{16}(2^{15})$ array in Taguchi and Wu, 1982). The best two v values for each stage are given in Table 21. They are consistently better than the v value given by the analysis of marginal means. For each v value, we note the control array it belongs to. If the midlevels of this parental array give the best (respectively, second-best) v value among the previous runs, we say that this v value has "parent index" 1 (respectively, 2).

Three points are observed from Table 21.

1. Use of SEL in scheme V is effective. This is confirmed by the simulations reported later.

Table 21 Best v Values from Designs at Different Stages

Stage		Design	Cumulative sample size	Levels	Best two values		Parent index
I	1st	$L_{27}(3^{13})$	27	$(m-6, m, m+6)$	133.023,	179.074	
	2nd	2 × same	81	$(m-5, m, m+5)$	2.622,	5.954	2, 1
	3rd	2 × same	135	$(m-4, m, m+4)$	1.843,	1.890	1, 1
	4th	2 × same	189	$(m-3, m, m+3)$	1.037,	1.340	2, 1
	5th	2 × same	243	$(m-2, m, m+2)$	.857,	1.019	1, 1
	6th	2 × same	297	$(m-1, m, m+1)$	.833,	.847	1, 1
V	1st	$L_{27}(3^{13}) + L_{16}(2^{13})$	43	$(m-6, m, m+6)$	133.023,	179.074	
	2nd	2 × same	129	$(m-5, m, m+5)$	1.751,	2.622	2, 2
	3rd	2 × same	215	$(m-4, m, m+4)$	.845,	1.270	1, 1
	4th	2 × same	301	$(m-3, m, m+3)$	.759,	.776	1, 1
	5th	2 × same	387	$(m-2, m, m+2)$	.736,	.749	2, 1
	6th	2 × same	473	$(m-1, m, m+1)$	.684,	.690	2, 2

2. As v becomes smaller, some factor levels become more extreme. Consider, for example, the history of runs (Table 22) that lead to the smallest value .684 in scheme V. The levels of factors A, B, C move down while those of E, F, G, I, K move up.

3. For iterative search, it is good to keep two best combinations, not just one. Consider again the situation in Table 22. The first and the fifth generations that lead to the smallest value $v = .684$ are the second best in their respective generations. This practice is especially beneficial at the beginning stages when optimum values are yet to be found.

Five design schemes (see Table 23) are compared in the simulation study. Schemes I and V were described before. In Table 23, "$L_{27} + L_{16}$" means "$L_{27}(3^{13})$ followed by $L_{16}(2^{13})$ whose levels are chosen by SEL"; the multiplier 2 denotes the two best combinations chosen at each stage. Both SEL(mean) and SEL(mini) are considered. In the simulations, the columns of $L_{27}(3^{13})$ (respectively, $L_{16}(2^{13})$) are generated from random permutations of the 13 columns of the $L_{27}(3^{13})$ array in Appendix B (respectively, the first 13 columns of the array $L_{16}(2^{15})$ in Taguchi and Wu (1982, Appendix).

Based on 100 simulations, the percentiles of the 100 values of v are given in Table 24 for each design scheme. In the original treatment by Zhang and Zheng (1983), two iterative schemes are considered. One requires 676 runs and yields the best v value 1.54. The other requires 648 runs and yields the best v value 1.34. The percentages of simulations with $v < 1.34$ are also reported in the table.

Important points in Table 24 are summarized below.

1. The greatest gain comes from the use of SEL at the initial stage. Note the improvement of scheme II over I. Use of SEL at later stages (e.g., in schemes III to V) continues to give improvement, though of a smaller magnitude.

2. SEL(mini) outperforms SEL(mean). The difference is more significant at the beginning stages.

3. Designs using SEL such as schemes II to V are superior to the schemes considered by Zhang and Zheng (1983). For SEL(mini), it takes 183 runs (scheme III) to 215 runs (schemes IV and V) to have at least 50% of $v \leq 1.34$ in the simulations. For SEL(mean), it takes 205 runs (scheme II) to 301 runs (scheme V) to have at least

Table 22 History of Runs Leading to $v = .684$

Generation	Factor level													v	Parent index
	A	B	C	D	E	F	G	H	I	J	K	L	M		
1st	57	64	56	69	63	53	39	47	51	51	51	57	45	179.074	
2nd	52	59	56	74	68	53	44	52	56	56	51	52	40	1.751	2
3rd	48	55	52	70	68	57	48	48	60	56	55	56	40	.845	1
4th	45	52	49	67	68	60	51	45	63	56	58	59	40	.759	1
5th	47	54	47	69	68	58	51	45	65	54	56	61	42	.749	1
6th	46	53	46	68	68	59	52	44	66	54	57	62	42	.684	2

Table 23 Design Schemes and Their Cumulative Sample Sizes (C.S.S.)

Stage	I	C.S.S.	II	C.S.S.	III	C.S.S.	IV	C.S.S.	V	C.S.S.
1st	L_{27}	27	$L_{27} + L_{16}$	43	Same as II		Same as III		Same as IV	
2nd	$2 \times L_{27}$	81	$2 \times L_{27}$	97	$2 \times (L_{27} + L_{16})$	129				
3rd	$2 \times L_{27}$	135	$2 \times L_{27}$	151	$2 \times L_{27}$	183	$2 \times (L_{27} + L_{16})$	215		
4th	$2 \times L_{27}$	189	$2 \times L_{27}$	205	$2 \times L_{27}$	237	$2 \times L_{27}$	269	$2 \times (L_{27} + L_{16})$	301
5th	$2 \times L_{27}$	243	$2 \times L_{27}$	259	$2 \times L_{27}$	291	$2 \times L_{27}$	323	$2 \times (L_{27} + L_{16})$	387
6th	$2 \times L_{27}$	297	$2 \times L_{27}$	313	$2 \times L_{27}$	345	$2 \times L_{27}$	377	$2 \times (L_{27} + L_{16})$	473

Design scheme

Table 24 Selected Percentiles (5th, 25th, 50th, 75th, 95th) of Simulation Distributions and Percentages (in Parenthesis) of Simulations with $v \leq 1.34$ for Nine Schemes (Based on 100 Simulations)

Design scheme (SEL(mini) for $L_{27} + L_{16}$)

Stage	I	II	III	IV	V
1st	418, 470, 514, 583, 834	19, 38, 63, 81, 158	.82, 1.1, 2.3, 4.5, 8.4		
2nd	20.0, 29.9, 38.4, 51.9, 69.7	.88, 1.8, 3.8, 6.6, 19.5	.77, .93, 1.2, 2.8, 5.7 (54%)	.63, .86, 1.0, 2.5, 5.3 (64%)	.56, .78, .92, 1.4, 3.7 (74%)
3rd	2.4, 4.0, 6.3, 8.0, 17.3 (0%)	.79, 1.00, 1.4, 3.1, 7.8 (44%)	.60, .83, .97, 1.5, 4.4 (72%)	.58, .80, .93, 1.5, 4.5 (74%)	.53, .74, .86, 1.1, 3.1 (81%)
4th	1.3, 2.1, 3.3, 4.7, 13.8 (7%)	.70, .85, 1.1, 1.7, 4.7 (69%)	.60, .78, .93, 1.3, 3.8 (75%)	.55, .78, .90, 1.2, 3.1 (78%)	.50, .68, .81, .98, 2.0 (87%)
5th	1.1, 1.5, 2.2, 3.5, 12.8 (23%)	.61, .80, .96, 1.2, 3.6, (80%)	.55, .76, .88, 1.0, 2.7 (85%)	.50, .75, .85, .99, 2.5 (87%)	
6th	1.0, 1.2, 1.9, 2.8, 12.2 (27%)	.55, .77, .92, 1.1, 2.8 (84%)			

Design scheme (SEL(mean) for $L_{27} + L_{16}$)

Stage	II	III	IV	V
1st	26, 60, 90, 133, 275	.84, 1.5, 3.5, 4.8, 9.3		
2nd	.88, 2.1, 4.4, 8.0, 16.9	.75, .97, 1.8, 3.5, 4.8 (40%)	.73, .91, 1.5, 3.1, 4.6 (44%)	.55, .81, 1.1, 2.2, 4.2 (60%)
3rd	.81, 1.0, 1.9, 3.5, 5.7 (37%)	.67, .83, 1.1, 2.0, 4.0 (58%)	.59, .82, 1.1, 2.3, 4.2 (59%)	.53, .80, .94, 1.5, 3.5 (71%)
4th	.77, .88, 1.2, 1.9, 4.0 (63%)	.61, .80, .98, 1.5, 3.3 (70%)	.53, .80, .96, 1.6, 3.5 (68%)	.44, .76, .88, 1.3, 2.6 (79%)
5th	.61, .81, 1.0, 1.4, 3.2 (73%)	.53, .78, .91, 1.3, 2.8 (78%)	.53, .79, .90, 1.3, 2.7 (77%)	
6th	.56, .79, .93, 1.2, 2.7 (82%)			

50% of $v \leq 1.34$ in the simulations. Recall that 1.34 is the best v value out of 648 runs in Zhang and Zheng's study.

4. Our overall recommendation is schemes II and III with SEL(mini).

A more limited study of an iterative search procedure for Example B also confirms the above conclusions. To save space, the results are not given.

6   CONCLUDING REMARKS

Sequential elimination of levels (SEL) is a method that enables experimenters to focus quickly on regions of promise by eliminating factor levels that give poor responses. Its general utility is supported by our study of the three examples. When the current guess is far from optimum and when search over a wide grid is desired, SEL(mini) is especially effective in accelerating the approach to optimum. It is more effective when used at the beginning stages of the experimentation. To reduce the number of iterations and the size of arrays at subsequent stages, several levels may be eliminated at one time. $\text{SEL}(x)$, with x other than mean or minimum, may be promising and deserves further study.

We do not intend to suggest that the "analysis of marginal means" (ANOMM) should be totally rejected. Its simplicity appeals to users, especially when presented graphically (see, for example, the many case studies reported at the Supplier Symposiums on "Taguchi Method" organized by the American Supplier Institute). However, the method should be used with prudence. Interactions should be considered and confounding patterns be noted. The setting determined by ANOMM is often inappropriate as shown in our study. Further experimentation may be conducted in its neighborhood in the hope of getting better settings. Confirmatory runs should be based on information from the analysis of experimental data as well as from engineering judgment. SEL provides a convenient way of doing this by retaining only one to three levels (for each factor) around the setting chosen by ANOMM.

Perhaps the most striking feature of SEL is its deviation from traditional design strategies based on the concept of averaging (e.g., main effects and interactions are computed as signed averages of observations) and on empirical model building (e.g., response surface methodology, Box and Draper, 1986). When the response surface is too rugged or when search over a wide grid is desired, an empirical model that fits the data well may not be useful for predicting where better settings can be found. SEL works by eliminating poor factor levels and focusing on the neighborhood of the best current settings for iterations. Comparisons of these strategies deserve further investigation.

Other possibilities should also be considered. For example, all the iterative schemes in Section 5 select the best two combinations for the next iteration. Other methods for iterations such as directional search and the simplex method (Nelder and Mead, 1964) may be compared or incorporated. For mathematical models, there are better ways for choosing the control factor levels than the strict use of orthogonal arrays. See, for example, the discussion at the end of Section 3.

ACKNOWLEDGMENTS

This research was completed at the University of Wisconsin–Madison with the support of the National Science Foundation Grant DMS-8420968.

Appendix A $L_{18}(3^6)$

Run	I	II	III	IV	V	VI
1	1	1	3	2	2	1
2	2	1	1	1	1	2
3	3	1	2	3	3	3
4	1	2	2	1	2	3
5	2	2	3	3	1	1
6	3	2	1	2	3	2
7	1	3	1	3	1	3
8	2	3	2	2	3	1
9	3	3	3	1	2	2
10	1	1	1	1	3	1
11	2	1	2	3	2	2
12	3	1	3	2	1	3
13	1	2	3	3	3	2
14	2	2	1	2	2	3
15	3	2	2	1	1	1
16	1	3	2	2	1	2
17	2	3	3	1	3	3
18	3	3	1	3	2	1

Appendix B $L_{27}(3^{13})$

Level													

						Factor							
Level	A	B	C	D	E	F	G	H	I	J	K	L	M
1	1	1	3	2	1	2	2	3	1	2	1	3	3
2	2	1	1	1	1	1	3	3	2	1	1	2	1
3	3	1	2	3	1	3	1	3	3	3	1	1	2
4	1	2	2	1	1	2	2	2	3	1	3	1	1
5	2	2	3	3	1	1	3	2	1	3	3	3	2
6	3	2	1	2	1	3	1	2	2	2	3	2	3
7	1	3	1	3	1	2	2	1	2	3	2	2	2
8	2	3	2	2	1	1	3	1	3	2	2	3	3
9	3	3	3	1	1	3	1	1	1	1	2	1	1
10	1	1	1	1	2	3	3	1	3	2	3	3	2
11	2	1	2	3	2	2	1	1	1	1	3	2	3
12	3	1	3	2	2	1	2	1	2	3	3	1	1
13	1	2	3	3	2	3	3	3	2	1	2	1	3
14	2	2	1	2	2	2	1	3	3	3	2	3	1
15	3	2	2	1	2	1	2	3	1	2	2	2	2
16˙	1	3	2	2	2	3	3	2	1	3	1	2	1
17	2	3	3	1	2	2	1	2	2	2	1	1	2
18	3	3	1	3	2	1	2	2	3	1	1	3	3
19	1	1	2	3	3	1	1	2	2	2	2	3	1
20	2	1	3	2	3	3	2	2	3	1	2	2	2
21	3	1	1	1	3	2	3	2	1	3	2	1	3
22	1	2	1	2	3	1	1	1	1	1	1	1	2
23	2	2	2	1	3	3	2	1	2	3	1	3	3
24	3	2	3	3	3	2	3	1	3	2	1	2	1
25	1	3	3	1	3	1	1	3	3	3	3	2	3
26	2	3	1	3	3	3	2	3	1	2	3	1	1
27	3	3	2	2	3	2	3	3	2	1	3	3	2

REFERENCES

Box, G. E. P. and Draper, N. R. (1986). *Empirical Model Building and Response Surfaces*, Wiley, New York.

Chen, G. Y., Wang, D. Q., Jian, J. B., and Zhang, L. T. (1983). Mid-point Voltage of an OTL Circuit, *Three-Stage Design of Experiments with Known Transfer Functions* (edited by the Committee on Three-Stage Design, Chinese Applied Statistics Society), pp. 63–74 (in Chinese).

Cui, S. Z., Fan, Y. M., and Zhang, J. (1983). Outlet Temperature of a Heat Exchanger, *Three-Stage Design of Experiments with Known Transfer Functions* (edited by the Committee on Three-Stage Design, Chinese Applied Statistics Society), pp. 81–85 (in Chinese).

Leon, R., Shoemaker, A. C., and Kacker, R. N. (1987). Performance Measures Independent of Adjustment: An Explanation and Extension of Taguchi's Signal-to-Noise Ratios, *Technometrics*, 29: 253–285.

Leon, R. and Wu, C. F. J. (1989). A Theory of Performance Measures in Parameter Design, manuscript.

Nelder, J. A. and Mead, R. (1964). A Simplex Method for Function Minimization, *Computer Journal*, 7: 308–313.

Taguchi, G. (1986). *Introduction to Quality Engineering*, Asian Productivity Organization, Tokyo.

Taguchi, G. and Wu, Y. I. (1982). *Introduction to Off-Line Quality Control*, Central Japan Quality Association, Tokyo.

Wu, C. F. J., Mao, S. S., and Ma, F. S. (1987). An Investigation of OA-based Methods for Parameter Design Optimization, Report No. 24, Center for Quality and Productivity Improvement, University of Wisconsin–Madison.

Zhang, L. T. and Zheng, Y. Y. (1983), Voltage Stabilizer Circuit, *Three-Stage Design of Experiments with Known Transfer Functions* (edited by the Committee on Three-Stage Design, Chinese Applied Statistics Society), pp. 92–103 (in Chinese).

12

Modern Factorial Design Theory for Experimenters and Statisticians

Jagdish N. Srivastava   Colorado State University, Fort Collins, Colorado

1   ELEMENTS OF MATRIX THEORY

A rectangular array of numbers is called a *matrix*. If the matrix has m rows and n columns, then it is said to be a size $(m \times n)$. Thus, the matrix A below is of size (3×5):

$$A = \begin{bmatrix} 0 & 1 & 0 & 0 & 1 \\ 1 & 1 & 0 & 1 & 1 \\ 0 & 0 & 0 & 1 & 0 \end{bmatrix} \tag{1}$$

The above matrix is called a $(0,1)$ matrix since its elements are 0 and 1 only. The (i,j) cell of a matrix is the position in the matrix corresponding to the ith row and the jth column. Thus, for example, the above matrix A has no $(3,6)$ cell, but it does have a $(2,5)$ cell.

Two matrices can be added if and only if they are of the same size; if both are of size $(m \times n)$, then their sum (say, C) of size $(m \times n)$ is such

311

that if the (i, j) cell of A, B, and C, respectively, contains the numbers a_{ij}, b_{ij}, c_{ij}, then we have

$$c_{ij} = a_{ij} + b_{ij}, \qquad \text{for } i = 1, \ldots, m; \quad j = 1, \ldots, n \tag{2}$$

If C is the sum of A and B, then we write

$$C = A + B = B + A \tag{3}$$

Now, consider two matrices $G(m \times n)$ and $H(n \times p)$. Notice that we have chosen G and H in such a way that the number of columns in G is equal to the number rows in H; the product GH of the matrices G and H is defined if and only if this is the case. Notice that the order in which we write G and H in the product does matter; indeed, if $p \neq m$, then the product HG does not even exist. To define the product GH, let

$$G = \begin{bmatrix} g_{11} & g_{12} & \cdots & g_{1n} \\ g_{21} & g_{22} & \cdots & g_{2n} \\ \cdot & \cdot & \cdots & \cdot \\ g_{i1} & g_{i2} & \cdots & g_{in} \\ \cdot & \cdot & \cdots & \cdot \\ g_{m1} & g_{m2} & \cdots & g_{mn} \end{bmatrix},$$

$$H = \begin{bmatrix} h_{11} & h_{12} & \cdots & h_{1j} & \cdots & h_{1p} \\ h_{21} & h_{22} & \cdots & h_{2j} & \cdots & h_{2p} \\ \cdot & \cdot & \cdots & & \cdots & \cdot \\ h_{n1} & h_{n2} & \cdots & h_{nj} & \cdots & h_{np} \end{bmatrix} \tag{4}$$

The product GH of these matrices is then given by

$$GH = K \tag{5}$$

where K is an $(m \times p)$ matrix, which has the element k_{ij} in the cell (i, j), where $(i = 1, 2, \ldots, m)$ and $(j = 1, \ldots, p)$, and where

$$k_{ij} = g_{i1}h_{1j} + g_{i2}h_{2j} + \cdots + g_{in}h_{nj} \tag{6}$$

In other words, the element k_{ij} in the (i, j) cell of K is obtained by taking the elements in the ith row of G and the jth column of H and multiplying this row and column together elementwise and adding.

Consider now matrices $E(m \times n)$ and $F(n \times m)$. Notice that, in this case, both products EF and FE exist, but these two matrices are of different sizes. Also, even if $m = n$, we do not necessarily have $EF = FE$. (If $EF = FE$, then we say that E and F commute.)

The matrix which has only one row or only one column is called, respectively, a *row vector* or a *column vector*. Throughout this chapter, we shall express vectors as lowercase boldface letters. Matrices (and not vectors) will be denoted by capital letters. Matrices of size (1×1) are called scalars; these are treated as ordinary numbers and will be denoted by lowercase italic letters. Consider an $(m \times n)$ matrix E, and suppose that the element in the (i, j) cell of E is e_{ij}. Then, the $(n \times m)$ matrix whose (i, j) cell contains the element e_{ji} is called the *transpose* of E, and will be written E'. Thus, if A is given by (1), and if $\mathbf{x}' = (x_1, \ldots, x_n)$, then we have

$$
A' = \begin{bmatrix} 0 & 1 & 0 \\ 1 & 1 & 0 \\ 0 & 0 & 0 \\ 0 & 1 & 1 \\ 1 & 1 & 0 \end{bmatrix}, \qquad \mathbf{x} = \begin{bmatrix} x_1 \\ \vdots \\ x_n \end{bmatrix} \tag{7}
$$

If A is a square matrix such that $A = A'$, then A is said to be *symmetric*.

We now consider the important concept of a *determinant*, which is defined only for a square matrix. The determinant of the matrix A is denoted by $|A|$. If

$$
A_1 = [a_{11}], \qquad A_2 = \begin{bmatrix} a_{11} & a_{12} \\ a_{21} & a_{22} \end{bmatrix} \tag{8}
$$

then

$$
|A_1| = a_{11}, \qquad |A_2| = a_{11}a_{22} - a_{12}a_{21} \tag{9}
$$

We shall define the determinant of matrices of higher size by induction. Thus, suppose we know how to compute the determinant of every matrix of size $(n-1) \times (n-1)$. Let A be the $(n \times n)$ matrix as displayed below.

$$
A = \begin{bmatrix} a_{11} & a_{12} & \cdots & a_{1n} \\ a_{21} & a_{22} & \cdots & a_{2n} \\ \cdot & \cdot & \cdots & \cdot \\ a_{n1} & a_{n2} & \cdots & a_{nn} \end{bmatrix} \tag{10}
$$

Notice that the element in the (i, j) cell of A is a_{ij}. Let A_{ij} be the $(n-1) \times (n-1)$ submatrix of A obtained by deleting the ith row and the jth column of A; then A_{ij} is called the cofactor of a_{ij}. By assumption, we know how to compute the determinant $|A_{ij}|$. Then we have

$$
\begin{aligned}
|A| &= (-1)^{i+1} a_{i1} |A_{i1}| + (-1)^{i+2} a_{i2} |A_{i2}| + \cdots + (-1)^{i+j} a_{ij} |A_{ij}| + \cdots \\
&\quad + (-1)^{i+n} a_{in} |A_{in}| \\
&= (-1)^{j+1} a_{ij} |A_{1j}| + (-1)^{j+2} a_{2j} |A_{2j}| + \cdots + (-1)^{j+i} a_{ij} |A_{ij}| + \cdots \\
&\quad + (-1)^{j+n} a_{nj} |A_{nj}|
\end{aligned}
\tag{11}
$$

The first line of (11) gives the "expansion" of $|A|$ in terms of the ith row of A, whereas the second does the same in terms of the jth column of A. It can be proved that the determinant of A will come out to be the same, irrespective of which column or which row is used in (11) for the expansion. To exemplify, we have

$$
\begin{vmatrix} 1 & 0 & -3 & 8 \\ 0 & 4 & 0 & 7 \\ -6 & 3 & 3 & -4 \\ 0 & 1 & 0 & 5 \end{vmatrix} = (-1)^{2+2} \cdot 4 \begin{vmatrix} 1 & -3 & 8 \\ -6 & 3 & -4 \\ 0 & 0 & 5 \end{vmatrix}
$$

$$
+ (-1)^{2+4} \cdot 7 \begin{vmatrix} 1 & 0 & -3 \\ -6 & 3 & 3 \\ 0 & 1 & 0 \end{vmatrix}
\tag{12}
$$

$$
= (4)(-1)^{3+3} \cdot 5 \begin{vmatrix} 1 & -3 \\ -6 & 3 \end{vmatrix} + (7)(-1)^{3+2} \begin{vmatrix} 1 & -3 \\ -6 & 3 \end{vmatrix}
$$

$$
= 20(3 - 18) - 7(3 - 18) = 13 \times (-15) = -195
$$

Let c be a scalar and A be a matrix. Then, the product cA is the matrix obtained by multiplying each element of A by c.

Let $A(m \times n)$ and $B(p \times q)$ be any two matrices. Then, the left Kronecker product of A and B (written $A \otimes B$) is defined by

$$
A \otimes B = \begin{bmatrix} a_{11} B & a_{12} B & \cdots & a_{1n} B \\ \cdots & \cdots & \cdots & \cdots \\ a_{m1} B & a_{m2} B & \cdots & a_{mn} B \end{bmatrix}
\tag{13}
$$

where

$$A = \begin{bmatrix} a_{11} & \cdots & a_{1n} \\ \cdot & \cdots & \cdot \\ a_{m1} & \cdots & a_{mn} \end{bmatrix} \tag{14}$$

The matrix on the right-hand side (r.h.s.) of (13) is expressed as a set of mn submatrices, these being $a_{11}B$, $a_{12}B$, ..., $a_{mn}B$. Thus, the size of $A \otimes B$ is $(mp \times nq)$. To exemplify, let

$$A = \begin{bmatrix} 2 & -1 & 3 \\ 0 & 4 & 5 \end{bmatrix}, \qquad B = \begin{bmatrix} 1 & -4 \\ 6 & 0 \end{bmatrix} \tag{15}$$

Then,

$$A \otimes B = \begin{bmatrix} 2 & -8 & -1 & 4 & 3 & -12 \\ 12 & 0 & -6 & 0 & 18 & 0 \\ 0 & 0 & 4 & -16 & 5 & -20 \\ 0 & 0 & 24 & 0 & 30 & 0 \end{bmatrix}$$

Let H be a square matrix whose elements are 1 and (-1) such that the sum of products of the elements in any two distinct rows of H is zero; then H is called a "Hadamard matrix." An example is the (2×2) matrix H_1 given by

$$H_1 = \begin{bmatrix} + & + \\ + & - \end{bmatrix} \tag{16}$$

where, throughout this chapter

$$\text{``+'' means } (+1) \text{ and ``} - \text{'' means } (-1) \tag{17}$$

(The symbol 1 is omitted for brevity of presentation.) Let H_m be the $(2^m \times 2^m)$ matrix obtained by taking the Kronecker product of H_1 with itself m times. Then it can be checked that H_m is a Hadamard matrix. In particular, we have

$$H_2 = H_1 \otimes H_1 = \begin{bmatrix} + & + & + & + \\ + & - & + & - \\ + & + & - & - \\ + & - & - & + \end{bmatrix} \tag{18}$$

$$H_3 = H_1 \otimes H_1 \otimes H_1 = H_1 \otimes H_2 = \begin{bmatrix} + & + & + & + & + & + & + & + \\ + & - & + & - & + & - & + & - \\ + & + & - & - & + & + & - & - \\ + & - & - & + & + & - & - & + \\ + & + & + & + & - & - & - & - \\ + & - & + & - & - & + & - & + \\ + & + & - & - & - & - & + & + \\ + & - & - & + & - & + & + & - \end{bmatrix}$$

Let $(a_1, a_2, \ldots, a_n)$ and $(x_1, x_2, \ldots, x_n)$ be two vectors each with n elements. Then the "product" of these row vectors is defined to be equal to $(a_1 x_1 + a_2 x_2 + \cdots + a_n x_n)$. The expression $(a_1 x_1 + \cdots + a_n x_n)$ is said to be a linear function of x_1, x_2, ..., x_n. Similarly, it is also a linear function of a_1, a_2, ..., a_n. If p and q are scalars, then $\{p(a_1, a_2, \ldots, a_n) + q(x_1, x_2, \ldots, x_n)\}$ is said to be a linear combination of the vectors $(a_1, \ldots, a_n)$, and $(x_1, \ldots, x_n)$. The linear combinations of more than two vectors are similarly defined. Let n be any positive integer. Then the $(n \times n)$ matrix whose diagonal elements equal 1 and whose off-diagonal elements equal 0 is said to be the "identity matrix" of size $(n \times n)$ and is denoted by I_n. Also, J_{mn} and 0_{mn} will denote an $(m \times n)$ all of whose elements equal 1 or 0, respectively. Sometimes, when the sizes of the matrices are clear from the context, the suffixes n and m may be dropped.

Let A be a square matrix, such that $|A| \neq 0$. Then, A is said to be nonsingular. In this case, it can be proved that there exists a matrix $B(n \times n)$ such that

$$AB = BA = I_n \tag{19}$$

If A is nonsingular and (19) holds, then B is said to be the inverse of A, and we write

$$B = A^{-1} \tag{20}$$

If (20) holds, and if

$$A = \begin{bmatrix} a_{11} & \cdots & a_{1n} \\ \cdot & \cdots & \cdot \\ a_{n1} & \cdots & a_{nn} \end{bmatrix}, \qquad B = \begin{bmatrix} b_{11} & \cdots & b_{1n} \\ \cdot & \cdots & \cdot \\ b_{n1} & \cdots & b_{nn} \end{bmatrix} \tag{21}$$

then we have

$$b_{ij} = \frac{(-1)^{i+j}|A_{ji}|}{|A|} \tag{22}$$

where A_{ji} is the cofactor of a_{ji}.

In (23) below, we give an example:

$$A = \begin{bmatrix} 1 & 0 & 0 \\ 2 & -3 & 4 \\ 0 & 1 & 6 \end{bmatrix}, \quad |A| = (1)\begin{vmatrix} -3 & 4 \\ 1 & 6 \end{vmatrix} = (-22) \tag{23}$$

$$b_{11} = \frac{1}{(-22)}(-1)^{1+1}\begin{vmatrix} -3 & 4 \\ 1 & 6 \end{vmatrix} = 1, \quad b_{12} = \frac{1}{(-22)}(-1)^{1+2}\begin{vmatrix} 0 & 0 \\ 1 & 6 \end{vmatrix} = 0$$

$$b_{23} = \frac{1}{(-22)}(-1)^{2+3}\begin{vmatrix} 1 & 0 \\ 2 & 4 \end{vmatrix} = \frac{4}{22}, \quad \text{etc.}$$

Now, consider m vectors $\mathbf{x}_i$ ($i = 1, 2, \ldots, m$), where $\mathbf{x}_i' = (x_{i1}, x_{i2}, \ldots, x_{in})$. Also, let $\mathbf{y}' = [\equiv (y_1, y_2, \ldots, y_n)]$ be any vector. Then, we say that $\mathbf{y}$ is a linear combination of $\mathbf{x}_1$, $\mathbf{x}_2$, $\ldots$, $\mathbf{x}_m$, if and only if we have

$$\mathbf{y} = \lambda_1 \mathbf{x}_1 + \lambda_2 \mathbf{x}_2 + \cdots + \lambda_m \mathbf{x}_m \tag{24}$$

where λ_1, λ_2, $\ldots$, λ_m are some scalars. In case the relationship (24) is satisfied for any scalars λ, then $\mathbf{y}$ is said to be dependent on $\mathbf{x}_1$, $\mathbf{x}_2$, $\ldots$, $\mathbf{x}_m$; otherwise, $\mathbf{y}$ is said to be independent of $\mathbf{x}_1$, $\mathbf{x}_2$, $\ldots$, $\mathbf{x}_m$.

Let B be any $(n \times n)$ matrix and let $\mathbf{y}$ be any $(n \times 1)$ vector. Then the product $\mathbf{y}'B\mathbf{y}$ is a scalar and is said to be a quadratic form in the y's. If the value of $\mathbf{y}'B\mathbf{y}$ is nonnegative for all values of $\mathbf{y}$, then we say that the matrix B is positive semidefinite. If, on the other hand, we have $\mathbf{y}'B\mathbf{y} > 0$, for all $\mathbf{y} \neq \mathbf{0}$ (the zero vector), then the matrix B is said to be positive definite.

Let λ be a symbol which denotes a variable scalar. Then the matrix $[B - \lambda I]$ equals the matrix B with λ subtracted along the diagonal. The determinant $|B - \lambda I|$ is then a polynomial of degree n in λ. This polynomial is said to be the characteristic polynomial of B. The roots of the equation

$$|B - \lambda I| = 0 \tag{25}$$

are called the characteristic roots of B. It can be shown that if B is a symmetric positive definite matrix, then its roots are all positive, and if B is positive semidefinite, then its roots are all nonnegative.

Let $R(m \times n)$ be any matrix. Let k ($\leq m, n$) be the largest integer such that there is a $(k \times k)$ submatrix R_1 inside R such that $|R_1| \neq 0$; then R is said to have "rank" k. It can be shown that if R has rank k, then R has exactly k linearly independent rows, and also exactly k linearly independent columns.

2 FACTORIAL EXPERIMENTS

For ease of explanation, we start with an illustration. Consider an agricultural experiment in which we want to study the effect of three fertilizers [say, nitrogen (N), phosphorus (P), and potassium (K)], on the yield (say, y) of wheat. In this situation, y is said to be a "response" and the fertilizers N, P, and K are said to be the "factors." In general, a response is that variable which we are measuring on different units, and on which the effect of the factors is being studied. The factors are the variables which are independent and which we may control and whose effects we want to study on the responses. The experiment is said to be uniresponse (or univariate) if there is only one response; otherwise, the experiment is said to be multiresponse or multivariate.

A *factorial experiment* is an experiment in which we wish to study simultaneously the effect of several factors. Although, the case of a single factor is not commonly considered to be a topic under factorial experiments, it is necessary nevertheless to understand such a case.

Usually, factors are of three kinds: continuous (such as the amount of a drug to be given to a patient), discrete unordered (like three varieties of wheat), and discrete ordered (like the number of irrigations).

In this chapter, we shall restrict our attention to discrete factors, both ordered, and unordered. (The situation in which the factor is continuous—and we do not make it discrete—comes under the so-called theory of response surfaces, which is indeed the theory of factorial experiments when factors are continuous.)

Consider now a discrete factor at l levels. Usually, for convenience, these levels are denoted by the integers $0, 1, 2, \ldots (l-1)$. In general, any level can be represented by any of these integers. In order to study the effect of this factor on the response y, clearly, all one can do is to compare the value of y at different levels of the factor. For example, when $l = 2$, we may consider the difference of the values of y at the two levels.

Before we go further into such comparisons, it is important here to point out that in most experimentation, the result (i.e., the observed value of y) is subject to chance fluctuations. In other words, the value of y is influenced not only by the factors under study, but other factors that are not under study. These other factors are usually called "nuisance factors." If there is a nuisance factor whose levels are such that we can classify the experimental material according to the levels of this nuisance factor, then the effect of such a nuisance factor can be eliminated by using the so-called theory of block designs. Thus, every nuisance factor gives rise to a block system. If there are several nuisance factors, then there can be several cross-classified block systems.

Apart from the nuisance factors whose levels can be easily identified with respect to the experimental material, there are a host of other nuisance factors which are usually quite small, and which are such that the experimental material cannot be easily classified with respect to these nuisance factors. The combined effect of all these nuisance factors then causes an unknown amount of fluctuation in the value of y, and is measured as the variance of y.

Now, let $\mathbf{t}$ denote a level combination (i.e., a combination of levels of the different factors). For example, in the illustration that we began with, suppose that N has four levels (say, 0, 20, 40, 60 lb/acre), P has three levels (0, 30, 60 lb/acre), and K has two levels (0, 40 lb/acre). Without loss of generality, the four levels of N may be denoted by $\{0, 1, 2, 3\}$. Similarly, those of P and K may be denoted respectively by $\{0, 1, 2\}$ and $\{0, 1\}$. Thus, here $\mathbf{t}$ may have $4 \times 3 \times 2 = 24$ different values, which we may denote by (t_1, t_2, t_3), where $t_1 = 0, 1, 2, 3$; $t_2 = 0, 1, 2$; and $t_3 = 0, 1$. For example, if $\mathbf{t} = (3, 0, 1)$, then we have the level combination in which N, P, K are respectively at 60, 0, and 40 lb/acre. In an obvious notation, the above experiment is called a "$4 \times 3 \times 2$ factorial experiment." In general, an $s_1 \times \cdots \times s_m$ experiment is one in which we have m factors, the ith factor $(i = 1, \ldots, m)$ having s_i levels (which may be denoted by $\{0, 1, \ldots, s_i - 1\}$). If all the s_i are equal, then the experiment is said to be symmetrical; otherwise it is asymmetrical.

Throughout this chapter, the terms "level combination," "run," "treatment," and "assembly" will be used synonymously (as is the popular practice).

Consider an ideal condition in which there are no random fluctuations whatsoever. Suppose that the yield (under such ideal conditions) for a given combination $\mathbf{t}$ is denoted by $\phi(\mathbf{t})$. In actual practice, we shall not observe $\phi(t)$; suppose the "observed yield" is $y(t)$. Then, we can write

$$y(t) = \phi(t) + \epsilon(t) \tag{26}$$

where $\epsilon(t)$ denotes the total effect of the random fluctuations on the yield in the unit on which t was applied. In this chapter, we shall not go too much into the statistical theory dealing with $\epsilon(t)$, except that we shall always assume

$$E[\epsilon(t)] = 0, \qquad V[\epsilon(t)] = \sigma^2 \tag{27}$$

where E denotes expectation, and V denotes variance, and where σ^2 is a (usually unknown) constant.

In this chapter, the quantity $\phi(t)$ will usually be called the "true yield" or the "signal," and $\epsilon(t)$ the "noise."

Consider now a factor, say at l levels, and let $\phi(f_i)$ denote the true yield at the level f_i of this factor. Then, if we wish to compare two levels f_i and f_j, it is proper to consider estimating the value of $\{\phi(f_i) - \phi(f_j)\}$. In other words, given any two levels, to compare them, we merely consider the difference in the values of ϕ at the two levels. If there are more than two levels, then any two levels may be compared in this manner. It can be shown that if there are l levels, then we can have at most $(l-1)$ linearly independent comparisons. For example, consider the case where $l = 3$, and consider the comparisons $(\theta_0 - \theta_1)$, $(\theta_0 - \theta_2)$, $(\theta_2 - \theta_0)$, where $\theta_i = \phi(f_i)$, for $(i = 0, 1, 2,)$. Although each of these three comparisons is meaningful, these comparisons are not linearly independent because we can write

$$(\theta_2 - \theta_0) = -(\theta_1 - \theta_2) - (\theta_0 - \theta_1) \tag{28}$$

which shows that $(\theta_2 - \theta_0)$ is a linear function of $(\theta_1 - \theta_2)$ and $(\theta_0 - \theta_1)$, so that, given $(\theta_0 - \theta_1)$ and $(\theta_1 - \theta_2)$, we can determine $(\theta_2 - \theta_0)$.

Let $\boldsymbol{\theta}' = (\theta_0, \theta_1, \ldots, \theta_{l-1})$, and $\mathbf{a}' = (a_0, a_1, \ldots, a_{l-1})$. Then we say that $\mathbf{a}'\boldsymbol{\theta}$ $(\equiv (a_0\theta_0 + a_1\theta_1 + \cdots + a_{l-1}\theta_{l-1}))$ is a contrast between the θ_i (i.e., between the levels of the factor under consideration) if and only if we have

$$a_0 + a_1 + \cdots + a_{l-1} = 0 \tag{29}$$

Contrasts of the type $(\theta_i - \theta_j)$ are called "elementary." It can be shown that every contrast can be expressed as a linear combination of elementary contrasts.

Let $\mathbf{a}'\boldsymbol{\theta}$ and $\mathbf{b}'\boldsymbol{\theta}$ be any two contrasts, where $\mathbf{b}' = (b_0, \ldots, b_{l-1})$; then these are said to be orthogonal if and only if we have

$$\mathbf{a}'\mathbf{b} \equiv a_0 b_0 + a_1 b_1 + \cdots + a_{l-1} b_{l-1} = 0 \tag{30}$$

Although there are advantages to working with sets of contrasts that are not all mutually orthogonal (or even linearly independent), it is simpler from the theoretical angle to restrict attention to contrasts that are orthogonal. This will be done in the present paper. However, authors interested in contrasts that are not orthogonal, or not linearly independent, may refer to the book by Roy, Gnanadesikan, and Srivastava (1970).

For the convenience of the reader, for any l (≥ 2), we present a set of very useful orthogonal contrasts $\mathbf{a}_1'\boldsymbol{\theta}, \ldots, \mathbf{a}_{l-1}'\boldsymbol{\theta}$; these are given by

$$\mathbf{a}_1' = (1, -1, 0, \ldots, 0) \cdot \left[1/\sqrt{\{2.1\}} \right]$$

$$\mathbf{a}_2' = (1, 1, -2, 0, \ldots, 0) \cdot \left[1/\sqrt{\{3.2\}} \right]$$

$$\cdots\cdots \tag{31}$$

$$\mathbf{a}_i' = (1, 1, \ldots, 1, -i, 0, \ldots, 0) \cdot \left[1/\sqrt{\{(i+1)i\}} \right]$$

$$\mathbf{a}_{l-1}' = (1, \ldots, 1, -(l-1)) \cdot \left[1/\sqrt{\{l(l-1)\}} \right]$$

Notice that, in the above, the ith contrast compares the first levels with the $(i+1)$ level. Also, each of the contrasts is normalized, which means that the sum of squares of the elements in each contrast equals 1. Normalization provides certain theoretical conveniences; but it is not necessary and may sometimes be ignored for simplicity.

Consider any contrast $\mathbf{b}'\boldsymbol{\theta}$ that equals $(b_0 \theta_0 + \cdots + b_{l-1}\theta_{l-1})$. For simplicity, this will sometimes be written as $\phi(b_0 f_0 + \cdots + b_{l-1} f_{l-1})$.

3 SEVERAL FACTORS: MAIN EFFECTS AND INTERACTIONS

Consider now the situation where we have several (say, m) factors: F_1, F_2, $\ldots$, F_m, where F_i has s_i levels. First, let $m = 2$. Suppose F_1 has s_1 levels and F_2 has s_2 levels. Consider all treatments in which the level of F_2 is j (where $j = 0, 1, 2, \ldots, s_2 - 1$). Obviously, there are s_1 such treatments. Consider a contrast among the levels of the factor F_1 at a fixed level j of factor F_2. Let the value of this contrast be denoted by u_j. If we now add up the values u_j over all values of j, then we obtain what is known as a main effect of the factor F_1. In other words, a main effect of factor F_1 is a contrast between the levels of F_1, averaged over all levels of F_2. A main effect of F_2 can be similarly defined. On the other hand, consider a contrast between the values u_j, for $j = 0, 1, 2, \ldots, s_2 - 1$. This contrast denotes how the value of u_j changes with different levels j of the factor F_2. This contrast is said to belong to the two-factor interaction between the factors F_1 and F_2. To exemplify, suppose $s_1 = 3$, $s_2 = 4$. Suppose that the true yield of the different level combinations is shown as in Table 1.

Each cell of this table gives the yield for the corresponding level combination. Thus, for the level combination $(1, 2)$ the value of ϕ is 30. We consider the contrast $u_j \equiv [(\text{level } 0 \text{ of } F_1) + (\text{level } 2 \text{ of } F_1) - 2(\text{level of } 1 \text{ of } F_1)]$, for each level of F_2. The value of u_j is as given in the last line of the table. The total of these values equals 26, which (except for a scalar multiplier) represents a component of the main effect of the factor F_1. However, one could take a contrast among the values u_j (for different j). Suppose we take the contrast in which the first three levels of F_2 are compared with the last level. Taking this contrast among the values of u_j, we obtain the number (-114), which (except for a constant multiplier) represents a component of the two-factor interaction between the factors F_1 and F_2.

We can generalize the above. Suppose we have three factors. Consider a contrast between the levels of F_1 at any fixed levels of F_2 and F_3. If we average out the value of this contrast over all the different level combinations of F_2 and F_3, then (except for a constant multiplier), we obtain a contrast representing the main effect of the factor F_1. Similarly, as indicated above, we can obtain a contrast which belongs to the interaction

Table 1

		Level of F_2			
		0	1	2	3
Level	0	10	12	18	40
of	1	6	20	30	20
F_1	2	4	14	45	35
		2	-14	3	35

between F_1 and F_2 at any fixed level of F_3. If the same contrast is averaged out over all the levels of F_3, then in this three-factor setup, the value obtained will be a measure of a component of the interaction between the factors F_1 and F_2. However, if this interaction contrast (between factors F_1 and F_2) is computed for different values of F_3, and then a contrast is taken between the values of this contrast at different levels of F_3, then the quantity obtained will be said to belong to the three-factor interaction between the three factors F_1, F_2, and F_3. This process can be generalized to m factors in an obvious manner.

The above can be written simply in terms of the notation introduced earlier. Let the levels of F_j be denoted by $(f_{j0}, f_{j1}, \ldots, f_{j,s_j-1})$. Let

$$\mathbf{b}'_j = (b_{j0}, b_{j1}, \ldots, b_{j,s_j-1}) \tag{32}$$

Consider the linear combination $(b_{j0}f_{j0} + b_{j1}f_{j1} + \cdots + b_{j,s_j-1}f_{js_j-1})$. If

$$b_{j0} = b_{j1} = \cdots = b_{j,s_j-1} = 1 \tag{33}$$

then this linear combination represents a sum over all the levels of the jth factor. On the other hand, if

$$b_{j0} + b_{j1} + \cdots + b_{j,s_j-1} = 0 \tag{34}$$

then the said linear combination represents a contrast between the levels of the jth factor. We shall assume that either (33) or (34) always holds. Consider now the product p, where

$$p = \prod_{j=1}^{m}(b_{j0}f_{j0} + \cdots + b_{j,s_j-1}f_{j,s_j-1}) \tag{35}$$

Notice that p is a polynomial in symbols of the form $f_{1u_1}, f_{2u_2}, \ldots, f_{mu_m}$, where $(u_j = 0, 1, \ldots, s_j - 1; j = 1, \ldots, m)$. Also, notice that the term $f_{1u_1} \cdots f_{mu_m}$ can be interpreted as a level combination in which the jth factor is at level u_j. Consider p as a linear combination of all such possible level combinations. Let $\phi(p)$ denote the linear combination in which the symbol $f_{1u_1}, \ldots, f_{mu_m}$ is replaced by $\phi(f_{1u_1}, \ldots, f_{mu_m})$. Then $\phi(p)$ may be considered as a linear combination of the quantities $\phi(f_{1u_1}, \ldots, f_{mu_m})$. This linear combination $\phi(p)$ is then said to belong to an l-factor interaction between factors (say, $F_{i_1}, \ldots, F_{i_l}$), if and only if equation (34) holds for $j = i_1 \ldots i_l$, and (33) holds for $j \neq i_1 \ldots i_l$. If the equation (33) holds for all j, then $\phi(p)$ is denoted by μ (the grand total, which is also sometimes referred to as the "general mean," ignoring a constant multiplier).

We shall present an example from a $(3 \times 3 \times 2)$ factorial experiment. Suppose

$$\mathbf{b}_1 = (1, -2, 1)', \qquad \mathbf{b}_2 = (1, 0, -1)', \qquad \mathbf{b}_3 = (1, -1)' \qquad (36)$$

Then, we have

$$
\begin{aligned}
\phi(p) = {} & \phi(f_{10}f_{20}f_{30}) - \phi(f_{10}f_{20}f_{31}) - 2\phi(f_{11}f_{20}f_{30}) + 2\phi(f_{11}f_{20}f_{31}) \\
& + \phi(f_{12}f_{20}f_{30}) - \phi(f_{12}f_{20}f_{31}) - 2\phi(f_{10}f_{22}f_{30}) + \phi(f_{10}f_{22}f_{31}) \\
& + 2\phi(f_{11}f_{22}f_{30}) - \phi(f_{11}f_{22}f_{31}) - \phi(f_{12}f_{22}f_{30}) + \phi(f_{12}f_{22}f_{31}) (37)
\end{aligned}
$$

The above linear combination $\phi(p)$ represents a component of the 3-factor interaction between the factors F_1, F_2, and F_3. Notice that if, in (36), we had taken $\mathbf{b}_3 = (1, 1)'$, then, in the r.h.s. of (37), the sign of all terms standing at even-number places would change, and the resulting expression would then represent a component of the two-factor interaction between the factors F_1 and F_2.

In this chapter, we shall be primarily concerned with the 2^m type of experiments, in which there are m factors each at two levels. In this case, it is easy to see that the vectors b_j can take only two values, $(1, 1)$ and $(-1, 1)$, respectively. In this case, there are 2^m "factorial effects," which include μ, the main effects, and the interactions. For $1 \leq l \leq m$, let $F_{i_1} F_{i_2} \cdots F_{i_l}$ denote the l-factor interaction between factors # $i_1, i_2, \ldots, i_l$. Then, it is easy to see that we have

$$F_{i_1} F_{i_2} \cdots F_{i_l} = \sum_{j_1, \ldots, j_m = 0, 1} d(i_1, \ldots, i_l; j_1, \ldots, j_m) \phi(f_{1j_1} \cdots f_{mj_m}) \qquad (38)$$

where

$$d(i_1, i_2, \ldots, i_l; j_1, \ldots, j_m) = (-1)^u$$

and $\qquad\qquad\qquad\qquad\qquad\qquad\qquad\qquad\qquad\qquad\qquad\qquad (39)$

$$u = j_{i_1} + j_{i_2} + \cdots + j_{i_l}$$

It is easy to describe the main effects and interactions in terms of the ϕ values for the different level combinations, by using the Hadamard matrices and the so-called "Yates order." For $m = 3$, this relationship is given by (40), in which both the vector of factorial effects (on the l.h.s.), and the

vector of the true values (on the r.h.s.) are in the Yates order, and the
(8×8) matrix on the r.h.s. is a Hadamard matrix. We have

$$
\begin{bmatrix}
F(0) \\
F(1) \\
F(2) \\
F(12) \\
F(3) \\
F(13) \\
F(23) \\
F(123)
\end{bmatrix}
=
\begin{bmatrix}
+ & + & + & + & + & + & + & + \\
+ & - & + & - & + & - & + & - \\
+ & + & - & - & + & + & - & - \\
+ & - & - & + & + & - & - & + \\
+ & + & + & + & - & - & - & - \\
+ & - & + & - & - & + & - & + \\
+ & + & - & - & - & - & + & + \\
+ & - & - & + & - & + & + & -
\end{bmatrix}
\begin{bmatrix}
\phi(0) \\
\phi(1) \\
\phi(2) \\
\phi(12) \\
\phi(3) \\
\phi(13) \\
\phi(23) \\
\phi(123)
\end{bmatrix}
\tag{40}
$$

where for $(1 \leq l \leq m; 1 \leq i_1 < i_2 < \cdots < i_l \leq m)$, we shall always use
the notation

$$
F_{i_1} F_{i_2} \ldots F_{i_l} = F_{i_1 i_2 \cdots i_l} = F(i_1 i_2 \cdots i_l)
$$
$$
\phi(0) \equiv \mu, \qquad \phi(i_1, i_2, \ldots, i_l) = \phi(f_{i_1}, \ldots, f_{i_l})
\tag{41}
$$

where $(f_{i_1}, f_{i_2}, \ldots, f_{i_l})$ is an abbreviated form of the level combination in
which factors $(i_1, i_2, \ldots, i_l)$ are at level 0, and all the other factors are at
level 1.

The above is easily generalizable to larger m. For example, for $m = 4$,
we have $2^4 (= 16)$ level combinations, out of which the first 2^3 elements
are the same as those occurring in the vectors in (40). The remaining 2^3
elements involve, in order:

$$
(4), (14), (24), (124), (34), (134), (234), (1234).
\tag{42}
$$

Also, then, the matrix relating the F to the ϕ will be the $(2^4 \times 2^4)$ Hadamard
matrix H_4 defined earlier.

4   THE NATURE OF THE REALISTIC PROBLEMS IN FACTORIAL EXPERIMENTS

The basic objective behind a factorial experiment may be one or more of
the following: (1) to estimate $\phi(\mathbf{t})$, for all $\mathbf{t}$ as accurately as possible, (2) to
obtain the set of $\mathbf{t}$ for which $\phi(\mathbf{t})$ is large (or maximum), (3) to obtain
the set of $\mathbf{t}$ for which $\phi(t)$ is small (or minimum), or (4) to obtain the set

of t for which $\phi(t)$ lies within a certain range, etc. Objective 1 subsumes other objectives. Some people go for the estimation of a particular set of parameters; this, however, cannot be done unless we have a certain type of knowledge about the remaining parameters. (As we shall see, the knowledge of all parameters is equivalent to the knowledge of $\phi(t)$ for all t.) A deeper examination of scientific needs shows that in general it is wise to have objective 1.

Thus, the basic problem is to estimate all the parameters, i.e., μ, and the main effects and interactions. Interest does not usually lie in the estimation of the general mean μ; however, it turns out that in most situations we have to act as if this needs to be estimated anyway.

Because of the linear relationship between the set of the parameters and the set of the true effects of the treatments (i.e., the set of the values $\phi(t)$, for various treatments t), it is clear that if we had the values $\phi(t)$ for all t, the set of parameters could be easily determined. As explained earlier, there are chance fluctuations, so that, given any t, we observe a value $y(t)$, which may be different from $\phi(t)$. However, since the expected value of $y(t)$ is $\phi(t)$, we can estimate the parameters using the $y(t)$, though it will necessarily introduce some error in the estimate. It can be shown that if the error variance for any experimental unit is σ^2, then, in a factorial experiment in which each treatment t is replicated r times, the variance of the estimate of any parameter is σ^2/r (where we assume that the parameters are normalized).

Thus, if it was possible to do the whole experiment even once, there would not be much problem in estimating the parameters, and thus there will be no need of any complicated theory. However, it is easy to see that as m, the number of factors, increases, the total number of treatments increases very rapidly. For example, even if all the factors are at two levels, the number of treatments is 64 with only 6 factors. This is not a small number. If the factors have more than two levels, then the increase in the number of treatments is even faster. Thus, the problems of the theory of factorial designs emerge out of this basic fact.

The above fact has repercussions both when a nuisance factor is present and is not present. If one or more nuisance factors are present, then, even if one has the resources to replicate the whole experiment many times for a reasonably large value of r (say, $r = 3$ or 4), there are problems of accommodating all of the treatments at any given level of the nuisance factor. Usually, each nuisance factor may present two kinds of problems. The first one occurs when each level of the nuisance factor comes with a clearly specified set of experimental units. This would occur, for example, in an experiment on chicken, where the nuisance factor occurs in the form of different litters in which chicken are raised, and where every litter represents a particular level of the nuisance factor. In this case, if a particular litter has 8 chickens in it, then we can accommodate up to 8 treatments in this particular litter or block. On the other hand, it may happen that the nuisance factor may be such that we only broadly know which experimental units may correspond to the same level of the nuisance factor. This may occur, for example, in an agricultural experiment, where we may think that the fertility gradient is occurring in a particular direction in the field, and experimental units which lie perpendicular to that direction may generally be assumed to be at the same level of the nuisance factor. Even here, it is clear that the smaller the number of experimental units we put inside *each block* (i.e., *each level* of the nuisance factor), the more homogeneous the block will be, and hence a more accurate estimate will be possible. Indeed, this is what gives rise to the *theory of block designs*. This theory arises because we are unable to accommodate all treatments in each block of the nuisance factor. In the case of factorial experiments, it is obvious that in almost all cases it will be impossible to accommodate all the treatments at any particular level of the nuisance factor. This gives rise to a special case of the theory of block designs when the treatments have factorial structure, namely, the theory of "confounded design." Because of lack of space, the theory of confounding will not be treated in this paper. Furthermore, for simplicity, we shall assume throughout that there are nuisance factors.

Clearly, even if there are no nuisance factors, usually a factorial experiment is large enough so that it is prohibitive to include all of the different treatments in the experiment. From this it is obvious that if all the parameters were possibly unknown, and could possibly be nonzero, we could not estimate them. However, fortunately, a certain simplification arises because, in many natural situations, many of these parameters have been empirically found to be negligible. In any given experiment, though, we may not know which particular parameters are nonzero. Usually, many of the main effects and some of the two-factor interactions may be nonzero, and a few other higher interactions may also be nonzero. However, which parameters are actually nonzero is almost always not known. If many of the parameters are negligible, then one intuitively expects that it might be possible to conduct a smaller experiment involving a smaller number of runs. This expectation is also theoretically justifiable.

However, the experimenter has, in general, a considerable role to play. If the experimenter can accurately tell which parameters are indeed nonnegligible, then it is a relatively simple matter to design an experiment which would involve a reasonably small number of treatments, and which would provide a *nonsingular* design, i.e., a design which is able to estimate all the parameters that have not been assumed to be negligible. (By *design* we mean any particular set of treatments.) Actually, there are a very large number of nonsingular designs for any such situation. Among these, which design would be most accurate (in other words, which design will provide the most information in the sense of having a very small variance of the estimates of the various parameters) is a much harder problem to deal with. As the subject stands now, solutions to such problems are given in only relatively few cases.

Trouble usually arises, however, because it is very difficult for the experimenter, in general, to identify the set of parameters which we definitely expect to be negligible.

Let $\mathbf{L}^*$ be the set of parameters which is *actually nonnegligible*, and let $\mathbf{L}_0^*$ be the set of parameters which is *actually* negligible so that $\mathbf{L}^*$ and $\mathbf{L}_0^*$ together exhaust the set of all parameters. Thus, in almost all cases, $\mathbf{L}^*$ and $\mathbf{L}_0^*$ are *not* known. Let $\mathbf{L}$ be the set of parameters which are *assumed* to be *nonnegligible*, and $\mathbf{L}_0$ the set which are *assumed negligible*. Let $\mathbf{L}^0$ be the set of all parameters included in $\mathbf{L}_0$ which are actually nonnegligible (so that these are elements of $\mathbf{L}^*$). Then, of course, much of the success of the experiment may depend upon how accurate $\mathbf{L}$ is, i.e., how small is the set $\mathbf{L}^0$. The "theory of nonsingular designs" is concerned with the problem of obtaining a set of treatments T such that, given $\mathbf{L}$, it can be estimated by using the treatments in T. The theory of optimal designs is concerned with finding a set of treatments T^* which forms a nonsingular design, and which at the same time maximizes the information among all competing nonsingular designs T which belong to a certain class. If no restriction is placed on this class, then T^* is said to be optimal among the class of all designs. In studies on optimality, the class of competing designs is restricted so that all designs have the same number of treatments. The reason is that it can be shown in general that the "information content" of a design increases as the number of treatments in it are increased. If, in any design, every treatment is not used with equal frequency, then such a design is said to be a "fractional replicate." A fractional replicate design is said to be regular if the number of the set of all treatments is a multiple of the number of treatments N in the design; otherwise, it is said to be "irregular."

From the above discussion, it is clear that the theories of nonsingular and optimal designs are concerned with the situation where we try to estimate $\mathbf{L}$, assuming $\mathbf{L_0}$ to be negligible. However, as mentioned earlier, $\mathbf{L_0}$ hardly ever contains only the negligible parameters. Thus, there remains the further problem of identifying the set of parameters $\mathbf{L^0}$, and estimating these as well as the elements of $\mathbf{L}$. Designs which allow us to identify the elements of $\mathbf{L^0}$ and estimate them (along with estimating the elements of $\mathbf{L}$) will be touched upon in Sections 8 and 9, under the headings "Search Designs" and "Probing Designs."

Experimentation may usually proceed in several possible ways. One way is to conduct the whole experiment all in one stage. In other words, the set of all treatments is, in a sense, tried together or at the same time. Such experimentation is said to be single-stage experimentation. On the other hand, the term "multistage experimentation" is used for the situation where an experiment is done, and its results noted and used in the planning of the next experiment, which is used in the planning of the experiment after that, and so on. Thus, in multistage experimentation, the total experiment is done in several stages. Contrasted with these is the situation called "sequential experimentation," where there are a large number of stages, and where at each stage only one or two or just a few treatments are used.

The theory of search designs deals with the situation where the experimentation is conducted all in one stage, while probing designs are multistage or sequential. It is to be noted that the purpose of both search designs and probing designs is to identify $\mathbf{L^0}$, and to estimate it along with $\mathbf{L}$.

As mentioned earlier, unless $\mathbf{L}$ is taken to be a rather large set (in most cases, unreasonably large), it is extremely unlikely that $\mathbf{L}$ would contain $\mathbf{L^*}$. But the larger the size of $\mathbf{L}$ is allowed to be, the larger the experiment, and hence the bigger the expenditure of resources. Hence, the basic problem in the theory of factorial designs boils down to this. Choose $\mathbf{L}$ in a clever way such that it is not too large, and yet there is a reasonably good chance that most of the elements of $\mathbf{L^*}$ are contained in it, so that $\mathbf{L^0}$ is a rather small set. No hard-and-fast rules for choosing $\mathbf{L}$ can be prescribed. If the inclusion of various treatments in the experiment is relatively inexpensive, $\mathbf{L}$ should be taken large enough so that $\mathbf{L^0}$ is a rather small set (preferably, having less than five elements). (Again, one can never accurately predict this. Thus, what we mean is that one should go by the best guesses possi-

then **L** should be chosen to be rather small, and perhaps multistage or even sequential experimentation might be desirable in order to keep the size of the total experiment as small as possible.

Of course, if there are nuisance factors present, then all of the above becomes one step more complex. However, it must be added that even when there are no nuisance factors, the problems involved in identifying **L**⁰ and estimating all the nonnegligible parameters while at the same time keeping the experiment small in size, are *extremely nontrivial.*

Certain sciences are noisier than others. In other words, in certain sciences, the random fluctuations ϵ [of equation (26)] are relatively large, resulting in a large σ^2. However, in other sciences, the opposite is true. Generally speaking, there is much more noise in the agricultural, biological, and social sciences. In the field of nutrition and in medical sciences, as well as in engineering sciences, the noise level is relatively less. In the more exact sciences, like physics and chemistry, the noise level is far less. Although, most of the results are pertinent and important for all sciences, some procedures (such as those in Section 9) are suitable for the situation where noise is small.

5   SOME USEFUL COMBINATORIAL STRUCTURES

In the last section, it was pointed out that one of the main purposes of the theory of factorial designs is to obtain a design (i.e., a set of treatments) T which allows us to identify the elements of **L**⁰ and estimate them along with **L**. In this section, we consider some discrete mathematical techniques, which have been found useful in generating such designs T.

We start with the concept of a *finite, or Galois, field*, denoted by $GF(p)$, where $(p > 1)$ is an integer. We shall assume that p is a prime number, i.e., p is divisible only by 1 and itself, and by no other integers. (For example, we can have $p = 2, 3, 5, 7, 11, 13, 17, \ldots$). Although, for lack of space, we shall assume p to be prime, the field $GF(p)$ exists even when p is a power of a prime (such as 4, 8, 9, 16, 25, 27, 32, etc.) Unless otherwise stated, most of the theorems included in this chapter in which $GF(p)$ is used hold for general p.

We first explain what is meant by doing computations or "arithmetic," $mod(p)$. The phrase "$mod(p)$" stands for modulo p, and basically it means that we will continue to multiply, add, divide, and subtract integers as usual, except that in doing so, we will add a multiple of p to the final result in such a way that we obtain an integer in the set $\{0, 1, \ldots, (p-1)\}$. For example, suppose we are doing arithmetic mod(7). Now, in ordinary arithmetic, the value of the expression $\{3 \times 4 - (5)^2\}$ is (-13). We need to add a multiple of 7 to (-13) so that the result is a number in the set $\{0, 1, \ldots, 6\}$. By simple checking, it is clear that 14 is the required multiple of 7. Thus, mod(7), the value of the expression $\{3 \times 4 - (5)^2\}$ is $(14 - 13) = 1$.

However, $GF(p)$ refers to our intention to work with only the integers $(0, 1, \ldots, (p-1)\}$, where it is understood that we will be working mod(p), when p is a prime number. The integers $(0, 1, \ldots, p-1)$ are said to be the elements of $GF(p)$. In this system, it can be shown that division can also be uniquely done. In other words, the ratio of two elements of $GF(p)$ (where the denominator is nonzero) can be shown to be a unique element of $GF(p)$. Thus, in $GF(7)$, the value of the expression $\{-(3/5)\} = \{(-3)/5 = \{(-3 + 4 \times 7)/5\} = 5$.

In this chapter, the field that will be used often is $GF(2)$, in which we work with only the integers $\{0, 1\}$, where 0 and 1 are multiplied and added as usual, except that we assume that $1 + 1 = 0$, since (mod 2), we have $2 = 0$.

We now consider linear equations "over" finite fields, which means that all constants occurring in the equations belong to such a field, and also that the variables can take values only in the field. Without loss of generality, we shall always take equations which are linearly independent (i.e., no one of them can be expressed as a linear combination of the others). As an example, consider the following equations over $GF(3)$:

$$x_1 + 2x_2 = 0$$
$$x_1 + 2x_3 = 0 \tag{43}$$

One solution of these equations is obviously $(x_1, x_2, x_3) = (0, 0, 0)$. Other solutions are easily seen to be $(1, 1, 1)$ and $(2, 2, 2)$. If the constants on the r.h.s. are 1 and 2, then the solutions are easily seen to be $(0, 2, 1)$, $(1, 0, 2)$ and $(2, 1, 0)$.

The following result is quite important.

Theorem 5.1 Let the matrix $A(k \times m)$ and the vector $\mathbf{c}(k \times 1)$ be over $GF(p)$. Suppose all the rows of A are linearly independent. Let $\mathbf{x}(m \times 1)$ be a vector of variables. Then, the equations

$$A\mathbf{x} = \mathbf{c} \tag{44}$$

have exactly p^{m-k} distinct solutions for $\mathbf{x}$.

The equations (43) above constitute an example of Theorem 5.1. with

$$p = 3, \quad m = 3, \quad k = 2, \quad A = \begin{bmatrix} 1 & 2 & 0 \\ 1 & 0 & 2 \end{bmatrix}, \quad \mathbf{c} = \begin{bmatrix} 0 \\ 0 \end{bmatrix} \quad (45)$$

As another example, consider the case where

$$p = 2, \quad m = 7, \quad k = 4,$$

$$A = \begin{bmatrix} 1 & 0 & 0 & 0 & 0 & 1 & 1 \\ 0 & 1 & 0 & 0 & 1 & 0 & 1 \\ 0 & 0 & 1 & 0 & 1 & 1 & 0 \\ 0 & 0 & 0 & 1 & 1 & 1 & 1 \end{bmatrix}, \quad \mathbf{c} = \begin{bmatrix} 0 \\ 0 \\ 0 \\ 0 \end{bmatrix} \quad (46)$$

For this case, it can be checked that the solutions are given by the columns of the (7×8) matrix T where

$$T = \begin{bmatrix} 0 & 0 & 1 & 1 & 1 & 1 & 0 & 0 \\ 0 & 1 & 0 & 1 & 1 & 0 & 1 & 0 \\ 0 & 1 & 1 & 0 & 0 & 1 & 1 & 0 \\ 0 & 1 & 1 & 1 & 0 & 0 & 0 & 1 \\ 0 & 1 & 0 & 0 & 1 & 1 & 0 & 1 \\ 0 & 0 & 1 & 0 & 1 & 0 & 1 & 1 \\ 0 & 0 & 0 & 1 & 0 & 1 & 1 & 1 \end{bmatrix} \quad (47)$$

Next, we define the concept of an *orthogonal array* (OA). An OA with s symbols, m constraints, N assemblies, strength t, and index λ is a matrix $T(m \times N)$ whose elements belong to the set of integers $\{0, 1, 2, \ldots, (s-1)\}$, such that in every submatrix T^* (inside T), all the s^t column vectors of the form $(a_1, a_2, \ldots, a_t)'$, where $[a_1, a_2, \ldots, a_t = 0, 1, 2, \ldots, (s-1)]$, appear as a column exactly λ times. Such an array is denoted as $\text{OA}(s, m, N, t)$. As an example, the matrix T at (47) is an $\text{OA}(2, 7, 8, 2)$. It may be verified that if T^* is any (2×8) submatrix of T at (47), then all column vectors of the form $(a_1, a_2)'$, where $(a_1, a_2 = 0, 1)$ occur as a column of T^* exactly $\lambda = 8 \cdot 2^{-2} = 2$ times.

Notice that the OA at (47) is actually a solution of (44) for the special case at (46). This shows that orthogonal arrays may sometimes be obtained as solutions of the equation (44). Indeed, this is true in general, if the matrix A obeys certain restrictions. We now proceed to consider this in detail.

Consider a matrix $A(k \times m)$. Then, A is said to have the property Q_t if every nonzero row vector $(a_1, a_2, \ldots, a_m)$ of size $(1 \times m)$, which is dependent on the rows of A, is of weight more than t. (Throughout this chapter, the weight of a matrix or a vector will be defined as the number of nonzero elements in the matrix or the vector.)

Theorem 5.2   Let T be an $(m \times N)$ matrix where $N = p^{m-k}$, whose columns constitute the set of all distinct solutions of the equations (44), where we assume that the matrix A in the equation has property Q_t, for some integer t. Then T will be an orthogonal array with p symbols, m rows, p^{m-k} columns, strength t, and index p^{m-k-t}.

It should be remarked that the method of obtaining OA's, by solving (44) with appropriate A, is the principal method available for obtaining OA's.

If $\mathbf{c}$ is the zero vector, then the resulting T will be called a "standard array," and if $\mathbf{c}$ is nonzero, then T will be called a "coset array." Notice that the design so obtained has p^{m-k} treatments. Thus, if we want to reduce the size of the designs obtained by this process, then, for a given m and t, we must increase the value of k to a maximum. (Alternatively, for a given value of t and r where $r = m - k$, we wish to maximize m.) This is known as Bose's "packing problem."

Bose's packing problem is also connected with coding theory. Consider a matrix $A(k \times m)$ with property Q_t. Then, all row vectors dependent on the rows of A may be called "code words." Let $\mathbf{a}'$ and $\mathbf{b}'$ be two such code words; then we define the distance δ between them by

$$\delta(\mathbf{a}', \mathbf{b}') = \text{weight}(\mathbf{a}' - \mathbf{b}') \tag{48}$$

One important problem in coding theory is to obtain a set of code words (called the "code") such that the distance between any two code words inside the code is larger than t, where t is some positive integer. If t is of the form of $t = 2l$, where l is an integer, then the code is said to be l-error correcting, and if $t = 2l + 1$, then the code is said to be l-error correcting and $(l+1)$-error detecting. It can be shown that the distance between any two code words inside the code is t if and only if A has property Q_t.

Thus, there is a fundamental connection between the theory of factorial designs and coding theory, a connection which was discovered in Bose (1961). A great deal of work has been done on Bose's packing problem since he first formulated it in 1947. The problem has been solved in general for $t = 2$, and for some larger values of t (≥ 3) when the value of m is relatively small. However, for $t > 3$, the problem remains largely unsolved, though there are certain codes available in which the value of k is quite large though not the maximum. Examples of such codes include the well-known BCH codes, and Srivastava codes. These codes provide "good" matrices A, i.e., in which k is quite large (with m and t being general). These codes are described in most books on coding theory, one of which is Berlekamp (1968).

It is instructive to show how to solve equations (43) when $\mathbf{c} = \mathbf{0}$. We first consider the case where A is of the form

$$A = [I_k : B] \tag{49}$$

where B is a $[k \times (m - k)]$ matrix. Define $A^0(m \times (m - k))$ by

$$A^0 = \begin{bmatrix} -B \\ \cdots\cdots \\ I_{m-k} \end{bmatrix} \tag{50}$$

Then we have

Theorem 5.3

1. The set of all solutions to equations (44) where A is of the form (49) is given by $\mathbf{x}$ at (51), where

$$\mathbf{x} = A^0 \mathbf{u} + \begin{bmatrix} \mathbf{c} \\ \mathbf{0}_{m-k} \end{bmatrix} \tag{51}$$

and where $\mathbf{u}[(m - k) \times 1]$ takes all possible s^{m-k} distinct values. Let $T(m \times s^{m-k})$ be the matrix whose columns constitute such solutions.
2. Let $T_0(m \times s^{m-k})$ be the matrix T when $\mathbf{c} = \mathbf{0}$. Then, T is obtainable by adding $\begin{bmatrix} \mathbf{c} \\ \mathbf{0} \end{bmatrix}$ to each column of T_0.

Usually, A is given in the form of (49). If not, the form (49) is obtainable by premultiplying A by the inverse of a $(k \times k)$ nonsingular submatrix of A, and possibly rearranging the columns. Sometimes, the matrix A^0 may be given instead of A, in which case A can be written down very easily.

When $t = 2$, the solution to Bose's packing problem is easily given. We have

Theorem 5.4   Let A^0 be a matrix of size $(m \times (m - k))$ in which

$$m = \frac{p^r - 1}{p - 1}, \qquad k = m - r \tag{52}$$

where r is a positive integer, and where A^0 is given by (50), where the matrix $(-B)$ consists of all distinct nonzero vectors with elements from $\mathrm{GF}(p)$, which have 1 as the first nonzero element. Then, for every vector $\mathbf{c}$, the set of all values $\mathbf{x}$ given by (51) (for all possible values of $\mathbf{u}$) is an orthogonal array of strength t, with $t = 2$. Also, given r, the value of m at (52) is a maximum.

As an example, consider the case

$$t = 2, \qquad p = 3, \qquad r = 3, \qquad m = 13, \qquad k = 10 \tag{53}$$

The above gives rise to

$$A^{0\prime} = \begin{bmatrix} 0 & 0 & 1 & 1 & 1 & 1 & 1 & 1 & 1 & 1 & 1 & 0 & 0 \\ 1 & 1 & 0 & 0 & 1 & 2 & 1 & 1 & 2 & 2 & 0 & 1 & 0 \\ 1 & 2 & 1 & 2 & 0 & 0 & 1 & 2 & 1 & 2 & 0 & 0 & 1 \end{bmatrix}$$

and the OA$(3, 13, 27, 2)$ given by T where

$$T = \begin{bmatrix} 0 & 0 & 1 & 1 & 0 & 2 & 2 & 1 & 1 & 2 & 2 & 2 & 1 & 2 & 2 & 0 & 1 & 1 & 0 & 2 & 0 & 0 & 1 & 1 & 0 & 0 & 2 \\ 0 & 0 & 1 & 2 & 0 & 2 & 1 & 1 & 2 & 0 & 2 & 1 & 0 & 2 & 1 & 2 & 1 & 2 & 1 & 0 & 2 & 1 & 0 & 0 & 1 & 2 & 0 \\ 0 & 1 & 0 & 1 & 2 & 0 & 2 & 1 & 2 & 1 & 2 & 1 & 2 & 1 & 0 & 2 & 2 & 0 & 1 & 2 & 0 & 2 & 0 & 1 & 0 & 1 & 0 \\ 0 & 1 & 0 & 2 & 2 & 0 & 1 & 1 & 0 & 2 & 2 & 0 & 1 & 1 & 2 & 1 & 2 & 1 & 2 & 0 & 2 & 0 & 2 & 0 & 1 & 0 & 1 \\ 0 & 1 & 1 & 0 & 2 & 2 & 0 & 2 & 1 & 1 & 1 & 2 & 2 & 0 & 1 & 1 & 0 & 2 & 2 & 2 & 0 & 0 & 1 & 1 & 0 & 0 \\ 0 & 1 & 2 & 0 & 2 & 1 & 0 & 0 & 1 & 2 & 0 & 2 & 1 & 2 & 1 & 2 & 1 & 2 & 1 & 0 & 0 & 2 & 2 & 0 & 0 & 1 & 1 \\ 0 & 1 & 1 & 1 & 2 & 2 & 2 & 2 & 2 & 1 & 1 & 1 & 0 & 0 & 0 & 0 & 0 & 0 & 1 & 1 & 2 & 0 & 2 & 2 & 1 \\ 0 & 1 & 1 & 2 & 2 & 2 & 1 & 2 & 0 & 0 & 1 & 0 & 0 & 0 & 2 & 2 & 0 & 1 & 1 & 1 & 0 & 2 & 1 & 2 & 0 & 1 & 2 \\ 0 & 1 & 2 & 1 & 2 & 1 & 2 & 0 & 2 & 0 & 0 & 1 & 0 & 2 & 0 & 1 & 1 & 0 & 2 & 1 & 2 & 0 & 1 & 2 & 1 & 0 & 2 \\ 0 & 1 & 2 & 2 & 2 & 1 & 1 & 0 & 0 & 1 & 0 & 0 & 2 & 2 & 0 & 1 & 1 & 0 & 2 & 1 & 1 & 0 & 1 & 2 & 2 & 0 \\ 0 & 1 & 0 & 0 & 2 & 0 & 0 & 1 & 1 & 0 & 2 & 2 & 0 & 1 & 1 & 0 & 2 & 2 & 0 & 1 & 1 & 1 & 1 & 2 & 2 & 2 & 2 \\ 0 & 0 & 1 & 0 & 0 & 2 & 0 & 1 & 0 & 1 & 2 & 0 & 2 & 2 & 0 & 1 & 1 & 0 & 2 & 1 & 1 & 2 & 2 & 2 & 2 & 1 & 1 \\ 0 & 0 & 0 & 1 & 0 & 0 & 2 & 0 & 1 & 1 & 0 & 2 & 2 & 0 & 2 & 2 & 0 & 1 & 1 & 1 & 2 & 1 & 2 & 2 & 1 & 2 & 1 \end{bmatrix} \tag{54}$$

The following corollary of Theorem 5.4 will be useful.

Corollary 5.1   Let A^0 be an $(m \times (m - k))$ matrix with elements from GF(2), such that $m = (2^r - 1)$, $k = m - r$, where r is a positive integer, and where A^0 consists of all possible nonzero row vectors of size $(1 \times r)$ with elements from GF(2). Let T be the $(2^r - 1) \times 2^r$ matrix whose columns equal $A^0 \mathbf{u}$, and $\mathbf{u}(r \times 1)$ takes all possible 2^r values over GF(2). Then T is an orthogonal array of strength 2 with two symbols. Furthermore, suppose T^0 is obtained by adding a row of zeros to T so that T^0 is of size $(2^r \times 2^r)$. Let T^* be obtained from T^0 by changing 0 to $(+1)$ and 1 to (-1) everywhere in T^0. Then T^* is a Hadamard matrix of size $(2^r \times 2^r)$. Also, T^* is the same as the Hadamard matrix of size $(2^r \times 2^r)$ discussed in Section 1, except for a possible permutation of rows and/or columns.

The above result shows the connection between Hadamard matrices and orthogonal arrays of strength 2 with two symbols. The following result is, in a sense, the reverse of the last one.

Theorem 5.5   Let H be a Hadamard matrix of size $(N \times N)$; i.e., H is such that (1) the elements of H are $+1$ or -1, (2) the product of any two rows of H is 0. Without loss of generality, we will assume that one row of H consists entirely of $(+1)$'s. (If a matrix H_0 is given in which this is not the case, then by multiplying the appropriate columns of H_0 with (-1), we can reduce H_0 to a new Hadamard matrix H in which the first row will have only $(+1)$.) Now, suppose the row of H which consists of $(+1)$ everywhere is deleted so that the resulting matrix H_1 is of size $(N - 1) \times N$. Then H_1 is an orthogonal array of strength 2 and two symbols.

It must be remarked here that Hadamard matrices of size $(N \times N)$ are possible only if N is of the form $N = 4N_0$, where N_0 is a positive integer, or $N_0 = \frac{1}{2}$. At the time of this writing, methods of obtaining Hadamard matrices are known for all N_0, up to a certain very large value. A good deal of very useful work on Hadamard matrices was done by Paley (1933). This work is more mathematical, and will not be reproduced here. However, it is important to note that Hadamard matrices with $N = 12, 20, 24, 28$, etc. (which are not powers of 2), are quite useful in the theory of factorial designs. For the benefit of the reader, a Hadamard matrix of size (12×12)

is reproduced below.

$$T = \begin{bmatrix} 1 & 1 & 1 & 1 & 1 & 1 & 1 & 1 & 1 & 1 & 1 & 1 \\ 1 & 1 & 0 & 1 & 1 & 1 & 0 & 0 & 0 & 1 & 0 & 0 \\ 1 & 0 & 1 & 0 & 1 & 1 & 1 & 0 & 0 & 0 & 1 & 0 \\ 1 & 0 & 0 & 1 & 0 & 1 & 1 & 1 & 0 & 0 & 0 & 1 \\ 1 & 1 & 0 & 0 & 1 & 0 & 1 & 1 & 1 & 0 & 0 & 0 \\ 1 & 0 & 1 & 0 & 0 & 1 & 0 & 1 & 1 & 1 & 0 & 0 \\ 1 & 0 & 0 & 1 & 0 & 0 & 1 & 0 & 1 & 1 & 1 & 0 \\ 1 & 0 & 0 & 0 & 1 & 0 & 0 & 1 & 0 & 1 & 1 & 1 \\ 1 & 1 & 0 & 0 & 0 & 1 & 0 & 0 & 1 & 0 & 1 & 1 \\ 1 & 1 & 1 & 0 & 0 & 0 & 1 & 0 & 0 & 1 & 0 & 1 \\ 1 & 1 & 1 & 1 & 0 & 0 & 0 & 1 & 0 & 0 & 1 & 0 \\ 1 & 0 & 1 & 1 & 1 & 0 & 0 & 0 & 1 & 0 & 0 & 1 \end{bmatrix} \tag{55}$$

The next theorem gives certain techniques of obtaining orthogonal arrays of strength 3 with two symbols.

Theorem 5.6 Consider the case where $m = 2^{r-1}$, $k = m - r$, where r is a positive integer. Let $A^0(m \times r)$ consist of all the possible distinct row vectors of size $(1 \times r)$ which have odd weight. Let $T(m \times N)$, $N = 2^r$, be the (0,1) matrix whose columns consist of all the possible distinct vectors $\mathbf{x}$ obtained from (51), as $\mathbf{u}$ takes the set of all possible 2^r distinct values. Then the matrix T so obtained is an orthogonal array with two symbols and strength 3. Also, for a given r, 2^{r-1} is the maximum possible value of m.

Theorem 5.7 Let $T_1(m \times N_1)$ be an orthogonal array of strength 2, with two symbols 0 and 1. Let $T_2 = \bar{T}_1$; i.e., T_2 is obtained from T_1 by interchanging 0 and 1. Let $N = 2N_1$, and let $T((m+1) \times N)$ be the (0,1) matrix given by

$$T = \begin{bmatrix} \mathbf{J}' & \mathbf{0}' \\ T_1 & T_2 \end{bmatrix} \tag{56}$$

where $\mathbf{J}'$ and $\mathbf{0}'$ are respectively $(1 \times N_1)$ vectors containing 1 everywhere and 0 everywhere. Then, T is an orthogonal array of strength 3, with two symbols.

Thus, for example, using the above results, we can make an orthogonal array of strength 3, and of sizes (8×16) and (12×24). In the latter case, we use the matrix at (55).

It will be seen later (Section 7) that if an array is obtainable as a solution of the equations of the form (44), then, there is a certain ease in its use as a factorial design. In the above, we have given only a few elementary results. A large number of other results are available. However, in spite of these, the class of arrays of strength more than 3 is quite limited. In what follows, arrays which can be obtained as solutions of equations (44) will be said to be of "type A."

Because of the various needs in the theory of factorial designs which we shall consider later on, generalizations of the technique in Theorem 5.1 have been considered abundantly in the literature. We now consider these briefly.

Let n be a positive integer. Suppose we take n equations of the form (44), with the same matrix A, but with possibly changing vectors $\mathbf{c}$. Consider the equations

$$A\mathbf{x} = \mathbf{c}_i; \qquad i = 1, 2, \ldots, n \tag{57}$$

Let T_i $(i = 1, 2, \ldots, n)$ be the $(m \times p^{m-k})$ matrix whose columns contain all the distinct solutions of the equations (57) for a given value of i. Define the $(m \times p^{m-k} \cdot n)$ matrix T by

$$T = [T_1 : T_2 : \ldots T_n] \tag{58}$$

Then one may consider using T as a design for various purposes. In the context of orthogonal arrays, the following result is important.

Theorem 5.8 Consider the array T given by (58), which is obtained as a solution of the equations (57). Define

$$C = [\mathbf{c}_1, \mathbf{c}_2, \ldots, \mathbf{c}_n] \tag{59}$$

Then, a necessary and sufficient condition that T is an orthogonal array of strength t is that for every vector $\mathbf{u}'(1 \times k)$ over $\mathrm{GF}(p)$ such that $\mathbf{u}'A \neq 0_{1m}$, and $\mathrm{wt}(\mathbf{u}'A) \leq t$, we have that the $(1 \times n)$ matrix $\mathbf{u}'C$ is an OA of strength 1.

Now, suppose that T is an OA of the form (58). Then, T is said to be an OA of type B if and only if it is not of type A, and $k < m$. All orthogonal arrays which are neither of type A nor B are said to be of type C; such arrays trivially correspond to the equations (58) but with

$k = m$. Theorem 5.8 was originally discovered in unpublished work by Srivastava, and was first mentioned in Srivastava and Chopra (1975).

Designs of type T at (58) are usually known as "parallel-flats" (or "parallel-planes," or "partial pencil") designs since each design T_i can be interpreted as a hyperplane in a "finite Euclidean space." We shall not go into the mathematical aspects of this here. However, these designs constitute a valuable generalization of designs of type A. Such designs were first introduced in Connor and Young (1959). Subsequently, many authors made contributions to these; notable among these are S. Addleman, C. Daniel, P. W. M. John, and M. S. Patel. To illustrate their work, some references to their work are included in the bibliography at the end. In the middle 1960's, the author tried to develop a general mathematical theory of the parallel-flats designs, which culminated in a long paper. The work was added to in the 1970's by Anderson and Mardekian, and later again by the author, culminating in the paper by Srivastava, Anderson, and Mardekian (1984). This work was further advanced in Srivastava (1987a). The last two papers are devoted to the study of the information matrices for designs of this type. We shall discuss this topic later on. Here, it is important to note that the main feature of these designs is that the number of assemblies N is a multiple of p^{m-k}. Thus, we have a larger flexibility in the choice of the values of N compared to designs for which $n = 1$.

When the experimental costs for including different treatments in the experiment are quite high, then it is important that, for any given situation, we should be able to provide a design for which N has any particular value that we desire. Obviously, this characteristic is not possessed by the parallel-plane designs of type A or B since N has to be a multiple of p^{m-k} and since, for nontriviality, we require $k < m$. For these and other reasons to be explained later, balanced designs were introduced in Srivastava (1961), wherein some of the detailed algebraic and other statistical properties were developed. We now proceed to consider this work, which has been extended and also generalized in many directions.

We first define a *balanced array* (BA), which is a natural generalization of an orthogonal array. A balanced array with s symbols, m constraints, N assemblies, and strength t is a matrix (say, T) of size $(m \times N)$ such that the elements of T belong to the set $\{0, 1, \ldots, (s-1)\}$ and, furthermore, such that, for every submatrix $T_0(t \times N)$ contained in T, and for any two column vectors $\mathbf{a}$ and $\mathbf{b}$, each of size $(t \times 1)$, such that both $\mathbf{a}$ and $\mathbf{b}$ are obtainable from each other by permuting the elements, it is true that the number of times $\mathbf{a}$ occurs as a column of T_0 is the same as the number of times $\mathbf{b}$ occurs as a column of T_0. Furthermore, we require that the number of times $\mathbf{a}$ (or $\mathbf{b}$) occurs as a column of T_0 be the same for all T_0 of size $(t \times N)$ contained in T. Such an array is denoted as $BA(s, m, N, t)$.

As an example, the matrix T in (60) below is a $BA(3, 4, 11, 2)$:

$$T = \begin{bmatrix} 1 & 0 & 0 & 1 & 1 & 1 & 2 & 2 & 2 & 2 & 2 \\ 1 & 1 & 2 & 0 & 1 & 2 & 0 & 1 & 2 & 2 & 2 \\ 1 & 1 & 2 & 1 & 2 & 0 & 2 & 0 & 1 & 2 & 2 \\ 1 & 2 & 1 & 1 & 0 & 2 & 2 & 1 & 0 & 2 & 2 \end{bmatrix} \tag{60}$$

To illustrate the definition, let us take $T_0 (2 \times 11)$ as the submatrix obtained by taking the 2nd and 3rd rows of T. Let $\mathbf{a} = (0, 2)'$, and $\mathbf{b} = (2, 0)'$. Then $\mathbf{a}$ and $\mathbf{b}$ are obtainable from each other by permuting the elements. Notice that both $\mathbf{a}$ and $\mathbf{b}$ occur exactly once as a column T_0. Notice that this is true, irrespective of which two rows of T are chosen to constitute T_0. If we take $\mathbf{a} = (0, 0)'$, then there is no vector $\mathbf{b}$ which is a permutation of $\mathbf{a}$ and which is distinct from $\mathbf{a}$. Thus, given any individual T_0, the condition that $\mathbf{a}$ occurs as many times as any permutation of $\mathbf{a}$ as a column of T_0 is automatically satisfied. In the above case, it will be noticed that $\mathbf{a}$ does not occur in T_0, irrespective of how T_0 is chosen. The following theorem summarizes many elementary facts concerning balanced arrays.

Theorem 5.9

1. Every OA is a BA of the same strength, but every BA is not necessarily an OA.
2. A $BA(s, m, N, t)$ exists for every value of N.
3. Every BA of strength t is also a BA of strength $(t - 1)$, for every positive integer t, but the converse is not generally true.
4. Consider a BA with s symbols. Suppose we take a vector $\mathbf{a}(m \times 1)$ which contains some or all of the s symbols, where any symbol may appear 0, or 1, or more times. Let T be an array with m rows which contains all possible distinct permutations of the vector $\mathbf{a}$. Then T is a BA of strength m.
5. Suppose T_i $(i = 1, \ldots, g)$ are BA's of strength t whose elements come from the set of symbols $\{0, 1, \ldots, (s - 1)\}$. Let

$$T = [T_1 : T_2 : \ldots T_g] \tag{61}$$

   Then T is a BA of strength t with s symbols.
6. Let T be a $BA(s, m, N, t)$. Let T^* be made from T as follows. Let β and γ be any two symbols which appear in T. Then, replace β

with γ everywhere, and call the resulting array T^*. Then T^* is a $BA(s-1, m, N, t)$.

7. If T is a BA with s symbols, then T does not necessarily involve all the s symbols; indeed, it may involve only s' symbols, where $1 \le s' \le s$.

8. Consider BA's with two symbols. Let Ω_{mj} $(j = 0, 1, \ldots, m)$ be the $m \times \binom{m}{j}$ matrix whose columns constitute the set of all distinct $(m \times 1)$ vectors of weight j. Then Ω_{mj} is a $BA(2, m, \binom{m}{j}, m)$.

9. Let T be a $BA(s, m, N, t)$. Let $T_1(m \times N_1)$ be a submatrix of T which is also a $BA(s, m, N_1, t)$. Let $T_2(m \times N_2)$, where $N_2 = N - N_1$, be the matrix obtained by deleting T_1 from T. Then T_2 is a $BA(s, m, N_2, t)$.

10. Let T be a $BA(s, m, N, t)$, and let T^* be an $(m_1 \times N)$ submatrix of T where $m_1 \ge t$. Then T^* is a $BA(s, m_1, N, t)$.

A few explanatory remarks on the above theorem should be useful. Thus, for example, the matrix T at (54) which is an $OA(3, 13, 27, 2)$ is also a $BA(3, 13, 27, 2)$. The first column of this T, which consists of 0 everywhere, constitutes in itself a $BA(3, 13, 1, 13)$. If we take this first column out of T, then the remaining (13×26) matrix (say, T_1) is a $BA(3, 13, 26, 2)$, which is not an OA. Let T_3 be the (4×8) matrix obtained by deleting the first and the last two columns in T at (60). Then T_3 is a $BA(3, 4, 8, 2)$. If, in this BA, we replace 2 by 0 everywhere, then we get a $BA(2, 4, 8, 2)$ given by T_4 where

$$T_4 = \begin{bmatrix} 0 & 0 & 1 & 1 & 1 & 0 & 0 & 0 \\ 1 & 0 & 0 & 1 & 0 & 0 & 1 & 0 \\ 1 & 0 & 1 & 0 & 0 & 0 & 0 & 1 \\ 0 & 1 & 1 & 0 & 0 & 0 & 1 & 0 \end{bmatrix} \tag{62}$$

Notice that the column number 6 of T_4 constitutes Ω_{40}; removing this column from T_4 gives us a $BA(2, 4, 7, 2)$.

The BA's were first defined in Chakravarti (1956), where they were called "partially balanced arrays." In that paper, only the definition was given, and no further statistical or combinatorial developments were made.

In recent years, the term "partially balanced arrays" PBA has been used for another kind of array by Kuwada. (See, for example, Kuwada, 1988). The PBA's are a generalization of the BA's, and are defined as follows.

Consider a $(0, 1)$ matrix $T(m \times N)$. Let $m = m_1 + m_2$, and suppose T is divided into two submatrices $T_1(m_1 \times N)$ and $T_2(m_2 \times N)$. Let t_1, t_2

be positive integers. Then T is a PBA with N runs, $(m_1 + m_2)$ factors, and strength (t_1, t_2) if the following holds. Let $T_{j0}(t_j \times N)$ $(j = 1, 2)$ be any submatrices of T_j, and let $T_0 = \begin{bmatrix} T_{10} \\ T_{20} \end{bmatrix}$. Then the number of columns $\begin{pmatrix} \mathbf{u}_1 \\ \mathbf{u}_2 \end{pmatrix}$ of T_0 [where $\mathbf{u}_j$ is a column of T_{j0} $(j = 1, 2)$], which are such that $\mathbf{u}_j$ $(j = 1, 2)$ is of weight i_j $(j = 1, 2)$ (where $0 \leq i_j \leq t_j$), equals $\mu(i_1, i_2)$, and does not depend upon the T_0 with which we start so long as T_0 has the above structure. Another generalization of a BA called "G-balanced arrays," has been made in Bose and Iyer (1984).

The theory of BA's was systematically attacked in Srivastava (1972), Chopra and Srivastava (1975), and in many other papers of Srivastava and Chopra. Some other authors such as Seiden, Zemach, Rafter, and certain Japanese authors have also contributed to this field. Many of these papers are included in the references at the end.

In the next theorem, we give an upper bound on the value of m for fixed values of s, N, and t in order that an OA(s, m, N, t) may exist. These conditions are usually known as "Rao-inequality." They were later discovered independently in the field of coding theory by Hamming (1950), and so now are referred to as "Rao–Hamming bounds."

Theorem 5.10   A necessary condition that an OA(s, m, N, t) may exist is that the following inequalities are satisfied. Two cases arise according to whether t is odd or even. The inequalities in (63a) and (63b), respectively, refer to the cases $t = 2g$ and $t = 2g + 1$:

$$\binom{m}{0} + \binom{m}{1}(s-1) + \binom{m}{2}(s-1)^2 + \cdots + \binom{m}{g}(s-1)^g \leq N \qquad (63a)$$

$$\binom{m}{0} + \binom{m}{1}(s-1) + \binom{m}{2}(s-1)^2 + \cdots$$
$$+ \binom{m}{g}(s-1)^g + \binom{m-1}{g}(s-1)^{g+1} \leq N \qquad (63b)$$

Detailed work on BA's has been done in the above-mentioned papers. Some of the major techniques that have been found successful involve the so-called single, double, and triple linear diophantine equations (i.e., equations which have to be solved only in terms of positive integer values of the variables). Most of the work concerned deals with the case $s = 2$, although some work is also available for $s = 3$ [see, for example, Srivastava and Ariyaratna (1981) and Kuwada (1980)]. In this chapter, we shall be concerned mostly with factorial designs of the 2^m series. Because of this, it is useful to consider some of the facts for the case $s = 2$ in more detail.

Consider a BA$(2, m, N, t)$, denoted by T. Let μ_{ti} $(i = 0, 1, \ldots, t)$ denote the number of times any $(t \times 1)$ vector of weight i occurs as a column in any submatrix $T_0(t \times N)$ of T. Then this set of numbers $(\mu_{t0}, \mu_{t1}, \ldots, \mu_{tt})$ $(= \mu_t'$, say) is called the "index set" of the array T. Thus, for example, the index set of the array at (62) is (3,2,1). It is easy to check that the index set must satisfy the equation

$$N = \mu_{t0} \binom{t}{0} + \mu_{t1} \binom{t}{1} + \cdots + \mu_{tt} \binom{t}{t} \tag{64}$$

A balanced array with parameters (s, m, N, t) is said to be of "full strength" if $t = m$; such an array is also called "simple." Thus, the arrays Ω_{mj} introduced earlier are simple arrays. An important result by Kuriki says that certain BA's are always simple. In Srivastava (1972), necessary and sufficient conditions for the existence of a BA$(2, m, N, t)$, have been obtained for the case where $m \leq (t + 2)$.

6   INFORMATION MATRIX, OPTIMAL DESIGNS

We shall now present a brief discussion of the salient features of some of the important work done in the field of statistical estimation of parameters in a factorial experiment, and the related problem of obtaining a design which would maximize the accuracy of the estimates.

Consider an $s_1 \times s_2 \times \cdots \times s_m$ factorial experiment. Let $s_1 \times s_2 \times \cdots \times s_m = \nu_1$. Then, in this experiment, we have ν_1 treatments, and also ν_1 factorial effects. Let $\phi_1(\nu_1 \times 1)$ be the vector whose elements are $\phi(\mathbf{t})$, where $\mathbf{t}$ ranges over the set of all treatments. The order in which the treatments occur in ϕ_1 will be specified later in special situations to the extent needed. Similarly, let $\mathbf{F}_1(\nu_1 \times 1)$ denote the vector whose elements are the set of all distinct factorial effects. Broadly speaking, the first element of $\mathbf{F}_1$ will always be μ, the general mean. This will be followed by the main effects, then the two-factor interactions, then the three-factor interactions, and so on, up to the m-factor interaction. For the 2^m case, sometimes, the elements of $\mathbf{F}_1$ will be arranged in the Yates order explained earlier. For situations where some factors have more than two levels, and even for some situations of the 2^m type, the elements of $\mathbf{F}_1$ may be arranged in some other order. The details will be specified as needed.

The vectors F_1 and ϕ_1 are mutually related by the linear equations

$$F_1 = X'_{(1)}\phi_1, \qquad (65)$$

where X_1 is a $(\nu_1 \times \nu_1)$ matrix. The relationship (65) is simply a restatement of the definition of the elements of F_1. It turns out that any two rows of the matrix $X'_{(1)}$ are mutually orthogonal. Let $D_1(\nu_1 \times \nu_1)$ be a diagonal matrix whose ith diagonal element equals the square root of the sum of squares of the ith row of $X'_{(1)}$. Then, it is easy to verify that $D_1^{-1}X'_{(1)}$ is an orthogonal matrix. Because of this we can solve the equation (65) to obtain

$$\phi_1 = X_{(1)}D_1^{-2}F_1 \qquad (66)$$

Notice that the above equation implies that if all the elements of F_1 were known, we could obtain all the elements of ϕ_1. *Notice that the object of the experiment is to somehow estimate or determine the value of all the elements of ϕ_1 without doing the whole experiment.*

Suppose now that the vector $F_1(\nu_1 \times 1)$ is divided into two subvectors $L(\nu \times 1)$ and $L_0(\nu_0 \times 1)$, such that

$$\nu + \nu_0 = \nu_1 \qquad (67)$$

and such that L and L_0 are mutually disjoint; i.e., they have no common elements. Now, suppose we assume that the elements of L_0 are all negligible. Suppose that $D^{-2}(\nu \times \nu)$ is a submatrix of D_1^{-2} such that the columns of D_1^{-2} which correspond to the elements of L_0 are deleted, and the rows corresponding to the deleted columns are also deleted. Let $X_{(10)}(\nu_1 \times \nu_1)$ be the submatrix of $X_{(1)}$ in which the columns corresponding to L_0 are deleted. Then under the assumption that the elements of L_0 are negligible, we have

$$\phi_1 = X_{(10)}D^{-2}L \qquad (68)$$

Now, suppose that all the ν_1 treatments of the experiment are not tried equally often; in other words, each treatment may be tried 0, 1, or more times. Let N be the total number of treatments (not necessarily distinct) which are included in the experiment. Let $\phi(N \times 1)$ be a vector obtained from ϕ_1 by deleting those elements which correspond to treatments which are not tried, and repeating the remaining elements of ϕ_1 according to the number of times they occur in the experiment. Similarly, suppose $X(N \times \nu)$ is obtained from $X_{(10)}$ by deleting and/or repeating rows of $X_{(10)}$

corresponding to the elements of ϕ_1 which have been deleted or repeated. Then, the last equation gives

$$\phi = XD^{-2}\mathbf{L} \qquad (69)$$

Let $\mathbf{y}(N \times 1)$ denote the vector of observed yields, such that each element of $\mathbf{y}$ corresponds to a particular treatment which has been included in the experiment, and such that the elements of $\mathbf{y}$ correspond in order to the elements of ϕ so that we have

$$E(\mathbf{y}) = \phi \qquad (70)$$

where, for simplicity, we are assuming that there are no nuisance factors. Then this experiment gives rise to the linear model

$$E(\mathbf{y}) = XD^{-2}\mathbf{L}, \qquad V(\mathbf{y}) = \sigma^2 I_N \qquad (71)$$

Now, the statistical problem is to estimate $\mathbf{L}$, given that we have observed $\mathbf{y}$. The best linear unbiased (BLU) estimate of $\mathbf{L}$ is $\hat{\mathbf{L}}$, where $\hat{\mathbf{L}}$ satisfies the so-called normal equations given by

$$(D^{-2}X'XD^{-2})\hat{\mathbf{L}} = D^{-2}X'\mathbf{y} \qquad (72)$$

This leads to the following important result, on which all further statistical theory is based.

Theorem 6.1   In the experiment under consideration, $\mathbf{L}(\nu \times 1)$ is estimable if and only if $(X'X)$ is nonsingular. If $(X'X)$ is nonsingular, then the estimate of $\mathbf{L}$ is given by $\hat{\mathbf{L}}$, where

$$\hat{\mathbf{L}} = D^2(X'X)^{-1}X'\mathbf{y} \qquad (73)$$

Also, we have

$$V(\hat{\mathbf{L}}) = \left\{ D^2(X'X)^{-1}D^2 \right\} \sigma^2 \equiv V\sigma^2, \text{ say} \qquad (74)$$

The above shows that the accuracy of the estimate $\hat{\mathbf{L}}$ depends upon the matrix $M(\nu \times \nu)$ defined by

$$M = D^{-2}(X'X)D^{-2} \qquad (75)$$

The matrix M is usually called the "information matrix." Now, the variance of $\hat{\mathbf{L}}$ is proportional to M^{-1}, in case M is nonsingular. Thus, it is clear that first we want M to be nonsingular so that $\mathbf{L}$ is estimable, and second that in some sense M should be "large" so that its inverse (which is proportional to the variance) is "small." Given $\mathbf{L}$, the theory of nonsingular designs deals with the choice of a set of treatments T (which influences the rows of $X_{(1)}$ which are deleted or repeated in order to form the rows of X) so that M is nonsingular. The theory of optimal designs deals with the further problem of choosing T so that in some sense the matrix M is maximized. Usually, for this purpose, certain functions of M such as the determinant of M ($|M|$), the smallest characteristic root of $M(\mathrm{ch_{min}}(M))$, and the sum of the diagonal elements of $M^{-1}(\mathrm{tr}\,M^{-1})$ are selected. In response surface work, a criterion known as integrated mean square error (IMSE) is used by certain authors; an analogous criterion has been given in the book of Federov (1972). All of these (and many other) criteria are interrelated, and most designs turn out to be simultaneously optimal under the different criteria. Different authors have emphasized different criteria in their work. In the field of discrete factorial experiments, the author and others have used the trace criterion most often. A discussion of these criteria will be found in Srivastava and Anderson (1974). A large amount of theory dealing with the equivalence or otherwise of these criteria is due to Kiefer and other authors. Kiefer and certain other authors in the field of continuous optimal design theory usually use the determinant criterion.

Note that the number of choices of the design is very large, being ν_1^N. Designs with N treatments, which are optimal in this class of all possible designs, are known only for a few cases. Under certain situations, for example, orthogonal arrays are optimal. Similarly, under certain other situations, certain balanced arrays have been shown to be optimal [Cheng (1980)]. Most of the work on optimal design has been done by obtaining designs which are optimal in certain important subclasses. In the theory of discrete designs, the two large subclasses are the class of balanced designs (in which balanced arrays have been used) and the class of parallel-flats designs. The combinatorial structures corresponding to these were introduced in the last section, and now we shall look into the statistical aspects of the same.

Before we proceed further, it is important to comment on the nature of the vectors $\mathbf{L}$ which have been considered. Usually, the following kind of $\mathbf{L}$ is considered. We include in $\mathbf{L}$, the general mean μ and all factorial effects involving l or a smaller number of factors. Designs which allow the estimation of $\mathbf{L}$ of this kind are said to be of resolution $(2l + 1)$. Another class of designs has been considered where $\mathbf{L}$ consists of μ and all factorial effects involving $(l + 1)$ or a lesser number of factors, but interest lies in estimating a subset of parameters which consists of μ and effects involving only l or a lesser number of factors. Designs which allow us to do this are said to be of resolution $(2l + 2)$. The most important designs from the practical point of view are of resolution III, IV, and V.

As mentioned earlier, in actual practice, it has been found as an empirical fact that as l increases, the number of factorial effects which involve l factors and which are nonnegligible decreases. However, in a specific situation, it is difficult to say which interactions are negligible. For example, in a large number of situations, it may be that almost all interactions might be negligible, but to claim that absolutely all interactions are negligible may be quite farfetched. The available theories have been built more because of the mathematical ease of developing the corresponding designs than because of the correctness of the assumption behind **L**. Thus, it should be kept in mind that the objectives of the experiment ought not to be to assume some **L** to be nonnegligible, the remaining parameters to be negligible, and then to simply proceed aggressively with the estimation of **L**. Some people mistakenly believe that such a procedure may lead them to some small errors; however, this belief is quite unfounded. In situations where the results of the experiment are going to influence the quality of a product, the consequences of such mistakes could result in losses of millions of dollars because of the inferior quality of the resulting product. Thus, although it is important to develop theory in which we try to maximize the information matrix, it is far more important to make sure that all the nonnegligible parameters have been identified. The idea of reducing the bias is emphasized also in Box and Draper (1987), in the context of a response surface theory.

Besides the designs of resolution l for various values of l, there are certain other structures which also seem to be very important. One such structure is the tree structure, which we now define.

Consider a positive integer $l > 1$. Consider an l-factor interaction involving factors $i_1, i_2, \ldots, i_l$, where $1 \leq i_1, i_2, \ldots, i_l \leq m$, and where the i are all distinct. Then, **L** is said to have a tree structure if and only if whenever **L** has an element belonging to it which represents a component of the l-factor interaction involving factors $i_1, i_2, \ldots, i_l$ (where the i are as above), then, **L** also has another element in it which corresponds to an $(l - 1)$-factor interaction involving factors $j_1, \ldots, j_{l-1}$, where the j are all distinct and are a subset of the set $(i_1, \ldots, i_l)$. For example, suppose $m = 6$, and suppose **L** has a component which belongs to the interaction $F_2 F_4 F_5$. Then, if **L** has a tree structure, **L** should have at least one element which belongs to either the interaction F_2F_4, F_2F_5, or F_4F_5. It may be remarked here that the author noticed the tree structure in investigations on the experiments in the field of social sciences and psychology. This work was contained in a report by the author jointly with P. Srivastava (1977). After this, the author has found this structure to be present in experiments in many other sciences as well. Quite often, the structure is present but not always, there being a few exceptions. However, in almost all cases examined, the exceptions were very few indeed.

At the time of this writing, the author intuitively believes that, in most situations, the structure of L^* has two important features S_1 and S_2, where S_1 says that there is essentially a tree structure, and S_2 says that the number of three-factor and higher interactions is not too large. Also, the author believes that, in most cases, the ultimate objective of the experiment must be to estimate $\phi(\mathbf{t})$, for all the $(s_1 \times \cdots \times s_m)$ runs $\mathbf{t}$ with which the experiment is concerned, Thus, because of these facts (concerning the nature of L^*, and the objective of the experiment), an appropriate statistical approach to the planning of the experiment would be as follows. Firstly, we should select an L which includes all factorial effects which the experimenter suspects to be nonnegligible. Next, this L should be augmented in a proper way to obtain a new L which would have S_1 (tree structure). Finally, we should choose a design which would estimate this L, and also identify and estimate L^0. At this point, it should be remembered that a single-stage design for this purpose will usually tend to have a relatively large number of runs, and, furthermore, that one cannot guarantee that L^0 will be identifiable and estimable. Thus, one should always be prepared for an experiment which may involve several stages.

From the above point of view, it is clear that the use of designs of resolution VI and higher would generally have two disadvantages. Firstly, the set L may be excessively large (compared to L^*), resulting in an unnecessarily large value of N. Secondly, of course, there is no provision for identifying the set L^0. Thus, they are excessively large, and yet miss a part of the problem which could be major.

Designs of resolution V may be excessively large in some cases, and may not be so in certain others. However, these also may miss a major part of the problem, because there is no provision for identifying L^0. Designs of resolution III are of course available in small sizes. However, in these cases, the set L^0 may be excessively large. Thus, such designs may be missing a very large class of nonnegligible parameters, and may indeed be quite *dangerous* in most situations.

The situation with respect to designs of resolutions IV appears to be more complicated. If $\mathbf{L}^0$ is large, and contains components of interactions involving an odd number of factors, then such designs could be very misleading. Even if most of the elements of $\mathbf{L}^0$ are components of interactions involving only an even number of factors, the designs of resolution IV could still be very inadequate. The reason is that the purpose of doing a factorial experiment in most cases is not to estimate the main effects; *the purpose is to estimate the true yield for every treatment of the experiment, without doing the experiment. Indeed, we need to do an experiment which is as small as possible, but, at the same time, is able to identify* $\mathbf{L}^*$, *and estimate it with sufficient accuracy.*

It was established by C. R. Rao that orthogonal arrays of strength $2l$ constitute designs of resolution $(2l+1)$, which are optimal in the class of all designs [of resolution $(2l + 1)$]. Because of the existence of the Hadamard matrices, resolution III designs (which are orthogonal arrays) are available for all values of N which is a multiple of 4. Furthermore, for other values N, Galil and Kiefer (1980) and others have filled many gaps. For resolution V and VII, optimal balanced designs have been obtained by the author and collaborators, and also by several Japanese statisticians. A number of references have been given at the end; these are only illustrative and not exhaustive.

Designs of the parallel-flats type (for both symmetrical and asymmetrical factorial experiments) have been developed for the situation where $\mathbf{L}$ corresponds to a design of resolution $(2l + 1)$, for $l = 1, 2$. Generally, the optimality of such designs in any particular class has not been established. However, there is a large body of theory of such designs where $\mathbf{L}$ is completely general and does not necessarily correspond to the designs of a particular resolution.

All of the above work addresses the situation where one desires to estimate an $\mathbf{L}$, assuming that the given parameters $\mathbf{L}_0$ are negligible. This work does not address itself to the problem of identifying $\mathbf{L}^0$ and estimating the same. As mentioned earlier, single-stage designs which address problems of this sort are known as "search designs." Sequential designs for similar purposes (where we do not necessarily start with an $\mathbf{L}$) have been recently introduced; these are called "probing designs."

We now discuss the 2^m designs of resolution $(2l + 1)$, which are balanced. By "balanced" we mean the following: suppose θ_1 and θ_2 are two (not necessarily distinct) elements of $\mathbf{L}$ such that θ_i $(i = 1, 2)$ involves m_i factors, and such that between θ_1 and θ_2, exactly m_3 factors are common.

Also, suppose $\hat{\theta}_1$ and $\hat{\theta}_2$ are the BLU estimates, respectively, of θ_1 and θ_2 [obtained in (73)]; then the value of $\text{Cov}(\hat{\theta}_1, \hat{\theta}_2)$ depends merely on the numbers m_1, m_2, and m_3 and does not depend otherwise on the θ_1 and θ_2 with which we start.

As an example in the context of balanced designs of resolution V, the above implies that the covariance matrix V [at (73)] has at most 10 distinct elements, say, $v_1, \ldots, v_{10}$, where

$$v_1 = V(\hat{\mu}), \qquad v_2 = V(\hat{F}_i), \qquad v_3 = V(\hat{F}_{ij}), \qquad v_4 = \text{Cov}(\hat{\mu}, \hat{F}_i)$$
$$v_5 = \text{Cov}(\hat{\mu}, \hat{F}_{ij}), \qquad v_6 = \text{Cov}(\hat{F}_i, \hat{F}_j), \qquad v_7 = \text{Cov}(\hat{F}_i, \hat{F}_{ij}) \quad (76)$$
$$v_8 = \text{Cov}(\hat{F}_i, \hat{F}_{jk}), \qquad v_9 = \text{Cov}(\hat{F}_{ij}, \hat{F}_{ik}), \qquad v_{10} = \text{Cov}((\hat{F}_{ij}, \hat{F}_{kl})$$

where, in the above, i, j, k, l, are all distinct and $1 \leq i, j, k, l \leq m$. The above means that we can express V as

$$V = \sum_{i=1}^{10} v_i M_i \tag{77}$$

where the M_i $(i = 1, \ldots, 10)$ are $(0,1)$ matrices. It turns out that the matrices M_i have important algebraic properties, being related to the multidimensional partially balanced association schemes and the related linearly associative algebras (which were introduced in Srivastava (1961) and Bose and Srivastava (1964b)). The following result is a major consequence of this important algebra of Bose and Srivastava (which is a generalization of the earlier defined "Bose–Mesner algebra").

Theorem 6.2 If V has the structure as at (77), then M has the same structure and is given by

$$M = \sum_{u=1}^{10} m_u M_u \tag{78}$$

where m_u are certain real numbers.

Results similar to equations (77) and (78) hold for all balanced designs of resolution $(2l+1)$ for all positive integers l, although, in this chapter, we shall consider only the case $l = 2$ in detail. From the results of the above type, the following theorem can be readily derived.

Theorem 6.3 Let $T(m \times N)$ be a $(0,1)$ matrix whose columns represent treatments from a 2^m factorial experiment. Then T is a balanced design of resolution $(2l + 1)$ only if T is a BA of strength $2l$. Also, if T is a BA of strength $2l$, and if the information matrix corresponding to the design T is nonsingular, then T is a balanced design of resolution $(2l + 1)$. Furthermore, if $T(m \times N)$ is a BA of strength $2l$, and if T is used as a design [of resolution $(2l + 1)$] for a 2^m factorial experiment, then every element of the information matrix corresponding to T will be a function only of m and the index set of T. If the information matrix of T is nonsingular, then the covariance matrix of $\hat{\mathbf{L}}$ will also be a function only of m and the index set of T.

Now, suppose that, given an even integer l, we want to obtain a design $T(m \times N)$ which is balanced and of a resolution of $(l+1)$, Then, in view of the above results, we need to choose an index set, say, $\boldsymbol{\mu}_l' \equiv (\mu_{l0}, \ldots, \mu_{ll})$ satisfying (64), such that the corresponding information matrix M is nonsingular. As the above theorem tells, the elements of M are functions of m and $\boldsymbol{\mu}_l$. Using the Bose–Srivastava algebra, one can compute the characteristic roots of the matrix M. Now, a necessary and sufficient condition that M is nonsingular is that each of its roots be positive. This then gives valuable conditions on $\boldsymbol{\mu}_l$ so that M is nonsingular. Thus, to obtain a nonsingular balanced design T, we simply choose a $\boldsymbol{\mu}_l$ satisfying the last-mentioned condition.

Next, we consider optimal balanced designs of resolution $(l + 1)$, with l even. We consider, proceeding as above, index sets $\boldsymbol{\mu}_l$ such that M is nonsingular. Using the theory referred to in the last paragraph, for each of these index sets, one may obtain the different characteristic roots of M. Now, the determinant of M ($|M|$) is the product of the characteristic roots. Similarly, $\operatorname{tr} V$ is the sum of the reciprocals of these characteristic roots. Suppose that we are working with $\operatorname{tr} V$. Then, given any index set $\boldsymbol{\mu}_l$ (which satisfies the condition that M is nonsingular), we obtain $\operatorname{tr} V$. The index set for which $\operatorname{tr} V$ is minimized is then considered.

Next, we consider making a BA of size $(m \times N)$ which has the index set just obtained. This work may turn out to be anywhere from moderately easy to excessively difficult, or impossible. Our experience shows that, in general, this work is quite nontrivial. If we succeed in making the required BA, then our problem of obtaining the optimal balanced design is solved. On the other hand, if we prove that it is impossible to make a BA with the given index set, we try to find the index set which gives the next-larger value of $\operatorname{tr} V$ (i.e., an index set for which $\operatorname{tr} V$ is minimum for all possible index sets except the one for which we have proved that no BA exists). If we can make a BA with the new index set, our problem is solved. On the other hand, if we prove that no BA exists with this second index set, then we proceed as before to obtain a third index set, and so on. We continue until we find an index set for which a BA exists. This BA will then be the optimal balanced design for the given value of N and t. Sometimes, this procedure can be shortened by showing that certain relatively large classes of BA's do not exist combinatorially.

Optimal balanced designs of resolution V for values of m in the range $4 \leq m \leq 8$ are obtained in a series of papers by Srivastava and Chopra. Some further work for the cases $m = 9$ and 10 has been done by Chopra. Some of these papers are listed at the end, and their titles reveal the cases that are covered. (For each m, designs have been given for *every* value of N in a large range of values that might actually arise). In almost all cases, the above optimal balanced designs turn out to be simple BA's. Because of this, it is easy to describe them. Now, suppose an array $T(m \times N)$ is simple. Then, it is easy to see that T consists of a few copies of Ω_{mj} $(j = 0, 1, \ldots, m)$ taken together, where for any j, Ω_{mk} may be totally left out or may be repeated one or more times. Thus, suppose, for every permissible j, Ω_{mj} is repeated λ_j times in T. Then, T can be described by the vector $\boldsymbol{\lambda}' \equiv (\lambda_m, \lambda_{m-1}, \ldots, \lambda_1, \lambda_0)$. Thus, for illustration, the simple array T for $m = 4$, and $\boldsymbol{\lambda}' = (1, 0, 0, 1, 2)$ is given by

$$
T = \begin{bmatrix} 0 & 0 & 1 & 0 & 0 & 0 & 1 \\ 0 & 0 & 0 & 1 & 0 & 0 & 1 \\ 0 & 0 & 0 & 0 & 1 & 0 & 1 \\ 0 & 0 & 0 & 0 & 0 & 1 & 1 \end{bmatrix} \tag{79}
$$

Also, in view of (76), it is clear that the covariance matrix of each optimal balanced balanced design of resolution V will have at most 10 distinct elements. Thus, it is easy to describe the covariance matrix of each such design by simply giving the vector of these ten numbers.

One important quantity to take into account while planning an experiment is the "absolute efficiency" of a design. Let T be a design with N treatments, for a 2^m factorial experiment. Suppose that $\mathbf{L}$ is of size $(\nu \times 1)$, so that a total of ν parameters are being estimated. Suppose that we are interested in the trace criterion; in other words, we measure the goodness of a design by the trace of the covariance matrix of the estimates of the parameters. Corresponding to the design T, let V_T denote the covariance matrix of the estimate of the parameters. Then we define

Absolute efficiency of T (under trace criterion) $= (\nu/N)/(\operatorname{tr} V_T)$ (80)

Similarly for the determinant criterion we shall have

Absolute efficiency of T (under determinant criterion) $= (1/N)(|V_T|))^{1/\nu}$
(81)

As illustration, in Table 2, we present for $m = 5$ and 6 and, for various values of N, the values of $\boldsymbol{\lambda}'$ corresponding to optimal designs, and also their absolute efficiency E with respect to the trace criterion.

We now briefly consider the use of Table 2. First, the following result is useful in the sequel.

Theorem 6.4

1. For any given positive integer m, and any nonnegative integer j such that $0 \le j \le m$, Ω_{mj} is a BA of strength l ($0 \le l \le m$), with index set given by

$$\mu_{li} = \binom{m-l}{j-i}; \qquad i = 0, 1, \ldots, l \tag{82}$$

2. Consider a simple array $T(m \times N)$ for which the parameter $\boldsymbol{\lambda}' = (\lambda_m, \lambda_{m-1}, \ldots, \lambda_1, \lambda_0)$; then, the index set of T (considered as an array of strength l) is given by

$$\mu_{li} = \sum_{j=0}^{m} \lambda_j \binom{m-l}{j-i}, \qquad \text{for } i = 0, 1, \ldots, l \tag{83}$$

3. Consider a design T as in part (2) above. Then, in the notation of (78) above, the elements of the information matrix M_T corresponding to the design T (considered as a design of resolution V) are given by

$$m_1 = m_2 = m_3 = N = \sum_{j=0}^{m} \lambda_j \binom{m}{j}$$

$$m_4 = m_7 = \mu_{44} + 2\mu_{43} - 2\mu_{41} - \mu_{40}$$

$$m_5 = m_6 = m_9 = \mu_{44} - \mu_{42} + \mu_{40} \tag{84}$$

$$m_8 = \mu_{44} - 2\mu_{43} + 2\mu_{41} - \mu_{40}$$

$$m_{10} = \mu_{44} - 4\mu_{43} + 6\mu_{42} - 4\mu_{41} + \mu_{40}$$

Table 2

N	E	λ5	λ4	λ3	λ2	λ1	λ0	N	E	λ6	λ5	λ4	λ3	λ2	λ1	λ0
		$m = 5$								$m = 6$						
16	100	0	1	0	1	0	1	22	61	1	1	0	0	1	0	0
17a	97	0	1	0	1	0	2	23	85	0	1	0	0	1	0	2
17b		1	0	1	0	1	1	24	82	1	1	0	0	1	0	2
18	95	1	1	0	1	0	2	25	79	2	1	0	0	1	0	2
19a	91	1	1	0	1	0	3	26	77	2	1	0	0	1	0	3
19b		2	0	1	0	1	2	27	84	1	0	0	1	0	1	0
20	87	2	1	0	1	0	3	28	83	2	0	0	1	0	1	0
21a	90	0	1	0	1	1	1	29	81	2	0	0	1	0	1	1
21b		1	0	1	0	2	0	30	79	3	0	0	1	0	1	1
22a	89	1	1	0	1	1	1	31	94	1	0	1	0	1	0	0
22b		2	0	1	0	2	0	32	100	1	0	1	0	1	0	1
23a	87	1	1	0	1	1	2	33	99	1	0	1	0	1	0	2
23b		2	0	1	0	2	1	34	98	2	0	1	0	1	0	2
24a	85	2	1	0	1	1	2	35	95	3	0	1	0	1	0	2
24b		3	0	1	0	2	1	36	94	3	0	1	0	1	0	3
25a	82	2	1	0	1	1	3	37	95	1	0	1	0	1	1	0
25b		3	0	1	0	2	2	38	95	2	0	1	0	1	1	0
26a	90	0	1	0	2	0	1	39	94	2	0	1	0	1	1	1
26b		1	0	1	1	1	0	40	93	3	0	1	0	1	1	1
27a	90	0	1	0	2	0	2									
27b		1	0	1	1	1	1									
28a	91	1	1	0	2	0	2									
27b		1	0	1	1	1	1									
29a	89	1	1	0	2	0	3									
29b		2	0	1	1	1	2									
30	91	0	1	1	1	1	0									
31a	97	0	1	1	1	1	1									
31b		1	0	2	0	2	0									
32a	100	0	2	0	2	0	2									
32b		1	1	1	1	1	1									

Now, consider any design given in Table 2. To fix ideas, we will take the design with $m = 6$, $N = 23$. For this design, the value of $\boldsymbol{\lambda}' = (0\ 1\ 0\ 0\ 1\ 0\ 2)$, and the value of the index is $\boldsymbol{\mu}_4' = (3\ 2\ 1\ 1\ 2)$. The values of $\boldsymbol{\mu}_4'$ and $\boldsymbol{\lambda}'$ can be checked to satisfy equations (83), with $m = 6$, and $t = 4$. It is easy to see that for the given value of λ', the T is given by

$$T = \begin{bmatrix} 1\,1\,1\,1\,1\,0 & 1\,1\,1\,1\,0\,1\,1\,1\,0\,1\,1\,0\,1\,0\,0 & 0\,0 \\ 1\,1\,1\,1\,0\,1 & 1\,1\,1\,0\,1\,1\,1\,0\,1\,1\,0\,1\,0\,1\,0 & 0\,0 \\ 1\,1\,1\,0\,1\,1 & 1\,1\,0\,1\,1\,1\,0\,1\,1\,0\,1\,1\,0\,0\,1 & 0\,0 \\ 1\,1\,1\,0\,1\,1 & 1\,1\,0\,1\,1\,1\,0\,1\,1\,0\,1\,1\,0\,0\,1 & 0\,0 \\ 1\,1\,0\,1\,1\,1 & 1\,0\,1\,1\,1\,0\,1\,1\,1\,0\,0\,0\,1\,1\,1 & 0\,0 \\ 1\,0\,1\,1\,1\,1 & 0\,1\,1\,1\,1\,0\,0\,0\,0\,1\,1\,1\,1\,1\,1 & 0\,0 \\ 0\,1\,1\,1\,1\,1 & 0\,0\,0\,0\,0\,1\,1\,1\,1\,1\,1\,1\,1\,1\,1 & 0\,0 \end{bmatrix} \qquad (85)$$

For this design T, the elements of the information matrix using (84) turn out to be: $m_1 = m_2 = m_3 = 23$, $m_4 = m_7 = -3$, $m_5 = m_6 = m_9 = 3$, $m_8 = 1$, and $m_{10} = -1$.

We have not given the elements of the covariance matrix. They can be obtained from the papers. However, the reader could, with a simple calculator, easily obtain these elements to any degree of precision by using the following simple procedure:

Let

$$\begin{bmatrix} N & m_4 & m_5 \\ m m_4 & N + (m-1)m_5 & 2m_4 + (m-2)m_8 \\ \imath(m-1)m_5 & (2m-2)m_4 + (m^2 - 3m + 2)m_8 & (m^2 - 5m + 6)m_{10}(m-2)m_5 + 2N \end{bmatrix}$$

$$= \begin{bmatrix} b_{11} & b_{12} & b_{13} \\ b_{21} & b_{22} & b_{23} \\ b_{31} & b_{32} & b_{33} \end{bmatrix} \qquad (86)$$

$$\begin{bmatrix} N - m_5 & m_4 - m_8 \\ (m-2)(m_4 - m_8) & N - 2m_5 + m_{10} + (m-2)(m_5 - m_{10}) \end{bmatrix}^{-1}$$

$$= \begin{bmatrix} c_{11} & c_{12} \\ c_{21} & c_{22} \end{bmatrix} \qquad (87)$$

$$\begin{bmatrix} c_{11} & c_{12} \\ c_{21} & c_{22} \end{bmatrix} \begin{bmatrix} m_4 & m_5 & m_8 \\ (m-1)m_5 & m_4 + (m-2)m_8 & (m-3)m_{10} + 2m_5 \end{bmatrix}$$

$$\times \begin{bmatrix} b_{11} & b_{12} & b_{13} \\ b_{21} & b_{22} & b_{23} \\ b_{31} & b_{32} & b_{33} \end{bmatrix}$$

$$= \begin{bmatrix} d_{11} & d_{12} & d_{13} \\ d_{21} & d_{22} & d_{23} \end{bmatrix} \tag{88}$$

Then, we have

$$v_1 = b_{11}$$
$$v_2 = c_{11} - d_{12}$$
$$v_3 = (n - 2m_5 + m_{10})^{-1}[1 - 2b_{13}m_5 - 2c_{12}(m_4 - m_8) + 4d_{13}(m_4 - m_8)$$
$$\quad - 2m_8 b_{33} - 2c_{22}(m_5 - m_{10}) + 4d_{23}(m_5 - m_{10}) - 2m_{10}b_{33}]$$
$$v_4 = b_{12} = (-d_{11})$$
$$v_5 = 2b_{13} \tag{89}$$
$$v_6 = (-d_{12})$$
$$v_7 = c_{12} - 2d_{13}$$
$$v_8 = (-2d_{13})$$
$$v_9 = (N - 2m_5 + m_{10})^{-1}[-2b_{13}m_5 - c_{12}(m_4 - m_8) + 4d_{13}(m_4 - m_8)$$
$$\quad - 2m_8 b_{23} - c_{22}(m_5 + m_{10}) + 4d_{23}(m_5 - m_{10}) - 2m_{10}b_{33}]$$
$$v_{10} = (N - 2m_5 + m_{10})^{-1}[-2b_{13}m_5 + 4d_{13}(m_4 - m_8)$$
$$\quad - 2m_8 b_{33} + 4d_{23}(m_5 - m_{10}) - 2m_{10}b_{33}]$$

The work on designs of resolution VII is very similar to the above; the interested reader is referred to the Japanese authors (e.g., Shirakura (1976b)). For designs of resolution III, as mentioned before, orthogonal arrays are available if N is a multiple of 4. Now, very often, one needs a few degrees of freedom for the estimation of error (σ^2). Also, one may need to do a multistage sequential design for the estimation of the L^0 (which is very likely to be large in case we begin with a resolution III design). Because of this, it appears that the case where N is not a multiple of 4 is not very important from the factorial design point of view.

In statistical work such as estimation of parameters, testing of hypotheses, or putting confidence intervals on parameters, the most important and cumbersome computation that needs to be done is to obtain the variance matrix of the estimate of the parameters. For balanced designs, as we have seen, this computation is relatively easy. However, for other designs, such as those based on partially balanced arrays, the computation becomes much more complicated. No doubt, for PBA's as well, the information matrix has a certain pattern. But this pattern is excessively complicated. It appears easier to directly obtain M from (75) and to invert it on the computer. The same is true for the designs which we discuss in the next section. There, the information matrix breaks down into several matrices of much smaller size. These matrices do not have much pattern. But because of their small size, they are relatively simple to invert.

7 DESIGNS BASED ON PARALLEL PLANES

Suppose s is a number such that GF(s) exists. Analogously to (58) and (59), consider the design $T(m \times n \cdot s^{m-k})$, where

$$T = [T_1 : \cdots : T_n] \tag{90}$$

and where $T_i(m \times s^{m-k})$ contains the solutions to the equations

$$Ax = c_i \tag{91}$$

where A and c_i $(i = 1, \ldots, n)$ are over GF(s). Let

$$C = [c_1, \ldots, c_n] \tag{92}$$

We now consider the information matrix for T. We will assume throughout that A is of the form

$$A = [I_k : B] \tag{93}$$

where B is of size $k \times (m - k)$. Define

$$\tilde{A} = [-B' : I_{m-k}] \tag{94}$$

At this point, two cases arise according to whether $s > 2$ or $s = 2$. Because there is a certain complication in the former case, we shall treat it separately. The case $s = 2$ is simpler, and we treat it first.

Consider $\mathbf{L}(\nu \times 1)$. This is a vector of factorial effects which we want to estimate, assuming that $\mathbf{L}_0$ is negligible. Now, for each of the 2^m possible factorial effects, we shall define the corresponding "defining vector." For the general mean μ, the defining vector will be 0_{m1}. Now, consider any factorial effect, say $\{F_{i_1} F_{i_2} \ldots F_{i_l}\}$, where $1 \le i_1 < i_2 < \cdots < i_l \le m$, $1 \le l \le m$; the defining vector for this factorial effect will be the vector

$\mathbf{e} = (e_1, \ldots, e_m)'$, where

$$e_i = \begin{cases} 1, & \text{if } i \in (i_1, i_2, \ldots, i_l) \\ 0, & \text{otherwise} \end{cases} \tag{95}$$

Let $\tilde{L}(m \times \nu)$ be the matrix whose ith $(i = 1, \ldots, \nu)$ column represents the defining vector of the ith element of $\mathbf{L}$.

We now consider a reordering of $\tilde{L}$; this is being done for convenience. This reordering of the columns of $\tilde{L}$ will induce a corresponding reordering of the elements of $\mathbf{L}$, and hence of the rows and columns of the matrices M, D, X, etc. *We assume that this has been done.* This reordering is being done for convenience, and is being done without loss of generality. To explain the reordering, we consider the matrix $(\tilde{A}\tilde{L})$ of size $(m - k \times \nu)$; we shall arrange the elements of $\tilde{L}$ in such a way that all columns of $\tilde{A}\tilde{L}$ which are equal to each other occur adjacent to each other, and all columns which equal the zero vector occur first in the matrix $\tilde{A}\tilde{L}$. We shall now group the elements of $\mathbf{L}$ and given them an additional name. All elements of $\mathbf{L}$ which are such that they give rise to identical columns in $\tilde{A}\tilde{L}$ are said to be in the same group, called an "alias-set." We shall assume that there are $(g + 1)$ alias-sets, where g is a positive integer. These sets or groups will be numbered $0, 1, \ldots, g$ where the 0th alias-set will correspond to all elements of $\mathbf{L}$ which give rise to the zero column in $\tilde{A}\tilde{L}$. Notice that the 0th alias-set could be empty; however, philosophically the author believes that μ should always be a member of $\mathbf{L}$ and, consequently, the 0th alias-set will not be empty. On the other hand, such an assumption about μ being in $\mathbf{L}$ does not have any adverse repercussions on the rest of the theory. Indeed, the theory becomes much easier if the 0th alias-set is either empty or consists of μ alone.

Let u_j $(j = 0, 1, \ldots, g)$ be the number of elements in the jth alias-set. Then, the following fundamental result of parallel-planes design holds.

Theorem 7.1 Suppose that $\tilde{L}$, $\mathbf{L}$, and M, etc., have been rearranged as above. Then, M is a direct sum of matrices M_j $(j = 0, 1, \ldots, g)$, i.e.,

$$M = M_0 + M_1 + \cdots + M_g \tag{96}$$

where M_0 is nonexistent if $u_0 = 0$, and where M_j $(j = 0, 1, \ldots, g)$ is of size $(u_j \times u_j)$.

Thus, the above important theorem says that the estimates of parameters which belong to different alias-sets are uncorrelated with each other.

We now describe how to obtain the matrices M_j. For this purpose we shall rename the elements of $\mathbf{L}$ in such a way that the hth $(h = 1, 2, \ldots, u_j)$ element of the jth $(j = 0, 1, \ldots, g)$ alias-set is denoted by $\mathbf{e}_{jh}$. Notice that although we rearranged the columns of $\tilde{L}$ and hence the elements of $\mathbf{L}$, etc., we have not given a complete description of the rearrangement. We have only grouped together the elements of the same alias-set. This grouping is sufficient for our purposes. Thus, any ordering of the elements within the same alias-set will be all right. For each permissible h and j, let $\mathbf{e}_{jh}$ denote the defining vector of the element e_{jh}; thus, $\mathbf{e}_{jh}$ is a column within the jth group of columns (corresponding to the jth alias-set) inside $\tilde{L}$. Let $e_{01} \equiv \mu$, so that $\mathbf{e}_{01}$ is the zero vector. Now, $\mathbf{e}_{jh}$ (for all permissible j and h) is an $(m \times 1)$ vector; let $\mathbf{e}_{jh}(1)$ be the $(k \times 1)$ subvector of $\mathbf{e}_{jh}$ which contains in order the first k elements of $\mathbf{e}_{jh}$. Thus, for all $i\ (= 1, \ldots, n)$, j $(= 1, \ldots, g)$, and $h\ (= 1, \ldots, u_j)$, define $q(j, h, i)$ by

$$q(j, h, i) = [\mathbf{e}'_{jh}(1)]\mathbf{c}_i \qquad (97)$$

Also, for all $h\ h' \in (1, 2, \ldots, u_j)$, and $j = 0, 1, \ldots, g$, define the nonnegative integer $n_r(j, h, h')$ to be the number of values of i in the set $(1, 2, \ldots, n)$ such that we have

$$q(j, h', i) - q(j, h, i) = r \qquad (98)$$

where $r \in (0, 1)$. Also, for $r \in (0, 1)$, and $j = 0, 1, \ldots, g$, define the $(u_j \times u_j)$ matrix N_{jr} to be such that its (h, h') cell contains the element $n_r(j, h, h')$. Then we have

Theorem 7.2   Under the definition of factorial effects as at (38) and (39), the information submatrices are given by

$$M_j = 2^{-2m}(N_{j0} - N_{j1}) = 2^{-2m}(2N_{j0} - nJ_{u_j u_j}), \qquad \text{for } j = 0, 1, \ldots, g \qquad (99)$$

We illustrate this by

Example 7.1   Consider a 2^7 experiment and let

$$\mathbf{L}' = \{\mu, F_1, F_2, F_5, F_6, F_7, F_{12}, F_{15},$$
$$F_{23}, F_{24}, F_{27}, F_{124}, F_{235}, F_{257}, F_{1234}, F_{1235}\}$$

Notice that this $\mathbf{L}'$, of size (1×16), has tree structure; such an $\mathbf{L}$ is quite likely to arise in practice. We try to obtain a nonsingular design for this case. We take A to be given by

$$A = [I_4 : B] \tag{101a}$$

where

$$B' = \begin{bmatrix} 0 & 1 & 1 & 1 \\ 1 & 0 & 1 & 1 \\ 1 & 1 & 0 & 1 \end{bmatrix} \tag{101b}$$

For this case, it is easy to check that

$$\tilde{A} = \begin{bmatrix} 0 & 1 & 1 & 1 & 1 & 0 & 0 \\ 1 & 0 & 1 & 1 & 0 & 1 & 0 \\ 1 & 1 & 0 & 1 & 0 & 0 & 1 \end{bmatrix},$$

$$\tilde{A}\tilde{L} = \begin{bmatrix} 0 & 0 & 1 & 1 & 0 & 0 & 1 & 1 & 0 & 0 & 1 & 0 & 1 & 0 & 1 & 1 \\ 0 & 1 & 0 & 0 & 1 & 0 & 1 & 1 & 1 & 1 & 0 & 0 & 1 & 0 & 1 & 0 \\ 0 & 1 & 1 & 0 & 0 & 1 & 0 & 1 & 1 & 0 & 0 & 1 & 1 & 0 & 1 & 0 \end{bmatrix} \tag{102}$$

This gives a rearranged vector $\mathbf{L}'$ given by

$\mathbf{L}'$ (rearranged)

$\quad = F\{0, 257; 1, 23; 2; 5, 27, 1235; 6, 24; 7, 124; 12; 15, 235, 1234\}$ (103)

It is thus clear that we have eight alias-sets, which are separated in $\mathbf{L}'$ at (103) by semicolons.

Now, let us take

$$C = \begin{bmatrix} 0 & 0 & 0 \\ 0 & 1 & 0 \\ 0 & 0 & 1 \\ 1 & 1 & 0 \end{bmatrix} \tag{104}$$

and consider obtaining $\mathbf{e}'_{jh}(1)\mathbf{c}'_i$ for different j, h, and i. Notice that $\mathbf{e}'_{jh}(1)$ in this case will involve only the first four elements of $\mathbf{e}'_{jh}$; this means that, in $\mathbf{L}'$ as given at (103), we can ignore the factors number 5, 6, and 7. Now, e_{01} is the zero vector, and hence $\mathbf{e}'_{01}\mathbf{c}_i = 0$, for each i. Next, $e_{02} = F(257)$, so that $\mathbf{e}'_{02}(1) = (0100)$, since we are ignoring factors 5, 6, and 7. Thus,

$e'_{01}c_i = 0, 1, 0$, for $i = 1, 2, 3$. Clearly, we have

$$u_0 = 2, \quad u_1 = 2, \quad u_2 = 1, \quad u_3 = 3,$$
$$u_4 = 2, \quad u_5 = 2, \quad u_6 = 1, \quad u_7 = 3 \qquad (105)$$

We now present the information matrix for this example. The actual information matrices equal 2^{-14} times the information matrices presented here [in equation (107) below]; this is being done for simplicity. Consider the quantities $e'_{jh}(1)c_i$ for various permissible values of i, j, h. These values are given in the matrix (106) below. Here the rows correspond, respectively, to $i = 1, 2, 3$ and the columns to the pairs (j, h), where $h = 1, \ldots,$ u_j, and $j = 0, 1, \ldots, 7$:

$$\begin{bmatrix} 0\,0 & 0\,0 & 0 & 0\,0\,0 & 0\,1 & 0\,1 & 0 & 0\,0\,1 \\ 0\,1 & 0\,1 & 1 & 0\,1\,1 & 0\,0 & 0\,0 & 1 & 0\,1\,0 \\ 0\,0 & 0\,1 & 0 & 0\,0\,1 & 0\,0 & 0\,0 & 0 & 0\,1\,1 \end{bmatrix} \qquad (106)$$

In the above display the columns corresponding to the different alias-sets have been separated by a space, for the reader's convenience. Now, consider, for example, the case $j = 0$. Firstly, from (99), it is clear that each M_j will have the number n $(= 3)$ on the diagonal. Now, for the 0th alias-set, we see from (106) that the number of common elements in these two columns is 2; thus the off-diagonal element in $N_{00}(2 \times 2)$ is 2. From (99), we thus get M_0 as exhibited below. Similarly, other information matrices are obtained:

$$M_0 = M_4 = M_5 = \begin{bmatrix} 3 & 1 \\ 1 & 3 \end{bmatrix}, \quad M_1 = \begin{bmatrix} 3 & -1 \\ -1 & 3 \end{bmatrix}, \quad M_2 = M_6 = [3]$$

$$M_3 = \begin{bmatrix} 3 & 1 & -1 \\ 1 & 3 & 1 \\ -1 & 1 & 3 \end{bmatrix}, \quad M_7 = \begin{bmatrix} 3 & -1 & -1 \\ -1 & 3 & -1 \\ -1 & -1 & 3 \end{bmatrix} \qquad (107)$$

From (107) it is clear that all the information submatrices are nonsingular, the reason being that in each case the diagonal elements exceed the sum of the absolute value of the remaining elements in the same row (or column). Thus, all the 16 parameters are estimable by our design, which contains 24 treatments.

It is clear from the above that the parallel-planes approach constitutes a powerful method for creating designs. Indeed, it is the most powerful method known at present. The method is good because most of the estimates are mutually orthogonal. If some singularity is present in one or more of the information submatrices, it is easy to change the matrix C, and to assess the effect of that on the different information (sub)matrices, since the computations involved are rather minor, and may usually be done without even the need of a hand calculator. It is also clear that those who want to restrict themselves to orthogonal arrays will have to work with very large designs in order to be able to estimate this set of parameters. Otherwise, they will miss certain parameters which are nonnegligible, and will come up with bad estimates which (as we will see later on) could be very misleading. Approaches built on BA's or PBA's etc., will also involve large cumbersome matrices to work with. Thus the current approach is also the most practical one. The approach based on balanced arrays, for example, is useful when someone wants to estimate all interactions up to a given order, assuming that the rest of the parameters are negligible. However, these kinds of situations do not arise too often in real life. (On the other hand, there are some other advantages to approaches based on balanced arrays as we shall explain in Section 9.)

We now consider the s^m experiment, where s is a prime number large than 2. In this case, there are s^m factorial effects. However, for $s \geq 3$, one can define two types of factorial effects, respectively called "geometric" and "analytic." The definition of factorial effects given at (35) (following Bose and Srivastava, 1964a), for example, is analytic. The analytic sets have good physical interpretation. The geometric effects (Bose and Kishen, 1940) do not usually have such interpretation. However, they are useful in obtaining appropriate designs. The following result is important.

Theorem 7.3 Consider an s^m experiment, where s is a power of a prime number. Also, consider the r-factor interaction between any r-factors (say, $i_1, \ldots, i_r$; $1 \leq i_1 < \cdots < i_r \leq m$). Then, this r-factor interaction will have $(s-1)^r$ linearly independent components (both under the analytic and geometric definition of factorial effects). Let $\mathbf{G}(i_1, \ldots, i_r)$ be a column vector with $(s-1)^r$ elements containing all the distinct geometric components of the r-factor interaction under consideration. Similarly, let $\mathbf{A}(i_1, \ldots, i_r)$ be a column vector of size $(s-1)^r \times 1$ whose elements are the distinct components of the r-factor interaction under consideration, under the analytic definition. Then we have

$$\mathbf{G}(i_1, \ldots, i_r) = U_r \mathbf{A}(i_1, \ldots, i_r) \tag{108}$$

where U_r is orthogonal of size $(s-1)^r \times (s-1)^r$.

Below, we define the geometric effects. Since the analytic effects have been defined before, one could consider obtaining the matrix U_r; however, we shall not go into this here. The above theorem is presented, however, to help the reader appreciate the fact that, for any fixed r-factor interaction, the set of $(s-1)^r$ components under one definition is related to the set of $(s-1)^r$ components under the other definition. Thus, a theory built under one definition can be easily transferred to a theory under the other definition.

Before discussing the geometric definition (for $s \geq 3$), we would like to caution the reader that because of its inclusive nature, some readers may find it a bit complicated. It is suggested that during a first reading, they should ignore any material not needed in the understanding of the statements made in Theorems 7.4–7.7.

For any treatment combination $\mathbf{t}$, recall, from Section 3, that $\phi(\mathbf{t})$ denotes the true effect of the treatment $\mathbf{t}$. Then we define

$$\mu = s^{-m/2} \sum \phi(\mathbf{t}) \tag{109}$$

where the $\sum$ in (109) runs over all the s^m distinct treatments $\mathbf{t}$.

Let $\mathbf{e}'(1 \times m)$ be a vector over GF(s). Then $\mathbf{e}'$ will be considered to be the defining vector for a geometric factorial effect which will be denoted by $F(\mathbf{e})$. Since $\mathbf{e}$ can take s^m values, there are s^m factorial effects. If $\mathbf{e}$ is the zero vector, then the corresponding $F(\mathbf{e})$ is identical with μ. Thus, we need to define $F(\mathbf{e})$ when $\mathbf{e}$ is not the zero vector. For all nonzero $\mathbf{e}$, and for all α over GF(s), let $\phi(\mathbf{e}, \alpha)$ denote the sum of the true effects of all the s^{m-1} distinct level combinations $\mathbf{t}$ which satisfy

$$\mathbf{e}'\mathbf{t} = \alpha \tag{110}$$

Also, let ζ be a $s \times (s-1)$ matrix whose columns are mutually orthogonal, and the sum of squares of the elements of each column equals 1. The elements of ζ will be denoted by $\zeta_{r'r}$ $[r' = 0, 1, \ldots, (s-1); r = 1, \ldots, (s-1)]$. We shall also assume that the sum of elements in every column of ζ is zero. Thus we have

$$\zeta'\zeta = I_{s-1}, \qquad \zeta'J_{s1} = \mathbf{0}_{s-1} \tag{111}$$

Next, let p_e denote a permutation of the set $\{\alpha_0, \alpha_1, \ldots, \alpha_{s-1}\}$ corresponding to the effect whose defining vector is $\mathbf{e}$, where $\mathbf{e}$ is nonzero. For example, suppose $m = 5$, $s = 5$, $e' = (0, 2, 1, 0, 3)$. Then GF(s) has the elements $\{0, 1, 2, 3, 4\}$, and p_e is merely any permutation of the elements of GF(5). If, for example, $p_e = (0\ 4\ 2\ 3\ 1)$, then we have $p_e(0) = 0$, $p_e(1) = 4$, etc. Thus, for each nonzero $\mathbf{e}$, we choose a permutation p_e. Then we define $F(\mathbf{e})$ by

$$F(\mathbf{e}) = \left\{ \sum_{r'=0}^{s-1} \zeta_{r'r} \phi[\mathbf{e}, p_e(\alpha_r \bullet \alpha_{r'})] \right\} s^{-(n-1)/2} \tag{112}$$

where $\alpha_r = e_{(1)}$, where $e_{(1)}$ is the first nonzero element of $\mathbf{e}$. As it stands, the definition is slightly complicated. However, it is very useful, and is easy to master. To help in this, we illustrate with the case $s = 3$, $m = 5$, $e = F_2^2 F_3^2 F_4^2 F_5$. This factorial effect e has the defining vector given by $e' = (0\ 2\ 2\ 2\ 1)$, so that we have $e_{(1)} = 2 = \alpha_r$, so that $r = 2$. (Here we are denoting the elements of GF(3) in a dual way; one by the ordinary integers, 0, 1, 2, and the other by the symbols already adopted, namely, $\alpha_0, \alpha_1, \alpha_2$.) Next, for this effect e, suppose we take the permutation $p_e = \left(\begin{smallmatrix} 0 & 1 & 2 \\ 2 & 0 & 1 \end{smallmatrix} \right)$. This means that this permutation permutes the vector $(0\ 1\ 2)$ to the vector $(2\ 0\ 1)$, so that 0 is changed to 2, 1 is changed to 0, and 2 is changed to 1. Also, throughout this paper, for $s = 3$, we shall take the matrix ζ to be given by

$$\zeta(\text{for } s = 3) = \begin{bmatrix} 1/\sqrt{2} & 1/\sqrt{6} \\ -1\sqrt{2} & 1/\sqrt{6} \\ 00 & -2/\sqrt{6} \end{bmatrix} = \begin{bmatrix} 1 & 1 \\ -1 & 1 \\ 0 & 2 \end{bmatrix} \begin{bmatrix} 1/\sqrt{2} & 0 \\ 0 & 1\sqrt{6} \end{bmatrix} \tag{113}$$

Thus, for $r' = 0, 1, 2$, we have $\alpha_{r'} \cdot \alpha_r = 0, 2, 1$ so that $p_e(\alpha_r \cdot \alpha_r) = 2, 1, 0$. Thus, equation (112) gives

$$F(e) = 3^{-2} \cdot (1/\sqrt{6}) \{\phi(\mathbf{e}, \alpha_2) + \phi(\mathbf{e}, \alpha_1) - 2\phi(\mathbf{e}, \alpha_0)\} \tag{114}$$

where $\phi(\mathbf{e}, \alpha_i)$ (for $i = 0, 1, 2$) equals $\sum_i \phi(\mathbf{t})$, where $\sum_i$ runs over all $\mathbf{t}$ ($= (t_1, t_2, t_3, t_4, t_5)$), such that $2t_2 + 2t_3 + 2t_4 + t_5 = \alpha_i$.

The reader acquainted with this subject should note that our p_e above correspond to the notation p_e^{*-1} in the SAM paper. The geometric interactions were introduced in Bose and Kishen (1940) and Bose (1947); therein, and in various papers before the SAM paper, traditionally, one took p_e to be the identity permutation, i.e., the permutation which leaves everything unchanged. However, in the SAM paper it was shown that if we allow ourselves to choose p_e in a certain manner, then a certain simplification in the information matrix is obtainable. This simplification in the information matrix leaves the characteristic roots of the information matrix unchanged.

Now, consider $(s-1)$ distinct factorial effects e_u $(u = 1,\ldots,s-1)$ with defining vectors e'_u, such that these defining vectors are (nonzero) multiples of each other. Then this set of $(s-1)$ effects, also called $(s-1)$ "degrees of freedom," are said to form a "pencil." Thus, besides μ, the rest of the factorial effects are divided into $(s^m - 1)/(s-1)$ pencils of factorial effects, each pencil carrying $(s-1)$ degrees of freedom. This terminology is fairly standard. As in the SAM paper, we shall assume throughout that $\mathbf{L}$ is such that, given any pencil, it is either totally contained in $\mathbf{L}$, or it is totally excluded from $\mathbf{L}$. Because of this assumption, and the introduction of the permutation p_e, a great simplification occurs in the information matrix, which we now discuss. Recall (65), and $\mathbf{F}_1$. For convenience, we shall assume that the factorial effects are arranged in $\mathbf{F}_1$ as explained before (65), and furthermore that all factorial effects which belong to the same pencil occur adjacently in $\mathbf{F}_1$. Consider the matrix $X_{(1)}$; let $x(e,\mathbf{t})$ be the element of this matrix in the row corresponding to this factorial effect e and the column corresponding to the treatment t. Then it is clear that

$$x(\mu,\mathbf{t}) = s^{-m/2} \tag{115a}$$

$$x(e,\mathbf{t}) = s^{-(m-1)/2}\zeta_{r'r} \tag{115b}$$

where $e \neq \mu$, and r' is obtained using the equation

$$\mathbf{e}'\mathbf{t} = \alpha_r \cdot \alpha_{r'} \tag{116}$$

where α_r is the first nonzero element of $\mathbf{e}$. Now, let A and $\tilde{A}$ be as at (93), (94). Now, consider $\mathbf{L}$. For every factorial effect $e \in \mathbf{L}$, compute (as for the case $s = 2$) the vector $\tilde{A}\mathbf{e}$, where $\mathbf{e}$ is the defining vector of e. Also, if e_1, e_2 are any two distinct factorial effects in $\mathbf{L}$, and $\mathbf{e}_1$ and $\mathbf{e}_2$ are the corresponding defining vectors, then put e_1 and e_2 in the same alias-set if and only if $(\tilde{A}\mathbf{e}_1) = \gamma(\tilde{A}\mathbf{e}_2)$, where γ is a nonzero constant. Thus, suppose $(g+1)$ alias-sets are obtained. We shall assume that the 0th alias-set has μ in it. Also, we shall denote the effects belonging to the jth $(j = 1,\ldots,g)$ alias-set by e_{jhr}, where $(h = 1,\ldots,u_j)$, and $[r = 1,\ldots,(s-1)]$, where the jth set contains u_j pencils, each carrying $(s-1)$ effects. Now, the 0th alias-set may or may not exist; however, for philosophical reasons we shall assume that μ always belongs to $\mathbf{L}$, so that the 0th alias-set does exist. If there are other effects in $\mathbf{L}$, besides μ, then they may be divided into $(u_0 - 1)$ pencils, each carrying $(s-1)$ effects. Traditionally, we may consider μ to be an exception in the sense that it belongs to a pencil which has only one effect in it. Thus, the 0th alias-set may be considered to have u_0 pencils in it. The individual effect, denoted by e_{jhr}, is the rth individual effect inside the hth pencil inside the jth alias-set; where, the rth effect is the one whose defining vector has r as the first nonzero element in it. With this convention, let $\mathbf{e}_{jh}$ be the defining vector of the effect e_{jh1}, for all permissible j and h. Now consider $\mathbf{L}$; we assume that the elements are so arranged that all effects which belong to the same alias-set occur adjacently, and similarly all effects which belong to the same pencil in any given alias-sets occur adjacently to each other in order of the values of $r = 1, \ldots, s-1$. Now, recall $X(N \times \nu)$, and assume that its columns have been rearranged so as to correspond to the new ordering of the elements of $\mathbf{L}$. Then X can be partitioned as

$$X = [X_0 : X_1 : \ldots : X_g] \tag{117}$$

where X_j $(j = 1, \ldots, g)$ is an $[N \times u_j(s-1)]$ matrix, and where X_0 is of size $N \times \{1 + (u_0 - 1)(s-1)\}$. Also, recall that $N = n \cdot s^{m-k}$. The following result is important.

Theorem 7.4   The information matrix $X'X$ can be broken up as a direct sum of the matrices $X_j'X_j$ so that we have

$$X'X = X_0'X_0 \dotplus X_1'X_1 \dotplus \cdots \dotplus X_g'X_g \tag{118a}$$

where

$$X_j'X_j = s^{-k}Z_j'Z_j, \qquad \text{for } j = 1, \ldots, g \tag{118b}$$

where Z_j is an $N(s-1) \times u_j(s-1)$ matrix. (Here we shall not give the matrix Z_j because of lack of space. However the matrix $Z_j'Z_j$ will be described. Also, the matrix $X_0'X_0$ will be described separately.)

Now, let the rth $[r = 0, 1, \ldots, (s-1)]$ row of ζ be denoted by ζ_r', where we shall now assume that $r \in \mathrm{GF}(s)$. Then we define the important matrix Q_0 of size $(s-1) \times (s-1)$ by

$$Q_0 = \sum_{r=0}^{s-1} \zeta_r \zeta_{r+1}' \tag{119}$$

The matrix Q_0 has very interesting properties in general. For example, when $s = 3$, we have

$$Q_0 = \begin{bmatrix} -1/2 & \sqrt{3}/2 \\ -\sqrt{3}/2 & -1/2 \end{bmatrix}, \qquad Q_0 Q_0' = I_2,$$

$$Q_0^2 = \begin{bmatrix} -1/2 & -\sqrt{3}/2 \\ -\sqrt{3}/2 & -1/2 \end{bmatrix} = -I_2 - Q_0 = Q_0' \qquad (120)$$

For $j = 1, \ldots, g$, we now introduce certain matrices Λ_j of size $(n \times u_j)$, which are related to the matrices Z_j. The matrices Λ_j involve complex numbers. [A complex number is a number of the form $a + bi$, where $i = \sqrt{-1}$, and where a and b are real numbers. The "conjugate" of the complex number $(a + bi)$ is the complex number $(a - bi)$. Thus, the conjugate of the number $(\sqrt{2} - 3i)$ is the number $(\sqrt{2} + 3i)$]. Now, for $j = 1, \ldots, g$, let $\mathbf{a}_j'[1 \times (m - k)]$ be a vector such that we have

$$\mathbf{a}_j' \tilde{A} \mathbf{e}_{j1} \neq 0 \qquad (121)$$

Define the $(n \times 1)$ vector $\mathbf{w}_{jh}$ $(j = 1, \ldots, g; h = 1, \ldots, u_j)$ by

$$\mathbf{w}_{jh} = (\mathbf{a}_j' \tilde{A} \mathbf{e}_{j1})^{-1} \mathbf{e}_{jh} \qquad (122)$$

Also, partition these vectors by

$$\mathbf{w}_{jh}' = [\mathbf{w}_{jh}'(1) : \mathbf{w}_{jh}'(2)] \qquad (123)$$

where the $\mathbf{w}_{jh}'(1)$ are of size $(1 \times k)$. Now, for $(i = 1, \ldots, n; j = 1, \ldots, g; h = 1, \ldots, u_j)$, let

$$q(j, h, i) = \mathbf{e}_{jh}' \mathbf{w}_{jh}(1) \qquad (124)$$

Notice that the q are elements of GF(s). Now, for $[r = 0, 1, \ldots, (s-1); j = 1, \ldots, g; h, h' = 1, 2, \ldots, u_j]$, define the nonnegative integer $n_r(j, h, h')$ to be the number of values of i in the set $(1, 2, \ldots, n)$ such that we have

$$q(j, h', i) - q(j, h, i) = r \qquad (125)$$

and let N_{jr} be the $(u_j \times u_j)$ matrix whose (h, h') element is $n_r(j, h, h')$.

Next, consider the sth root of unity given by $e^{2\pi i/s}$, which equals $\{\cos(2\pi/s) + i\sin(2\pi/s)\}$, where now i denotes the imaginary number

$\sqrt{-1}$. Let

$$\omega_s \equiv e^{2\pi i/s} \qquad (126)$$

Clearly, ω_s is an sth root of unity. Now, consider (121); the quantity on the left-hand side is obviously an element of GF(s). However, we can also consider this quantity to be a nonnegative integer; thus, let θ_j ($j = 1, \ldots, g$) denote the nonnegative integer which is identical to the element of GF(s) represented by the left-hand side of (121). Then, we define another sth root of unity by ω_{sj}, where

$$\omega_{sj} = \omega_s^{\theta_j} = e^{2\pi i \theta_j/s}, \qquad \text{for } j = 1, \ldots, g \qquad (127)$$

Then, we have the following major result of the theory of parallel-planes designs.

Theorem 7.5 For $j = 1, \ldots, g$, there exists a choice of the permutations p_e (for the different effects e) such that the following results are true.

1. The matrix $Z_j[n(s-1) \times u_j(s-1)]$ is related to a matrix $\Lambda_j(n \times u_j)$, whose elements are complex numbers (being actually powers of ω_s). Let Λ_j^* denote the "conjugate transpose" of Λ_j; i.e., $\Lambda_j^*(u_j \times n)$ is the matrix obtained by first taking the transpose of Λ_j, and then replacing each element by its conjugate. Then, we have

$$\Lambda_j^* \Lambda_j = \sum_{r=0}^{s-1} \omega_{sj}^r N_{rj} \qquad (128)$$

2. The matrix $Z_j' Z_j$ of size $u_j(s-1) \times u_j(s-1)$ is obtainable form the $u_j \times u_j$ matrix $\Lambda_j^* \Lambda_j$ by replacing ω_s everywhere by the $(s-1) \times (s-1)$ matrix Q_0.

3. The determinant $|\Lambda_j^* \Lambda_j|$ is a real number; let its value be l_j. Then we have

$$|X_j' X_j| = s^{-k} |Z_j' Z_j| = s^{-k} l_j^{s-1} \qquad (129)$$

(Notice that, in (128), we need to evaluate N_{rj} only for $r = 0, \ldots, (s-2)$, since we always have

$$1 + \omega_{sj} + \cdots + \omega_{sj}^{s-1} = 0 \qquad (130)$$

provided that $\omega_{sj} \neq 1$, which is true in our case.)

Now, we consider the 0th alias-set, where, without loss of generality, we assume that μ belongs to it. Suppose that the jth ($j = 0, 1, \ldots, g$) alias-set has ν_j factorial effects in it. Then, from the previous discussion, we have

$$\nu = \sum_{j=0}^{g} \nu_j, \qquad \nu_0 = 1 + (u_0 - 1)(s - 1), \qquad \nu_j = u_j(s - 1), \qquad \text{(for } j > 0)$$

(131)

Let e_{0hr} $[h = 1, \ldots, (u_0 - 1)]$ be the factorial effect inside the hth pencil inside the 0th alias-set, whose defining vector (say $\mathbf{e}_{0hr}$) has r as its first nonzero element.

Now, $\mathbf{e}_{0hr}$ (for all permissible h and r) is an $(m \times 1)$ vector; let $\mathbf{e}_{0hr}(1)$ be the subvector of $\mathbf{e}_{0hr}$ consisting of the first k elements. Let

$$\gamma_{0hi} = r^{-1} \mathbf{c}_i' \mathbf{e}_{0hr}(1); \qquad h = 1, \ldots, (u_0 - 1); \qquad i = 1, \ldots, n \qquad (132)$$

where, it is easy to check that γ_{0hi} does not depend upon r. Now, for any given e_{0hr}, let r' ($\in \mathrm{GF}(s)$) be such that

$$p_e(r \cdot r') = \gamma_{0hi} \qquad (133)$$

where in (133) e stands for e_{0hr}, and where obviously r' will depend upon h and i. Next, for $i \in (1, \ldots, n)$, let $\mathbf{x}_{0i}'$ be a vector of size $(1 \times \nu_0)$ defined as follows. The first elements of $\mathbf{x}_{0i}'$ (which corresponds to μ) equals $(1/\sqrt{s})$. Also, if, as above, e_{0hr} is any effect in the 0th alias-set, then, corresponding to this effect, the element of $\mathbf{x}_{0i}'$ is $\zeta_{r'r}$, where r' is obtained from (133). Then we have

Theorem 7.6 The information matrix for the 0th alias-set is given by

$$X_0' X_0 = s^{-k+1} \sum_{i=1}^{n} \mathbf{x}_{0i} \mathbf{x}_{0i}' \qquad (134)$$

We now present some necessary conditions for the nonsingularity of the information matrix for all prime numbers s.

Theorem 7.7 The following conditions are necessary in order that the information matrix be nonsingular:

1. For $s = 2$, we must have

$$n \geq u_j; \qquad j = 0, 1, \ldots, g \qquad (135a)$$

2. For $s \geq 3$, we must have

$$n \geq u_j, \qquad \text{for } j = 1, \ldots, g \quad n \geq \nu_0 \qquad (135b)$$

3. For all $j > 0$, when $s > 2$, and for all j when $s = 2$ and, furthermore, for all h, $h' \in (1, 2, \ldots, u_j)$, there does not exist a constant $\beta_j \in$ GF(s), such that we have

$$q(j, h', i) - q(j, h, i) = \beta_j, \qquad \text{for all } i = 1, \ldots, n \qquad (136)$$

where the $q(j, h, i)$ are defined at (97) for $s = 2$, and (124) for $s > 2$.

We now illustrate the above results with an example. Let us take

$$\mathbf{L}' = \{\mu, F_2, F_2^2, F_4, F_4^2, F_5, F_5^2, F_1 F_4, F_1^2 F_4^2, F_3 F_4, F_3^2 F_4^2,$$
$$F_4 F_5^2, F_4^2 F_5, F_2 F_3 F_4, F_2^2 F_3^2 F_4^2, F_2 F_3 F_4 F_5^2, F_2^2 F_3^2 F_4^2 F_5\} \quad (137)$$

As for the case $s = 2$, we can write down the matrix $\tilde{L}$ containing the defining vector of the elements of $\mathbf{L}'$; this is given by

$$\tilde{L} = \begin{bmatrix} 0 & 0 & 0 & 0 & 0 & 0 & 0 & 1 & 2 & 0 & 0 & 0 & 0 & 0 & 0 & 0 & 0 \\ 0 & 1 & 2 & 0 & 0 & 0 & 0 & 0 & 0 & 0 & 0 & 0 & 0 & 0 & 0 & 1 & 2 \\ 0 & 0 & 0 & 0 & 0 & 0 & 0 & 0 & 0 & 1 & 2 & 0 & 0 & 1 & 2 & 1 & 2 \\ 0 & 0 & 0 & 1 & 2 & 0 & 0 & 1 & 2 & 1 & 2 & 1 & 2 & 1 & 2 & 1 & 2 \\ 0 & 0 & 0 & 0 & 0 & 1 & 2 & 0 & 0 & 0 & 0 & 2 & 1 & 1 & 2 & 2 & 1 \end{bmatrix} \quad (138)$$

Let

$$A = \begin{bmatrix} 1 & 0 & 0 & 0 & 1 \\ 0 & 1 & 0 & 1 & 1 \\ 0 & 0 & 1 & 1 & 2 \end{bmatrix}, \qquad C = \begin{bmatrix} 0 & 1 \\ 0 & 2 \\ 0 & 2 \end{bmatrix} \qquad (139)$$

Then we have

$$\tilde{A} = \begin{bmatrix} 0 & 2 & 2 & 1 & 0 \\ 2 & 2 & 1 & 0 & 1 \end{bmatrix} \qquad (140)$$

Thus, we get

$$\tilde{A}\tilde{L} = \begin{bmatrix} 0 & 2 & 1 & 1 & 2 & 0 & 0 & 1 & 2 & 0 & 0 & 1 & 2 & 0 & 0 & 2 & 1 \\ 0 & 2 & 1 & 0 & 0 & 1 & 2 & 2 & 1 & 1 & 2 & 2 & 1 & 2 & 1 & 2 & 1 \end{bmatrix} \quad (141)$$

By inspection of the last equation, we find that the alias-sets are given in the rearranged form of $\mathbf{L}'$ below, where the different alias-sets are separated by semicolons:

$$\mathbf{L}' \text{ (rearranged)} = \{\mu; F_2, F_2^2, F_2 F_3 F_4 F_5^2, F_2^2 F_3^2 F_4^2 F_5, F_4, F_4^2,$$

$$F_2 F_3 F_4, F_2^2 F_3^2 F_4^2; F_5, F_5^2, F_3 F_4, F_3^2 F_4^2; F_1 F_4, F_1^2 F_4^2, F_4 F_5^2, F_4^2 F_5\} \quad (142)$$

First, consider the 0th alias-set; here it consists of μ alone, so that we have $\nu_0 = 1$. Clearly, we have $n = 2$; thus, for $i = 1, 2$, the vector $\mathbf{x}'_{0i}$ is of size (1×1) and consists of the element $(1/\sqrt{3})$. For the next alias-set, we have $j = 1$, and $\mathbf{e}_{11} = (0\ 1\ 0\ 0\ 0)'$, $\mathbf{e}_{12} = (0\ 1\ 1\ 1\ 2)'$. We take $\mathbf{a}'_1 = (1,1)$. Thus, we have $\tilde{A}\mathbf{e}_{1h} = (2,2)'$, for $h = 1, 2$, so that $\mathbf{a}'_1 \tilde{A}\mathbf{e}_{j1} = 1$. Thus, $\mathbf{w}_{j1} = \mathbf{e}_{j1}$, $\mathbf{w}_{j2} = \mathbf{e}_{j2}$, and $\mathbf{w}'_{j1}(1) = (0\ 1\ 0)$, $\mathbf{w}'_{j2}(1) = (0\ 1\ 1)$. Let q_j (for all permissible $j > 0$) be the $(n \times u_j)$ matrix whose (i, h) element $(i = 1, \ldots, n; h = 1, \ldots, u_j)$ is $q(j, h, i)$. Then, we have

$$q_1 = \begin{bmatrix} 0 & 0 \\ 2 & 1 \end{bmatrix} \quad (143)$$

We proceed in exactly the same manner for the other alias-sets. We define $\mathbf{a}'_2 = (1, 0)$, $\mathbf{a}'_3 = (0, 1)$, and $\mathbf{a}'_4 = (1, 0)$. Then we obtain

$$g = 4, \quad u_0 = 1, \quad u_1 = u_2 = u_3 = u_4 = 2,$$

$$\nu_0 = 1, \quad \nu_1 = \nu_2 = \nu_3 = \nu_4 = 4, \quad \nu = 17$$

$$q_2 = \begin{bmatrix} 0 & 0 \\ 0 & 1 \end{bmatrix}, \quad q_3 = \begin{bmatrix} 0 & 0 \\ 0 & 2 \end{bmatrix}, \quad q_4 = \begin{bmatrix} 0 & 0 \\ 1 & 0 \end{bmatrix} \quad (144)$$

Now, from (143) and the earlier definition, we have

$$M_0 = X'_0 X_0 = [1/3]3^{-3+1} = 3^{-3}$$

$$N_{10} = \begin{bmatrix} 2 & 1 \\ 1 & 2 \end{bmatrix}, \quad N_{11} = \begin{bmatrix} 0 & 0 \\ 1 & 0 \end{bmatrix}, \quad N_{12} = \begin{bmatrix} 0 & 1 \\ 0 & 0 \end{bmatrix} \quad (145)$$

It can be checked that for all $j > 0$, we shall have $\theta_j = 1$, and hence $\omega_{sj} = \omega_s = \omega_3 = e^{2\pi i/3} = (\cos 2\pi/3 + i \sin 2\pi/3) = \{(-\frac{1}{2}) + i(\sqrt{3}/2)\} = \omega$, say. Hence, from Theorem 7.5, we obtain

$$\Lambda_1^* \Lambda_1 = \begin{bmatrix} 2 & 1 + \omega^2 \\ 1 + \omega & 2 \end{bmatrix}, \quad |\Lambda_1^* \Lambda_1| = 3, \quad |M_1| = 3^{-3}(3)^{3-1} = 3^{-1}$$

$$(146)$$

since $(1 + \omega + \omega^2) = 0$, and $\omega^3 = 1$. Similarly, for other cases, we obtain

$$\Lambda_2^* \Lambda_2 = \begin{bmatrix} 2 & 1 + \omega \\ 1 + \omega^2 & 2 \end{bmatrix}, \qquad \Lambda_3^* \Lambda_3 = \begin{bmatrix} 2 & 1 + \omega^2 \\ 1 + \omega & 2 \end{bmatrix},$$

$$\Lambda_4^* \Lambda_4 = \begin{bmatrix} 2 & 1 + \omega \\ 1 + \omega^2 & 2 \end{bmatrix} \tag{147}$$

which shows that the determinant in each case equals 3. The information matrix corresponding to the first alias-set is, from (146), given by

$$X_1' X_1 = 3^{-3} \begin{bmatrix} 2I_2 & I_2 + Q_0^2 \\ I_2 + Q_0 & 2I_2 \end{bmatrix} = 3^{-3} \begin{bmatrix} 2I_2 & -Q_0 \\ -Q_0' & 2I_2 \end{bmatrix} \tag{148}$$

where Q_0 and Q_0^2 are given by (120). The other cases are similar.

With A and C as at (139), the design matrix $T(5 \times 18)$ for the above example is given by

$$T = [T_1 : T_2] = \begin{bmatrix} 0\,2\,0\,2\,0\,1\,1\,1\,2 & 1\,0\,1\,0\,1\,2\,2\,2\,0 \\ 2\,2\,0\,1\,1\,1\,2\,0\,0 & 1\,1\,2\,0\,0\,0\,1\,2\,2 \\ 2\,1\,0\,0\,1\,2\,0\,1\,2 & 1\,0\,2\,2\,0\,1\,2\,0\,1 \\ 1\,0\,0\,1\,2\,0\,2\,1\,2 & 1\,0\,0\,1\,2\,0\,2\,1\,2 \\ 0\,1\,0\,1\,0\,2\,2\,2\,1 & 0\,1\,0\,1\,0\,2\,2\,2\,1 \end{bmatrix} \tag{149}$$

The above design T with 18 runs enables us to estimate the 17 parameters given in $\mathbf{L}$ at (138). As for the case $s = 2$, it will be noted that once A and C are selected, the computations for obtaining the information matrices and their determinants are quite simple. In this particular case, it will also be found that not every matrix C will work. If any particular C does not work, by changing some of its elements, we can easily assess the effect on the information matrices through the intermediate matrices which involve ω. In any given situation, of course, the first problem is to choose the matrix A. This should be chosen by taking an appropriate matrix B. Even before doing this, the factors should be numbered in such a way that the effects which we want to estimate are not aliased with μ. As is clear from (135), the design may have to be made unnecessarily large if the 0th alias-set is too big.

We shall now make some brief comments on the designs of the type for $2^{m_1} \times 3^{m_2}$, in which there are m_1 factors at two levels, and m_2 at three levels. Quite a bit of work has been done on these types of designs, when

L consists of μ, the main effects, and the two-factor interactions. A rather large number of such designs are available, for instance, in the work of Connor and Young (1959), Patel (1961), and Addleman (1964), and the unpublished work of Srivastava (1967). No attempt at any general theory has been made. The main idea behind making these designs is to give a nonsingular design which is not too large. The basic technique for making this design is as follows. First, we make a parallel-flats designs for the 2^{m_1} case in which the matrix C has n columns. Let the n sets of runs so obtained be denoted as usual by $T_{11}, T_{12}, \ldots, T_{1n}$. Similarly, for the 3^{m_2} experiment as well, we choose a C matrix which has n columns, leading to m sets of runs, say, $T_{21}, \ldots, T_{2n}$. Then the design is given by

$$T = T_{11} \otimes T_{21} + T_{12} \otimes T_{22} + \cdots + T_{1n} \otimes T_{2n} \qquad (150)$$

where the meaning of the last equation is as follows. Consider $T_{1i} \otimes T_{2i}$; suppose each T_{1i} $(i = 1, \ldots, n)$ has $2^{m_1 - k_1}$ runs. Similarly, suppose that each T_{2i} $(i = 1, \ldots, n)$ has $3^{m_2 - k_2}$ runs in it. Then $T_{1i} \otimes T_{2i}$ is the set of $(2^{m_1 - k_1} \times 3^{m_2 - k_2})$ runs obtained by adjoining each run of the two-level factors in T_{1i} with each run of the three-level factors in T_{3i}. Then, (150) says that we should consider a design T which involves $(n \times 2^{m_1 - k_1} \times 3^{m_2 - k_2})$ runs obtained by taking all the sets $T_{1i} \otimes T_{2i}$ together. The interested reader is referred to the above documents. We shall not go into further details here for lack of space.

8 SEARCH DESIGNS

In the last two sections, we studied the problem of obtaining (nonsingular) designs T which allow estimation of a given set of parameters $\mathbf{L}$, assuming that the remaining parameters $\mathbf{L}_0$ are nonnegligible. However, as we discussed earlier, this approach suffers from the flaw that we assume the parameter set $\mathbf{L}^0$ to be negligible, when it is actually not the case. This causes two kinds of error: (1) the estimates of the parameters in $\mathbf{L}$ have unknown amounts of biases, which can lead to serious misjudgments, and (2) without knowing $\mathbf{L}^*$ (the vector of all nonnegligible effects), it is not possible for us to know $\mathbf{F}_1$ and hence $\phi(\mathbf{t})$ for every run $\mathbf{t}$, and so the main purpose of the experiment may not be accomplished.

The popular approach is to assume some $\mathbf{L}_0$ to be negligible and try to estimate the remaining set of parameters $\mathbf{L}$. As we have mentioned many times, this approach is faulty. We now consider the bias involved in this approach. Recall the discussion in Section 6 before Theorem 6.1. Consider (66). Notice that (69) holds only because of our assumption that L_0 is zero. We now consider the value of ϕ without such an assumption. Let $X_{(11)}(N \times \nu_1)$ be obtained from $X_{(1)}(\nu_1 \times \nu_1)$ by deleting, or repeating, rows of X according to whether a particular element of ϕ_1 is deleted, or repeated, to obtain ϕ. Then, it is clear that we shall have

$$\phi = X_{(11)}D_1^{-2}\mathbf{F}_1 \tag{151}$$

We assume that the elements of $\mathbf{F}_1$ are so rearranged that we have

$$\mathbf{F}_1' = [\mathbf{L}' : \mathbf{L}_0] \tag{152a}$$

This induces a partitioning of the diagonal matrix $D_1^{-2}(\nu_1 \times \nu_1)$ in the form

$$D_1^{-2} = \mathrm{diag}(D^{-2} : D_0^{-2}) \tag{152b}$$

where $D^{-2}(\nu \times \nu)$ is defined earlier, and $D_0^{-2}(\nu_0 \times \nu_0)$ is the matrix corresponding to $\mathbf{L}_0$. Then, (151) is equivalent to

$$\phi = X_{(11)} \begin{bmatrix} D^{-2} & L \\ D_0^{-2} & L_0 \end{bmatrix} \tag{153}$$

which then is the equation which should replace (69) in case no assumptions are made on $\mathbf{F}_1$. From (70), (73), and (153) we obtain the actual expected value of $\hat{\mathbf{L}}$ which is given by

$$E(\hat{\mathbf{L}}) = D^2(X'X)^{-1}X'X_{(11)} \begin{bmatrix} D^{-2} & L \\ D_0^{-2} & L_0 \end{bmatrix} \tag{154}$$

Clearly, the matrix $X(N \times \nu)$ consists of the first ν columns of $X_{(11)}$; thus, let us write

$$X_{(11)} = [X : X_{(0)}] \tag{155}$$

where $X_{(0)}$ is an $(N \times \nu_0)$ matrix. Then, (154) gives

$$E(\hat{\mathbf{L}}) = \mathbf{L} + \mathbf{L}_b \tag{156}$$

where the bias vector $\mathbf{L}_b$ will be given by

$$\mathbf{L}_b = D^2(X'X)^{-1}(X'X_0)D_0^{-2}\mathbf{L}_0 = B_0\mathbf{L}_0, \text{ say} \tag{157}$$

Experimenters should program the last equation for any design T and any given vector $\mathbf{L}$; this will then enable them to examine the bias vector $\mathbf{L}_b$, and how it depends upon the elements of $\mathbf{L}_0$. This should be considered a very important part of the analysis of the data from the experiment.

It is useful to comment on the nature of $\mathbf{L}_b$ for the two major classes of designs. Firstly, consider the class of balanced designs of resolution, say, $(2l + 1)$ of the 2^m series. We shall assume that the design is a simple BA (which is that situation in almost all cases, at least when $l = 2$). For such designs, it turns out that the matrix $B_0(\nu \times \nu_0)$ at (157) is also "balanced" in the following sense: Let π_1 and π_2, respectively, be l_1-factor and l_2-factor interactions chosen out of $\mathbf{L}$ and $\mathbf{L}_0$. Also, let l_3 be the number of common factors between π_1 and π_2. Then, the element of B_0, which is in the row corresponding to π_1 and the column corresponding to π_2, depends only upon l_1, l_2, l_3, and is otherwise independent of π_1 and π_2 with which we start. A matrix B_0 of this kind is relatively easy to examine; this then constitutes another implicit advantage in using BA's.

A certain simplicity, though of a different type, occurs also for the case of parallel-planes designs. Basically, in the study of bias, we are replacing (69) by (151); i.e., we are replacing $\mathbf{L}$ by the full vector $\mathbf{F}_1$. Consider a design T based on a matrix A as at (93). If we replace $\mathbf{L}$ by $\mathbf{F}_1$, then it is easily seen that we will obtain s^{m-k} alias-sets, each containing exactly s^k factorial effects. These s^{m-k} alias-sets could be divided into two classes I and II, where class I is the class of all alias-sets which contain at least one member of $\mathbf{L}$, and class II contains the remaining alias-sets. Clearly, factorial effects which are in alias-sets of class II are being totally ignored, whereas those in class I are being estimated assuming that $\mathbf{L}_0$ is zero, and the design T is nonsingular. For factorial effects in class I, the expected value of the BLU estimate does not involve any effect which is not in the same alias-set. Using (157), one can write down the bias vectors for any factorial effect included in $\mathbf{L}$ for any given matrix C.

It should be emphasized here that even if $\mathbf{L}_b$ is known accurately, this in itself may not fully meet the purposes of the experiment, since $\mathbf{L}^0$ may still not be known. Only when we know $\mathbf{L}^*$, we know $\mathbf{F}_1$ and hence $\phi(\mathbf{t})$, for all $\mathbf{t}$.

The above suggests that, in a lot of situations, it would be important not simply to examine $\mathbf{L}_b$, but to identify $\mathbf{L}^0$. This is the objective of "search designs," which we now consider. These designs have as their objective the identification of $\mathbf{L}^0$, and its estimation along with that of $\mathbf{L}$. Because of the extra problem of identification involved, these designs are based on a generalization of a linear model, known as a "search linear model," which we now define.

Consider a vector $\mathbf{z}(N \times 1)$ containing random variables, whose expected value and variance is given by

$$E(\mathbf{z}) = E_1 \boldsymbol{\xi}_1 + E_2 \boldsymbol{\xi}_2, \qquad V(\mathbf{y}) = \sigma^2 I_N \tag{158}$$

where $E_1(N \times p_1)$ and $E_2(N \times p_2)$ are known matrices, $\boldsymbol{\xi}_1(p_1 \times 1)$ is an unknown vector of parameters, σ^2 is a positive constant, and $\boldsymbol{\xi}_2(p_2 \times 1)$ is partially known in the following sense. It is known that there is an integer p_3 such that at most p_3 elements of $\boldsymbol{\xi}_2$ are nonnegligible, the remaining elements being negligible; however, it is not known which set of elements of $\boldsymbol{\xi}_2$ is actually nonnegligible and what the values of these nonnegligible elements are. A model such as (70) is called a "search linear model." The problem under the search linear model is this. Let $\boldsymbol{\xi}_{20}(p_4 \times 1)$ (where $0 \le p_4 \le p_3$) be the subvector of elements from $\boldsymbol{\xi}_2$ which is nonnegligible. Then, we wish to identify the vector $\boldsymbol{\xi}_{20}$, and estimate it along with $\boldsymbol{\xi}_1$. Note that the number p_3 may or may not be known.

As an example, consider a factorial experiment. Suppose that $\mathbf{z}$ constitutes the observations on a set of treatments T. Also, suppose that $\boldsymbol{\xi}_1$ contains the general mean, the main effects, and the two-factor interactions, and $\boldsymbol{\xi}_2$ contains the three-factor and the four-factor interactions, it being known that interactions involving more than four factors are negligible. Suppose it is believed that, at most, five elements of $\boldsymbol{\xi}_2$ are nonnegligible. Then, we have $p_3 = 5$, and we wish that the design T and hence $\mathbf{z}$ are such that the nonnegligible elements of $\boldsymbol{\xi}_2$ (p_4 in number, where $p_4 \le 5$) are identifiable and also estimable along with $\boldsymbol{\xi}_1$.

The model (158) was introduced by the author (1975), wherein the following fundamental theorem was established.

Theorem 8.1

1. Consider (158). A necessary condition that $\boldsymbol{\xi}_{20}$ can be identified, and estimated along with $\boldsymbol{\xi}_1$ is that the following fundamental rank condition:

$$\text{Rank}[E_1 : E_{20}] = p_1 + 2p_3 \tag{159}$$

is satisfied for every ($N \times 2p_3$) submatrix E_{20} in E_2.

2. In the noiseless case (i.e., when $\sigma^2 = 0$), the condition at (159) is also sufficient for the identification of $\boldsymbol{\xi}_{20}$; in this case both $\boldsymbol{\xi}_{20}$ and $\boldsymbol{\xi}_1$ are estimable with variance zero.

The case where σ^2 is very small or negligible is usually called the "noiseless case," the rest being the "noisy case." In the more exact sciences, σ^2 may actually be very small. However, the noiseless case is important in all sciences. The reason is that so far as the design aspect of this whole theory is concerned, we only need to work with the noiseless case. The noisy case does not have any impact on the design problem. If σ^2 is not negligible (in other words, when it may be moderately, or even very, large), the same procedure for identification for $\boldsymbol{\xi}_{20}$ and its estimation along with that of $\boldsymbol{\xi}_1$ are used; the only difference is that now the probability that $\boldsymbol{\xi}_{20}$ is correctly identified may be not only less but much less than 1, and also both $\boldsymbol{\xi}_{20}$ and $\boldsymbol{\xi}_1$ will be estimated with variance which will be a multiple of σ^2, which may be quite large.

Now, the question arises: given the observations $\mathbf{z}$ and the matrices E_1 and E_2, and the number p_3 (which we assume to be known), how do we carry out the identification of $\boldsymbol{\xi}_{20}$? Many procedures were suggested in Srivastava (1975); the one which appears to be most useful is described below. Let $\boldsymbol{\xi}_{21}$ be any $(p_3 \times 1)$ subvector of $\boldsymbol{\xi}_2$ and let $E_{21}(N \times p_3)$ be the submatrix of E_2 whose columns correspond to $\boldsymbol{\xi}_{21}$. Consider the model

$$E(\mathbf{z}) = E_1\boldsymbol{\xi}_1 + E_{21}\boldsymbol{\xi}_{21}, \qquad V(\mathbf{z}) = \sigma^2 I_N \qquad (160)$$

then, we notice that this will be the correct model in case it so happens that

$$\boldsymbol{\xi}_{20} = \boldsymbol{\xi}_{21} \qquad (161)$$

Under the model (160), let $s_e^2(\boldsymbol{\xi}_{21})$ denote the "sum of squares due to error." From the statistical theory of linear models, it is well known that we shall have

$$s_e^2(\boldsymbol{\xi}_{21}) = \mathbf{z}'\mathbf{z} - \mathbf{z}'[E_1 : E_{21}] \begin{bmatrix} E_1'E_1 & E_1'E_{21} \\ E_{21}'E_1 & E_{21}'E_{21} \end{bmatrix}^{-1} \begin{bmatrix} E_1' \\ E_{21}' \end{bmatrix} \mathbf{z} \qquad (162)$$

We shall assume that (159) is satisfied; this will also insure that the matrix whose inverse is involved in (162) will be nonsingular. Now, we compute $s_e^2(\boldsymbol{\xi}_{21})$ for all possible $\binom{p_2}{p_3}$ values of $\boldsymbol{\xi}_{21}$. Let $\boldsymbol{\xi}_{20}^*$ be a value of $\boldsymbol{\xi}_{21}$ such that

$$s_e^2(\boldsymbol{\xi}_{20}^*) = \min s_e^2(\boldsymbol{\xi}_{21}) \qquad (163)$$

where the minimum is taken over all possible values of $\boldsymbol{\xi}_{21}$. Then, our decision rule says that we take $\boldsymbol{\xi}_{20}^*$ to be a set of parameters which contains all the nonnegligible parameters in it.

It can be shown that, in the noiseless case, $\boldsymbol{\xi}_{20}^*$ will contain as a subset all possible parameters which could be nonnegligible. Also, (159) will insure that the vector $[\boldsymbol{\xi}_1' : \boldsymbol{\xi}_{20}^{*\prime}]$ will be estimable with variance 0; examination of the estimates will then reveal exactly how many and which elements of $\boldsymbol{\xi}_{20}^*$ (and hence of $\boldsymbol{\xi}_2$) are nonzero. In the noisy case, the above procedures may, of course, be followed. In this case, the probability that $\boldsymbol{\xi}_{20}^*$ will include all the nonnegligible parameters will usually be less than 1, and the estimates of the $\boldsymbol{\xi}_1$ and the other nonnegligible parameters will have nonzero variance. This, however, is a common feature of all statistical work. Monte Carlo work (Mallenby, 1977) has shown that the above procedure has quite a reasonably high probability of correct identification of the nonnegligible parameters.

In applications, usually p_2 is much large than p_3. If the classical approach (based on the ordinary statistical linear model rather than the search linear model) is used, then either we can estimate both $\boldsymbol{\xi}_1$ and $\boldsymbol{\xi}_2$ (in which case we shall have $N \geq p_1 + p_2$), or we can ignore $\boldsymbol{\xi}_{20}$ and work only with $\boldsymbol{\xi}_1$ (in which case the parameters $\boldsymbol{\xi}_{20}$ will introduce unknown biases in the estimate of $\boldsymbol{\xi}_1$). Furthermore, even if $\boldsymbol{\xi}_1$ is accurately estimated, the true value of $E(\mathbf{z})$ will still remain unknown). Because of the condition (159), it can be shown that

$$N \geq p_1 + 2p_3 \tag{164}$$

However, in most cases, we could identify $\boldsymbol{\xi}_{20}$ with a relatively small value of N (much smaller than $p_1 + p_2$).

Let $\tilde{\boldsymbol{\xi}}_1$ be the estimate of $\boldsymbol{\xi}_1$ if, in (158), we ignore $\boldsymbol{\xi}_{20}$ (and hence $\boldsymbol{\xi}_2$). Then we have

$$\tilde{\boldsymbol{\xi}}_1 = (E_1'E_1)^{-1}E_1'\mathbf{z}$$
$$E(\tilde{\boldsymbol{\xi}}_1) = \boldsymbol{\xi}_1 = (E_1'E_1)^{-1}(E_1'E_{20})\boldsymbol{\xi}_{20} \tag{165}$$

Now, let $\hat{\boldsymbol{\xi}}_1$ be the BLU estimate of $\boldsymbol{\xi}_1$ obtained under the search linear model (158) and, using the above decision rule, for the identification of $\boldsymbol{\xi}_{20}$. Then, the author intuitively believes that even if σ^2 is rather large, and even if p_3 is not known, and although $\boldsymbol{\xi}_{20}^*$ may fail to contain most of the nonnegligible parameters, usually the estimate $\hat{\boldsymbol{\xi}}_1$ will still be much more accurate than $\tilde{\boldsymbol{\xi}}_1$. (This remark will usually not hold if $p_4 = 0$.)

Besides the above decision rule for identifying ξ_{20}, there are many other rules described in Srivastava (1975). Another, different, class of rules is described in Srivastava and Mallenby (1985).

The model (158) is called a pure search model if $p_1 = 0$. Many times, it is convenient to reduce a general model at (158) to a pure search model, as we shall now see.

Suppose that, in (158), $\mathbf{z}(N \times 1)$, $E_1(N \times p_1)$, and $E_2(N \times p_2)$ are partitioned as below:

$$\mathbf{z} = \begin{bmatrix} \mathbf{z}_1 \\ \mathbf{z}_2 \end{bmatrix}, \qquad E_1 = \begin{bmatrix} E_{11} \\ E_{21} \end{bmatrix}, \qquad E_2 = \begin{bmatrix} E_{12} \\ E_{22} \end{bmatrix} \tag{166}$$

where $\mathbf{z}_1$, E_{11}, and E_{12} have N_1 rows each, and Z_2, E_{21}, and E_{22} have N_2 rows each, where

$$N = N_1 + N_2 \tag{167}$$

We assume that $E_{11}(N_1 \times p_1)$ has rank p_1. (In view of (159), this can always be done by rearranging the elements of $\mathbf{z}$, and hence the rows of E_1 and E_2.) Thus, we have $N_1 \geq p_1$. The case when $(N_1 = p_1)$ is more important, since the procedure is more sensitive; notice that we can always achieve $(N_1 = p_1)$ by rearranging the elements of $\mathbf{z}$ in an appropriate manner. In what follows, we shall use the "property P_l" of a matrix M, which we now define. A matrix M is said to have the property P_l (with respect to columns) if and only if every set of l columns of M are linearly independent. The following result is sometimes very useful in obtaining search designs.

Theorem 8.2 Define

$$Q = E_{22} - E_{21}(E_{11}'E_{11})^{-1}E_{11}'E_{12} \tag{168}$$

Then, a sufficient condition that (159) holds is that we have

$$\text{Rank}(E_{11}) = p_1 \tag{169}$$

and Q has property P_{2p_3}. Furthermore, if E_{11} is a square matrix (i.e., $N_1 = p_1$), then the conditions (169) are also necessary in order that (159) may hold.

A search design T is a design such that if we take observation $\mathbf{z}$ corresponding to the runs in T, such that (158) holds, then (159) also holds. In other words, a search design T allows the identification of ξ_{20} and its estimation (along with that of $\boldsymbol{\xi}_1$), such that, in the noiseless case, the identification is possible with certainty, and the estimation is possible with zero variance. Thus, in the attempt to obtain search designs, we need to check the condition (159). The last theorem simplifies this checking by reducing it to the investigation of the matrix Q in which we need to consider $2p_3$ columns at a time, which contrasts with the matrix $(E_1 : E_2)$ in which we need to consider at a time p_1 columns of E_1, and $2p_3$ columns of E_2. We now give a second decision rule for identifying $\boldsymbol{\xi}_{20}$, which uses the above kind of reduction for a pure search model.

Let $R(N_2 \times N_1)$ and $G(N_2 \times N)$ be defined by

$$R = E_{21}(E_{11}' E_{11})^{-1} E_{11}' = E_{21} E_{11}^{-1} \qquad \text{(if E_{11} is square)}$$
$$G = [-R, I_{N_2}] \tag{170}$$

Multiplying both sides of (158) by G, we get the reduced model

$$E(\mathbf{z}_0) = (K^{-1}Q)\boldsymbol{\xi}_2, \qquad V(\mathbf{z}_0) = \sigma^2 I_{N_2} \tag{171}$$

where $K(N_2 \times N_2)$ is a matrix such that

$$KK' = GG', \qquad \mathbf{z}_0 = K^{-1}G\mathbf{z} \tag{172a}$$

It can be shown that such a matrix K always exists. Clearly, the decision rule for identifying $\boldsymbol{\xi}_{20}$ can be used on the pure model (171). Thus, suppose that we take $\boldsymbol{\xi}_{2(1)}(p_3 \times 1)$ to be a subvector of $\boldsymbol{\xi}_2$, and temporarily assume that it contains all the nonnegligible parameters. Then, under this assumption we shall have

$$E(\mathbf{z}_0) = (K^{-1}Q)\boldsymbol{\xi}_{2(1)} \tag{172b}$$

Let $s_e^{*2}(\boldsymbol{\xi}_{2(1)})$ be the sum of squares of error for the model (172b). Then, the following result holds.

Theorem 8.3   We have

$$s_e^{*2}(\boldsymbol{\xi}_{2(1)}) = \mathbf{y}'H\mathbf{y} - \mathbf{y}'HE_{2(1)}[E_{2(1)}' HE_{2(1)}]^{-1} E_{2(1)}'H\mathbf{y} \tag{173a}$$

where

$$H = G'(GG')^{-1}G \tag{173b}$$

and where $E_{2(1)}(N \times p_3)$ is the submatrix of $E_2(N \times p_2)$, whose columns correspond to the elements of $\boldsymbol{\xi}_{2(1)}$.

Notice that R and, hence, G and H are defined in terms of E_1, and $E_{2(1)}$. Thus, in the computation of the sum of squares due to error by using (173a), there is no need for determining the matrix K. Thus, our second decision rule for identification is this. We compute s_e^{*2} for each of the $\binom{p_2}{p_3}$ choices of $(\boldsymbol{\xi}_{2(1)})$ inside $\boldsymbol{\xi}_2$, using (173a), and then proceed as for the decision rule given earlier. The advantage in using this decision rule is that a matrix of a much smaller size $(p_3 \times p_3)$ needs to be inverted under the formula (173a). This contrasts with the matrix of size $(p_1 + p_3) \times (p_1 + p_3)$ which needs to be inverted if we use the earlier decision rule. In applications, p_3 is usually quite small, and therefore this procedure can be carried out with relative ease.

The above decision rules can be used for general diagnostics. Indeed, we should not use them very strictly. We should look at the value of $s_e^{*2}(\boldsymbol{\xi}_{2(1)})$ for various sets $\boldsymbol{\xi}_{2(1)}$. In particular, we should examine whether there are many sets $\boldsymbol{\xi}_{2(1)}$ for which the value of the s_e^{*2} is quite small, though it may not be the smallest. In other words, we should look at the cluster of the values of s_e^{*2}. If there is a cluster of small values, which is separated by most of the other values, then this has an important message. The message is that perhaps there are many nonnegligible parameters which are contained in one or more of the sets $\boldsymbol{\xi}_{2(1)}$ for which the value of s_e^{*2} is small. We should then try to see whether the design permits the estimation of all of the parameters which are indicated to be possibly nonzero by these small values of s_e^{*2}. If not, a second-stage design should be used to estimate these parameters. These remarks apply even to designs which are not, strictly speaking, search designs. In fact, these decision rules can be used on any design. If a design is not a search design, certain sets of parameters will give the *same* value of s_e^{*2} because (159) does not hold. These then should be examined with respect to the possible nonnegligibility of effects. These procedures can be followed with simplicity both when the design is a parallel-planes design and a simple B-array.

We now discuss search designs. We shall consider these designs mainly in the context of 2^m factorial experiments. For experiments of the type $s_1 \times s_2 \times \cdots \times s_m$, in which some of the s_i are different from 2, the following idea can be used. Firstly, suppose that each s_i is a power of 2. Then, if $s_i = 2^{n_i}$, we can consider a search design for the 2^n experiment where $n = n_1 + n_2 + \cdots + n_m$. On the other hand, if s_i is not a power of 2 for some value of i, still we can "embed" this experiment in a 2^m type of experiment. For example, suppose $s_1 = 7$; then we can consider another experiment in which $s_1 = 8$ and one level is a "dummy" level, which could be the same as one of the other levels. We shall not go into the details of the embedding here, except that it should be emphasized that this approach appears to be simpler at this time for making search designs than many other available approaches.

We now consider a 2^m factorial design of resolution $\{(2l+1):p\}$. This means that the design corresponds to the model (158), where ξ_1 has the general mean, the main effects, and all interactions involving l or lesser numbers of factors. Also, ξ_2 contains all the interactions involving $(l+1)$ factors or more. Also, ξ_2 may contain up to p nonnegligible parameters. Thus,

$$p_1 = 1 + \binom{m}{1} + \cdots \binom{m}{l}, \qquad p_2 = 2^m - p_1, \qquad p_3 = p \qquad (174)$$

In other words, in the context of the discussions in the previous sections, this corresponds to a situation where $\mathbf{L}$ has p_1 elements corresponding to ξ_1, and $\mathbf{L}_0$ corresponds to ξ_2 with p_2 elements and, furthermore, $\mathbf{L}^0$ has p_3 elements and corresponds to ξ_{20} of model (158). Thus, for example, a search design of resolution 3.2 is a design in which we have $\xi_1' = \mathbf{L}' = \{\mu,$ main effects$\}$, $\xi_2' = \{$all interactions$\}$, and $p_3 = 2$. A design of resolution $0 \cdot p$ (also called a pure search design of order p) is a design in which $p_1 = 0$, ξ_2 consists of all possible factorial effects, and $p_3 = p$. Take any arbitrary situation, i.e., one where $[\xi_1', \xi_2'] = \mathbf{F}_1'$, and where ξ_1 and ξ_2 are otherwise arbitrary. Also, let $p_3 = p$. Then it can be shown that this design is also a pure search design of order p. This shows the general importance of pure search designs of order p. The following result for the case $p = 1, 2$, was contained in the original paper of Srivastava (1975) where the theory of search designs was introduced.

Theorem 8.4 Consider a design $T(m \times N)$ for a 2^m experiment. Then, T can be considered as a $(0,1)$ matrix whose elements belong to GF(2). Without loss of generality, we assume that T contains the run J_{m1} in which all factors are at level 1. The following results hold.

1. T is a design of resolution 0.1 (i.e., a pure search design of order 1) if and only if rank$(\bar{T}) = m$, where $\bar{T}$ is obtained from T by interchanging 0 and 1.
2. A necessary and sufficient condition that T is a pure search design of order 2 is the following. For every matrix $R(2 \times m)$, of rank 2, and every vector $\mathbf{r}(2 \times 1)$, there is at least one treatment $\mathbf{t}(m \times 1)$

in $\bar{T}$ (i.e., there is at least one column **t** in $\bar{T}$) such that

$$Rt = r \tag{175}$$

3. A necessary condition that T is a pure search design of order 2^u (where $u > 1$ is a positive integer) is the following. For every matrix $R(u \times m)$, of rank u, and every vector $r(u \times 1)$, there is at least one treatment **t** in $\bar{T}$ such that (175) is satisfied, where now in (175) we assume that R and **r** have u rows.

4. Let $m(N, 4)$ be the maximum value of m such that there exists a design $T(m \times N)$ of order 2. Then we have

$$\lim_{N \to \infty} \frac{1}{N} m(N, 4) \le 0.2835 \tag{176}$$

5. We have

$$m(N, 4) = 3 \quad \text{for } 6 \le N \le 9, \quad m(10, 4) = 4 \tag{177}$$

6. Let $N(m, 4)$ be the minimum value of N such that, for a given m, there exists a design $T(m \times N)$ of order 2. Then we have

$$N(3, 4) = 7, \qquad N(4, 4) = 10, \qquad N(5, 4) = 14 \tag{178}$$

7. Let $T(m \times N)$ be a design of order 2. Also let $K(m \times m)$ be a nonsingular matrix with elements from GF(2). Let T^* be the $(m + 1) \times (N + m)$ matrix given by

$$T^* = \left[\begin{array}{c|c} T & \bar{K} \\ \hline J_{1N} & O_{1m} \end{array} \right] \tag{179}$$

Then T^* is of order 2 for the 2^{m+1} experiment. (It may be remarked that sometimes, with an appropriate choice of K in (179), one can delete some treatments T^* from the part $\binom{T}{J_{1N}}$ in such a way that the remaining set of treatments still constitutes a design of order 2.

The above results are from Srivastava (1975) and Katona and Srivastava (1983).

Because of the importance of the case when $p_3 = 2$, many studies have been made on the functions $m(N, 4)$ or $N(m, 4)$ which are closely interrelated. These include the Ph.D. dissertation of Jain (1980), under the guidance of R. C. Bose, and unpublished work of Katona and his associates, and also of the author. Pure search designs of order p are not likely to be used directly; however, as mentioned above, they are important in general since a general design with a general value of (p_1, p_2, p_3) has also to be a design of order p_3.

We now consider designs of resolution 3.1 which are quite important from the practical angle. The smallest value of m of interest is $m = 4$. The following theorem (due to Srivastava and Arora, 1988b) is important.

Theorem 8.5 Consider the 2^4 factorial experiment. Suppose $T(4 \times N)$ is a design of resolution 3.1. Then, we have the following:

1. $N \geq 10$.
2. There are exactly six nonisomorphic designs T, where $T = T^{(i)}$; $i = 1, \ldots, 6$, where the $T^{(i)}$ are given below.

$$T^{(1)} = \{1, 2, 3, 4, 12, 13, 34, 123, 124, 1234\}$$

$$T^{(2)} = \{1, 2, 13, 24, 34, 123, 124, 134, 234, 1234\}$$

$$T^{(3)} = \{\phi, 1, 2, 3, 13, 34, 123, 124, 134, 1234\}$$

$$T^{(4)} = \{\phi, 3, 13, 14, 24, 34, 123, 124, 134, 1234\}$$

$$T^{(5)} = \{1, 3, 12, 13, 23, 24, 34, 123, 124, 1234\}$$

$$T^{(6)} = \{\phi, 2, 3, 4, 13, 14, 23, 34, 123, 1234\} \tag{180}$$

In (180), for each $T^{(i)}$, the treatments are represented in the following way. For $1 \leq l \leq 4$, and $1 \leq i_1 < i_2 < \cdots < i_l \leq 4$, the symbol $(i_1, i_2, \ldots, i_l)$ represents the treatment (t_1, t_2, t_3, t_4), in which $t_j = 1$ if $j \in (i_1, i_2, \ldots, i_l)$, and $t_j = 0$, otherwise. This symbolism will be used in some later theorems as well.]

We now present a result which is central to the development of the theory in several papers of several authors. This result is due to Srivastava and Gupta (1979).

Consider the partitioning of the vector $\mathbf{z}$ and the matrices E_1 and E_2 as at (166). Let $T_1(m \times N_1)$ be a set of treatments which correspond to $\mathbf{z}_1$, and similarly let $\bar{z}_2$ correspond to the treatments in $T_2(m \times N_2)$. Now, suppose $T_1 = \Omega_{mm} + \Omega_{m1}$, where the symbol Ω_{mj} is to be recalled from Section 6. Let $T = [T_1 : T_2]$ so that T is $(m \times N)$, where $N = N_1 + N_2$. We now consider T as a candidate for a design of resolution 3.1. Notice that, following this partitioning of T, the partitioning induced on E_1 and E_2 will be such that E_{11} will be a square nonsingular matrix. Consider Q

as at (168). Then Q is $(N_2 \times p_2)$, where $p_2 = 2^m - m - 1$. The rows of Q correspond to the treatments $\mathbf{t}$ in T_2, and the columns of Q correspond to the various two-factor effects. Let $q(\mathbf{t}, e)$ be the element of the matrix Q in the row corresponding to the treatment $\mathbf{t}$ (in T_2), and in the column corresponding to the factorial effect e. Let e be the interaction denoted by $e = F(i_1 i_2, \ldots, i_l)$, where $2 \le l \le m$, and $1 \le i_1 < i_2 < \cdots < i_l \le m$. For any pair $(\mathbf{t}, e)$, let $\theta(\mathbf{t}, e) = \sum_{r=1}^{l} t_{i_r}$. Also, for any positive integer g, define the ζ function by

$$\zeta(g) = \begin{cases} g - 1, & \text{if } g \text{ is odd} \\ g, & \text{if } g \text{ is even} \end{cases} \tag{181}$$

Theorem 8.6   We have

$$q(\mathbf{t}, e) = 2(-1)^l \left\{ \left[1 - \left(\sum_{i=1}^{m} t_i \right) \right] (m-1)^{-1} \zeta(p) + \zeta(\theta) \right\} \tag{182}$$

Now, let T_1 be replaced with $T_1^* = \Omega_{mm} + \Omega_{m,m-1}$; and let the rest of the notation remain the same as for the previous theorem. Because of this change from T_1 to T_1^*, let T change to T^*, Q to Q^*, and $q(\mathbf{t}, e)$ to $q^*(\mathbf{t}, e)$. Then the following result (Srivastava and Arora, 1988c) is worth noting because of its simplicity and potential importance.

Theorem 8.9   We have

$$q^*(\mathbf{t}, e) = 2\zeta(g) \tag{183}$$

where

$$g = l - \sum_{r=1}^{l} t_{i_r} = l - \theta(\mathbf{t}, e) \tag{184}$$

Gupta and his co-workers considered the situation of Theorem 8.8, with the further assumption that T_2 contains the treatment O_{m1}. They also considered the situation where $T_1 = \Omega_{mm} + \Omega_{m1} + \Omega_{m0}$; notice that this means that the corresponding E_{11} is not square, so that the first expression for Q in (168) will have to be used. With this latter T_1, Gupta and Carvajal (1986) gave a 2^5 design of resolution 3.1 with $N_2 = 7$, $N = 14$, and Gupta (1988) gave a similar design for the 2^6 case with $N = 19$.

On 2^m designs of resolution 3.2, the following results were obtained in Srivastava and Arora (1988a, c).

Theorem 8.10

1. Consider the design $T(4 \times n)$, with $N = 11$, and $T = \Omega_{44} + \Omega_{42} + \Omega_{41}$. Then T is a 2^4 design of resolution 3.2.

2. Consider the 2^m experiment with $m \geq 3$. Consider the design $T(m \times N)$, where

$$N = 2 + m + m(m-1)/2$$
$$T = \Omega_{mm} + \Omega_{m2} + \Omega_{m1} + \Omega_{m0} \tag{185}$$

Then T is a 2^m design of resolution 3.2.

We now consider designs of resolution 5.1. These have been considered principally by Srivastava and Ghosh, and Ohnishi and Shirakura. The following result by Srivastava and Ghosh (1976), gives a general series of designs of this class.

Theorem 8.11   Let $T(m \times N)$ be a design for a 2^m factorial experiment, such that

$$T = \Omega_{mm} + \Omega_{m,m-2} + \Omega_{m1} + \Omega_{m0} \tag{186a}$$

with

$$N = \binom{m}{m} + \binom{m}{m-2} + \binom{m}{1} + \binom{m}{0} = 2 + m + m(m-1)/2 \tag{186b}$$

Then T is a resolution 5.1 plan.

Indeed, the above gives an infinite series of designs, each of which is a BARE 5.1 (balanced resolution 5.1) plan. More work on this line was done by Srivastava and Ghosh (1977). They considered simple BA's for values of the pair (m, N) in a selected range, such that each of these constitutes a design of resolution V. Next, each of these BA's was tested as to whether it was of resolution 5.1. Values of (m, N) in the following range were considered: $(m = 4,\ 13 \leq N \leq 15,\ p_1 = 11)$, $(m = 5,\ 18 \leq N \leq 31,\ p_1 = 16)$, $(m = 6,\ 24 \leq N \leq 40,\ p_1 = 22)$, $(m = 7,\ 31 \leq N \leq 68,\ p_1 = 29)$, and $(m = 8,\ 39 \leq N \leq 59,\ p_1 = 37)$. For the convenience of the reader, for each value of m we are giving the value of p_1 $(= 1 + \binom{m}{1} + \binom{m}{2})$. In (187) below, we present, for selected values of (m, N), the values of λ'

(the index set of the simple BA, which constitutes a BARE 5.1 plan) of Srivastava and Ghosh:

$$
\begin{aligned}
m &= 4, & N &= 15, & \lambda' &= (1\ 1\ 1\ 1\ 0), & p_1 &= 11 \\
m &= 5, & N &= 21, & \lambda' &= (1\ 1\ 1\ 0\ 1\ 0), & p_1 &= 16 \\
m &= 5, & N &= 26, & \lambda' &= (0\ 1\ 1\ 1\ 0\ 1), & p_1 &= 16 \\
m &= 6, & N &= 28, & \lambda' &= (1\ 1\ 1\ 0\ 0\ 1\ 0), & p_1 &= 22 \\
m &= 6, & N &= 37, & \lambda' &= (0\ 1\ 1\ 0\ 1\ 0\ 1), & p_1 &= 22 \\
m &= 7, & N &= 36, & \lambda' &= (1\ 1\ 1\ 0\ 0\ 0\ 1\ 0), & p_1 &= 29 \\
m &= 7, & N &= 50, & \lambda' &= (0\ 1\ 0\ 1\ 0\ 0\ 1\ 1), & p_1 &= 29 \\
m &= 8, & N &= 45, & \lambda' &= (1\ 1\ 1\ 0\ 0\ 0\ 0\ 1\ 0), & p_1 &= 37 \quad (187)
\end{aligned}
$$

Since, in view of (159), we must have $N \geq p_1 + 2p_3$, for each of the above cases, the minimum value of N would be $(p_1 + 2)$. However, it is not known whether resolution 5.1 designs exist for $N = p_1 + 2$. In this situation then, the next-smallest value of N will generally be important from the practical viewpoint. From the tables of Srivastava and Ghosh, we have given in (187) the design for some such values. We have also given a few other values of N in case one experimenting with a large design later decides to examine the results, considering the design to be a search design.

Srivastava (1977) considered combining the concepts of "search" and "optimality." A new optimality criterion (called "AD-optimality") was advanced. Essentially, this consists of taking the average of the determinants

$$
\begin{vmatrix}
E_1' E_1 & E_1' E_{21} \\
E_{21}' E_1 & E_{21}' E_{21}
\end{vmatrix}
$$

over all choices $E_{21}(N \times p_3)$ in E_2. Each E_{21} (i.e., each ξ_{21}) is given a certain probability of being the correct and nonnegligible set of parameters. These probabilities should be decided upon by the experimenter. Considering these probabilities, a simplified value of the AD-optimal criterion was obtained. Later on, Ohnishi and Shirakura (1985) worked in this field and obtained designs which are AD-optimal in certain classes of balanced designs for various values of (m, N) in a practical range.

Various authors, particularly Ghosh, Gupta, and Shirakura, have considered various other situations with respect to the parameter sets ξ_1' and ξ_2'. For example, the case where ξ_1' consists of μ and the main effects, but

ξ_2' is restricted to interactions up to order 2 or 3 or 4, has been considered. These studies are important from the practical point of view. However, it would also be useful in future studies to consider ξ_2 of a type which corresponds to a tree structure in the set of parameters.

Various authors have also considered search designs for the 3^m series. Some of the important work in this field is due to authors such as Anderson and Thomas (1980), Chatterjee and Mukherjee (1986), and Ghosh and Zhang (1987). Besides the 3^m case, some authors, such as Chatterjee and Mukherjee, have also considered search designs for mixed factorials. However, we omit a discussion of these results because of lack of space.

Before closing this section, it is important to discuss briefly the so-called *factor-screening* problems. In 1988, *Communications in Statistics* recognized the importance of this area by devoting an entire issue to it, edited by Patel.

Briefly, the factor-screening problem is this. Consider a 2^m experiment in which m is possibly large and the chance that interactions are nonnegligible is very small. In other words, we have a situation like that of the design of resolution III. We want to estimate the main effects, quite a few of which are possibly negligible. Our problem is to screen out the large number of ineffective factors and find the important factors. Of course, our earlier theory says that the number of points in a design of resolution III (which is what we need here) is $(m + 1)$. However, we do not want to use large designs because m is large. This apparent paradox is sometimes resolved because, in most factor-screening problems, the experimenter knows the sign (positive or negative) of each main effect F_i $(i = 1, \ldots, m)$. This then helps in simplifying the problem and carrying it toward solution.

In Srivastava (1976), a connection was made between this field and the field of search designs, and it was shown that the factor-screening problem could be approached in a convenient way as a special case of the search design theory. Further contributions in this field have been made, for example, in Ghosh and Avila (1985). Even in factor-screening problems, because m is large, at least a few interactions are nonnegligible. Thus, a more accurate approach would need to include this in the formulation (perhaps under the assumption of the tree structure). Generally, it may be that the experimenter has an idea of the sign of any possible interaction that may be nonnegligible.

Some of the recent work on the search design theory shows that in many cases the value of N turns out to be rather large. This is the case particularly for mixed factorials. However, as mentioned earlier, it might be possible to help ourselves by embedding the mixed factorials in the two-level case. However, for two-level situations, m may tend to be large. Because of this, we need to develop further the theory of "weak search designs," where the condition (159) is not totally satisfied, but is satisfied for "important" values of ξ_{21}. In other words, the sets ξ_{21} for which the condition (159) is not satisfied should be such that the chance that they are nonnegligible is considered to be minimum.

9 PROBING DESIGNS: THE REVEALING POWER OF DESIGNS

In this section, we continue the study of the identification and estimation problem. Here, we present a brief discussion of the situation where $\mathbf{L}^*$ is to be identified; i.e., $\mathbf{L}$ is nonexistent, and $\mathbf{L}^0 = \mathbf{L}^*$. This corresponds to the search designs, under the pure search model. However, the search designs are single-stage, whereas we shall now consider multistage or sequential designs. It may be emphasized here that the study of the sequential aspect is not in the sense of Abraham Wald's sequential analysis.

Before we proceed further, it is important to emphasize the nature of situations in which the methods of this section may work. These methods are suited primarily to situations in which trying new runs is expensive and in which σ^2 is either quite small to start with or can be "made" small. To exemplify, consider the situation in a factory where different factors correspond to different structures, and the levels of a factor correspond to different settings of the structure. Now, to try a new treatment (or level combination), we shall have to set the different structures at settings that correspond to this particular level combination. This may mean closing down the factory for a while so that the different structures are appropriately set, which may be quite expensive. However, once the structures are set to correspond to a particular level combination $\mathbf{t}$ "production" can be restarted, and we shall obtain a number (say r) of values of $y(\mathbf{t})$. Now, as production is continued, r increases. The average of the values of $y(\mathbf{t})$ for r observations will have variance σ^2/r, which goes to zero as r becomes large. It is in this sense that we say that σ^2 can be "made" small. Thus, for each $\mathbf{t}$ that we try, we will obtain a fairly accurate value of $\phi(\mathbf{t})$, but trying different treatments $\mathbf{t}$ is very expensive.

The ideas of this section are based on several pieces of work. For example, in Srivastava (1984), where certain foundational questions in statistical design theory were examined, the concept of revealing power of a design was introduced and studied along with the concept of "sensitivity." These two concepts and the concept of optimality were examined in general.

Hussein (1986) considered multistage designs for 3^m factorial experiments, where he considered designs based on parallel flats. At each stage, one or more flats, i.e., one or more columns of the matrix C at (91), are used, where the total number of columns of C (which is n) is now a variable. In this work, generally, two assumptions were made: (1) all interactions involving three or more factors are negligible, and (2) a linear combination of factorial effects (which are nonnegligible and distinct) is zero, only with probability zero.

In Srivastava (1987b), the concept of revealing power was more explicitly developed. It was shown through example that the customary orthogonal arrays present difficulties as far as the question of revealing nonnegligible elements of $\mathbf{L}$ is concerned. (It was pointed out that these difficulties appeared to occur largely because, by using an orthogonal array, we estimate a number of linear combinations of disjoint sets of elements out of $\mathbf{F}_1$. Indeed, each linear combination is a linear combination of elements in one alias-set). New ground was broken concerning the problem of multistage or sequential identification of $\mathbf{L}^*$. Certain new concepts and original decision rules (called "sieves") were introduced. Thus, ground was laid for the identification and estimation of nonnegligible factorial effects, and the general problem of factorial experimentation was addressed more completely, without unrealistic and artificial assumptions concerning the negligibility of higher-order interactions. At the same time, an attempt was made not to lose sight of the fact that, in most situations, we need designs in which the number of runs is of a desirable magnitude. Some of these ideas are further developed in Srivastava and Hveberg (1988), where the case of a 2^m experiment, with $m = 3$, was examined at length, under the assumption of the tree structure. The concept of the "average sample number" and other useful concepts were introduced. It will be difficult to give a detailed discussion of these developments, important though they are. However, we present an example to illustrate the nature of the developments.

Consider a 2^4 experiment, for which

$$\mathbf{L}^* = (\mu, F_2, F_{24}, F_{124})' \tag{188}$$

where

$$\mu = 1, \qquad F_2 = 5, \qquad F_{24} = -4, \qquad F_{124} = 2 \tag{189}$$

where the remaining parameters (which are not in $\mathbf{L}^*$) are assumed to be negligible. For this situation, using (38), (39), and (66), we can determine the value of ϕ_1 from $\mathbf{F}_1$ (which is known since $\mathbf{L}^*$ is known and the value of its elements are known). We find

$$\phi_1' = \{4, 0, -2, 2, 4, 0, -2, 2, 8, 12, -6, -10, 8, 12, -6, -10\} \qquad (190)$$

In equation (190), $\phi_1' = \phi(\mathbf{t})$, where $\mathbf{t}$ runs over the set of all possible treatments, which are arranged in the Yates order as explained at (40) and (42).

Now, of course, the knowledge of (188)–(190) is not available to us; the purpose of the experiment is indeed to discover (190). We plan to do the same by determining $\mathbf{L}^*$, and its value as at (189).

For the above purpose, we first consider an approach based on orthogonal arrays. This will show the shortcomings of such an approach, which are not well known to many experimenters. Besides the shortcomings that have been well known since OA's were defined by Rao, there are new difficulties that were pointed out in Srivastava (1987b). Basically, an approach based on OA's gives us the estimates of a certain number of linear combinations of sets of factorial effects, where these sets are mutually disjoint. (Indeed, usually, these sets are nothing but alias-sets, which are mutually disjoint.) As will be seen, this feature precludes further examination of the parameters, and the search for the elements of $\mathbf{L}^*$.

We start with $N = 4$ runs. Now, the OA-users usually go for the estimation of μ and the main effects. This assumption is not realistic at all, since nature hardly ever comes in a totally additive fashion (which is equivalent to saying that all interactions are negligible). Sometimes, these workers suggest that we include in our $\mathbf{L}$ certain interactions, and use the OA to estimate them. Now, it is unreasonable to assume (for most situations) that $\mathbf{L}^*$ would be known. Because of this, as we shall see, misleading estimates of the parameters are obtained.

To make an orthogonal array with $N = 4$ runs, consider the equation $A\mathbf{x} = \mathbf{c}_1$, where $A(2 \times 4)$ of rank 2, is given by

$$A = \begin{bmatrix} 1 & 0 & 1 & 1 \\ 0 & 1 & 1 & 1 \end{bmatrix} \qquad (191)$$

It is easy to verify that, for this case, the main effects of F_1 and F_2 are confounded, However, it can also be verified that there does not exist a matrix A of this size and rank, which will avoid the confounding of at least two main effects. However, the above matrix A has another problem associated with it. It can be seen that there does not exist a vector c_2 such that the eight runs which satisfy the equation $Ax = c_i$ $(i = 1, 2)$ generate an OA of strength 3, which is what we need in order that main effects be estimable free from the two-factor interactions.

It is verifiable that if we want to start with an $A(2 \times 4)$ such that by using two vectors c_1 and c_2, we can produce an OA of strength 3, then it is necessary to use the A given by

$$A = \begin{bmatrix} 1 & 0 & 1 & 0 \\ 0 & 1 & 0 & 1 \end{bmatrix} \tag{192}$$

Unfortunately, this A leads to the confounding of the main effects F_1 and F_2, and also of the main effects of F_3 and F_4. Since there are basically only these two types of A matrices, we have to make a choice. Since $\mathbf{L}^*$ is not known, it is clear that an OA with only four runs is unlikely to be able to identify and estimate of the elements of $\mathbf{L}^*$. Thus, it seems prudent to assume that we will need an OA with at least eight runs. Because of this, we consider the OA of strength 3, whose runs constitute the solutions of the equations

$$t_1 + t_2 + t_3 + t_4 = 0 \tag{193}$$

Now, this array provides the estimates of the following linear combinations of the parameters: $F(0 + 1234)$, $F(1 + 234)$, $F(2 + 134)$, $F(3 + 124)$, $F(4 + 123)$, $F(12 + 34)$, $F(13 + 24)$, $F(14 + 23)$, where, throughout, we follow the notation

$$F(0) \equiv \mu \tag{194}$$
$$F(\pi_1(i_1 i_2 \ldots i_l) + \pi_2(j_1 j_2 \ldots j_{l'})) = \pi_1 F(i_1 \ldots i_l) + \pi_2 F(j_1 \ldots j_{l'})$$

where the notation is in accordance with (41), and where π_1 and π_2 are any real numbers. Thus, $F(0 + 1234) = \mu + F_{1234}, F(14 + 23) = F_{14} + F_{23}$. (Notice that, in both of these cases, we have $\pi_1 = \pi_2 = 1$.) Thus, we see that the 16 factorial effects have become divided into eight sets of two each and, for each set, we are able to estimate the sum of the two effects. Sticking to the tradition of "estimating" the main effects, and any interactions that might seem to be nonzero, we take the following estimates:

$$\hat{\mu} = 1, \qquad \hat{F}_1 = 0, \qquad \hat{F}_2 = 5,$$

$$\hat{F} - 3 = 2, \qquad \hat{F}_4 = 0, \qquad \hat{F}_{13} + \hat{F}_{24} = -4 \qquad (195)$$

Two cases arise according to whether we believe that F_{13} or F_{24} is nonnegligible. Thus, assuming that $\hat{\mu}_1$, $\hat{F}_2$, $\hat{F}_3$, and $\hat{F}_{13}$ are nonnegligible, and the remaining parameters are negligible, and using (66), we obtain

$$\phi_1' = \{4, 12, -6, 2, 8, 0, -2, -10, 4, 12, -6, 2, 8, 0, -2, -10\} \qquad (196)$$

Similarly, assuming that F_{24} is nonnegligible, and $F_{13} = 0$, and continuing with the same assumptions as above on the other parameters, we obtain

$$\phi_1' = \{4, 4, 2, 2, 0, 0, -2, -2, 12, 12, -6, -6, 8, 8, -10, -10\} \qquad (197)$$

Comparing (196) and (197) with (190), we find that there is quite a bit of difference in the value of $\phi(\mathbf{t})$ in the three cases. For example, suppose we want to choose $\mathbf{t}$ so that $\phi(\mathbf{t})$ is maximized. In the first case, assuming (196) to be true, we find that the maximum occurs for $\mathbf{t} = (0111)$ and (0110). Notice that the treatment (0110) does have a maximum as shown by (190). However, the experimenter does not know this. Thus, imagine an industrial experiment where, for some reason, an experimenter finds the treatment (0111) to be more appealing than the treatment (0110) on certain grounds. Since both give the same maximum, suppose the treatment (0111) is chosen. Then, the quality of the product will be quite inferior since, from (190), we find that $\phi(0111) = 0$. Similar misleading results may be arrived at starting from (197), and for the cases where we wish to minimize $\phi(\mathbf{t})$.

Thus, even in this simple case where the number of nonnegligible parameters is quite small and we use an OA of strength 3 (rather than 2, which is customarily used), we may end up with a very wrong conclusion if we depend only on the classical approach based on OA's.

As mentioned earlier, in spite of the above observation about the misleading nature of the results that we may reach if we use OA's disregarding the assumptions behind them, no appropriate technique was available to get around this problem. However, the techniques introduced in Srivastava (1987b) offer a breakthrough. We now briefly describe these. For this purpose, it is necessary first to introduce certain concepts.

Any set of linear combinations of factorial effects will be called a "viewfield." An "invariant viewfield" (IVF) is a viewfield which is such that if any linear function θ of factorial effects belongs to it, then so does $p(\theta)$ where $p(\theta)$ is the linear function obtained by permuting the factor symbols according to any permutation whatsoever. Thus, for example, suppose a particular IVF contains the linear combination $F_1 - 2F_4$ (assuming $m = 4$). Then, if the viewfield is invariant, all linear combinations of the type $F_i - 2F_j$ will belong to it, where i and j are distinct and where $1 \le i$, $j \le 4$.

The new techniques involve analysis of the data as the experimentation proceeds. Certain techniques that are to be used during the analysis of the data are now introduced. One of them is called the "temporary elimination procedure" (TEP). The TEP says the following: Let $(\pi_1 h_1 + \cdots + \pi_k h_k)$ be a linear combination of factorial effects, where the π are real numbers, and where the h are factorial effects. Suppose that this linear combination is such that we cannot express it in terms of the sum of two linear combinations, where the value of one of the linear combinations in this sum is known. Now, furthermore, suppose we know that $\pi_1 h_1 + \cdots + \pi_k h_k = 0$. Then the TEP says that, during the analysis of the data, we may temporarily assume that $h_1 = \cdots = h_k = 0$. In other words, the TEP says that, given a number of factorial effects, it is *unlikely* that we will come across a linear combination whose value will be zero. Note that the rule says that it is "unlikely" rather than "impossible."

The TEP is an example of a "tentative decision rule" (this is a rule that the experimenter is allowed to modify, in certain circumstances, which may not be known in advance).

Another example of a tentative decision rule is the "intersection sieve" (IS). Let θ_1, θ_2, and θ_3, be three linear combinations of factorial effects, such that the set of factorial effects involved in the linear combination θ_i is disjoint with the set of effects involved in θ_j if i and j are distinct. Then, if, during the intermediate analysis of the data (as the experimentation proceeds), it turns out that we have $\theta_1 + \theta_2 = \theta_1 + \theta_3 = \pi$, where π is a real number, then the intersection sieve says that it is very likely that we have $\theta_1 = \pi$, and $\theta_2 = \theta_3 = 0$.

There is a third tentative decision rule, called the "majority sieve." Here, we consider an IVF. We consider the class of those linear combinations in the IVF for which the estimated value is numerically relatively large. Then, we find the frequency of each factorial effect within this last set of linear combinations. Factorial effects which occur in a relatively large number of linear combinations are then suspected of being nonnegligible. On the other hand, factorial effects which occur in linear combinations whose value is very small are suspected of being negligible.

Now, we illustrate the use of the above tentative rule in the example with which we started.

We start the experiment with the simple BA consisting of the five runs in $(\Omega_{44} + \Omega_{43})$.

Recall the notation from Section 6. From (190), we obtain

$$\phi(1111) = 4, \qquad \phi(01111) = 0, \qquad \phi(1011) = -2,$$
$$\phi(1101) = 4, \qquad \phi(1110) = 8 \tag{198}$$

Define, for all treatments **t**,

$$\alpha(\mathbf{t}) = \frac{1}{2}\{\phi(1111) - \phi(\mathbf{t})\} \tag{199}$$

Then, using (38), (39), and (198), we obtain

$$\alpha_1 \equiv \alpha(0111) \equiv A(1 + 12 + 13 + 123 + 14 + 124 + 134 + 1234) = 2 \tag{200}$$
$$\alpha_2 \equiv \alpha(1011) \equiv A(2 + 12 + 23 + 123 + 24 + 124 + 234 + 1234) = 3 \tag{201}$$
$$\alpha_3 \equiv \alpha(1101) \equiv (3 + 13 + 23 + 123 + 34 + 134 + 234 + 1234) = 0 \tag{202}$$
$$\alpha_4 \equiv \alpha(1110) \equiv A(4 + 14 + 24 + 124 + 34 + 134 + 234 + 1234) = -2 \tag{203}$$

Using TEP, (202) suggests that all factorial effects involving the factor 3 are negligible. In other words, it suggests that

$$\mathbf{L}^* \subset A(0, 1, 2, 4, 12, 14, 24, 124) \tag{204}$$

Considering the quantity $(\alpha_1 + \alpha_2 - \alpha_4)$, and using (204), we obtain

$$A(1 + 2 - 4 + (2)12 + 124) = 7 \tag{205}$$

where (2) denotes the real number 2. We need to have more information in order to sort out the effects. Let us try (0000), and let $\alpha_5 = \alpha(0000)$. This gives

$$A(1 + 2 + 4 + 124) = 7 \tag{206}$$

in view of (204). Now, taking $\theta_1 = A(1 + 2 + 124)$, $\theta_2 = (2)A(-4 + 12)$, $\theta_3 = A(4)$, and using the IS, we obtain

$$A(1 + 2 + 124) = 7, \qquad A(4) = A(12) = 0 \tag{207}$$

Using this in (201) and (203), we obtain

$$A(2 + 24 + 124) = 3, \qquad A(14 + 24 + 124) = -2, \qquad A(2 - 14) = 5 \tag{208}$$

Next, we try the run (1010). Looking at its α value, simplifying, and using (208), we obtain

$$A(2) = 5, \qquad A(14) = 0 \tag{209}$$

Clearly, we need to separate $A(24)$ and $A(124)$. We can consider trying the runs (1001) or (1000). However, it will be seen that these runs do not give us any new information. Thus, a design procedure which involves using these runs at this stage will obviously have less "revealing power." On the other hand, it we tried the run (0011), then, in view of the above, its α value leads us to

$$A(1 + 24) = 1 \tag{210}$$

Now, from (208) and (209), we get

$$A(24 + 124) = 0 \tag{211}$$

Finally, from (200), (210), and (211), we obtain

$$A(1) = 0, \qquad A(24) = -4 \tag{212}$$

The value of μ is now found from the first equation in (198), which gives

$$\mu = 1 \tag{213}$$

At this stage, *tentatively*, we have identified $\mathbf{L}^*$, and obtained the values of the various parameters. However, this analysis has been based on a tentative procedure which, obviously, would not always give correct results. At this stage, it is suggested that the experimenter should randomly select a treatment which has not been used so far. If the yield for this treatment "matches" the yield predicted by the results obtained so far, then it will show that our work is perhaps finished. (It should be emphasized that unless we do the whole experiment, we can never be absolutely certain about the value of $\phi(\mathbf{t})$ for all $\mathbf{t}$. However, our objective is to conduct the experiment (including the intermediate analysis of the data) in such a way that the chance is very good that our results are correct.

Notice that the above procedure is much more successful than the one based on the OA. The OA leads to the estimation of disjoint sets of parameters. If we try a few more runs, and perhaps use the tentative rules, we can obtain $\mathbf{L}^*$ and its estimate. However, such a procedure would require more observations and thus would have less revealing power.

The above example illustrates the nature of the approach that has been adopted. These approaches are new, and quite different from some sequences of parallel-planes designs discussed earlier in Addleman (1969) and Daniel (1962).

In the example that we considered, the number of nonnegligible parameters was rather small (although, relative to the total number of parameters, it was not too small). In general, when the number of nonnegligible parameters is not too large, it is best to start with the BA consisting of $\Omega_{mm} + \Omega_{m,m-1}$, as we did in our example. This may be helpful in situations where we expect few main effects, and only a few interactions to be nonnegligible. However, when several main effects, two-factor interactions, and higher interactions are expected to be nonnegligible, it is better to start with the BA's consisting of $\Omega_{mm} + \Omega_{m,m-1}$ or $\Omega_{mm} + \Omega_{m,m-1} + \Omega_{m,m-2}$. Both of these have certain advantages. In Srivastava (1987b), the use of the latter was illustrated. However, the former appears to have more revealing power in certain situations.

ACKNOWLEDGEMENTS

This work has been supported by AFOSR grant 830080. I wish to express my thanks to Pat Key for her excellent and painstaking typing of this manuscript.

REFERENCES

Addleman, S. (1969). Sequences of two-level fractional factorial plans. *Technometrics* 11, 477–509.

Addleman, S. (1964). Techniques for constructing fractional replicate plans. *JASA* 58, 45–71.

Addleman, S. (1962a). Orthogonal main-effect plans for asymmetrical factorial experiments. *Technometrics* 4, 21–46.

Addleman, S. (1962b). Symmetrical and asymmetrical fractional factorial plans. *Technometrics* 4, 47–58.

Addleman, S. (1961). Irregular fractions of the 2^n factorial experiments. *Technometrics* 3, 479–496.

Addleman, S. and Kempthorne, O. (1961). Some main effect plans and orthogonal arrays of strength two. *Ann. Math. Stat.* 32, 1167–1176.

Anderson, D. A. and Thomas, A. M. (1980). Weakly resolvable IV.3 search designs for the p^n factorial experiment. *JSPI* 4, 299–312.

Anderson, D. A. and Mardekian, J. (1979). Parallel flats fractions for the 3^k factorial, abstract. *IMS Bull.* 8(4), 45.

Anderson, D. A. and Thomas, A. M. (1979). Near minimal resolution IV designs for the s^n factorial experiment. *Technometrics* 21(30), 331–336.

Berlekamp, E. R. (1968). *Algebraic Coding Theory.* McGraw–Hill, New York.

Bose, R. C. and Iyer, H. K. (1984) Balanced E-optimal designs of resolution III for the $2^m \times 3^n$ series. *JSPI* 10, 345–364.

Bose, R. C. and Srivastava, J. N. (1964a). Analysis of irregular factorial fractions. *Sankhya Ser. A* 26, 117–144.

Bose, R. C. and Srivastava, J. N. (1964b). Multidimensional partially balanced designs and their analysis with applications to partially balanced factorial fractions. *Sankhya Ser. A* 26, 117–144.

Bose, R. C. (1961). On some connections between the design of experiments and information theory. *Bull. Inst. Intern. Stat.* 38, 4e, 257–271.

Bose, R. C. (1959). Notes on linear models. North Carolina Institute of Statistics.

Bose, R. C. and Bush, K. A. (1952). Orthogonal arrays of strength two and three. *Ann. Math. Stat.* 23, 502–524.

Bose, R. C. (1947). Mathematical theory of the symmetrical factorial design. *Sankhya* 8, 107–166.

Bose, R. C. and Kishen, K. (1940). On the problem of confounding in the general symmetrical factorial design. *Sankhya* 5, 21–36.

Box, G. E. P. and Draper, N. R. (1987). *Empirical Model-Building and Response Surfaces.* Wiley, New York.

Box, G. E. P. (1986). Studies in quality improvement: Signal to noise ratios, performance criteria and statistical analysis: Part I. Report No. 11. Center for Quality and Productivity Improvement, Univ. of Wisconsin, Madison.

Box, G. E. P. and Fung, C. A. (1986). Studies in quality improvement: Minimizing transmitted variation by parameter design. Report No. 8. Center for Quality and Productivity Improvement. Univ. of Wisconsin, Madison.

Box, G. E. P. and Meyer, R. D. (1986). Studies in quality improvement: Dispersion effects from fractional designs. Report No. 1. Center for Quality and Productivity Improvement, Univ. of Wisconsin, Madison.

Box, G. E. P. and Hunter, J. S. (1961). The 2^{k-p} fractional factorial designs, Parts I and II. *Technometrics* 3, 311–351, 449–458.

BuHamra, S. (1987). Theory of parallel flat fractions for s^n factorial experiments. Ph.D. thesis, Department of Statistics, Univ. of Wyoming, Laramie.

Chakravarti, I. M. (1963). Orthogonal and partially balanced arrays and their application in design of experiments. *Metrika* 7, 231–243.

Chakravarti, I. M. (1956). Fractional replication in asymmetrical factorial designs and partially balanced arrays. *Sankhya* 17, 143–164.

Chatterjee, K. and Mukherjee, R. (1986). Some search designs for symmetric and asymmetric factorials. *JSPI* 13, 357–363.

Cheng, C. S. (1985). Run orders of factorial designs. *Proceedings of the Berkeley Conference in Honor of Jerzy Neyman and Jack Kiefer*, Vol. II, (Lucien M. LeCam and R. A. Olshen, eds.), Wadsworth, Belmont, Calif., pp. 619–633.

Cheng, C.-S. (1980). Optimality of some weighing and 2^m fractional-factorial designs. *Ann. Stat.* 8, 436–446.

Chopra, D. V. (1975a). Balanced optimal 2^8 fractional factorial designs of resolution V, $52 \leq N \leq 50$. In: *A Survey of Statistical Designs and Linear Models* (J. N. Srivastava, ed.), North–Holland, Amsterdam, 91–99.

Chopra, D. V. (1975b). Optimal balanced 2^8 fractional factorial designs of resolution V, with 60–65 runs. *Proceedings of the International Statistics Institute* (Warsaw), pp. 164–168.

Chopra, D. V. (1975c). Some investigations on trace optimal balanced designs of resolution V and balanced arrays. Wichita State Univ. Bull. L1–4, Univ. Stud. 105.

Chopra, D. V. and Srivastava, J. N. (1977). Optimal balanced 2^7 fractional factorial designs of resolution V, $43 \leq N \leq 48$. *Sankhya Ser. B* 4, 429–447.

Chopra, D. V. and Srivastava, J. N. (1975). Optimal balanced 2^7 fractional factorial designs of resolution V, $43 \leq N \leq 48$. *Sankhya Ser. B* 37, 429–447.

Chopra, D. V. and Srivastava, J. N. (1974). Optimal balanced 2^8 fractional factorial designs of resolution V, $37 \leq N \leq 51$. *Sankhya Ser. A* 36, 41–52.

Chopra, D. V. and Srivastava, J. N. (1973a). Optimal balanced 2^7 fractional factorial designs of resolution V with $N \leq 42$. *Ann. Inst. Stat. Math.* 25–3, 587–604.

Chopra, D. V. and Srivastava, J. N. (1973b). Optimal balanced 2^7 fractional factorial designs of resolution V, $49 \leq N \leq 55$. *J. Comm. Stat.* 2–1, 59–84.

Connor, W. S. (1960). Construction of the fractional factorial designs of the mixed $2^m \times 3^n$ series. In: *Contributions to Probability and Statistics—Essays in Honor of Harold Hotelling.* Stanford Univ. Press, Stanford, Calif., pp. 168–181.

Connor, W. S. and Young, S. (1959). Fractional factorial designs for experiments with factors at two and three levels. Applied Mathematical Series, National Bureau of Standards, U. S. Government Printing Office, Washington, D.C.

Daniel, C. (1962). Sequential fractional replicates in the 2^{p-q} series. *JASA* 57, 403–429.

Ehlich, H. (1964). Determinantenabschätzungen für binäre Matrizen mit $n = 3 \bmod 4$. *Math. Z.* 84, 438–447.

Federov, V. V. (1972). *Theory of Optimal Experiments.* Academic Press, New York.

Finney, D. J. (1945). The fractional replication of factorial experiments. *Ann. Eugenics* 12, 291–310.

Fisher, R. A. (1960). *The Design of Experiments.* Hafner, New York.

Fisher, R. A. (1945). A system of confounding for factors with more than two alternatives giving completely orthogonal cubes and higher powers. *Annals of Eugenics* 12, 283–290.

Fisher, R. A. (1943). The theory of confounding in factorial experiments in relation to the theory of groups. *Ann. Eugenics* 12, 291–312.

Fisher, R. A. (1942). The theory of confounding in factorial experiments in relation to the theory of groups. *Ann. Eugenics* 11, 341–353.

Galil, Z. and Kiefer, J. (1980). D-optimum weighing designs. *Ann. Stat.* 8, 1293–1306.

Ghosh, S. and Zhang, X. D. (1987). Two new series of search designs for 3^m factorial experiments. *Utilitas Math.*, 32, 245–254.

Ghosh, S. (1987). Non-orthogonal designs for measuring dispersion effects in sequential factor screening experiments using search linear model. *Comm. Stat. Theory Meth.* 16(10), 2839–2850.

Ghosh, S. and Avila, D. (1985). Some new factor screening designs using the search linear model. *JSPI* 11, 259–266.

Ghosh, S. (1981). On some new search designs for 2^m factorial experiments. *JSPI* 5, 381–389.

Ghosh, S. (1980). On main effects plus one plan for 2^m factorials. *Ann. Stat.* 8, 922–930.

Graybill, F. A. (1976). *Theory and Application of the Linear Model.* Duxbury Press, North Scituate, Mass.

Gupta, B. C. (1988). A bound connected with factorial search designs of resolution III · 1. *Comm. Stat.* (to appear).

Gupta, B. C. and Carvajal, S. S. R. (1986). A lower bound for number of treatments in a main effect plus one plan for 2^5 factorial. *Utilitas Math.* (to appear).

Gupta, B. C. and Carvajal, S. S. R. (1984a). A necessary condition for the existence of main effect plus one plan for 2^m factorials. *Comm. Stat. Theor. Meth.* 13(5), 567–580.

Gupta, B. C. and Carvajal, S. S. R. (1984b). A lower bound for number of treatments in a main effect plan for 2^4 factorials. *Utilitas Math.*, 259–67.

Gupta, B. C. and Sheult, A. (1975). Some results in search designs with special reference to 2^m factorial experiments. *Metron* 33, 379–388.

Gupta, S. and Mukherjee, R. (1989). A calculus for factorial arrangements. (unpublished).

Hamming, R. W. (1950). Error detecting and error correcting codes. *Bell System Tech. J.* 29, 147–160.

Hedayat, A. and Wallis, W. D. (1978). Hadamard matrices and their applications. *Ann. Stat.* 6, 1184–1238.

Hoke, A. T. (1975). The characteristic polynomial of the information matrix for second-order models. *Ann. Stat.* 3, 780–786.

Hoke, A. T. (1974). Economical second-order designs based on irregular fractions of the 3^n factorial. *Technometrics* 16, 375–384.

Hussein, A. H. (1986). Sequential designs for the 3^m factorial. Ph.D. dissertation, Univ. of Wyoming, Laramie.

Hveberg, R. (1988). Sequential factorial probing designs for the 2^3 case, when the three factor interaction may not be negligible. *Proc. Intern. Statist. Inst.*, Paris, 1989.

Jain, N. C. (1980). Investigations on 2-covering of a finite Euclidean space based on GF(2). Ph.D. dissertation, Colorado State Univ., Fort Collins.

John, P. W. M. (1962). Three quarter replicates of 2^n designs. *Biometrics* 18, 172–184.

Kageyama, S. (1975). Note on the construction of partially balanced arrays. *Ann. Ins. Stat. Math.* 27, 177–180.

Kakkar, R. N. (1985). Offline quality control, parameter design and the Taguchi method (and discussion). *J. Quality Tech.* 17, 176–209.

Kakkar, R. N. and Shoemaker, A. C. (1986). Robust design: A cost effective method for improving manufacturing process. *AT&T Tech. J.* 65, 39–50.

Katona, G. and Srivastava, J. N. (1983). Minimal 2-coverings of a finite affine space based on GF(2). *JSPI* 8, 375–88.

Kempthorne, O. (1973). *The Design and Analysis of Experiments.* Robert E. Krieger Publ., New York.

Kempthorne, P. (1947). A simple approach to confounding the fractional replications in factorial experiments. *Biometrika* 34, 255–272.

Khuri, A. and Cornell, J. A. (1987). *Response Surfaces.* Marcel Dekker, New York.

Kiefer, J. (1961). Optimum design in regression problems. *Ann. Math. Stat.* 32, 298–325.

Kiefer, J. C. (1959). Optimum experimental designs. *J. Roy. Stat. Soc.* B 21, 273–319.

Kiefer, J. and Wolfowitz, J. (1959). Optimum design in regression problems. *Ann. Math. Stat.* 30, 270–294.

Kishen, K. and Srivastava, J. N. (1960). Mathematical theory of confounding in symmetrical and a symmetrical factorial designs. *J. Ind. Soc. Agric. Stat.* 11, 73–110.

Kishen, K. (1948). On fractional replication of the general symmetrical factorial design. *J. Ind. Soc. Agric. Stat.* 1, 91–106.

Kuriki, S. (1988). Existence of 2-symbol arrays of strength t and $(t + 2)$ constraints. *JSPI* 20, 225–228.

Kuriki, S. and Yamamoto, S. (1984). Nonsimple 2-symbol balanced arrays of strength t and $t + 2$ constraints. *TRU Math.* 20(2), 249–263.

Kuwada, M. (1988a). A-optimal partially balanced $2^{m_1+m_2}$ factorial designs of resolution V, with $4 \le m_1 + m_2 \le 6$. *JSPI* 18, 177–193.

Kuwada, M. (1988b). Analysis of variance and hypotheses testing of balanced fractional 3^n factorial designs of resolution V. Technical Report 225, Hiroshima Univ., Hiroshima, Japan.

Kuwada, M. (1981). Characteristic polynomial of the information matrix of balanced fractional 3^n factorial designs of resolution V. *JSPI* 5, 189–209.

Kuwada, M. (1980). Balanced arrays of strength 4 and balanced fractional 3^m factorial designs. *JSPI* 4(1).

Kuwada, M. (1979). Optimal balanced fractional 3^m factorial designs of resolution V and balanced third-order designs. *Hiroshima Math. J.* 347–450.

Mallenby, D. W. (1977). Inference for the search linear method. Ph.D. dissertation. Colorado State Univ., Fort Collins.

Mardekian, J. (1979). Parallel flats fractions for the 3^k factorial. Unpublished Ph.D. Thesis, Department of Statistics, Univ. of Wyoming, Laramie.

Margolin, B. H. (1969). Results on factorial designs of resolution IV for the 2^n and $2^n 3^n$ series. *Technometrics* 11, 421–444.

Meyer, R. D. (1986). Studies in quality improvement: Analysis of fractional factorials. Report No. 16. Center for Quality and Productivity Improvement. Univ. of Wisconsin, Madison.

Myers, R. H. (1976). *Response Surface Methodology.* Edward Brothers, Ann Arbor, Mich.

Nair, K. R. and Rao, C. R. (1948). Confounding in asymmetrical factorial experiments. *J. Roy. Stat. Soc. Series B* 10, 109–131.

Nishii, R. (1981). Balanced fraction $r^m \times s^n$ factorial designs and their analysis. *Hiroshima Math. J.* 11, 379–413.

Ohnishi, T. and Shirakura, T. (1985). Search designs for 2^m factorial experiments. *JSPI* 11, 241–245.

Paley, R. E. A. C. (1933). On orthogonal matrices. *J. Math. Phys.* 122, 311–320.

Patel, M. S. (ed.) (1988). Special Issue on Factor Screening. *Comm. Stat.* 16.

Patel, M. S. (1961). Investigations in factorial designs. Ph.D. Thesis, Department of Statistics, Univ. of North Carolina, Chapel Hill.

Phadke, M. S. (1986). Design optimization case studies. *AT&T Tech. J.* 65, 51–68.

Phadke, M. S., Kakkar, R. N., Speeney, D. V., and Grieco, M. J. (1983). Off-line quality control for integrated circuit fabrication using experimental design. *Bell System Techn. J.* 62, 1273–1309.

Plackett, R. L. and Burman, J. P. (1946). The design of optimum multifactorial experiments. *Biometrika* 33, 305–325.

Rafter, J. A. and Seiden, E. (1974). Contributions to the theory and construction of balanced arrays. *Ann. Stat.* 2, 1256–1273.

Raghavarao, D. (1971). *Construction and Combinatorial Problems in Design of Experiments.* Wiley, New York.

Raktoe, R. L., Hedayat, A., and Federer, W. T. (1981). *Factorial Designs.* Wiley, New York.

Rao, C. R. (1973a). Some combinatorial problems of arrays and applications to design of experiments. In: *A Survey of Combinatorial Theory,* (J. N. Srivastava, ed.), North–Holland, Amsterdam, pp. 349–359.

Rao, C. R. (1973b). *Linear Statistical Inference.* Wiley, New York.

Rao, C. R. (1950). The theory of fractional replications in factorial experiments. *Sankhya* 10, 81–85.

Rao, C. R. (1947). Factorial arrangements derivable from combinatorial arrangements of arrays. *Suppl. J. Roy. Stat. Soc.* 9, 128–139.

Rao, C. R. (1946). Hypercubes of strength d leading to confounded designs in factorial experiments. *Bull. Calcutta Math. Soc.* 38, 67–73.

Rechtschaffner, R. L. (1967). Saturated fractions of 2^n and 3^n factorial design. *Technometrics* 9, 568–575.

Roy, S. N., Gnanadesikan, R., and Srivastava, J. N. (1970). *Analysis and Design of Multiresponse Experiments.* Pergamon Press, New York.

Seiden, E. and Zemach, R. (1966). On orthogonal arrays. *Ann. Math. Stat.* 37, 1355–1370.

Shirakura, T. (1987). Main effects plus k plans for 2^m factorials. Technical Report #305, Hiroshima Univ., Hiroshima, Japan.

Shirakura, T. and Ohnishi, T. (1985). Search designs for 2^m factorials derived from balanced arrays of strength $2(l+1)$ and AD optimal search designs. *JSPI* 11, 247–258.

Shirakura, T. (1980). Necessary and sufficient condition for a balanced array of strength $2l$. *Austra. J. Stat.* 22, 69–74.

Shirakura, T. (1979). Optimal balanced fractional 2^m factorial designs of resolution IV derived from balanced arrays of strengths four. *J. Japan Stat. Soc.* 9, 19–27.

Shirakura, T. (1977). Contributions to balanced fractional 2^m factorial designs derived from balanced arrays of strength $2l$. *Hiroshima Math. J.* 7, 217–285.

Shirakura, T. (1976a). Balanced fractional 2^m factorial designs of even resolution obtained from balanced arrays of strength $2l$ with $u_l = 0$. *Ann. Stat.* 4, 723–735.

Shirakura, T. (1976b). Optimal balanced fractional 2^m factorial designs of resolution VII, $6 \leq m \leq 8$. *Ann. Stat.* 4, 515–531.

Shirakura, T. and Kuwada, M. (1976). Covariance matrices of the estimates for balanced fractional 2^m factorial designs of $2l+1$. *J. Japan Stat. Soc.* 6, 27–31.

Shirakura, T. (1975). Optimum balanced fractional 2^m factorial designs of resolution VII, $6 \leq m \leq 8$, *Ann. Stat.* 4, 515–531.

Shirakura, T. and Kuwada, M. (1975). Note on balanced fractional 2^m factorial designs of resolution $2l + 1$. *Ann. Inst. Stat. Math.* 27, 377–386.

Srivastava, J. N. and Arora, S. (1988a). The minimal resolution $3 \cdot k$ $(k = 1, 2)$ plans for the 2^4 factorial experiment. In: *Coding Theory and Design Theory* (Ed: D. K. Raychandhari) Springer-Verlag. (To be published).

Srivastava, J. N. and Arora, S. (1988b). An infinite series of resolution $3 \cdot 2$ designs for the 2^m factorial experiment. (Submitted).

Srivastava, J. N. and Arora, S. (1987). A minimal search design of resolution $3 \cdot 2$ for the 2^4 factorial experiment. *Ind. Jour. Math.* 29, 309–320.

Srivastava, J. N. and Hveberg, R. (1988). Sequential factorial probing designs for identifying and estimating nonnegligible factorial effects for the 2^m experiment under the tree structure. *Jour. Stat. Plan. Inf.* (To appear).

Srivastava, J. N. (1987a). Advances in the general theory of factorial designs based on partial pencils in Euclidean n-space. *Utilitas Math.* 32, 75–94.

Srivastava, J. N. (1987b). On the inadequacy of the customary orthogonal arrays in quality control and scientific experimentation, and the need of probing designs. *Comm. Stat.* 16, 2901–2941.

Srivastava, J. N. and Mallenby, D. M. (1985). On a decision rule using dichotomies for identifying nonnegligible parameters in certain linear models. *J. of Multivariate Anal.* 16, 318–334.

Srivastava, J. N. (1984). Sensitivity and revealing power: Two fundamental statistical criteria other than optimality arising in discrete experimentation. In: *Experimental Designs, Statistical Models, and Genetic Statistics Models, and Genetic Statistics* (K. Hinkelmann, ed.), Marcel Dekker, New York, pp. 95–117.

Srivastava, J. N., Anderson, D. A., and Mardekian, J. (1984). Theory of factorial designs of the parallel flats type 1: The coefficient matrix. *JSPI* 9, 229–252.

Srivastava, J. N. (1983). On certain structures arising in the set of nonnegligible effects in factorial experiments in the field of psychology. (Unpublished).

Srivastava, J. N. and Ariyaratna, W. M. (1982). Inversion of information matrices of balanced 3^m factorial designs of resolution V, and optimal designs. In: *Statistics and Probability* (G. Kallianpur and P. Krishnaiah, eds.). North–Holland, Amsterdam.

Srivastava, J. N. and Ariyaratna, W. M. (1981). Balanced arrays of strength t with three symbols and $(t+1)$ rows. *J. Comb. Inform. Sys. Sci.* 6, 335–355.

Srivastava, J. N. and Ghosh, S. (1980). Enumeration and representation of nonisomorphic bipartite graphs. *Ann. Disc. Math.* 6, 315–332.

Srivastava, J. N. and Gupta, B. C. (1979). Main effect plan for 2^m factorials which allow search and estimation of one nonnegligible effect. *JSPI* 3, 259–265.

Srivastava, J. N. (1978a). A review of some recent work on discrete optimal factorial designs for statisticians and experiments. *Devel. Stat.* 1, 267–329.

Srivastava, J. N. (1978b). On the linear independence of sets of 2^q columns of certain $(1,-1)$ matrices with a group structure, and its connection with finite geometries. In: *Combinatorial Mathematics* (D. A. Holton and J. Seberry, eds.). Springer–Verlag, Berlin.

Srivastava, J. N. (1977). Optimal search designs or designs optimal under bias-free optimality criteria. In: *Statistical Decision Theory and Related Topics* (S. S. Gupta, and D. S. More, eds.), Vol. II, 375–409.

Srivastava, J. N. and Ghosh, S. (1977). Balanced 2^m factorial designs of resolution V which allow search and estimation of one extra unknown effect, $4 \le m \le 8$. *Comm. Stat. Theor. Meth.* A6(2), 141–166.

Srivastava, J. N. and Srivastava, P. (1977). Examination of results of experiments in social sciences using factorial designs. (Unpublished report).

Srivastava, J. N. (1976). Some further theory of search linear models In: *Contributions to Applied Statistics*, Swiss–Australian Region of Biometry Society, pp. 249–256.

Srivastava, J. N. and Ghosh, S. (1976). A series of 2^m factorial designs of resolution V which allow search and estimation of one extra unknown effect. *Sankhya Ser. B* 38, 280–289.

Srivastava, J. N. (1975). Designs for searching nonnegligible effects. In: *A Survey of Statistical Designs and Linear Models* (J. N. Srivastava, ed.), pp. 507–519. North Holland, Amsterdam.

Srivastava, J. N. and Anderson, D. A. (1974). A comparison of the determinant, trace, and largest root optimality criteria. *Comm. Stat.* 3, 933–940.

Srivastava, J. N. and Chopra, D. V. (1974). Balanced trace-optimal 2^7 fractional factorial designs of resolution V with 56 to 68 runs. *Utilitas Math.* 5, 263–279.

Srivastava, J. N. and Chopra, D. V. (1973). Balanced arrays and orthogonal arrays. In: *A Survey of Combinatorial Theory* (J. N. Srivastava et al., eds.), North–Holland Amsterdam, pp. 411–428.

Srivastava, J. N. (1972). Some general existence conditions for balanced arrays of strength t and 2 symbols. *J. Comb. Theory* 13, 198–206.

Srivastava, J. N. and Anderson, D. A. (1971). Factorial subassembly association scheme and multidimensional partially balanced designs. *Ann. Math. Stat.* 42, 1167–1181.

Srivastava, J. N. and Chopra, D. V. (1971a). Balanced optimal 2^m fractional factorial designs of resolution V, $m \leq 6$. *Technometrics* 13, 257–269.

Srivastava, J. N. and Chopra, D. V. (1971b). On the characteristic roots of the information matrix of balanced fractional 2^m factorial designs of resolution V, with applications. *Ann. Math. Stat.* 42, 722–734.

Srivastava, J. N. and Chopra, D. V. (1971c). On the comparison of certain classes of balanced 2^8 fractional factorial designs of resolution V, with respect to the trace criterion. *J. Ind. Soc. Agric. Stat.* 23, 124–131.

Srivastava, J. N. (1970). Optimal balanced 2^m fractional factorial designs. In: *Essays in Probability and Statistics* (R. C. Bose et al., Eds.), Univ. of North Carolina Press, Chapel Hill, pp. 689–706.

Srivastava, J. N. (1970). Optimal balanced 2^m fractional factorial designs. In: *S. N. Roy Memorial Volume*, Univ. of North Carolina and Indian Stat. Inst., pp. 689–706.

Srivastava, J. N. and Anderson, D. A. (1970). Optimal fractional factorial plans for main effects orthogonal to two-factor interactions: 2^m series. *JASA* 65, 828–843.

Srivastava, J. N. (1967). Investigations on the basic theory of $2^m 3^n$ fractional factorial designs of resolution V and related orthogonal arrays (Abstract). *Ann. Math. Statist.* 38, 637.

Srivastava, J. N. (1961). Contributions to the construction and analysis of designs. Institute of Statistics, Univ. of North Carolina, Chapel Hill, Mimeo Ser. No. 301.

Taguchi, G. and Wu, Y. (1979). Introduction to off-line quality control. Central Japan Quality Control Association. Available in U.S.A. from American Supplier Institute Center for Taguchi Methods, Dearborn, Mich.

Vijayan, K. (1976). Hadamard matrices and submatrices. *J. Austral. Math. Soc.* A 22 469–475.

Webb, S. R. (1965). Expansible and contractible factorial designs and the application of linear programming to combinatorial problems. Aerospace Research Laboratories Technical Report, 65–116, Part I.

Whitwell, J. D. and Morbey, G. K. (1961). Reduced designs of resolution V. *Technometrics* 3, 459–477.

Williamson, J. (1946). Determinants whose elements are 0 and 1. *Amer. Math. Monthly* 53, 427–434.

Yamamoto, S., Kuriki, S., and Natori, S. (1984). Some nonsimple 2-symbol balanced arrays of strength t and $t + 2$ constraints. *TRU Math.* 20(2), 225–228.

Yamamoto, S., Shirakura, T., and Kuwada, M. (1976a). Balanced arrays of strength $2l$ and balanced fractional 2^m factorial designs. *Ann. Inst. Stat. Math.* 27, 143–147.

Yamamoto, S., Shirakura, T., and Kuwada, M. (1976b). Characteristic polynomials of the information matrices of balanced fractional 2^m factorial designs of higher $(2l+1)$ resolution. In: *Essays in Probability and Statistics* (S. Ikeda et al., eds.), 73–94. Shinko Tsusho, Ltd., Tokyo.

Yamamoto, S., Shirakura, T., and Kuwada, M. (1975). Balanced arrays of strength $2l$ and balanced fractional 2^m factorial designs. *Ann. Inst. Stat. Math.* 27, 143–157.

Yamamoto, S., Fujii, Y., and Hamada, N. (1965). Composition of some series of association algebras. *J. Sci. Hiroshima Univ. Ser. A-1,* 29, 181–215.

Yang, C. H. (1966). Some designs for maximal $(+1,-1)$-determinant of order $n \equiv 2(\mathrm{mod}\,4)$. *Math. of Comp.* 20, 147.

Yang, C. H. (1968). On designs of maximal $(+1,-1)$-matrices of order $n \equiv 2(\mathrm{mod}\,4)$. *Math. of Comp.* 22, 174.

Yates, F. (1937). The design and analysis of factorial experiments. *Imp. Bur. Soil Sci. Tech. Comm.* 35.

13

New Properties of Orthogonal Arrays and Their Statistical Applications

A. S. Hedayat   University of Illinois at Chicago, Chicago, Illinois

1   INTRODUCTION

Factorial designs are among the most utilized statistical designs in industrial laboratories. Indeed, such designs are considered to be the workhorses of modern investigations and quality control studies. To minimize cost and to speed up the course of investigations, many researchers rely on factorial experiments in which certain level combinations are excluded from the study. Clearly, such deletions should not be done on an ad hoc basis. Otherwise, the price we may have to pay later may be more than what we can afford.

It is customary to refer to any factorial design in which one or more level combinations are excluded from the study as a fractional factorial design. To recommend a fraction for a particular problem is usually hard, and it is often impossible to identify the "best fraction." In a series of three papers, Rao (1946, 1947a, 1947b) identified special types of fractions with a great deal of symmetry and many desirable statistical properties. These special fractions are now known as orthogonal arrays. Numerous beautiful results have been obtained by both statisticians and mathematicians on this topic. Earlier contributions include Bush (1952), Bose and Bush (1952), Seiden

(1954), and Seiden and Zemach (1966). These results are published in a wide variety of journals and presented in many different styles. In their forthcoming book, Hedayat and Stufken (1990) have presented the various results on this fascinating subject in a unified and comprehensive way.

In the remainder of this chapter, we will have to assume that the reader has some familiarity with the standard concepts in factorial experiments and the theory of ANOVA. For a quick review of these concepts, useful references are Raktoe, Hedayat, and Federer (1981) and McLean and Anderson (1984).

Throughout, we shall concentrate on quantitative factors, that is, factors whose values can be measured on a numerical scale such as temperature, pressure, time, and so forth.

The chapter is divided into six sections. In Section 2, we will review the basic definitions and terminology of the subject. In Section 3, we will present an efficient way of obtaining design and information matrices associated with factorial designs in general and orthogonal arrays in particular. In Section 4, we establish how orthogonal arrays of strength t can be used to estimate model parameters orthogonally, including other parameters than those usually mentioned in the literature. The concept of orthogonal arrays of strength $t+$ is introduced in Section 5. These are arrays that are somewhere between orthogonal arrays of strength t and $t + 1$. The usefulness of these latter arrays is indicated. Throughout, we use examples to demonstrate ideas and concepts. The results in this chapter are summarized and highlighted via some concluding remarks in Section 6.

2 PRELIMINARIES

We begin by giving some of the basic definitions and terminology. Let S be a set of s symbols, coded by 0, 1, ..., $s - 1$. A $k \times N$ array A with entries from S is called an orthogonal array, denoted by OA(N, k, s, t), if each $t \times N$ subarray of A has the property that every possible $t \times 1$ vector with entries from S appears equally often (say λ times) in the columns of the chosen subarray. The integer λ is called the index of the array, while N, k, s, and t are said to be the parameters of the array. If $\lambda = 1$, the array is said to be of index unity. The relation $N = \lambda s^t$ is an immediate consequence of the definition.

The reader may verify that the following array is an OA$(27, 4, 3, 3)$, an orthogonal array of index unity.

$$0\ 1\ 2\ 0\ 1\ 2\ 0\ 1\ 2\ 0\ 1\ 2\ 0\ 1\ 2\ 0\ 1\ 2\ 0\ 1\ 2\ 0\ 1\ 2\ 0\ 1\ 2$$
$$0\ 1\ 2\ 1\ 2\ 0\ 2\ 0\ 1\ 1\ 2\ 0\ 2\ 0\ 1\ 0\ 1\ 2\ 2\ 0\ 1\ 0\ 1\ 2\ 1\ 2\ 0$$
$$0\ 1\ 2\ 2\ 0\ 1\ 1\ 2\ 0\ 1\ 2\ 0\ 0\ 1\ 2\ 2\ 0\ 1\ 2\ 0\ 1\ 1\ 2\ 0\ 0\ 1\ 2$$
$$0\ 0\ 0\ 0\ 0\ 0\ 0\ 0\ 0\ 1\ 1\ 1\ 1\ 1\ 1\ 1\ 1\ 1\ 2\ 2\ 2\ 2\ 2\ 2\ 2\ 2\ 2$$

The terminology used in the literature for the parameters of an orthogonal array is as follows: N refers to the size of the array or the number of runs, assemblies, level combinations, or treatments; k to the number of constraints or factors in the array; s to the number of levels or symbols; and t to the strength of the array.

For fixed values of the strength t and the index λ, the number of factors k that can be accommodated in an OA$(\lambda s^t, k, s, t)$ is bounded from above. Although for any t and λ, an upper bound for k is available, the maximum value of k for which an OA$(\lambda s^t, k, s, t)$ exists is generally unknown. Determining this value and providing constructions of orthogonal arrays with this maximum number of factors, or at least a number that comes close to it, form the most challenging problems. It easily follows that:

1. Any array obtained from an orthogonal array by permuting columns, rows, or symbols in one or more rows will again be an orthogonal array with the same parameters.
2. Any $k' \times \lambda s^t$ subarray of an OA$(\lambda s^t, k, s, t)$ is an OA$(\lambda s^t, k', s, t')$, with $t' = \min\{k', t\}$.
3. Any orthogonal array of strength t is an orthogonal array of strength t', with $t' \leq t$.
4. Combining columns of OA$(\lambda_i s^t, k, s, t)$, $i = 1, 2$, leads to an OA$(\lambda s^t, k, s, t)$, where $\lambda = \lambda_1 + \lambda_2$.

3 FRACTIONAL FACTORIAL DESIGNS UNDER ORTHOGONAL POLYNOMIAL MODELS

As fractional factorial designs orthogonal arrays are highly efficient if orthogonal polynomial models are postulated for the response under study. It was indeed in this context that Rao introduced the concept of orthogonal arrays in the literature. The purpose of this section is twofold, one of which is to present the orthogonal polynomial models associated with orthogonal arrays so that our results in Sections 4 and 5 will be easily understood.

Second, we present an efficient way, borrowed from Hedayat and Stufken (1990), for generating the design and hence the information matrices associated with any arbitrary fractional factorial design. This method is very handy for analyzing data obtained under fractional factorial designs and for studying their efficiencies.

Assume that our interest is in an experiment based on k controllable factors, each of which can be set at s different levels. The levels are coded by 0, 1, ..., $s - 1$. A level combination is an experimental condition to which each factor contributes a level. For example, $(i_1, \ldots, i_j, \ldots, i_k)$ is a level combination with the jth factor at level i_j. Thus, there are s^k distinct level combinations. These can be exhibited columnwise in a $k \times s^k$-array, which we will denote by ρ and refer to as a minimal complete factorial based on k factors as s levels each.

Let g be a level combination and y_g the observed response under the experimental condition g. We will assume that y_g is a random variable,

$$y_g = f'(g)\theta + \epsilon_g$$

where $f'(g)$ is a row vector of real known functions of g, and θ is a column vector of unknown parameters. The random error component ϵ_g will be assumed to have mean zero and unknown variance σ^2.

The structure of $f'(g)$ is an important consideration in practice. If all factors are quantitative factors, a popular way of modeling y_g is to structure $f'(g)$ via the orthogonal polynomial model (see, for example, Chapter 4 of Raktoe, Hedayat, and Federer, 1981). We will use this model from here on. If we perform the experiment using all s^k treatment combinations, the resulting vector of observations, Y_ρ, can be expressed as

$$Y_\rho = X_\rho \beta_\rho + \epsilon_\rho \tag{1}$$

where X_ρ is the design matrix and β_ρ is the vector of general mean, main effects, and interactions. The entries in β_ρ, following the common notations, are

$$\phi_1^{j_1} \phi_2^{j_2} \ldots \phi_k^{j_k}, \qquad j_i \in \{0, 1, \ldots, s - 1\}$$

We will now present an easy way to find the entries of X_ρ under a full-order orthogonal polynomial model. Let $x_0, x_1, \ldots, x_{s-1}$ be the actual, uncoded levels of the first factor. Form the following matrix:

$$
X_1 = \begin{bmatrix}
1 & x_0 & x_0^2 & \cdots & x_0^{s-1} \\
1 & x_1 & x_1^2 & \cdots & x_1^{s-1} \\
\vdots & \vdots & \vdots & & \vdots \\
1 & x_{s-1} & x_{s-1}^2 & \cdots & x_{s-1}^{s-1}
\end{bmatrix}
$$

Orthogonalize the columns of X_1 from left to right, and call the resulting matrix M_1. Similarly, obtain $M_2, \ldots, M_k$ corresponding to the other $k-1$ factors. Denote the entry in the ith row and jth column of M_l by $n_{ij}^{(l)}$, $0 \le i$, $j \le s-1$, $1 \le l \le k$. Then, the entry in X_ρ corresponding to the treatment combination $(i_1, i_2, \ldots, i_k)$ and parameter $\phi_1^{j_1} \phi_2^{j_2} \ldots \phi_k^{j_k}$ is given by

$$
n_{i_1 j_1}^{(1)} n_{i_2 j_2}^{(2)} \ldots n_{i_k j_k}^{(k)} \tag{2}
$$

At times, a full-order orthogonal polynomial model may not be needed for an adequate explanation of the variation in the response. Use of a second- or third-order model may, for example, be sufficient. One way to obtain the design matrix of such a model is by finding that of the full-order orthogonal polynomial model and deleting the columns corresponding to the undesired terms. An advantage of this is that any model can be obtained in this way, but a disadvantage is that we will have to go through the same computations as for the full-order model. A computationally simpler method is to delete in the computation of each M_i one or more columns of the corresponding matrix X_i, whatever is appropriate. Thus, for example, for a second-order orthogonal polynomial model, we would only have to construct each M_i and then obtain the design matrix as described in the preceding.

If the s levels of the factors are equally spaced, the orthogonalized matrices M can be obtained from the available literature. Table 4.1 in Raktoe, Hedayat, and Federer (1981) lists such matrices for $2 \le s \le 7$, which is sufficient for most practical purposes. The same table can be used for $s = 8$ or 9 if a polynomial model of degree ≤ 5 is fitted for each factor. We demonstrate some of the above ideas in the following example.

Example 3.1

Let $s = 3$, $k = 2$. Under equally spaced levels, coded by 0, 1, 2, we have

$$\rho = \begin{array}{cccccccccc} 0 & 0 & 0 & 1 & 1 & 1 & 2 & 2 & 2 \\ 0 & 1 & 2 & 0 & 1 & 2 & 0 & 1 & 2 \end{array}$$

$$M_1 = M_2 = \begin{array}{c} \\ 0 \\ 1 \\ 2 \end{array} \begin{array}{ccc} 0 & 1 & 2 \\ \left[\begin{array}{ccc} 1 & -1 & 1 \\ 1 & 0 & -2 \\ 1 & 1 & 1 \end{array} \right] \end{array}$$

$$\beta'_\rho = (\phi_1^0 \phi_2^0, \phi_1^0 \phi_2^1, \phi_1^0 \phi_2^2, \phi_1^1 \phi_2^0, \phi_1^1 \phi_2^1, \phi_1^1 \phi_2^2, \phi_1^2 \phi_2^0, \phi_1^2 \phi_2^1, \phi_1^2 \phi_2^2)$$

Thus for example, the entry in X_ρ corresponding to treatment combination $(1,2)$ and parameter $\phi_1^2 \phi_2^0$ is, from (2), equal to

$$n_{12}^{(1)} n_{20}^{(2)} = (-2) \cdot (1) = -2$$

Completing the example, we obtain

$$\begin{bmatrix} Y_{00} \\ Y_{01} \\ Y_{02} \\ Y_{10} \\ Y_{11} \\ Y_{12} \\ Y_{20} \\ Y_{21} \\ Y_{22} \end{bmatrix} = \begin{bmatrix} 1 & -1 & 1 & -1 & 1 & -1 & 1 & -1 & 1 \\ 1 & 0 & -2 & -1 & 0 & 2 & 1 & 0 & -2 \\ 1 & 1 & 1 & -1 & -1 & -1 & 1 & 1 & 1 \\ 1 & -1 & 1 & 0 & 0 & 0 & -2 & 2 & -2 \\ 1 & 0 & -2 & 0 & 0 & 0 & -2 & 0 & 4 \\ 1 & 1 & 1 & 0 & 0 & 0 & -2 & -2 & -2 \\ 1 & -1 & 1 & 1 & -1 & 1 & 1 & -1 & 1 \\ 1 & 0 & -2 & 1 & 0 & -2 & 1 & 0 & -2 \\ 1 & 1 & 1 & 1 & 1 & 1 & 1 & 1 & 1 \end{bmatrix} \beta_\rho + \epsilon_\rho$$

4   NEW PROPERTIES OF ORTHOGONAL ARRAYS

It is often impractical or impossible because of other restrictions, such as money and time, to use all possible level combinations in an experiment. Clearly, we would like to select a fraction of the s^k treatment combinations, where we allow that some of the selected combinations are used more than once in the experiment. If A is an $OA(N, k, s, t)$ based on $S = \{0, 1, \ldots, s - 1\}$, then the columns of A form such a fraction of the minimal complete factorial.

Under the orthogonal polynomial model (1), possibly neglecting some of the parameters in β_ρ, we can write

$$Y_A = X_A \beta_A + \epsilon_A \tag{3}$$

Here β_A is the vector of parameters obtained from β_ρ be deleting those parameters that can be neglected, while X_A is the design matrix obtained as in (2), but now using only the treatment combinations in A and the vector of parameters β_A.

Example 4.1

Let A be the following OA(8, 4, 2, 3):

$$
\begin{array}{cccccccc}
0 & 1 & 1 & 1 & 1 & 0 & 0 & 0 \\
1 & 0 & 1 & 1 & 0 & 1 & 0 & 0 \\
1 & 1 & 0 & 1 & 0 & 0 & 1 & 0 \\
1 & 1 & 1 & 0 & 0 & 0 & 0 & 1 \\
\end{array}
$$

Let β_A consist of the general mean, the main effects, and the two-factor interactions between factors 1 and 2, 1 and 3, and 1 and 4. In our notation, this means

$$
\beta_A = (\phi_1^0\phi_2^0\phi_3^0\phi_4^0, \phi_1^1\phi_2^0\phi_3^0\phi_4^0, \phi_1^0\phi_2^1\phi_3^0\phi_4^0, \phi_1^0\phi_2^0\phi_3^1\phi_4^0,
$$
$$
\phi_1^0\phi_2^0\phi_3^0\phi_4^1, \phi_1^1\phi_2^1\phi_3^0\phi_4^0, \phi_1^1\phi_2^0\phi_3^1\phi_4^0, \phi_1^1\phi_2^0\phi_3^0\phi_4^1)
$$

Since

$$
M_1 = M_2 = M_3 = M_4 = \begin{bmatrix} 1 & -1 \\ 1 & 1 \end{bmatrix}
$$

we obtain, by (2),

$$
\begin{bmatrix}
Y_{0111} \\
Y_{1011} \\
Y_{1101} \\
Y_{1110} \\
Y_{1000} \\
Y_{0100} \\
Y_{0010} \\
Y_{0001}
\end{bmatrix}
=
\begin{bmatrix}
1 & -1 & 1 & 1 & 1 & -1 & -1 & -1 \\
1 & 1 & -1 & 1 & 1 & -1 & 1 & 1 \\
1 & 1 & 1 & -1 & 1 & 1 & -1 & 1 \\
1 & 1 & 1 & 1 & -1 & 1 & 1 & -1 \\
1 & 1 & -1 & -1 & -1 & -1 & -1 & -1 \\
1 & -1 & 1 & -1 & -1 & -1 & 1 & 1 \\
1 & -1 & -1 & 1 & -1 & 1 & -1 & 1 \\
1 & -1 & -1 & -1 & 1 & 1 & 1 & -1
\end{bmatrix}
\beta_A + \epsilon_A
$$

It is interesting to observe that, in the above example, $X'_A X_A = 8I$, implying that the general mean, the main effects, and the 3 two-factor interactions can be estimated orthogonally, assuming that other parameters can be neglected. This property is not based on coincidence. We will soon see a result that explains this property. In the meantime, the reader is invited to verify that the same property holds if we replace the interaction between factors 1 and 4 in β_A by the interaction between factors 2 and 3.

We will proceed by giving some of the statistical properties of orthogonal arrays as fractional factorial designs under the orthogonal polynomial model but, for a concise formulation, we need one more concept. If r is even, then a fractional factorial design is said to be of resolution r if all factorial effects involving $(r-2)/2$ or fewer factors are estimable, assuming that all factorial effects involving $(r+2)/2$ or more factors can be neglected. If r is odd, the design is said to be of resolution r if all effects involving $(r-1)/2$ or fewer factors are estimable, assuming that the remaining effects can be neglected. The usual convention about the general mean is that, for odd r, it is among the parameters to be estimated while, for even r, it is not among those to be estimated nor among those that can be neglected. So, for example, a resolution III design is one for which the general mean and main effects can be estimated, assuming that all other effects can be neglected. A resolution IV design allows us to estimate all main effects in the presence of the general mean and all two-factor interactions, while all other effects are neglected.

Remark   The reason for including the general mean for estimation in odd resolution designs but not in even resolution designs are:

1. In odd resolution designs, we estimate all parameters not assumed to be zero. Thus, if we can also estimate the general mean, we will have a predictive model. However, in even resolution designs, we do not estimate some of the nonzero parameters because they are not of interest to us. Therefore, if we estimate the general mean, we cannot build a predictive model anyway.
2. If we include the general mean as a part of the parameters not to be estimated, we will end up with some less cumbersome mathematics in even resolution designs.

The importance of orthogonal arrays in the above-described statistical context is due mainly to the following well-known result.

Theorem 4.1   An orthogonal array of strength t is a design of resolution $t+1$. Moreover, the associated estimated effects are orthogonally estimable, while the general mean can always be included in the estimable effects both for even and odd t.

A proof of this result can, for example, be found in Theorem 13.1 of Raktoe, Hedayat, and Federer (1981). This is a very nice and powerful result. It tells us, for example, that in an OA(8,4,2,3) the general mean and main effects can be estimated orthogonally, assuming that three- and four-factor interactions are absent. However, a property as in Example 4.1 cannot be explained by this result.

For this purpose, we formulate here a more general result. Let x be a k-dimensional $(0,1)$ vector, i.e., $x' = (x_1, x_2, \ldots, x_k)$ with $x_i \in \{0,1\}$. With factorial effects corresponding to x, we will mean all those effects $\phi_1^{j_1} \phi_2^{j_2} \ldots \phi_k^{j_k}$ with $j_i = 0$ if and only if $x_i = 0$, $1 \leq i \leq k$. The set of effects for an s^k factorial associated with x' is designated by $E_x(k,s)$. Clearly, $E_x(k,s)$ contains $(s-1)^t$ effects whenever t components of x' are nonzero.

Example 4.2

Consider a 3^5 factorial. The set $E_x(5,3)$ associated with $x' = (1,1,0,1,0)$ consists of the following eight effects:

$$\phi_1^1 \phi_2^1 \phi_3^0 \phi_4^1 \phi_5^0, \qquad \phi_1^1 \phi_2^1 \phi_3^0 \phi_4^2 \phi_5^0, \qquad \phi_1^1 \phi_2^2 \phi_3^0 \phi_4^1 \phi_5^0, \qquad \phi_1^1 \phi_2^2 \phi_3^0 \phi_4^2 \phi_5^0$$
$$\phi_1^2 \phi_2^1 \phi_3^0 \phi_4^1 \phi_5^0, \qquad \phi_1^2 \phi_2^1 \phi_3^0 \phi_4^2 \phi_5^0, \qquad \phi_1^2 \phi_2^2 \phi_3^0 \phi_4^1 \phi_5^0, \qquad \phi_1^2 \phi_2^2 \phi_3^0 \phi_4^2 \phi_5^0$$

Now, assume that B and C are two disjoint sets of k-dimensional $(0,1)$ vectors. The set C is allowed to be the empty set. Suppose B and C are constructed such that

$$\text{If } x, y \in B, \text{then } |\{i \in \{1,2,\ldots,k\} : x_i = 1 \text{ or } y_i = 1\}| \leq t \qquad (4)$$

$$\text{If } x \in B, y \in C, \text{ then } |\{i \in \{1,2,\ldots,k\} : x_i = 1 \text{ or } y_i = 1\}| \leq t \quad (5)$$

In words, conditions (4) and (5) require that the vector $x + y$ should have no more than t nonzero components.

Example 4.3

Let $k = 4$. Then the following sets B and C satisfy conditions (4) and (5) for $t = 3$.

$$B = \{(0,0,0,0),(1,0,0,0),(0,1,0,0),$$
$$(0,0,1,0),(0,0,0,1),(1,1,0,0),(1,0,1,0)\}$$
$$C = \{(1,0,0,1)\}$$

Within the context of s^k-factorial, we denote the set of effects associated with a set X of k-dimensional (01,) vectors by $E_X(k,s)$.

Using the preceding concept and notation, we are now ready to show that an orthogonal array of strength t can be more than a fractional factorial design of resolution $t + 1$.

Theorem 4.2   If A is an OA(N,k,s,t), then with B and C satisfying conditions (4) and (5), all effects in $E_B(k,s)$ can be estimated orthogonally, assuming that effects in $E_C(k,s)$ are the only other ones that are not neglected. The general mean can always be chosen to be in $E_B(k,s)$.

A proof of this theorem is given in the Appendix. Here we elucidate and demonstrate the practical meaning of Theorem 4.2 by two examples.

Example 4.4

If we take $B = \{(0,0,0,0),\ (1,0,0,0),\ (0,1,0,0),\ (0,0,1,0),\ (0,0,0,1),$ $(1,1,0,0),\ (1,0,1,0),\ (1,0,0,1)\}$ and C to be the empty set, then, for $k = 4$, $s = 2$, and $t = 3$, we obtain an explanation for Example 4.1 without going through the computations. Another possible choice for sets B and C that could be applied to the design in Example 4.1 is

$$B = \{(0,0,0,0),(1,0,0,0),(0,1,0,0),$$
$$(0,0,1,0),(0,0,0,1),(1,0,1,0),(1,0,0,1)\}$$
$$C = \{(1,1,0,0),(0,0,1,1)\}$$

An expert in the area of factorial experiments could conclude the results of Examples 4.1 and 4.4 by observing that the fraction given in Example 4.1 has indeed a "defining relation" and, as a consequence, our conclusions for those examples could have been drawn without the support of Theorem 4.2. We like to point out, however, that not all orthogonal arrays have defining relations. Therefore, the traditional approach via defining relations is not applicable to them. Here is an example.

Example 4.5

Consider the following fraction from a 2^5 factorial:

$$OA(24, 5, 2, 3)$$

$$0\ 0\ 0\ 0\ 0\ 0\ 0\ 0\ 0\ 0\ 0\ 0\ 1\ 1\ 1\ 1\ 1\ 1\ 1\ 1\ 1\ 1\ 1\ 1$$
$$1\ 1\ 0\ 1\ 1\ 1\ 0\ 0\ 0\ 1\ 0\ 0\ 0\ 0\ 1\ 0\ 0\ 0\ 1\ 1\ 1\ 0\ 1\ 1$$
$$1\ 0\ 1\ 0\ 1\ 1\ 1\ 0\ 0\ 0\ 1\ 0\ 0\ 1\ 0\ 1\ 0\ 0\ 0\ 1\ 1\ 1\ 0\ 1$$
$$1\ 0\ 0\ 1\ 0\ 1\ 1\ 1\ 0\ 0\ 0\ 1\ 0\ 1\ 1\ 0\ 1\ 0\ 0\ 0\ 1\ 1\ 1\ 0$$
$$1\ 1\ 0\ 0\ 1\ 0\ 1\ 1\ 1\ 0\ 0\ 0\ 0\ 0\ 1\ 1\ 0\ 1\ 0\ 0\ 0\ 1\ 1\ 1$$

This fraction cannot have a defining relation. However, based on Theorem 4.2, we can orthogonally estimate the general mean, all five main effects, and the two-factor interactions between factors 1 and 2, 1 and 3, and 2 and 3, assuming that there are no other effects in the model. For this example,

$$B = \{(0,0,0,0,0), (1,0,0,0,0), (0,1,0,0,0), (0,0,1,0,0),$$
$$(0,0,0,1,0), (0,0,0,0,1), (1,1,0,0,0), (1,0,1,0,0), (0,1,1,0,0)\}$$

while the set C is empty.

5 ORTHOGONAL ARRAYS OF STRENGTH $t+$ AND THEIR PRACTICAL APPLICATIONS

In Section 4, we learned that orthogonal arrays can be used to fit various models and estimate their corresponding parameters orthogonally and efficiently. We reached these conclusions solely on the knowledge of their strength. However, we did not *need* to utilize any additional information concerning the structure of the arrays. By taking such information also into account, we can arrive at stronger conclusions. Instead of a general formulation, we shall illustrate the idea by an example.

Example 5.1

Let A be the following orthogonal array:

Factor								
1	0 0 0 0 1 1 1 1
2	0 0 1 1 0 0 1 1
3	0 1 0 1 0 1 0 1
4	0 1 0 1 1 0 1 0
5	0 1 1 0 1 0 0 1

This is an $OA(8, 5, 2, 2)$. We claim that the following eight effects can be estimated orthogonally, assuming that no other effects are present: the general mean, the main effects, and the two-factor interactions between factors 1 and 2 and 1 and 5. We cannot conclude this from our previous result. An explanation for the validity of our statement is as follows: Select any two effects to be estimated, and look at the rows in A corresponding to the factors involved. Each time these rows form one or more copies of a minimal complete factorial. For example, if we select the main effect of 4 and the interaction between 1 and 5, we see that the rows corresponding to factors 1, 4, and 5 form one copy of a minimal complete factorial. From this property, the validity of our statement follows. We cannot, however, reach the same conclusion for factors 2, 4, and 5. As seen below, restricted to these three factors, the array is not a minimal complete factorial.

Factor								
2	0 0 1 1 0 0 1 1
4	0 1 0 1 1 0 1 0
5	0 1 1 0 1 0 0 1

This lead us to the following definition.

Definition 5.1 An $OA(N, k, s, t+)$ is an $OA(N, k, s, t)$ that is not of strength $t+1$ but has one or more subarrays that form an $OA(N, k', s, t+1)$.

Thus, for example, the array in Example 5.1 is indeed an $OA(8, 5, 2, 2+)$. In the language of fractional factorial, an $OA(N, k, s, t+)$ consists of N level combinations from an s^k factorial with the following property. Restricted to every t factors, the design consists of one or more replications of the corresponding minimal complete factorial. The same property holds when the design is restricted to some, but not all, sets of $t + 1$ factors.

In general, we like to have arrays with high strength, but this can be costly since the number of runs needed tends to increase with the strength of the array. In many practical situations, we need to fit more parameters than an orthogonal array of strength t (resolution $t + 1$) allows but do not need all the parameters associated with an array of strength $t+1$ (resolution $t + 2$). In many such situations, we can reduce the cost of experimentation by running our experiment in the form of arrays of strength $t+$. Results on the existence and construction of $OA(N, k, s, t+)$ are reported in Hedayat (1989).

6 CONCLUDING REMARKS

The real art of design of experiments is to gather maximum information per unit cost. In this direction, orthogonal arrays can serve as extremely useful fractional factorial designs in industrial experimentations. We demonstrated that we can both fit and test many models if we use orthogonal arrays. In a sense, orthogonal arrays are model-robust and model-efficient. We also pointed out that, by a judicious selection and construction of such arrays, we can even benefit more by using such arrays. This practical consideration led us to introduce a new concept called orthogonal arrays of strength $t+$, which are somewhere between arrays of strength t and $t + 1$.

The subject of orthogonal arrays has received a great deal of attention by both mathematicians and statisticians. Many results that are useful to practitioners have been obtained. Many of these results based on two or three levels have been used by Taguchi and his followers in industrial experimentation. It is unfortunate that these users have not given credit to those researchers who originally discovered these results and that they have not used the most efficient arrays in terms of number of runs, and so forth. In their forthcoming book, Hedayat and Stufken (1990) have given a full account of the entire subject of orthogonal arrays. This includes the results of some of the earlier contributors mentioned in this article. Some recent discoveries can be found in Hedayat and Stufken (1988), (1989a), and (1989b) and references given there.

APPENDIX

Below we provide a proof of Theorem 4.2

Proof Let β_B and β_C be the vectors of factorial effects specified by the sets B and C. Then the orthogonal polynomial model associated with the orthogonal array A can be expressed as

$$Y_A = X_A\beta_A + \epsilon_A = X_B\beta_B + X_C\beta_C + \epsilon_A \qquad (A1)$$

The information matrix associated with A under model (3) can be expressed as

$$X_A'X_A = \begin{pmatrix} X_B'X_B & X_B'X_C \\ X_C'X_B & X_C'X_C \end{pmatrix} \qquad (A2)$$

Our claim is established if we show that $X_B'X_B$ is a full-rank diagonal matrix and $X_B'X_C$ is a null matrix. To establish these, let $\phi_1^{j_1}\phi_2^{j_2}\ldots\phi_k^{j_k}$ and $\phi_1^{j_1'}\phi_2^{j_2'}\ldots\phi_k^{j_k'}$ be two effects corresponding to $x = (x_1,\ldots,x_k)$ and $y = (y_1,\ldots,y_k)$. We assume that x is a vector in B and y is a vector in B or in C. Without loss of generality and for the ease of presentation, we may assume that $j_1 \neq j_1'$ and

$$\{i \in \{1,\ldots,k\} : x_i = 1 \text{ or } y_i = 1\} = \{1,2,\ldots,l\}, \qquad l \leq t \qquad (A3)$$

Let $(i_1,\ldots,i_k)$ be a treatment combination, which is a column of A. Then, by (2), the coefficients of $\phi_1^{j_1}\phi_2^{j_2}\ldots\phi_k^{j_k}$ and $\phi_1^{j_1'}\phi_2^{j_2'}\ldots\phi_k^{j_k'}$ corresponding to $(i_1,\ldots,i_k)$ in X_A are

$$n_{i_1j_1}^{(1)}n_{i_2j_2}^{(2)}\ldots n_{i_kj_k}^{(k)}, \qquad n_{i_1j_1'}^{(1)}n_{i_2j_2'}^{(2)}\ldots n_{i_kj_k'}^{(k)} \qquad (A4)$$

respectively. By (A3),

$$j_{l+1} = j_{l+2} = \cdots = j_k = 0, \qquad j_{l+1}' = j_{l+2}' = \cdots = j_k' = 0$$

and, consequently [see (2)],

$$n^{(l+1)}_{i_{l+1}j_{l+1}} = n^{(l+1)}_{i_{l+1}0} = 1, \ldots,$$

$$n^{(k)}_{i_k j_k} = n^{(k)}_{i_k 0} = 1$$

$$n^{(l+1)}_{i_{l+1}j'_{l+1}} = n^{(l+1)}_{i_{l+1}0} = 1, \ldots,$$

$$n^{(k)}_{i_k j'_k} = n^{(k)}_{i_k 0} = 1$$

Now we show that the column vectors in X_A corresponding to $\phi_1^{j_1} \phi_2^{j_2} \ldots \phi_k^{j_k}$ and $\phi_1^{j'_1} \phi_2^{j'_2} \ldots \phi_k^{j'_k}$ are orthogonal. The inner product of these two columns is

$$\sum_{(i_1,\ldots,i_k)} \left(n^{(1)}_{i_1 j_1} n^{(2)}_{i_2 j_2} \ldots n^{(k)}_{i_k j_k} \right) \left(n^{(1)}_{i_1 j'_1} n^{(2)}_{i_2 j'_2} \ldots n^{(k)}_{i_k j'_k} \right)$$

$$= \sum_{i_1=0}^{s-1} \left(n^{(1)}_{i_1 j_1} n^{(1)}_{i_1 j'_1} \right) \left\{ \sum_{(i_2 \ldots i_k)} \left(n^{(1)}_{i_2 j_2} \ldots n^{(l)}_{i_l j_l} \right) \left(n^{(2)}_{i_2 j'_2} \ldots n^{(l)}_{i_l j'_l} \right) \right\} = 0$$

since A is of strength t, and then the summation term inside the above $\{\ \}$ is independent of i_1 and

$$\sum_{i_1=0}^{s-1} \left(n^{(1)}_{i_1 j_1} n^{(1)}_{i_1 j'_1} \right) = 0$$

because M_1 is columnwise orthogonal.

ACKNOWLEDGMENTS

This research has been sponsored by AFOSR grant 89-0221.

REFERENCES

Bose, R. C. and Bush, K. A. (1952). Orthogonal arrays of strength two and three, *Ann. Math. Statist.*, *23*: 508–524.

Bush, K. A. (1952). Orthogonal arrays of index unity, *Ann. Math. Statist.*, *23*: 426–434.

Hedayat, A. (1989). Orthogonal arrays of strength $t+$ and their statistical applications. Technical Report, Statistical Laboratory, University of Illinois, Chicago.

Hedayat, A. and Stufken, J. (1988). Two-symbol orthogonal arrays, *Optimal Design and Analysis of Experiments* (Y. Dodge, V. V. Fedorov and H. P. Wynn, eds.), North–Holland, Amsterdam, pp. 47–58.

Hedayat, A. and Stufken, J. (1989a). On the maximum number of constraints in orthogonal arrays. *Ann. Statist. 17*: 448–451.

Hedayat, A. and Stufken, J. (1989b). On the maximum number of factors in two construction methods for orthogonal arrays. In *Recent Developments in Statistical Data Analysis and Inference* [A Volume in Honor of C. R. Rao] (Y. Dodge, ed.), North–Holland, Amsterdam, pp. 33–40.

Hedayat, A. and Stufken, J. (1990). *Orthogonal Arrays*, to appear.

McLean, R. A. and Anderson, V. L. (1984). *Applied Factorial and Fractional Designs*, Marcel Dekker, New York.

Raktoe, B. L., Hedayat, A., and Federer, W. T. (1981). *Factorial Designs*, Wiley, New York.

Rao, C. R. (1946). Hypercubes of strength d leading to confounded designs in factorial experiments, *Bull. Calc. Math. Soc., 38*: 67–78.

Rao, C. R. (1947a). Factorial experiments derivable from combinatorial arrangements of arrays, *J. Roy. Statist. Soc. 9*: 128–139.

Rao, C. R. (1947b). On a class of arrangements, *Proc. Edinburgh Math. Soc., 8*: 119–125.

Seiden, E. (1954). On the problem of construction of orthogonal arrays, *Ann. Math. Statist., 25*: 151–156.

Seiden, E. and Zemach, R. (1966). On orthogonal arrays, *Ann. Math. Statist., 37*: 1355–1370.

Taguchi, G. and Wu, Y. (1979). *Introduction to Off-line Quality Control*, Central Japan Quality Control Association, Tokyo.

14

Construction of Run Orders
of Factorial Designs

Ching-Shui Cheng   University of California, Berkeley, Berkeley, California

1   INTRODUCTION

The purpose of this chapter is to discuss the construction of systematic
run orders of factorial designs. Suppose a factorial experiment is to be
run in a time sequence with one observation taken at a time, and the
treatment combinations to be observed have already been chosen. Then
the experimenter has to decide in which order to observe these treatment
combinations. A common practice is to randomize. However, sometimes
randomization may lead to an undesirable sequence, and a systematic run
order may be preferred. For instance, in industrial experiments, it is often
difficult, time-consuming, and expensive to change factor levels or, after
they have been changed, it may take a long time for the system to return to
steady state; then the experimenter may want to have as few level changes
as possible. Another consideration is that when observations are taken
over time, they may be affected by some uncontrollable variables that are
highly correlated with time. Then the usual estimates of factorial effects
(main effects or interactions) may become very inefficient. An experimenter
who has some knowledge about the nature of the time trend should put

this knowledge to use and construct a run order in which the estimates of factorial effects are little disturbed by the presence of the time trend.

Previous research on the construction of run orders of factorial designs includes Hill (1960), Daniel and Wilcoxon (1966), Draper and Stoneman (1968), Dickinson (1974), Daniel (1976, Chapter 15), Joiner and Campbell (1976), Cheng (1985), Coster and Cheng (1988), John (1986), and Cheng and Jacroux (1988). In all these, time dependence is represented by a polynomial. The objective is to construct a run order such that the estimates of the important factorial effects are orthogonal or nearly orthogonal to the postulated (typically linear or quadratic) polynomial trend. Draper and Stoneman (1968), Dickinson (1974), Joiner and Campbell (1976), Cheng (1985), Coster and Cheng (1988) also considered costs of level changes. Another way to model time dependence is to consider correlated errors. Steinberg (1988) proposed a flexible class of models representing the time trend as an ARIMA time series model.

In this chapter, we shall review and discuss a simple and flexible method of constructing run orders of factorial designs for such different purposes as to eliminate polynomial time trends, to guard against correlated errors, or to save the cost of level changes. Various forms of this method have appeared in Daniel and Wilcoxon (1966), Cheng (1985), Coster and Cheng (1988), John (1986) and Cheng and Jacroux (1988).

For simplicity, we shall discuss only factorial designs in which all the factors have two levels. Also we shall not consider blocking. Many of the results can easily be extended to the case in which the number of levels is more than two and to the case in which there is a common trend within each block.

We assume that there are n factors, each occurring at two levels. The two levels, say, high and low levels, can be represented by 1 and 0, respectively. Then the 2^n different treatment combinations can be represented as n tuples of 1s and 0s. Given two treatment combinations $(x_1, x_2, \ldots, x_n)$ and $(y_1, y_2, \ldots, y_n)$, we can form another combination through component-wise addition modulo 2 : $(x_1, x_2, \ldots, x_n) + (y_1, y_2, \ldots, y_n) = (x_1 + y_1, x_2 + y_2, \ldots, x_n + y_n) \bmod 2$. Another convenient way to denote a treatment combination is to use a string of lowercase letters. Suppose we associate letter a_i with the ith factor. Then a particular treatment combination is represented as a string consisting of a subset of the letters $a_1, a_2, \ldots, a_n$, where letter a_i is present if the ith factor is at the high level and it is absent if the ith factor is at the low level. In particular, the treatment combination in which all the factors are at low levels is denoted by 1, wherein none of the letters appears. For instance, suppose there are six factors, which we call A, B, C, D, E, and F. Then acf represents the treatment combination where factors A, C, F are at high levels and the other three factors are at low levels, i.e., the treatment combination $(1,0,1,0,0,1)$. In this notation, the addition of treatment combinations defined earlier becomes symbolic multiplication subject to the rule that $a_i^2 = 1$; i.e., two identical letters cancel themselves out. A set of treatment combinations are said to be *independent* if none of these combinations can be expressed as a product (or sum when denoted as n tuples of 1s and 0s) of any others in the set. In the rest of the paper, we find it more convenient to denote treatment combinations as strings of lowercase letters. Therefore, multiplication instead of addition will be used unless it is stated otherwise.

The factorial effects are usually denoted by strings of capital letters. Letter A_i denotes the main effect of factor i, and a combination of t capital letters $A_{i_1} A_{i_2} \dots A_{i_t}$ represents a t-factor interaction. For instance, in an experiment with six factors, ACF represents the interaction of the three factors A, C, and F. We can also perform multiplications on factorial effects subject to the rule $A_i^2 = 1$. Independence of a set of factorial effects is similarly defined. For convenience, sometimes each string of letters (representing a factorial effect or a treatment combination) is called a *word*, and the number of letters it contains is called its *length*.

An experiment in which all the 2^n treatment combinations are observed is called a 2^n complete experiment or design. In the following, we shall consider complete, as well as fractional, factorial designs. Generally speaking, a fractional factorial design is such that only a subset of the 2^n treatment combinations is to be observed. In this chapter, however, we shall consider only those fractions in which the treatment combinations form a *group* under the operation of multiplication defined earlier. Readers who are not familiar with group theory should not be concerned. What this means is that there exist $k < n$ independent treatment combinations, called *generators*, such that the design consists of all the 2^k products that can be formed by multiplying these k independent generators in all possible ways. Let $p = n - k$. Then this represents a $1/2^p$ fraction of the complete factorial experiment and is often referred to as a 2^{n-p} fractional factorial design.

An ordered sequence of the 2^k combinations is called a *run order*. Given a run order of a 2^{n-p} design, for $i = 1, 2, \dots, n$, let $\mathbf{u}_i = (u_{1i}, \dots, u_{2^{n-p},i})'$ be the $2^{n-p} \times 1$ vector defined by $u_{si} = 1$ or -1, depending on whether, in the sth run, factor i is at its high or low level. Then we say that $\mathbf{u}_i$ is the contrast representing or defining the main effect of the ith factor. The contrast representing interaction $A_{i_1} \ldots A_{i_t}$ is the componentwise product of $\mathbf{u}_{i_1}, \ldots, \mathbf{u}_{i_t}$. We refer the readers to a standard textbook on experimental design such as John (1971) for other terminology and notations that will be used in this paper, particularly the notion of resolution, defining relation, aliasing, etc.

In Section 2, we give a general description of the method that will be used to construct run orders of two-level factorial designs in this paper. This method is then applied in later sections to deal with polynomial trends (Section 3), cost saving (Section 5), and correlated errors (Section 6). Section 4 discusses the relationship between this method and a method first proposed by Daniel and Wilcoxon (1966).

2   FOLDOVER METHOD

The method described below was used by Cheng (1985) to construct run orders of some factorial designs in which all the main effects are orthogonal to a linear trend and was formally defined in Coster and Cheng (1988). In Section 4, we shall relate it to a method first proposed by Daniel and Wilcoxon (1966) and more fully developed by Cheng and Jacroux (1988).

Suppose a given fractional factorial design is generated by k independent treatment combinations $\mathbf{x}_1, \mathbf{x}_2, \ldots, \mathbf{x}_k$, where $\mathbf{x}_i$ is a string of lowercase letters. A simple way to generate an order of the 2^k runs can be described by induction as follows. We start with 1, the combination in which all the factors are at low levels. It is then followed by $\mathbf{x}_1$. Suppose 2^s treatment combinations, $s < k$, have been generated. Then we follow these 2^s combinations by their products with $\mathbf{x}_{s+1}$ in the same order.

As an example, consider the 2^3 complete factorial design and the three generators ab, abc, ac. According to the above procedure, we start with 1, followed by ab, then abc ($= 1 \cdot abc$), c ($= ab \cdot abc$), and finally ac ($= 1 \cdot ac$), bc ($= ab \cdot ac$), b ($= abc \cdot ac$), a ($= c \cdot ac$). This produces the run order

$$1, ab, abc, c, ac, bc, b, a \qquad (1)$$

Of course the above procedure depends on the generators chosen and the order in which they appear. This will be dictated by the goal of the experimenter, who may want to guard against possible correlations between consecutive runs, to have estimates that are resistant to certain polynomial trends, or to have few level changes. The properties of the generators that are needed to achieve the various objectives will be discussed in later sections. For ease of reference, we shall call the method described above the *foldover* method. The generators, together with their ordering, are called a *generator sequence*. Thus, for example, the so-called *standard order* (the one used in applying Yates' algorithm) of a 2^n complete factorial design can be obtained by applying the foldover method to the generator sequence $a_1, a_2, \ldots, a_n$. For the complete 2^3 experiment, this leads to the run order 1, a, b, ab, c, ac, bc, abc.

Another equivalent way to apply the foldover method is the following: Given a sequence of generators, say, $\mathbf{y}_1, \mathbf{y}_2, \ldots, \mathbf{y}_k$. Again, we start with 1, followed by $\mathbf{y}_1$. But after 2^s treatment combinations, $s < k$, have been generated, they are followed by their products with $\mathbf{y}_{s+1}$ in *reverse* order. We shall call this the *reverse foldover* method. For instance, applying the reverse foldover method to the three generators ab, c, a of the complete 2^3 design, we obtain the run order 1, ab, abc ($= ab \cdot c$), c ($= 1 \cdot c$), ac ($= a \cdot c$), bc ($= abc \cdot a$), b ($= ab \cdot a$), a ($= 1 \cdot a$). This is the same as the one constructed earlier by applying the foldover method to the generator sequence ab, abc, ac. In general, the two methods lead to the same run order if

$$\mathbf{x}_1 = \mathbf{y}_1, \qquad \text{and} \qquad \mathbf{x}_i = \mathbf{y}_{i-1}\mathbf{y}_i \qquad \text{for all } i > 1 \qquad (2)$$

From (2), we have

$$\mathbf{y}_i = \prod_{j=1}^{i} \mathbf{x}_j \qquad \text{for all } i = 1, 2, \ldots, k \qquad (3)$$

The reverse foldover method has some advantages, which will be evident later when we describe the results on run orders that have a minimum number of level changes and those that are robust against correlated errors.

3 POLYNOMIAL TRENDS

Suppose the observations are equally spaced in time. Assume the usual uncorrelated homoscedastic model, but the observations are affected by a systematic time trend that is an mth-degree polynomial of the time. Under a certain run order of a factorial design, if the least-squares estimator of a factorial effect is the same as when the time trend is not present, that effect is said to be orthogonal to a polynomial time trend of degree m. If an effect is orthogonal to all polynomial time trends of degrees less than or equal to m, then it is called *m-trend-free*. Coster and Cheng (1988) derived a simple condition for a main effect to be m-trend-free in a run order generated by the foldover method described in Section 2. It turns out that, in such a run order, *the main effect of a given factor is m-trend-free if the corresponding letter appears at least $m + 1$ times in the generator sequence*. This actually is also a necessary condition. As an example, consider the run order (1) obtained from the generator sequence ab, abc, ac. Here each letter appears at least twice in the generator sequence, so that all the three main effects are linear-trend-free. Since a appears three times, the main effect of factor A is also quadratic-trend-free.

Compare the following two run orders of the 2^4 complete factorial design:

$$1, ab, abc, c, bcd, acd, ad, bd, d, abd, abcd, cd, bc, ac, a, b \qquad (4)$$

and

$$1, abd, abc, cd, acd, bc, bd, a, bcd, ac, ad, b, ab, d, c, abcd \qquad (5)$$

Run order (4) is obtained by applying the foldover method to the generator sequence ab, abc, bcd, d, while (5) is derived from the generator sequence abd, abc, acd, bcd. By counting the number of times each letter appears in the two generator sequences, we can easily see that, in (4), the main effects of factors A, C, and D are linear-trend-free and B is quadratic-trend-free while, in (5), the main effects of all the three factors are quadratic-trend-free.

Another method of constructing trend-free run orders of factorial designs was originally due to Daniel and Wilcoxon (1966). They discovered that some of the ordered contrasts associated with the 2^n complete factorial experiment are orthogonal to linear and quadratic trends (see their paper and the discussion below). If these contrasts are used to represent main effects (or other effects of interest), a run order in which these effects are orthogonal to linear and quadratic trends can be obtained. This idea was carried out on 16- and 32-run designs in Daniel and Wilcoxon (1966). Extending Daniel and Wilcoxon's result, Cheng and Jacroux (1988) showed that *in the standard order of a complete 2^n factorial design, any t-factor interaction is orthogonal to a $(t - 1)$-degree polynomial trend*. Therefore, starting from the standard order, by redesignating some higher-order interaction terms as main effects (or other effects of interest), one can derive another run order of the complete 2^n design in which these effects of interest are orthogonal to high-degree polynomial trends. More precisely, suppose we would like to construct a run order of the complete 2^n design in which the main effect A_i of the ith factor is m-trend-free. Then we choose an $(m+1)$-factor interaction (of course, this is possible only if $n > m$), say, $A_{i_1} \ldots A_{i_{m+1}}$. We know that $A_{i_1} \ldots A_{i_{m+1}}$ is m-trend-free in the standard order. Therefore, if we can order the runs so that the contrast originally defining $A_{i_1} \ldots A_{i_{m+1}}$ in the standard order now represents A_i, then A_i would be m-trend-free in the new order. This can be achieved by reassigning the high or low level to factor A_i on the sth run, $1 \leq s \leq 2^n$, depending on whether the sth component of the contrast representing $A_{i_1} \ldots A_{i_{m+1}}$ in the standard order is 1 or -1, respectively. To get a complete replicate of the 2^n design, $A_{i_1} \ldots A_{i_{m+1}}$ should not be an interaction involving only the factors other than the ith.

As an example, we consider again the complete 2^4 design. Suppose we would like to make the main effect of factor A quadratic-trend-free and we choose to redesignate the three-factor interaction ABC as A. We first write down the standard order of complete 2^4:

$$1, a, b, ab, c, ac, bc, abc, d, ad, bd, abd, cd, acd, bcd, abcd \tag{6}$$

In this order, ABC is represented by the following contrast:

$$(-1, 1, 1, -1, 1, -1, -1, 1, -1, 1, 1, -1, 1, -1, -1, 1) \tag{7}$$

which is orthogonal to a quadratic trend. Now, assign the low level to factor A wherever -1 appears in (7), but do not change the levels of the other factors. Then, from (6), we obtain the following run order:

$$1, a, ab, b, ac, c, bc, abc, d, ad, abd, bd, acd, cd, bcd, abcd$$

in which (7) represents main effect A and, therefore, A is quadratic-trend-free.

For convenience, we shall call this method the D-W method. Suppose we would like to make all the main effects m-trend-free. Then we choose one $(m+1)$-factor or higher-order interaction for each main effect; but, in order to get a complete replicate, these interactions must be independent. Sometimes this may not be possible. In a complete 2^n system, there are exactly n $(n-1)$-factor interactions. Cheng and Jacroux (1988) showed that these interactions are independent if and only if n is even. So for even n, one can construct a run order of the complete 2^n design such that all the main effects are $(n-2)$-trend-free. Returning to our example of 2^4, if we redesignate the three-factor interactions ABC, ABD, BCD, and ACD as main effects A, B, C, and D, respectively, then, from the standard order, we obtain

$$1, abd, abc, cd, acd, bc, bd, a, bcd, ac, ad, b, ab, d, c, abcd \qquad (8)$$

which is the same as the run order (5) obtained by applying the foldover method to the generator sequence abd, abc, acd, bcd. We will return to discuss the relationship between the two methods in the next section.

After n independent contrasts have been redefined as main effects, the trend-free status of all the other interactions are automatically determined. For any interaction, say, $A_{i_1} \ldots A_{i_t}$, we can simply multiply together those words representing the interactions redesignated as main effects $A_{i_1}, \ldots,$ and A_{i_t}. If the resulting word has length s, then interaction $A_{i_1} \ldots A_{i_t}$ is $(s-1)$-trend-free in the new order. For example, in constructing (8), we redefined ABC and ABD as A and B, respectively. Therefore, in the new order, the two-factor interaction AB is represented by the contrast that represents $ABC \cdot ABD = CD$ in the standard order and, hence, is linear-trend-free. In fact, one can check that this holds for all two-factor interactions; so run order (8) is such that all the main effects are quadratic-trend-free and all the two-factor interactions are linear-trend-free. There are usually many ways to choose the contrasts to be redefined as main effects. The choice may be determined by some secondary goals. For instance, in the complete 2^4 design, another way to make all the main effects at least quadratic-trend-free is to identify A, B, and C with ABC, ABD, and BCD as before, but D is identified with the four-factor interaction $ABCD$. All these interactions are independent, so that the assignment is legitimate. This will actually make the main effect D *cubic*-trend-free. However, a by-product is that none of the interactions AD, BD, and CD is even linear-trend-free. Thus, an experimenter who wants the main effects and some interactions to be to a certain degree trend-free must choose the interactions to be redefined as main effects in such a way that their products are also of suitable lengths. Cheng and Jacroux (1988) showed that, for even (respectively, odd) $n \geq 4$, one can always construct a run order of the complete 2^n design in which all the main effects are $(n-2)$-trend-free [respectively, $(n-3)$-trend free] and all the two-factor interactions are at least linear-trend-free.

One can also apply the D-W method to construct trend-resistant run orders of fractional factorial designs. It is known that for any 2^{n-p} design, there exist $n-p$ of the n factors, say, factors $i_1, i_2, \ldots, i_{n-p}$, such that the design contains a complete replicate of these factors, and the levels of the other p factors on each run can be determined by the defining relation. Let factors $i_1, i_2, \ldots, i_{n-p}$ be called *basic factors*. One can apply the D-W method to the complete factorial design in the basic factors as before. Once the contrasts to represent main effects of the basic factors have been chosen, the trend-free status of not only the interactions of the basic factors but also those involving the other p factors are automatically determined. Thus, an added complication here is that if one wants all the main effects to be to a certain degree trend-free, checking the lengths of those words identified with factors $i_1, i_2, \ldots,$ and i_{n-p} is not enough; the defining relation must also be employed to check the trend-free properties of the other p main effects. For example, consider the $1/4$ replicate of 2^7 defined by $I = -ABCDE = -ABCFG = DEFG$. This consists of the 32 treatment combinations $(x_1, x_2, x_3, x_4, x_5, x_6, x_7)$, $x_i = 1$ or 0 that satisfy the following two equations (the last word $DEFG$ in the defining relation is the product of the first two, so that there are only two independent equations):

$$x_1 + x_2 + x_3 + x_4 + x_5 = 0, \qquad x_1 + x_2 + x_3 + x_6 + x_7 = 0 \qquad (9)$$

This is a resolution IV design; i.e., all the main effects are estimable if all the interactions involving more than two factors are negligible. This design contains a complete replicate of the five factors A, B, C, D, and F. The levels of factors E and G on each run can be determined from the levels of the other factors by $x_5 = x_1 + x_2 + x_3 + x_4$, and $x_7 = x_1 + x_2 + x_3 + x_6$, respectively. In other words, the main effects of E and G are aliased with the four-factor interactions $ABCD$ and $ABCF$, respectively. To construct a run order of this 2^{7-2} design in which all the main effects are at least linear-trend-free, we first construct one such run order for the complete factorial in A, B, C, D, and F. Suppose we identify the main effects A, B, C, D, and F with interactions DF, $ABDF$, ABF, BCF, and CDF, respectively. These all have lengths of at least two; so that, by the D-W method, we obtain the following run order in which all the five main effects A, B, C, D, and F are at least linear-trend-free:

$$\begin{aligned}
& ab,\ ac,\ acd(e),\ abd(e),\ abd(e)f(g),\ acd(e)f(g),\ acf(g),\ abf(g), \\
& f(g),\ bcf(g),\ bcd(e)f(g),\ d(e)f(g),\ d(e),\ bcd(e),\ bc,\ 1,\ cdf,\ bdf, \\
& b(e)f,\ c(e)f,\ c(eg),\ b(eg),\ bd(g),\ cd(g),\ abcd(g),\ ad(g),\ a(eg), \\
& abc(eg),\ abc(e)f,\ a(e)f,\ adf,\ abcdf
\end{aligned} \qquad (10)$$

In (10), the levels of factors E and G indicated within the parentheses are obtained from the defining relation. Now, since the products $(DF)(ABDF)(ABF)(BCF) = BC$ and $(DF)(ABDF)(ABF)(CDF) = CD$ have lengths of at least two, we conclude that the main effects of E and G are also linear-trend-free.

4   RELATIONSHIP BETWEEN FOLDOVER AND D-W METHODS

We have seen that run order (8) obtained by the D-W method is identical to (5), which was constructed by the foldover method. In fact, *every run order derived by the D-W method can be obtained by the foldover method.* The D-W method starts with the standard order, which can be constructed by applying the foldover scheme to the generator sequence a_1, a_2, ..., a_n. Identifying main effects with some interaction terms amounts to replacing a_1, a_2, ..., a_n with a suitable generator sequence, and therefore the D-W method is in essence a foldover method. Note that the D-W method does not always lead to a run order starting with 1, the treatment combination with all the factors at low levels. Run order (10) is one such example. But if we multiply all the treatment combinations by the first one in the sequence, then a run order starting with 1 is obtained. This does not affect the trend-free properties of all the effects. To facilitate the comparison with the foldover method, we shall assume that this step in making the run order starting with 1 is always carried out, although it is not necessary. Then for a run order of the complete 2^n design constructed by the D-W method, we have the following simple rule for deriving the corresponding generator sequence:

A letter appears in the ith generator if and only if the ith factor appears in the interaction redesignated as the main effect represented by that letter.   (11)

For fractional factorial designs, one can simply apply the preceding rule to the basic factors and then use the defining relation to determine the levels of the other factors in the generators.

For example, consider run order (8), which was obtained from the D-W method with the main effects A, B, C, and D identified with the three-factor interactions ABC, ABD, BCD, and ACD, respectively. To obtain the first generator for constructing the same run order by the foldover method, we notice that the first factor (i.e., A) appears in the interactions ABC, ABD, and ACD, which are redesignated as A, B, and D, respectively, but not in the interaction BCD identified with C. Therefore according to (11), the first generator is abd. Likewise, the second factor (i.e., B) appears in the interactions ABC, ABD, and BCD, which are redesignated as A, B, and C; hence, the second generator is abc. By the same rule, the third and fourth generators are acd and bcd. So, the resulting generator sequence is abd, abc, acd, and bcd, which is the same as the one used to construct run order (5).

Using the same method, one can show that after multiplying all the treatment combinations in run order (10) by ab to make it start with 1, the resulting ordering can be obtained by applying the foldover scheme to the generator sequence bc, $bcde$, $defg$, $abfg$, $abcdf$. For example, the second factor (B) appears in $ABDF$, ABF, and BCF, which are identified with B, C, and D, respectively. Therefore b, c, and d must appear in the second generator. From (9), when factors A and F are at low levels and B, C, D are at high levels, E must be at the high level, and G must be at the low level. Therefore, the second generator is $bcde$.

An important application of (11) that has practical significance is that it provides a simple method for obtaining the generator sequence for a run order to be constructed by the D-W method. Once the generator sequence is determined, the run order can be systematically constructed in a very simple manner. There is no need to write down contrasts such as (7) and go through the somewhat cumbersome process of reassigning levels of the various factors on each run.

Given a generator sequence in the foldover method, we can also use (11) to write down the interactions that are redesignated as main effects in the equivalent D-W method. One can easily translate from one method to the other. As a by-product, (11) can be combined with the result of Cheng and Jacroux (1988) to derive a simple method for checking the trend-free properties of all the *interactions* for run orders constructed by the foldover method. By (11), the length of the word (interaction) redesignated as the main effect of a factor in the D-W method is equal to the number of times that factor appears in the corresponding generator sequence. Therefore, if a factor appears in $m + 1$ generators, its main effect is identified with an $(m + 1)$-factor interaction in the D-W method and, hence, is m-trend-free. This gives an alternate proof of the result of Coster and Cheng (1988) mentioned earlier. Now, consider a t-factor interaction, say, $A_{i_1} \ldots A_{i_t}$. To determine the trend-free property of $A_{i_1} \ldots A_{i_t}$, we notice that the interaction term ultimately representing $A_{i_1} \ldots A_{i_t}$ in the equivalent D-W method is the product of the words redesignated as the main effects of factors $A_{i_1}, \ldots$ and A_{i_t}. The length of this product is equal to the number of letters, each of which appears in an odd number of these t words and, by (11), is the same as the number of generators, each of which contains an odd number of letters out of $a_{i_1}, \ldots, a_{i_t}$. Therefore, we conclude the following.

Theorem 4.1 Given a run order of the complete 2^n design constructed by the foldover method, a t-factor interaction $A_{i_1} \ldots A_{i_t}$ is m-trend-free if and only if there are at least $m + 1$ generators, each of which contains an odd number of the corresponding lowercase letters $a_{i_1}, \ldots, a_{i_t}$. For fractional factorial designs, the same result holds for estimable factorial effects (main effects and interactions).

For $t = 2$, we find that a two-factor interaction is m-trend-free if and only if there are at least $m + 1$ generators, in each of which exactly one of the two factors appears. This result was also obtained by Coster and Cheng (1988) by a different method. As an example, consider run order (5) obtained by the generator sequence abd, abc, acd, bcd. By examining the generators, from Theorem 4.1, we conclude that all the two-factor interactions are linear-trend-free. Now, since all the four generators contain three letters, we see that the four-factor interaction $ABCD$ is cubic-trend-free.

It follows from Theorem 4.1 that *the trend-free properties of the main effects and interactions do not depend on the order in which the generators appear*. This makes it possible to construct run orders that also have other properties in addition to providing trend-free estimates. Some examples will be given in the next section.

5 RUN ORDERS WITH MINIMUM NUMBER OF LEVEL CHANGES

When all the factor levels are equally expensive to change, minimizing the cost of level changes is the same as minimizing the total number of level changes. This is the criterion used by Draper and Stoneman (1968) and Dickinson (1974) in their computer searches. Based on the group structure of a fractional factorial design, Cheng (1985) gave a mathematical characterization of the run orders with a minimum number of level changes. The result is particularly simple when applied to run orders constructed by the foldover method.

Recall that each treatment combination is represented by a string of lowercase letters. We say that one treatment combination is shorter than another if it contains fewer letters, i.e., fewer factors are at high levels. From a given generator sequence x_1, x_2, ..., x_k, we get y_1, y_2, ..., y_k described in (3). It follows from Cheng's (1985) result that a run order constructed by applying the foldover method to x_1, x_2, ..., x_k (or applying the reverse foldover method to y_1, y_2, ..., y_k) minimizes the total number of level changes among *all possible run orders* (not just among the run orders obtainable by the foldover method) if and only if

> y_1 is the shortest treatment combination in the design, and for all i with $1 < i \leq k$, y_i is the shortest treatment combination among those that cannot be expressed as products of y_1, ..., y_{i-1}.

This makes it possible to write a computer routine to find generators for constructing run orders with a minimum number of level changes. Sometimes run orders can be found that minimize the total number of level changes and also have desirable trend-free properties; see the remark made at the end of Section 4. For example, run order (10) of the 2^{7-2} design defined by $I = -ABCDE = -ABCFG = DEFG$ is such that not only all the main effects are linear-trend-free but also it has the minimum number of level changes among all possible run orders. More examples of this kind can be found in Cheng (1985) and Coster and Cheng (1988). Coster and Cheng (1988) also examined the designs tabled in two National Bureau of Standards publications: Applied Mathematics Series 48 (1957) and 54 (1959). Many of these designs are blocked, and the designs in Series 54 have three levels, but the results we have described can easily be modified to handle these situations. Assuming that there is a common trend within each block when the experiment is blocked, Coster and Cheng (1988) found that, of the 125 two-level designs listed in Series 48, 96 can be so ordered that the number of level changes is minimized and all the main effects are at least linear-trend-free. Furthermore, for 63 of these 96 designs, one can also obtain quadratic trend-free run orders (for all main effects) with a minimum number of level changes. All the 41 three-level designs in Series 54 can be ordered to make all main effects linear-trend-free and, at the same time, minimize the number of level changes.

Of course, in practice, usually different factors have different costs of level changes. The readers are referred to Cheng (1985) for a solution that applies to more general cost functions.

Another important application of the result described in this section is that, in the rule given above, for constructing run orders with a minimum number of level changes, if we replace the word "shortest" with "longest," we obtain a generator sequence that leads to a run order with *maximum* number of level changes among all possible run orders of the given design. The reader may wonder why one wants to maximize the total number of level changes. If the main concern of the experimenter is possible *positive* correlation between adjacent runs, then maximizing the number of level changes is a natural thing to do. This will be discussed in the next section.

6   CORRELATED ERRORS

Construction of run orders for guarding against correlated errors is still under active research. The results are largely unpublished. We shall only sketch some important findings.

A flexible class of models in which the time trend is represented by an ARIMA time series model was proposed by Steinberg (1988). In an unpublished work presented in the 1985 Workshop on Efficient Data Collection held at UCLA, he also carried out a simulation study to identify important design properties that characterize robust designs. One important conclusion is that for *smooth* trends, one needs to change factor levels a large number of times. This and some of Steinberg's other conclusions were theoretically justified by the present author in an unpublished work, wherein simpler but mathematically more tractable models were investigated.

Consider the simple model in which there are no interactions and the correlation structure can be described by a first-order autoregressive (AR(1)) process with positive lag 1 correlation ρ. We shall assume that ρ is known. In practice, if ρ is unknown, it needs to be estimated from the data. For a given run order, one can compute the information matrix $\mathbf{C}$ of the generalized least-squares estimators of the main effects. Then the objective is to find a run order that minimizes a certain functional Φ of the information matrix $\mathbf{C}$ among all possible run orders. For instance, $\Phi(\mathbf{C}) = \det \mathbf{C}^{-1}$ and $\Phi(\mathbf{C}) = \operatorname{tr} \mathbf{C}^{-1}$ are the D and A criteria, respectively.

A D- (or A)-optimal run order minimizes the determinant (or trace) of the dispersion matrix of the generalized least-squares estimators of the main effects. According to a result by Kiefer (1975), if we can find a run order such that $\mathbf{C}$ is a multiple of the identity matrix and it also maximizes tr $\mathbf{C}$ among all possible run orders, then it is optimal with respect to a large class of optimality criteria including the A and D criteria. Such run orders actually do not exist. But Kiefer's result has two important applications here. First, the ideally optimal values can be used to obtain lower bounds on the efficiencies of a given run order under various criteria; so it is possible to judge the performance of an arbitrary run order. Second, it provides a useful guideline for constructing highly efficient, if not optimal, run orders. If we can find a run order such that $\mathbf{C}$ is close to a multiple of the identity matrix and it also maximizes the trace of $\mathbf{C}$, then it is expected to be close to an optimal run order. It turns out that run orders maximizing the trace of $\mathbf{C}$ can be found among those that *maximize* the total number of level changes. In particular, *if we use the foldover method as described in Section 5 to construct a run order with maximum number of level changes, then it always maximizes the trace of* $\mathbf{C}$. The requirement that $\mathbf{C}$ is close to a multiple of the identity matrix leads to some neighbor balance properties, which can be described in terms of the word pattern of the generator sequence. For instance, it can be shown that a run order obtained by applying the *reverse* foldover method to a generator sequence satisfying the following condition has an information matrix that is close to a *diagonal* matrix.

For any $1 \leq i \neq j \leq n$, after the last generator in which exactly one of a_i and a_j appears, either both a_i and a_j appear in all the subsequent generators or neither appears in any of them. $\qquad$ (12)

The property that the diagonal elements are close to each other amounts to requiring that the various factors change levels about the same number of times. This can also be described by the word pattern of the generator sequence, but it seems to be a less important property than maximizing the total number of level changes and (12).

For example, consider the run order of the complete 2^5 design obtained by applying the reverse foldover method to the generator sequence $abcde$, $abcd$, $abce$, $abde$, $acde$. It obviously maximizes the total number of level changes because, at each stage, we use a longest possible word. Property (12) can easily be checked. The following gives lower bounds on the A and

D efficiencies of the resulting run order for several values of ρ, where the A efficiency (e_A) and D efficiency (e_D) are defined by

$$e_A(\mathbf{C}) = \operatorname{tr}(\mathbf{C}_A)/\operatorname{tr}(\mathbf{C}), \qquad e_D(\mathbf{C}) = [\det(\mathbf{C})/\det(\mathbf{C}_D)]^{1/n}$$

with $\mathbf{C}_A$ and $\mathbf{C}_D$ being A- and D-optimal information matrices, respectively.

ρ	0.1	0.2	0.3	0.4	0.5	0.7	0.9
e_A	0.9990	0.9968	0.9944	0.9922	0.9903	0.9875	0.9858
e_D	0.9995	0.9984	0.9973	0.9962	0.9953	0.9939	0.9931

Note that the presented values are only *lower bounds*, not actual efficiencies. Thus, this is a highly efficient run order. One also expects it to have high efficiencies under other criteria.

If the correlation ρ is negative, then one needs to minimize the total number of level changes. It would be compatible with the criterion of cost saving.

The same theory carries over to the case where the error structure is a first-order moving average and ordinary least-squares estimators are used to estimate the factorial effects. It is also possible to generalize to more complicated models incorporating linear trends, two-factor interactions, etc. Results in this direction will be reported elsewhere.

7 SUMMARY

The foldover method is a simple method of systematically laying out a run order of a factorial design. It is also very flexible in that the experimenter can choose appropriate generator sequences to achieve different goals. As demonstrated in this paper, the properties that the generator sequences need to satisfy in order to achieve various design objectives, such as trend elimination and cost saving, can be described in terms of the word pattern of the generator sequence in a simple manner. This helps identify, and provides a useful tool for finding, good run orders.

ACKNOWLEDGMENT

This paper was written during a visit to the Institute of Mathematics, University of Augsburg. The research was supported by National Science Foundation Grants No. DMS-8502784 and DMS-8802640.

REFERENCES

Cheng, C. S. (1985). Run orders of factorial designs, *Proceedings of the Berkeley Conference in Honor of Jerzy Neyman and Jack Kiefer*, Vol. II, L. LeCam and R. A. Olshen, (eds.), Wadsworth, pp. 619–633.

Cheng, C. S., and Jacroux, M. (1988). On the construction of trend-free run orders of two-level factorial designs, *J. Am. Stat. Assoc.*, 83: 1152–1158.

Coster, D. C., and Cheng, C. S. (1988). Minimum cost trend-free run orders of fractional factorial designs, *Ann. Stat.*, 16: 1188–1205.

Daniel, C. (1976). *Applications of Statistics to Industrial Experimentation*, Wiley, New York.

Daniel, C., and Wilcoxon, F. (1966). Fractional 2^{p-q} plans robust against linear and quadratic trends, *Technometrics*, 8: 259–278.

Dickinson, A. W. (1974). Some run orders requiring a minimum number of factor level changes for the 2^4 and 2^5 main effect plans, *Technometrics*, 16: 31–37.

Draper, N. R., and Stoneman, D. M. (1968). Factor changes and linear trends in eight-run two-level factorial designs, *Technometrics*, 10: 301–311.

Hill, H. M. (1960). Experimental designs to adjust for time trends, *Technometrics*, 2: 67–82.

John, P. W. M. (1971). *Statistical Design and Analysis of Experiments*, Macmillan, New York.

John, P. W. M. (1986). Time trends and screening experiments, unpublished manuscript.

Joiner, B. L. and Campbell, C. (1976). Designing experiments when run order is important, *Technometrics*, 18: 249–259.

Kiefer, J. (1975). Optimality and construction of generalized Youden designs, *A Survey of Statistical Designs and Linear Models*, J. N. Srivastava, ed., North–Holland, Amsterdam, pp. 333–353.

National Bureau of Standards (1957). *Fractional Factorial Experiment Design for Factors at Two Levels*, Applied Mathematics Series 48.

National Bureau of Standards (1959). *Fractional Factorial Experiment Design for Factors at Three Levels*, Applied Mathematics Series 54.

Steinberg, D. M. (1988). Factorial experiments with time trends, *Technometrics*, 30: 259–269.

15

Methods for Constructing Trend-Resistant Run Orders of 2-Level Factorial Experiments

Mike Jacroux Washington State University, Pullman, Washington

1 INTRODUCTION

In this paper we consider experimental situations in which a factorial experiment having all factors at two levels is to be conducted over time or space and where there may be unknown or uncontrollable variables influencing the experimental process that are highly correlated with the order in which the observations are obtained. For example, if a batch of material is created at the beginning of an experiment and treatments are to be applied to experimental units formed from the material over time, then there could be an unknown effect due to aging of the material which influences the observations obtained. Other variables that can often affect observations obtained in some specific order are equipment wear-out, learning, fatigue, etc. In such situations, the usual recommendation made to experimenters is to randomize the order in which treatments are applied to experimental units so as to "average" out the effects of the unknown trends. However, it may be that randomization will lead to a run order that is undesirable. Daniel and Wilcoxon (1966) studied the adverse results of linear and quadratic trends on the estimates of main effects arising from various orderings in 2-level factorial and fractional factorial designs. Based upon their findings, Daniel and Wilcoxon (1966) pointed out the advisability of choosing particular run orders of 2-level factorial designs to avoid the adverse effects of time or space trends.

Cox (1951) was the first to study the construction of systematic designs for the efficient estimation of treatment effects in the presence of a smooth polynomial trend. However, his investigation was carried out in the context of variety trials. Since then, a number of other authors including Daniel and Wilcoxon (1966) have studied the problem of constructing run orders of 2^n designs, not only with respect to trend resistance, but also with respect to other criteria. For example, another important consideration when conducting an experiment is cost. One important source of cost in sequential factorial experiments is changing the levels of the factors being studied between successive runs. Assuming that all level changes are equally expensive, this source of cost is minimized by minimizing the total number of level changes. Draper and Stoneman (1968) and Dickinson (1974) considered simultaneously both polynomial trend elimination and cost reduction through the minimization of the number of level changes. Joiner and Campbell (1976) took an approach in which each factor changes levels from one run to the next with a given probability. More expensive factors were assigned smaller probabilities of changing levels. Cheng (1985) gave a theoretical description of the cost structure in 2-level factorial designs and provided some examples of runs optimal with respect to the elimination of trend and reduction of cost. Coster and Cheng (1987) generalized many of Cheng's results to the case of arbitrary factorial designs having n factors occurring at s levels each, where s is a prime or prime power.

It is the purpose of this paper to describe some methods for constructing run orders of complete and fractional factorial designs having all factors at two levels that yield estimates for main effects and in certain cases 2-factor interactions that have a high degree of trend resistance against polynomial trend effects. Most of the construction methods described here come from corresponding results given in Cheng and Jacroux (1987) and Jacroux and Saha Ray (1988). In Section 2, we give a summary of the notation and terminology that is used throughout the paper. In Section 3, some methods are given for constructing trend-resistant complete 2^n designs. Some comments are also made with regard to constructing minimum cost trend-resistant run orders of complete 2^n designs. In Section 4, some methods are given for constructing trend-resistant sequential 2^{n-p} fractional factorial designs and, finally, in Section 5, we consider the construction of run orders of complete 2^n designs that must be blocked in 2^p blocks of size 2^{n-p} and where there may be polynomial trends within the various blocks used.

2 PRELIMINARY NOTATION AND DEFINITIONS

Throughout this paper we shall be considering factorial experiments involving n factors and where each of the n factors being studied occurs at two levels. We shall refer to one of the two levels of each factor as the high level and the other as the low level. With the factor i, we shall associate the letter A_i and call A_i the main effect of factor i. The product $A_{i_1} \cdots A_{i_t}$ of main effect letters shall be called the t-factor interaction for factors i_1, $i_2, \ldots, i_t$. An experiment in which each of the 2^n possible combinations of factor levels occurs once as a treatment shall be called a complete 2^n design. To denote the various treatment combinations that can occur in a complete 2^n design, we shall use strings of lowercase letters. In particular, with factor i, we also associate the letter a_i. A specific treatment combination is then represented as the product of some subset of letters out of $a_1, \ldots, a_n$, with the presence of a_i in the product indicating that factor i occurs at its high level and the absence of a_i indicating that factor i occurs at its low level. We use (1) to denote that treatment having all factors occurring at their low levels. If we define $A_i^2 = A_i^0 = 1$ ($a_i^2 = a_i^0 = 1$), the multiplicative identity, then the set of all 2^n products that can be formed by taking products of letters out of $A_1, \ldots, A_n$ ($a_1, \ldots, a_n$) form an abelian group that we denote by A^n ($\mathcal{A}^n$).

We shall also consider fractions of a complete 2^n design. In this paper, any design based on n factors which has as its treatments some subset of the 2^n possible treatments occurring once is called a fractional factorial design. We shall denote some fraction of a 2^n design by $d(n, m)$ where $n = $ the number of factors and $m = $ the number of treatments in $d(n, m)$. Any application of a treatment in $d(n, m)$ to an experimental unit is called a run and any ordered application of the treatments in $d(n, m)$ to experimental units is called a run order.

Under a given design $d(n, m)$, let $\mathbf{y} = (y_1, \ldots, y_m)'$ denote the vector of ordered observations obtained after applying the treatments in $d(n, m)$ to experimental units. The model assumed here for analyzing the data obtained in $d(n, m)$ is the standard linear model

$$\mathbf{y} = \mathbf{X}\boldsymbol{\beta} + \boldsymbol{\epsilon} = \mathbf{X}_1\boldsymbol{\beta}_1 + \mathbf{X}_2\boldsymbol{\beta}_2 + \boldsymbol{\epsilon}. \tag{1}$$

where $\boldsymbol{\epsilon}$ is an $m \times 1$ vector of independent error terms having expectation zero and constant variance σ^2. The parameters in $\boldsymbol{\beta}_1$ correspond to the usual factorial main effects and interactions and the parameters in $\boldsymbol{\beta}_2$ correspond to possible blocking effects and possible smooth unknown polynomial trend effects. Let $X = (\mathbf{x}_0, \mathbf{x}_1, \ldots, \mathbf{x}_t) = (x_{ij})$. Then we shall assume that the first column of X corresponds to a general mean, the next n columns to main effects A_i, the next $\binom{n}{2}$ columns to 2-factor interactions $A_i A_j$, $i \neq j$, etc. Notice that the sth entry of column i corresponding to main effect A_i is 1 or -1, depending on whether factor i occurs at its high or low level on the sth run of $d(n, m)$. The column of X_1 corresponding to interaction $A_{i_1} \cdots A_{i_t}$ is obtained by taking the componentwise product of columns $\mathbf{x}_{i_1}, \ldots, \mathbf{x}_{i_t}$; i.e., the sth entry of the column of X_1 corresponding to $A_{i_1} \cdots A_{i_t}$ is given by

$$\prod_{j=1}^{t} x_{s, i_j} \qquad \text{for } s = 1, \ldots, m. \tag{2}$$

If blocking is used in $d(n, m)$, then we shall assume throughout the sequel that the first k rows of X correspond to the ordered set of runs from the first block, the next k rows to the ordered set of runs from the second block, etc. There is one column in X_2 for each block and for any block column $\mathbf{x}_j$, $x_{ij} = 1$ if run i is in block j; otherwise, $x_{ij} = 0$. The remaining columns of X_2 correspond to some space or time trend. If blocking is used in $d(n, m)$, we shall assume the trend within the ith block can be approximated by

$$\text{trend effect} = \alpha_{i0} + \alpha_{i1}x + \alpha_{i2}x^2 + \cdots + \alpha_{it}x^t \tag{3}$$

and the values that x assumes correspond to the equally spaced positions at which observations are obtained within the ith block. If no blocking is used in $d(n, m)$, then we shall still assume a polynomial trend effect of the form given in (3) over all m-ordered observations in $\mathbf{y}$.

Throughout this paper we shall assume that all third and higher order interactions in model (1) are negligible. Within this context, we shall also restrict our attention to orthogonal designs. A design $d(n, m)$ is said to be orthogonal if the columns of X_1 corresponding to the factorial parameters of interest are orthogonal. In an orthogonal design, the columns of X_1 corresponding to main effects and interactions are called contrasts. If $d(n, m)$ is an orthogonal design, then in order for the least squares estimates of the factorial effects of interest to be free of bias that might be introduced from the unknown trend effects in β_2 of model (1), the columns of X_1 (excluding $\mathbf{x}_0$) must be orthogonal to the columns of X_2.

Definition 2.1   Let $d(n, m)$ be an orthogonal design in which treatments are applied to experimental units arranged in b blocks of size k and let $\mathbf{x}_j$ be the column of X_1 in model (1) corresponding to main effect A_j. Let $\mathbf{T}_x = (1^x, \ldots, k^x, 1^x, \ldots, k^x, \ldots, 1^x, \ldots, k^x)'$ be an $m \times 1$ vector. We say that $\mathbf{x}_j$ or A_j is t-trend free or t-trend resistant *across* blocks if

$$\mathbf{x}_j' \mathbf{T}_x = 0 \qquad \text{for } x = 0, 1, \ldots, t. \tag{4}$$

We say that $\mathbf{x}_j$ or A_j is t-trend free or t-trend resistant *within* blocks if

$$\sum_{i=1}^{k} x_{yk+i,j} i^z = 0 \qquad \text{for } y = 0, 1, \ldots, b-1 \text{ and } z = 1, \ldots, t. \tag{5}$$

If $\mathbf{x}_j$ is the column of X_1 in model (1) corresponding to interaction term $A_{i_1} \cdots A_{i_s}$, we say $\mathbf{x}_j$ or $A_{i_1} \cdots A_{i_s}$ is t-trend free or t-trend resistant *across* blocks if $\mathbf{x}_j$ satisfies (4) and we say $\mathbf{x}_j$ or $A_{i_1} \cdots A_{i_s}$ is t-trend free or t-trend resistant within blocks if $\mathbf{x}_j$ satisfies (5).

In Definition 2.1, if no blocking is used in a given design $d(n, m)$, then we say that the column $\mathbf{x}_j$ of X_1 in model (1) corresponding to main effect A_j or interaction $A_{i_1} \cdots A_{i_s}$ is t-trend free or t-trend resistant if $\mathbf{x}_j$ satisfies (4).

In the remaining sections of this paper, we describe some methods for constructing run orders of complete and fractional 2^n designs that yield estimates for main effects and in certain cases two-factor interactions that are not biased by the effects of unknown trends of the form given in (3).

3   RUN ORDERS IN COMPLETE 2^n DESIGNS

In this section we describe a method for constructing run orders of complete 2^n designs that yield estimates for main effects and in certain cases 2-factor interactions that are trend resistant. We will henceforth use "o" to denote the componentwise multiplication between two vectors $\mathbf{x}$ and $\mathbf{y}$, i.e., if $\mathbf{x} = (x_1, \ldots, x_m)'$ and $\mathbf{y} = (y_1, \ldots, y_m)'$, then $\mathbf{x} \circ \mathbf{y}$ is the vector given by $(x_1 y_1, \ldots, x_m y_m)'$. We also use $\prod_{j \in A} \mathbf{x}_j$ to denote the componentwise product between a set of vectors having subscripts in the set A. In a complete 2^n design, we note that if we let $m = 2^n$, then the columns of X_1 are mutually orthogonal, and if we let $\mathbf{1}_m$ denote the $m \times 1$ vector of 1's, then the columns of X_1 form an abelian group under the operation of componentwise multiplication with $\mathbf{1}_m$ serving as the group identity.

To begin our discussion, we note that there are a great many run orders for which none of the contrasts for main effects or inter- actions are even 1-trend free. For example, the two run orders of a complete 2^3 design given by $(1, a_1, a_2, a_3, a_2 a_3, a_1 a_2, a_1 a_3, a_1 a_2 a_3)$ and $(1, a_1, a_1 a_2, a_1 a_2 a_3, a_3, a_1 a_3, a_2 a_3, a_2)$ yield no contrasts for main effects or interactions that are 1-trend free. Daniel and Wilcoxon (1966) give numerical examples that demonstrate the adverse effects that unknown linear and quadratic trends can have on least squares estimates for main effects when non-trend–resistant run orders are applied over time or space to experimental units. In particular, the usual estimates for main effects contain a substantial amount of bias. For example, if or- dered observations are taken in a complete 2^3 design using the run order $((1), a_1, a_2, a_1 a_2, a_3, a_1 a_3, a_2 a_3, a_1 a_2 a_3)$, then the main effect contrast for A_3 is $y_1 + y_2 + y_3 + y_4 - y_5 - y_6 - y_7 - y_8$. Clearly, if an unknown lin- ear or quadratic run order trend is present, then the main effect estimate for A_3 will be biased due to the unknown trend. Thus if certain run or- ders of a 2^n design do not provide any protection against unknown trend effects, then we might ask just how much trend resistance is possible to attain in a complete 2^n design. This last question has been partially an- swered by the following lemma that is given in Jacroux and Saha Ray (1988).

Lemma 3.1

1. In a complete 2^n design the maximal degree of trend resistance that a contrast can have is $n - 1$.

2. In a run order of a complete 2^n design, the maximal number of contrasts that can be p-trend free is

$$2^n - 1 - \sum_{x=1}^{p} \binom{n}{x} \qquad \text{for } p = 1, \ldots, n - 1.$$

With Lemma 3.1 in mind, we now describe a method for constructing trend-free run orders of complete 2^n designs. The basic idea behind the construction process is introduced in Daniel and Wilcoxon (1966) and then more fully developed in Cheng and Jacroux (1987). The basic idea, as quoted from Daniel and Wilcoxon (1966), is that "certain of the ordered contrasts appearing in the 2^n system are orthogonal to linear and to quadratic trends.... The design problem is, then, to choose those sets of ordered contrasts that provide efficient estimation of all desired effects." To more fully describe this idea, we recall what it means for the runs in a complete 2^n design to be written down in standard order. In particular, the standard ordering is given by

$$(1), a_1, a_2, a_1 a_2, a_3, a_1 a_3, a_2 a_3, a_1 a_2 a_3 \ldots, a_n, a_1 a_n, \ldots, a_1 a_2 \cdots a_n.$$

Now let s_1, s_2, $\ldots$, s_n denote the main effect contrasts that would be derived from this standard ordering, i.e.,

$$s_1 = (-1, 1, -1, 1, \ldots, -1, 1)',$$
$$s_2 = (-1, -1, 1, 1, -1, -1, 1, 1, \ldots, -1, -1, 1, 1,)', \ldots,$$
$$s_n = (-1, -1, -1, \ldots, -1, 1, 1, 1, \ldots, 1)'.$$

We note that $s_1, \ldots, s_n$ are the same as the n columns of X_1 corresponding to main effects $A_1, \ldots, A_n$ that would be derived under model (1) from the standard ordering. Cheng and Jacroux (1987) call $s_1, \ldots, s_n$ the standard main effect contrasts and denote the group generated by $s_1, \ldots, s_n$ under the operation of componentwise multiplication by S^n. It is interesting to note that none of these standard main effect contrasts are even 1-trend free. However, Cheng and Jacroux (1987) do give the following lemma.

Lemma 3.2   Let $s_1, \ldots, s_n$ be the standard main effect contrasts as defined above. Then $s_{j_1} \circ \cdots \circ s_{j_p}$ is $(p-1)$-trend free for any subscripts $1 \leq j_1 < \cdots < j_p \leq n$.

From this last lemma, we see that from the standard ordering of runs in a complete 2^n design we can derive $\binom{n}{x}$ contrasts that are $(x-1)$-trend free for $x = 1, \ldots, n$. Thus the standard ordering of runs in a complete 2^n design produces contrasts that satisfy the bounds given in Lemma 3.1. Of course, having none of the main effects 1-trend free, as is the case in the standard ordering of runs, is not desirable since they are usually the factorial effects of primary interest. However, Lemma 3.2 does provide the basis for our construction process. Before actually giving the construction process, we note that in a complete 2^n design, once the main effect contrasts have been determined, all other interaction contrasts are obtained by componentwise multiplication of the appropriate main effect contrasts. Thus in order for a set of n contrasts out of S^n to serve as main effect contrasts in a complete 2^n design, they must generate S^n; i.e., all possible contrasts in S^n must be obtainable by taking appropriate products between contrasts out of the set. With this in mind, we now give a method for constructing a run order of a complete 2^n design that yields contrasts for main effects that are p_1-trend free and contrasts for 2-factor interactions that are p_2-trend free:

1. Find a set of n contrasts out of S^n which generate S^n and which involve products containing at least $p_1 + 1$ elements out of s_1, ..., s_n and whose pairwise products contain at least $p_2 + 1$ elements out of s_1, ..., s_n.
2. With each contrast obtained in step 1, identify exactly one of the main effects $A_1, A_2, ..., A_n$.   (6)
3. Write down the run order of the 2^n complete design indicated by the assignment of main effects to contrasts given in step 2; i.e., on run t, factor j occurs at its high or low level, depending upon whether the tth entry of the contrast identified with A_j is 1 or -1.

Example 3.1

Suppose we wish to obtain a run order of a 2^3 design where the main effects are at least 1-trend free. If we assign A_1, A_2 and A_3 to $s_2 \circ s_3$, $s_1 \circ s_3$ and $s_1 \circ s_2 \circ s_3$, respectively, we obtain the run order $(a_1 a_2, a_1 a_3, a_2 a_3, (1), a_3, a_2, a_1, a_1 a_2 a_3)$. To obtain the interaction contrasts, we take the appropriate products between the contrasts assigned to A_1, A_2 and A_3. This run order yields estimates for A_1, A_2 and A_3 that are 1-, 1- and 2-trend free, respectively. The estimate for the $A_1 A_2$ interaction is 1-trend free since its corresponding contrast is $(s_2 \circ s_3) \circ (s_1 \circ s_3) = s_1 \circ s_2$. However, the estimates for the interactions $A_1 A_3$, $A_2 A_3$ and $A_1 A_2 A_3$ are not even 1-trend free since their corresponding contrasts are s_1, s_2 and s_3, respectively.

The implementation of the technique described in (6) for finding run orders for arbitrary values of p_1 and p_2 is in general not easy. The main problem comes in finding a set of product contrasts that generate S^n and that satisfy the conditions of (6). However, if one is only interested in finding a run order that gives trend-resistant contrasts for main effects, then the task is easier. From Lemma 3.1, it follows that all main effects cannot be $(n-1)$-trend free in a complete 2^n design. However, using Lemma 3.2, we can construct a run order having all main effect contrasts at least $(n-2)$ trend free as follows:

1. Let $s_1, \ldots, s_n$ be the standard main effect contrasts from a complete 2^n design. For $i = 1, \ldots, n-1$, assign A_i to $\prod_{\substack{j=1 \\ j \neq i}}^{n} s_j$ and assign A_n to $\prod_{j=1}^{n} s_j$. (7)
2. Write down the run order corresponding to the assignment of main effects to contrasts given in step 1.

Example 3.1 provides an illustration of the construction technique described in (7) for the case of a 2^3 design. While the method of construction given in (7) provides a run order of a 2^n design that is at least $(n-2)$-trend free for main effects, some of the 2-factor interaction contrasts will not even be 1-trend free. This is also illustrated in Example 3.1. However, it is possible to construct 2^n designs for even (odd) values of $n \geq 4$ in which all the main effect contrasts are $(n-2)$-trend free $[(n-3)$-trend free] and all 2-factor interaction contrasts are at least 1-trend free. Such a construction process is given in Cheng and Jacroux (1987) and proceeds as follows:

1. Let $s_1, \ldots, s_n$ be the standard main effect contrasts in a complete 2^n design.

   a. If n is even, then for $i = 1, \ldots, n$, assign A_i to the contrast corresponding to $\prod_{\substack{j=1 \\ j \neq i}}^{n} s_j$. (8)

   b. If n is odd, then for $i = 1, \ldots, n-1$, assign A_i to the contrast corresponding to $\prod_{\substack{j=1 \\ j \neq i}}^{n-1} s_j$ and assign A_n to $\prod_{j=1}^{n} s_j$.

2. Write down the run order corresponding to the assignment of main effects to contrasts given in step 1 above.

Example 3.2

Assume $n = 4$ and let s_1, s_2, s_3 and s_4 be the standard main effect contrasts. Upon assigning A_1, A_2, A_3 and A_4 to $s_2 \circ s_3 \circ s_4$, $s_1 \circ s_3 \circ s_4$, $s_1 \circ s_2 \circ s_4$ and $s_1 \circ s_2 \circ s_3$, respectively, we obtain the following run order:

$((1), a_2 a_3 a_4, a_1 a_3 a_4, a_1 a_2, a_1 a_2 a_4, a_1 a_3, a_2 a_3,$

$a_4, a_1 a_2 a_3, a_1 a_4, a_2 a_4, a_3, a_3 a_4, a_2, a_1, a_1 a_2 a_3 a_4).$

In this run order, all main effects are 2-trend free and all 2-factor interactions are 1-trend free since the contrasts corresponding to A_1A_2, A_1A_3, A_1A_4, A_2A_3, A_2A_4 and A_3A_4 are $s_1 \circ s_2$, $s_1 \circ s_3$, $s_1 \circ s_4$, $s_2 \circ s_3$, $s_2 \circ s_4$ and $s_3 \circ s_4$, respectively.

As mentioned previously, the problem of finding a run order of a complete 2^n design that is at least p_1-trend free for main effects and at least p_2-trend free for 2-factor interactions is usually a nontrivial problem when considered for arbitrary values of p_1 and p_2. In general, the higher the degree of trend resistance that a run order provides for main effects, the lower the degree of trend resistance the run order provides for 2-factor interactions. For further information on the construction of run orders of complete 2^n designs, the reader is referred to Cheng and Jacroux (1987).

Another important factor in any experiment is cost. In most experiments, there is usually some flexibility in the actual construction of an appropriate run order. When such flexibility is available, one should always use the least expensive run order that achieves the desired degree of trend resistance for main effects and 2-factor interactions. There are primarily two sources of cost. The first is that of measuring the observations and is usually to some extent fixed. The second source of cost comes from making level changes between successive runs and can to some degree be controlled as indicated by the following lemma.

Lemma 3.3 Suppose $s_1, \ldots, s_n$ are the standard main effect contrasts from a complete 2^n design and suppose $\mathbf{A}_i$ is assigned to contrast $s_{i_1} \circ s_{i_2} \circ \cdots \circ s_{i_t}$ where $i_1 < i_2 < \cdots < i_t$ in a given run order. Then the number of level changes of factor i in the given run order is

$$(2^{n-i_1+1} - 1) - (2^{n-i_2+1} - 1) + (2^{n-i_3+1} - 1) + \cdots + (-1)^{t-1}(2^{n-i_t+1} - 1).$$

In view of Lemma 3.3, we make several observations. In general, the higher the degree of trend resistance that a main effect contrast provides, the larger is the number of level changes required for that factor (though not in all cases). Once the desired degree of trend resistance is determined for the factors being studied, there may be some factors whose level changes are more expensive than others. Thus, using Lemma 3.3, one should assign those factors whose levels are the most expensive to change to those contrasts that provide the required degree of trend resistance but also require the fewest level changes. For example, if the levels of factor 1 are the most expensive to change and a run order is desired that provides a

p_1-trend-free estimate for A_1, then A_1 should be assigned to the contrast among the available $\binom{n}{p_1+1}$ p_1-trend-free contrasts available that requires the fewest level changes. In the event that all level changes are equally expensive, then one should find a run order that provides the desired degree of trend resistance for main effects and minimizes the total number of level changes. For further discussion of this latter problem, the reader is referred to Coster and Cheng (1986).

4   FRACTIONAL FACTORIALS

One of the problems with running complete 2^n designs is that as n increases, the number of treatments and experimental units required to conduct the experiment increases rapidly. One solution to this problem when n is large is to use a fractional factorial design, i.e., take some subset of the runs that make up a complete 2^n design, apply them to experimental units, and then use the observations obtained to elicit information about the main effects and 2-factor interactions of interest. For the remainder of this section, we shall restrict our attention to the construction of fractions of 2^n designs where the number of experimental units available is 2^{n-p}. We shall call such fractions 2^{n-p} fractional factorial designs. One approach to the problem of constructing a run order of a 2^{n-p} fractional factorial design when the main objective is to obtain estimates for main effects that are p_1-trend free when $n \le \sum_{x=p_1+1}^{n-p} \binom{n-p}{x}$ is the following:

1. Let $\mathbf{s}_1, \ldots, \mathbf{s}_{n-p}$ denote the standard main effect contrasts of a complete 2^{n-p} design. Assign factors A_i, $i = 1, \ldots, n-p-1$ to the contrasts corresponding to $\prod_{\substack{j=1 \\ j \ne i}}^{n-p} \mathbf{s}_j$ and assign A_{n-p} to $\prod_{j=1}^{n-p} \mathbf{s}_j$. $\hspace{3cm}$ (9)
2. Assign factors $A_{n-p+1}, \ldots, A_n$ to any product of the standard main effect contrasts containing at least $p_1 + 1$ elements out of $\mathbf{s}_1, \ldots, \mathbf{s}_{n-p}$ that is not already assigned in step 1.
3. Write down the run order corresponding to the assignment of main effects to contrasts given in steps 1 and 2.

Example 4.1

Consider the construction of a 2^{6-2} fractional factorial design where it is desired to obtain estimates for all main effects that are at least 1-trend free. Following the technique described in (9), we could assign A_1, A_2, A_3,

A_4, A_5 and A_6 to $s_2 \circ s_3 \circ s_4$, $s_1 \circ s_3 \circ s_4$, $s_1 \circ s_2 \circ s_4$, $s_1 \circ s_2 \circ s_3 \circ s_4$, $s_1 \circ s_2 \circ s_3$ and $s_1 \circ s_2$, respectively, and obtain a run order where all main effect contrasts are at least 1-trend free. In fact, A_1, A_2, A_3 and A_5 are 2-trend free, A_4 is 3-trend free, and A_6 is 1-trend free.

Following the procedure outlined in (9), there are certain limitations. For instance, for a given value of p_1, we see from Lemma 3.1 that there are at most $\sum_{x=p_1+1}^{n-p} \binom{n-p}{x}$ contrasts in a complete 2^{n-p} design that are at least p_1-trend free. Thus if it is desired to obtain estimates for all main effects that are at least p_1-trend free, the number of factors n can be no larger than $\sum_{x=p_1+1}^{n-p} \binom{n-p}{x}$. Another and potentially more serious problem that can occur with the method given in (9) is that some main effect might be confounded with some low order interaction; i.e., the contrast estimates for the main effect and the interaction might be identical. When this happens, we say that the main effect and the interaction are confounded and are aliases of one another. In Example 4.1, the contrasts for A_6 and $A_1 A_2$ are confounded and are aliases of one another. While the confounding of main effects with 2-factor interactions is not a major problem if all 2-factor interactions are negligible, such confounding can be troublesome when trying to interpret main effect estimates in the presence of nonnegligible 2-factor interaction effects. With this in mind, we now consider the construction of run orders of "regular" or "traditional" 2^{n-p} fractional factorial designs that are obtained as in the following definition.

Definition 4.1   Choose a set of p interactions X_1, ..., X_p from A^n that are independent in the sense that no X_i is the product of any combination of other interactions in the set. Let $d(n, 2^{n-p})$ be the set of all 2^{n-p} runs from a complete 2^n design that have an even number of letters in common with each X_i, $i = 1, \ldots, p$. The runs in $d(n, 2^{n-p})$ are called a regular 2^{n-p} fractional factorial design.

The set of all 2^p products that can be generated from X_1, ..., X_p of Definition 4.1 are called the defining contrasts of $d(n, 2^{n-p})$. For any main effect or interaction Y that is not a defining contrast, the set of elements in A^n obtained by multiplying each defining contrast by Y is the aliasing set of Y and gives those elements in A^n that are confounded with Y. We now discuss the construction of run orders of regular 2^{n-p} fractional factorial designs.

Theorem 4.1   If $n - p \geq 5$, then there exists a run order of a regular 2^{n-p} fractional factorial design that yields estimates for all main effects that are $n-p-3$ trend free and has all main effects aliased with interactions of order 3 or higher, provided $2p \leq n - 1$.

Proof The proof of the Theorem is by construction. Let $s_1, \ldots, s_{n-p}$ be the standard main effect contrasts for a complete 2^{n-p} design. Now proceed as follows:

1. For $i = 1, \ldots, n-p-1$, assign A_i to the contrast corresponding to $\prod_{\substack{j=1 \\ j \neq i}}^{n-p} s_j$ and assign A_{n-p} to $\prod_{j=1}^{n-p} s_j$.
2. For $i = n-p+1, \ldots, n$, assign A_i to the contrast corresponding to $\prod_{\substack{j=1 \\ j \neq i-n+p}}^{n-p-1} s_j$. $\qquad(10)$
3. Write down the run order corresponding to the assignment of main effects to contrasts given in steps 1 and 2.

From Lemma 3.1, it follows that all the main effects are at least $(n - p - 3)$-trend free. Also, we see that if $n - p$ is even in the above assignment of main effects to contrasts, for $i = n - p + 1, \ldots, n$, A_i is aliased with $\prod_{\substack{j=1 \\ j \neq i-n+p}}^{n-p-1} A_j$ where as if $n - p$ is odd, for $i = n - p + 1, \ldots, n$, A_i is aliased with $\prod_{\substack{j=1 \\ j \neq i-n+p}}^{n-p-1} A_j \cdot A_{n-p}$. Using the assumption that $n - p \geq 5$ and $2p \leq n - 1$, it is easily verified that all the defining contrasts generated by the aliases associated with the A_i, $i = n-p+1, \ldots, n$, mentioned in the last sentence contain at least four letters; thus, we have the desired result.

Example 4.2

Suppose we wish to obtain a run order for a 2^{7-2} fractional factorial experiment that has all main effects at least 2-trend free and all main effects aliased with third or higher order interactions. Letting s_1, s_2, s_3, s_4 and s_5 denote the standard main effect contrasts from a complete 2^{7-2} design and following the procedure outlined in (10), we make the following assignment of main effects to contrasts:

$$A_1 \longrightarrow s_2 \circ s_3 \circ s_4 \circ s_5 \qquad A_5 \longrightarrow s_1 \circ s_2 \circ s_3 \circ s_4 \circ s_5$$

$$A_2 \longrightarrow s_1 \circ s_3 \circ s_4 \circ s_5 \qquad A_6 \longrightarrow s_2 \circ s_3 \circ s_4$$

$$A_3 \longrightarrow s_1 \circ s_2 \circ s_4 \circ s_5 \qquad A_7 \longrightarrow s_1 \circ s_3 \circ s_4$$

$$A_4 \longrightarrow s_1 \circ s_2 \circ s_3 \circ s_5$$

Given the above assignment of main effects to contrasts, we see that the defining contrasts are $A_2A_3A_4A_5A_6$, $A_1A_3A_4A_5A_7$ and $A_1A_2A_6A_7$. Thus all main effects are at least 2 trend free and aliased with third and higher order interactions.

Theorem 4.1 provides a means for finding run orders of regular 2^{n-p} fractional factorial designs that leave all main effects $(n-p-3)$-trend free and aliased with high order interactions. However, the designs that can be found using this technique leave various 2-factor interactions confounded with one another, and some 2-factor interactions that are not even 1-trend free. For instance, in Example 4.2 the A_1A_2 and A_6A_7 interactions are confounded with one another and all interactions A_iA_5, $i = 1, 2, 3, 4$, are not even 1-trend free. However, by using more elaborate aliasing schemes, it is possible to construct run orders of 2^{n-p} fractional factorials that have all main effects and 2-factor interactions aliased with third and higher order interactions and that give contrasts for main effects and 2-factor interactions that have varying degrees of trend freeness. However, these designs usually require larger numbers of experimental units. For more discussion concerning the construction of such designs, the reader is referred to Cheng and Jacroux (1987).

5   BLOCKING IN COMPLETE 2^n DESIGNS

In experimental settings where a complete 2^n design and blocking are simultaneously to be used, it is often the case that the number of experimental units available within each block is not large enough to accommodate all 2^n runs. Thus an incomplete block design must be used. In such situations, if runs within blocks are applied to experimental units over space or time, it is desirable to order the sequence of runs within blocks so as to protect against unknown trend effects which might be present within blocks. Coster and Cheng (1986) considered the problem of constructing run orders of complete 2^n designs in $b = 2^p$ blocks of size $k = 2^{n-p}$ that are resistant against polynomial trends across blocks. However, their constructions are based on the assumption that the polynomial trend within each block is the same. Here we describe a method for constructing run orders of complete 2^n designs in $b = 2^p$ blocks of size $k = 2^{n-p}$ which provide estimates for main effects that are not only resistant against polynomial trends across blocks when the trends within blocks are all the same but also provide estimates for main effects that are resistant against arbitrary polynomial trends within blocks of degree $n - p - 2$ or less. To describe the construction process, we use the Kronecker product. If $\mathbf{x}' = (x_1, \ldots, x_s)$ and $\mathbf{y}' = (y_1, \ldots, y_t)$, then the Kronecker product between $\mathbf{x}$ and $\mathbf{y}$, denoted by $\mathbf{x} \otimes \mathbf{y}$, is the $st \times 1$ vector given by $(x_1 \mathbf{y}', x_2 \mathbf{y}', \ldots, x_s \mathbf{y}')'$. Using the Kronecker product and assuming the ordering of observations described in model (1) with respect to blocking, we now describe the method obtained in Jacroux and Saha Ray (1988) for constructing a run order of a complete 2^n design in $b = 2^p$ blocks of size $k = 2^{n-p}$, $n - p \geq 3$, such that the main effect contrasts are resistant to arbitrary polynomial trends within blocks of degree $n - p - 2$ or less and are orthogonal to arbitrary polynomial trends across blocks when the polynomial trends within blocks are all the same.

1. Let $\mathbf{s}_1, \ldots, \mathbf{s}_{n-p}$ be the standard main effect contrasts from a complete 2^{n-p} design and let $\bar{s}_1, \ldots, \bar{s}_p$ be the standard main effect contrasts from a complete 2^p design. Now define

$$\mathbf{w}_i = \prod_{\substack{j=1 \\ j \neq i}}^{n-p} \mathbf{s}_j \quad \text{for } i = 1, \ldots, n-p-1 \text{ and } \mathbf{w}_{n-p} = \prod_{j=1}^{n-p} \mathbf{s}_j$$

and

$$\mathbf{z}_i = \prod_{\substack{j=1 \\ j \neq i}}^{p} \bar{s}_j \quad \text{for } i = 1, \ldots, p-1 \text{ and } \mathbf{z}_p = \prod_{j=1}^{p} \bar{s}_j. \tag{11}$$

2. For $i = 1, \ldots, n - p$, assign A_i to the contrast corresponding to $\mathbf{W}_i = \mathbf{z}_p \otimes \mathbf{w}_i$.
3. For $i = n - p + 1, \ldots, n - 1$, assign A_i to the contrast corresponding to $\mathbf{W}_i = \mathbf{z}_{i-n+p} \otimes \mathbf{w}_{n-p}$.
4. Assign A_n to $\left(\prod_{j=1}^{p-1} \bar{s}_j \right) \otimes \mathbf{w}_{n-p}$.
5. Write down the run order corresponding to the assignment of main effects to contrasts given in steps 2–4.

Example 5.1

Suppose we consider the case of blocking a 2^5 design in $b = 2^2$ blocks of size $k = 2^3$ such that all contrasts for main effects are at least 1-trend free within blocks. Following the procedure outlined in (11), we let $\mathbf{w}_1 = \mathbf{s}_2 \circ \mathbf{s}_3$, $\mathbf{w}_2 = \mathbf{s}_1 \circ \mathbf{s}_3$, $\mathbf{w}_3 = \mathbf{s}_1 \circ \mathbf{s}_2 \circ \mathbf{s}_3$, $\mathbf{z}_1 = \bar{s}_2$ and $\mathbf{z}_2 = \bar{s}_1 \circ \bar{s}_2$. Now, upon letting $\mathbf{W}_1 = \mathbf{z}_2 \otimes \mathbf{w}_1$, $\mathbf{W}_2 = \mathbf{z}_2 \otimes \mathbf{w}_2$, $\mathbf{W}_3 = \mathbf{z}_2 \otimes \mathbf{w}_3$, $\mathbf{W}_4 = \mathbf{z}_1 \otimes \mathbf{w}_3$, $\mathbf{W}_5 = \bar{s}_1 \otimes \mathbf{w}_3$, we get the following assignment of runs to blocks after assigning A_1, A_2, A_3, A_4 and A_5 to $\mathbf{W}_1$, $\mathbf{W}_2$, $\mathbf{W}_3$, $\mathbf{W}_4$ and $\mathbf{W}_5$, respectively.

$$B_1 = (a_1a_2a_3a_4, a_1a_3, a_2a_3, a_4a_5, a_3, a_2a_4a_5, a_1a_4a_5, a_1a_2a_3),$$
$$B_2 = (a_3a_4, a_2a_5, a_1a_5, a_1a_2a_3a_4, a_1a_2a_5, a_1a_3a_4, a_2a_3a_4, a_5),$$
$$B_3 = (a_3a_5, a_2a_4, a_1a_4, a_1a_2a_3a_5, a_1a_2a_4, a_1a_3a_5, a_2a_3a_5, a_4),$$
$$B_4 = (a_1a_2, a_1a_3a_4a_5, a_2a_3a_4a_5, (1), a_3a_4a_5, a_2, a_1, a_1a_2a_3a_4a_5).$$

We note that A_1, A_2, A_3, A_4 and A_5 are at least 1-trend free within blocks and that their corresponding contrasts are orthogonal to arbitrary polynomial trends across blocks when the polynomial trends within all blocks are the same.

For further results on the construction of run orders of complete 2^n designs in 2^p blocks of size 2^{n-p} under varying assumptions, the reader is referred to Coster and Cheng (1986) and Jacroux and Saha Ray (1988).

ACKNOWLEDGMENT

This research was supported by NSF Grant DMS-8700945.

REFERENCES

[1] Cheng, C. S. (1985). Run order of factorial designs. *Proceedings of the Berkeley Conference in Honor of Jerzy Neyman and Jack Kiefer*, Vol. II, 619-633. Lucien M. LeCam and R. H. Olshen, eds., Wadsworth, Belmont, California.

[2] Cheng, C. S. and Jacroux, M. (1988). On the construction of trend-free run orders of two-level factorial designs, *JASA*, 83, 1152–1158.

[3] Coster, C. D. and Cheng, C. S. (1988). Minimum cost trend free run orders of fractional factorial designs, *Ann. Stat.*, 16, 1188–1205.

[4] Cox, D. R. (1951). Some systematic experimental designs, *Biometrika*, 38, 312-323.

[5] Daniel, C. and Wilcoxon, F. (1966). Fractional 2^{p-q} plans robust against linear and quadratic trends, *Technometrics*, 8, 259-278.

[6] Dickinson, A. W. (1974). Some run orders requiring a minimum number of factor level changes for 2^4 and 2^5 main effects plans, *Technometrics*, 16, 31-37.

[7] Draper, N. R. and Stoneman, D. M. (1968). Factor changes and linear trends in eight-run two-level factorial designs, *Technometrics*, 10, 301-311.

[8] Hill, H. H. (1960). Experimental designs to adjust for time trends, *Technometrics*, 2, 67-82.

[9] Jacroux, M. and Saha Ray, R. (1988). Run orders of trend resistant 2-level factorial designs, submitted for publication.

[10] Joiner, B. L. and Campbell, C. (1976). Designing experiments when run order is important, *Technometrics*, 18, 249-259.

16

Measuring Dispersion Effects of Factors in Factorial Experiments

Subir Ghosh and Eric S. Lagergren   University of California, Riverside, Riverside, California

1   INTRODUCTION

An important problem in quality control studies is to find an optimum combination of levels of control factors in achieving stability against noise factors (Taguchi and Wu, 1985). Both "location" and "dispersion" effects of factors are pertinent to measure from the data in resolving this problem. This chapter considers the problem of measuring dispersion effects of factors in factorial experiments. The concept of dispersion effects in factorial experiments was considered in the work of G. Taguchi (Taguchi and Wu, 1985) for replicated factorial experiments and in the work of G. E. P. Box (Box and Meyer, 1986) for unreplicated factorial experiments. Factorial experiments may be complete or fractional factorial under completely randomized designs. Although, for clarity, we consider 2^m factorial experiments here, the ideas presented can be generalized easily to any symmetric or asymmetric factorial experiments. Kacker (1985), Phadke et al. (1983), and Nair (1986) made pioneering contributions to this area of research.

Ghosh (1987) used search linear models (Srivastava, 1975) to explain dispersion effects in factor screening experiments. Srivastava (1987) discussed briefly some aspects of dispersion modeling. Nair and Pregibon (1988) assumed a structured log-linear model of dispersions and dispersion effects. Carroll and Ruppert (1988) discussed the methods available in drawing inferences on location and dispersion parameters in regression analysis. We make a further contribution to this area of research.

We first assume that, for the fitted model to the data, there is no significant lack of fit. We then propose three sets of measures of dispersion effects of m factors. The dispersion effect of a factor depends on the dispersion at levels 0 and 1 of the factor. The dispersions at levels 0 and 1 based on the ordinary least-squares residuals are correlated in most situations. We introduce a method of adjusting residuals and then calculate the dispersions at levels 0 and 1 based on these adjusted residuals. The adjusted residuals at a particular level of the factor are uncorrelated with the residuals and the adjusted residuals at the other level of the factor.

Use of the proposed measures of dispersion effects in a factorial experiment will give a combination of levels of control factors that is optimum for reducing the process variability due to noise in the experiment. The reduction of the process variability due to noise factors is an important aspect in quality control studies.

2   DISPERSION EFFECTS

We consider a 2^m factorial experiment under a completely randomized design. Let $T(n \times m)$ be the design. The columns of T denote factors that are controllable at their lower and upper levels. The rows of T denote runs or treatment combinations. Runs are level combinations of control factors that are actually used in the experiment to collect observations or data. The design T is called an inner array for m control factors. The inner array T chosen for the experiment will play an important role in subsequent discussions. An important objective in quality control studies is to evaluate the sensitivity of the manufacturing process to noise. First, a list of noise factors likely to affect the process is made. Various level combinations of noise factors that provide a good representation of noise are then considered. The matrix representation of the level combinations of noise factors is called an outer array (Taguchi and Wu, 1985). Suppose that there are r (> 1) level combinations of noise factors in the experiment.

For every run in the inner array T, we collect r observations corresponding to r level combinations of the outer array. The r observations for a run in T are called r replicated observations. The variability in r replicated observations for a run in T is attributable to process variability due to noise in the experiment. Again, for simplicity, equal replication is considered for the experiment, and the idea can be easily extended to unequal replication.

Let y_{ij} be the jth observation for the ith run, $\bar{y}_i$ the mean of all observations for the run i, $i = 1, \ldots n$ and $j = 1, \ldots, r$, and $N (= nr)$ the total number of observations. The linear model assumed for the experiment is

$$E(\mathbf{y}) = X\boldsymbol{\beta} \tag{1}$$

$$\mathrm{Var}(\mathbf{y}) = \Sigma \tag{2}$$

$$\mathrm{Rank}\, X = p \tag{3}$$

where $\mathbf{y}(N \times 1)$ is the vector of observations and $\mathbf{y} = (y_{11}, \ldots, y_{1r}; \ldots; y_{n1}, \ldots, y_{nr})'$, $\boldsymbol{\beta}(p \times 1)$ is the vector of factorial effects considered in the experiment, $X(N \times p)$ is a known matrix that depends on the inner array T, and the matrix Σ is an unknown diagonal matrix with the diagonal elements for all observations on the ith run equal to σ_i^2, $i = 1, \ldots, n$. Model (1–3) in the special case $\sigma_1^2 = \cdots = \sigma_n^2 = \sigma^2$ is called the standard linear model. We denote $H = X(X'X)^{-1}X'$ and $R = (I - H)$. The vectors $\hat{\mathbf{y}} = H\mathbf{y}$ and $\mathbf{y} - \hat{\mathbf{y}} = R\mathbf{y}$ are the vector of ordinary least-squares fitted values and the vector of residuals, respectively. The fitted values for all observations corresponding to the ith run are identical and are denoted by $\hat{y}_i$, $i = 1, \ldots, n$. Suppose that, for the fitted model to the data, there is no significant lack of fit under the standard linear model. Again, under the standard linear model, the sum of squares of error is $\mathrm{SSE} = \sum_{i=1}^{n} \sum_{j=1}^{r}(y_{ij} - \hat{y}_i)^2$, the mean square of error is $\mathrm{MSE} = \mathrm{SSE}/(N - p)$, the sum of squares of pure error is $\mathrm{SSPE} = \sum_{i=1}^{n} \sum_{j=1}^{r}(y_{ij} - \bar{y}_i)^2$, and the mean square of pure error is $\mathrm{MSPE} = (\mathrm{SSPE}/n(r - 1))$. Note that both MSE and MSPE are measures of error variance σ^2 under the standard linear model. We now take MSE and MSPE as measures of noise under the assumed model (1–3). We then express MSPE as the weighted average of $(\mathrm{MSPE})_1$ and $(\mathrm{MSPE})_0$, where $(\mathrm{MSPE})_u$ is called the contribution of the level u $(u = 0, 1)$ of the factor to MSPE. Formal expressions of $(\mathrm{MSPE})_u$, $u = 0, 1$ are given in the next section. We do the same for MSE. Different levels of a factor may contribute differently to MSE and MSPE. In general, the contributions of levels of a factor to noise (measured by MSPE or MSE) are called the dispersions at levels of the factor. The dispersion effect of a factor is the ratio of the dispersion at level 1 and the dispersion at level 0 of the factor (see Box and Meyer, 1986). If the dispersion effect is greater than 1, then the dispersion at level 1 is more than the dispersion at level 0. We then prefer level 0 over level 1 in terms of smaller dispersion. Similarly, if the dispersion effect is less than 1, we prefer level 1 over level 0. Although we use the definition of dispersion effect presented in Box and Meyer (1986), we may also take the logarithm of the proposed ratio or the difference between dispersions at levels 1 and 0 of the factor. The use of any of these definitions will give the same conclusion since each of them compares the dispersions at levels 1 and 0. The main theme of this chapter is to investigate the possible ways of measuring dispersion effects of all factors. We first present the sample dispersion effects and then the corresponding population dispersion effects under the assumed model (1–3).

3   MEASURING DISPERSION EFFECTS

We take a single factor out of m factors and develop the methods of measuring the dispersion effect of the chosen factor. For simplicity of presentation, we do not introduce any notation for the chosen factor. The chosen factor appears at levels 0 and 1 in the n runs of the inner array T. We now introduce an indicator variable to identify the runs at which the factor appears at levels 0 and 1, respectively. Distinguishing between the level 0 and the level 1 runs for the factor will enable us to measure dispersions at level 0 and level 1 of the factor. We define for $i = 1, \ldots, n$,

$$\delta_i = \begin{cases} 1, & \text{if the level of the factor in the } i\text{th run is 1,} \\ 0, & \text{if the level of the factor in the } i\text{th run is 0,} \end{cases}$$

3.1   First Measure

We have

$$\text{SSPE} = \sum_{i=1}^{n} \sum_{j=1}^{r} \delta_i (y_{ij} - \bar{y}_i)^2 + \sum_{i=1}^{n} \sum_{j=1}^{r} (1 - \delta_i)(y_{ij} - \bar{y}_i)^2$$

Notice that $\sum_{j=1}^{r}(y_{ij} - \bar{y}_i)^2$ is the total corrected sum of squares of r observations for the ith run with the degrees of freedom $(r - 1)$. The first component in SSPE corresponds to level 1 of the factor and has degrees of freedom $(\sum_{i=1}^{n} \delta_i)(r-1)$. The second component corresponds to level 0 of the factor and has degrees of freedom $[\sum_{i=1}^{n}(1-\delta_i)](r-1)$. The set of measures of dispersions at levels 1 and 0 of the factor are

$$
\begin{aligned}
S_1^2(1) &= \frac{\sum_{i=1}^{n} \sum_{j=1}^{r} \delta_i(y_{ij} - \bar{y}_i)^2}{(\sum_{i=1}^{n} \delta_i)(r-1)} \\
S_0^2(1) &= \frac{\sum_{i=1}^{n} \sum_{j=1}^{r}(1 - \delta_i)(y_{ij} - \bar{y}_i)^2}{(\sum_{i=1}^{n}(1 - \delta_i))(r-1)}
\end{aligned} \tag{4}
$$

respectively. The first measure of the dispersion effect of the chosen factor is therefore $S_1^2(1)/S_0^2(1)$.

3.2 Second Measure

$$
\text{SSE} = \sum_{i=1}^{n} \sum_{j=1}^{r} \delta_i(y_{ij} - \hat{y}_i)^2 + \sum_{i=1}^{n} \sum_{j=1}^{r}(1 - \delta_i)(y_{ij} - \hat{y}_i)^2
$$

Note that $(y_{ij} - \hat{y}_i)$ is the residual for the jth observation on the ith run and $\sum_{j=1}^{r}(y_{ij} - \hat{y}_i)^2$ is the residual sum of squares for r observations on the ith run. The first component in SSE corresponds to level 1 of the factor and has V_1 degrees of freedom (df). The second component corresponds to level 0 of the factor and has V_0 df. The set of measures of dispersions at levels 1 and 0 of the factor are

$$
\begin{aligned}
S_1^2(2) &= \frac{\sum_{i=1}^{n} \sum_{j=1}^{r} \delta_i(y_{ij} - \hat{y}_i)^2}{V_1} \\
S_0^2(2) &= \frac{\sum_{i=1}^{n} \sum_{j=1}^{r}(1 - \delta_i)(y_{ij} - \hat{y}_i)^2}{V_0}
\end{aligned} \tag{5}
$$

respectively. The second measure of the dispersion effect of the chosen factor is $S_1^2(2)/S_0^2(2)$.

We now denote $\mathbf{y}_u$ as the vector of observations corresponding to the runs with the chosen factor at level u, $u = 0, 1$. Note that $\mathbf{y}$ consists of $\mathbf{y}_1$ and $\mathbf{y}_0$. Let X_u be the submatrix of X corresponding to $\mathbf{y}_u$, and let $\hat{\mathbf{y}}_u$ be the ordinary least-squares fitted values for $\mathbf{y}_u$. The elements $\hat{\mathbf{y}}_u$ and $\mathbf{y}_u - \hat{\mathbf{y}}_u$, $u = 0, 1$, are linear functions of the elements in $\mathbf{y}$. It can be seen that $\hat{\mathbf{y}}_u = X_u(X'X)^{-1}X'\mathbf{y}$, $u = 0, 1$. We denote $\mathbf{y}_u - \hat{\mathbf{y}}_u = r_u\mathbf{y}$, $u = 0,$

1. Then $V_u = \text{Rank}\, r_u$ and $V_u S_u^2(2) = \mathbf{y}' r_u' r_u \mathbf{y}$, $u = 0,\ 1$. Thus, $V_u S_u^2(2)$ is the sum of squares of the elements in $r_u \mathbf{y}$, $u = 0,\ 1$.

The second measure of the dispersion effect was in fact proposed in Box and Meyer (1986) for unreplicated experiments. We observe that $r_1 \mathbf{y}$ and $r_0 \mathbf{y}$ are correlated under the standard linear model and also under model (1– 3). In the following subsection, we present a method of adjusting the above residuals to make them uncorrelated. The purpose for this adjustment is to eliminate the overlap in the contribution of levels 1 and 0 of the factor to the overall noise measured by MSE.

3.3   Adjusted Residuals

We now present a vector of "adjusted residuals" at level 0 of the factor, adjusted w.r.t. $r_1 \mathbf{y}$ so that it is uncorrelated with $r_1 \mathbf{y}$ under the standard linear model. Let $r_{11}(V_1 \times N)$ be a submatrix of r_1 so that $\text{Rank}\, r_{11} = V_1$. We write

$$r_{0a} = r_0(I - r_{11}'(r_{11} r_{11}')^{-1} r_{11}) = r_0 - r_0 r_{11}'(r_{11} r_{11}')^{-1} r_{11} \qquad (6)$$

It can be seen that $\text{Cov}(r_{11}\mathbf{y}, r_{0a}\mathbf{y}) = 0$ and hence $\text{Cov}(r_1\mathbf{y}, r_{0a}\mathbf{y}) = 0$ under the standard linear model. In other words, $r_1\mathbf{y}$ and $r_{0a}\mathbf{y}$ are uncorrelated under the standard linear model. We call $r_{0a}\mathbf{y}$ the vector of "adjusted residuals" at level 0 of the factor and adjusted w.r.t. the residuals at level 1 of the factor. It can be checked that $\text{Rank}\, r_{0a} = [(N - p) - V_1] = V_{0a}$ (say). Let r_{01} be a $(V_0 \times N)$ submatrix of r_0 with $\text{Rank}\, r_{01} = V_0$. We write

$$r_{1a} = r_1(I - r_{01}'(r_{01} r_{01}')^{-1} r_{01}) = r_1 - r_1 r_{01}'(r_{01} r_{01}')^{-1} r_{01} \qquad (7)$$

Again, $\text{Cov}(r_0\mathbf{y}, r_{1a}\mathbf{y}) = 0$ under the standard linear model. In other words, $r_0\mathbf{y}$ and $r_{1a}\mathbf{y}$ are uncorrelated under the standard linear model. We call $r_{1a}\mathbf{y}$ the vector of "adjusted residuals" at level 1 of the factor and adjusted w.r.t. the residuals at level 0 of the factor. We have $\text{Rank}\, r_{1a} = [(N - p) - V_0) = V_{1a}$ (say). It can be checked that

$$\begin{aligned} r_{0a}\mathbf{y} &= r_0(I - r_{11}'(r_{11} r_{11}')^{-1} r_{11}) r_0' \mathbf{y}_0 \\ r_{1a}\mathbf{y} &= r_1(I - r_{01}'(r_{01} r_{01}')^{-1} r_{01}) r_1' \mathbf{y}_1 \end{aligned} \qquad (8)$$

Thus, for $u = 0,\ 1$, $r_{ua}\mathbf{y}$ depends on $\mathbf{y}$ only through $\mathbf{y}_u$ and, moreover, $\text{Cov}(r_{0a}\mathbf{y}, r_{1a}\mathbf{y}) = 0$, i.e., they are uncorrelated under the assumed linear model (1–3). We now present an illustrative example.

Example 1

We consider the example from page 20 of Box and Meyer (1986) and page 68 of Taguchi and Wu (1985). Daniel's normal probability plot indicates that, over the ranges studied, only factors B (period of drying) and C (welded materials) affect tensile location by amounts not readily attributed to noise (Box and Meyer, 1986). We now fit the following standard linear model to the data.

$$E[y(x_1, x_2)] = \mu + \alpha_1 B + \alpha_2 C$$

where $x_i = 0, 1$, $\alpha_i = (2x_i - 1)$, μ is the general mean, and B and C are the main effects of the factors. We can write the above model in the form (1–3). Notice that $N = 16$, $n = 4$, $p = 3$, $(\sum_{i=1}^{n} \delta_i) = [\sum_{i=1}^{n}(1 - \delta_i)] = 2$ for both factors B and C. The F value for the lack of fit test under the standard linear model and under the assumption of normality is 0.1971 (< 1), and we therefore conclude that there is no significant lack of fit under the standard linear model. Notice that this is an approximate test under (1–3). The inner array T is given by

$$T' = \begin{bmatrix} 1 & 1 & 0 & 0 \\ 1 & 0 & 0 & 1 \end{bmatrix}$$

Let us now choose the factor B. Recall that the vector of observations $\mathbf{y}$ consists of level 1 observations $\mathbf{y}_1$ and level 0 observations $\mathbf{y}_0$ of the factor B. We have in $\mathbf{y}_1$ four observations on each of the runs $(1, 1)$ and $(1, 0)$, and thus $\mathbf{y}_1' = (42.4, 42.4, 42.4, 42.5, 44.7, 45.9, 45.5, 46.5)$. We also have in $\mathbf{y}_0$ four observations on each of runs $(0, 0)$ and $(0, 1)$ and thus $\mathbf{y}_0' = (43.7, 42.2, 43.6, 44.0, 40.2, 40.6, 40.6, 40.2)$. We now present the matrices

$$X_1 = \begin{bmatrix} 1 & 1 & 1 \\ 1 & 1 & 1 \\ 1 & 1 & 1 \\ 1 & 1 & 1 \\ 1 & 1 & -1 \\ 1 & 1 & -1 \\ 1 & 1 & -1 \\ 1 & 1 & -1 \end{bmatrix}, \quad X_0 = \begin{bmatrix} 1 & -1 & -1 \\ 1 & -1 & -1 \\ 1 & -1 & -1 \\ 1 & -1 & -1 \\ 1 & -1 & 1 \\ 1 & -1 & 1 \\ 1 & -1 & 1 \\ 1 & -1 & 1 \end{bmatrix}$$

We write $\mathbf{y}' = [\mathbf{y}_1' : \mathbf{y}_0']$, $X' = [X_1' : X_0']$, $\hat{\mathbf{y}}_1 = X_1(X'X)^{-1}X'\mathbf{y}$, $\hat{\mathbf{y}}_0 = X_0(X'X)^{-1}X'\mathbf{y}$, $r_1\mathbf{y} = \mathbf{y}_1 - \hat{\mathbf{y}}_1$, and $r_0\mathbf{y} = \mathbf{y}_0 - \hat{\mathbf{y}}_0$. It can be checked that $X'X = 16I$ and the matrices $r_1(8 \times 16)$ and $r_0(8 \times 16)$ are given by

$$r_1 = [I : 0] - X_1(X'X)^{-1}X'$$

$$= \left[I - \frac{1}{16}X_1X_1' : \left(-\frac{1}{16}\right)X_1X_0'\right]$$

$$r_0 = [0 : I] - X_0(X'X)^{-1}X'$$

$$= \left[\left(-\frac{1}{16}\right)X_0X_1' : I - \frac{1}{16}X_0X_0'\right]$$

We now obtain

$$r_1 = \begin{bmatrix} I - \dfrac{3J}{16} & : & \left(-\dfrac{J}{16}\right) & : & \dfrac{J}{16} & : & \left(\dfrac{-J}{16}\right) \\[2mm] \left(-\dfrac{J}{16}\right) & : & I - \dfrac{3J}{16} & : & \left(-\dfrac{J}{16}\right) & : & \dfrac{J}{16} \end{bmatrix}$$

$$r_0 = \begin{bmatrix} \dfrac{J}{16} & : & \left(-\dfrac{J}{16}\right) & : & I - \dfrac{3J}{16} & : & \left(\dfrac{-J}{16}\right) \\[2mm] \left(-\dfrac{J}{16}\right) & : & \dfrac{J}{16} & : & \left(-\dfrac{J}{16}\right) & : & I - \dfrac{3J}{16} \end{bmatrix}$$

where J is a (4×4) matrix with all its elements unity. It can be checked that $V_1 = \text{Rank}\, r_1 = 7$ and $V_0 = \text{Rank}\, r_0 = 7$. Thus, $V_{0a} = V_{1a} = 6$. The matrix r_{11} is obtained from r_1 by deleting the last row. The calculations of r_{0a} from (6) and the vector of adjusted residuals $r_{0a}\mathbf{y}$ are straightforward. The calculations of r_{1a} and the vector of adjusted residuals $r_{1a}\mathbf{y}$ are similar.

3.4   Third Set of Measures

Let r_{ua1} be a $(V_{ua} \times N)$ submatrix of r_{ua} with $\text{rank}\, r_{ua1} = V_{ua}$, $u = 0$, 1. We now have the sum of squares of the sets of linear functions $r_{u1}\mathbf{y}$ and $r_{ua1}\mathbf{y}$ as

$$\begin{aligned} \text{SS}(r_{u1}\mathbf{y}) &= \mathbf{y}'r_{u1}'(r_{u1}r_{u1}')^{-1}r_{u1}\mathbf{y} \\ \text{SS}(r_{ua1}\mathbf{y}) &= \mathbf{y}'r_{ua1}'(r_{ua1}r_{ua1}')^{-1}r_{ua1}\mathbf{y} \end{aligned} \tag{9}$$

with df V_u and V_{ua}, respectively ($u = 0, 1$). The sum of squares for a set of linear functions of observations was defined by R. C. Bose and can be seen on page 127 of Scheffé (1959).

Table 1   Numerical Values of $\bar{y}_i, \hat{y}_i,$ and
$\sum_{j=1}^{r}(y_{ij} - \bar{y}_i)^2/(r-1), i = 1, 2, 3, 4$

i	Run	$\bar{y}_i$	$\hat{y}_i$	$\sum_{j=1}^{r}(y_{ij} - \bar{y}_i)^2/(r-1)$
1	11	42.425	42.4875	.0025
2	10	45.650	45.5875	.5700
3	00	43.375	43.4375	.6425
4	01	40.400	40.3375	.0533

We present the measures of dispersion and adjusted dispersion at level
$u, u = 0, 1,$ of the factor

$$S_u^2(3) = [\text{SS}(r_{u1}\mathbf{y})/V_u]$$
$$S_{ua}^2(3) = [\text{SS}(r_{ua1}\mathbf{y})/V_{ua}] \tag{10}$$

respectively.

We note that $S_0^2(3)$ and $S_{1a}^2(3)$, $S_1^2(3)$ and $S_{0a}^2(3)$ are uncorrelated and
hence independent under the standard linear model and under the assumption of normality. Furthermore, $S_{0a}^2(3)$ and $S_{1a}^2(3)$ are uncorrelated and
hence independent under the assumed model (1–3) and under the assumption of normality. The sample dispersions $S_0^2(3)$ and $S_1^2(3)$ are generally
correlated under the standard linear model and also under model (1–3).
The third set of measures of the dispersion effect of the chosen factor is
therefore $S_1^2(3)/S_0^2(3)$, $S_1^2(3)/S_{0a}^2(3)$, $S_{1a}^2(3)/S_0^2(3)$, and $S_{1a}^2(3)/S_{0a}^2(3)$.

Example 1 (cont.)

We present in Table 1 $\bar{y}_i$, $\hat{y}_i$, and $\sum_{j=1}^{r}(y_{ij} - \bar{y}_i)^2/(r-1)$, $i = 1, 2, 3,$
and 4, which are used in calculating the first and the second measures of

Table 2   Numerical Values of Measures of Dispersion for Factors B
and C

Factor	$S_1^2(1) = S_{1a}^2(3)$	$S_0^2(1) = S_{0a}^2(3)$	$S_1^2(2)$	$S_0^2(2)$	$S_1^2(3)$	$S_0^2(3)$
B	.2863	.3479	.2498	.3027	.2543	.3071
C	.0279	.6063	.0284	.5241	.0329	.5286

Table 3 Numerical Values of Measures of Dispersion
Effects for Factors B and C

Factor	$\dfrac{S_1^2(1)}{S_0^2(1)}$	$\dfrac{S_1^2(2)}{S_0^2(2)}$	$\dfrac{S_1^2(3)}{S_0^2(3)}$	$\dfrac{S_1^2(3)}{S_{0a}^2(3)}$	$\dfrac{S_{1a}^2(3)}{S_0^2(3)}$	$\dfrac{S_{1a}^2(3)}{S_{0a}^2(3)}$
B	.8229	.8252	.8281	.7310	.9323	.8229
C	.0460	.0542	.0622	.0543	.0528	.0460

dispersion effects. We write, for the factor B,

$$\hat{\mathbf{y}}_1 = (42.4875, 42.4875, 42.4875, 42.4875;$$
$$45.5875, 45.5875, 45.5875, 45.5875)'$$
$$\hat{\mathbf{y}}_0 = (43.4375, 43.4375, 43.4375, 43.4375;$$
$$40.3375, 40.3375, 40.3375, 40.3375)'$$

Both r_{1a1} and r_{0a1} can be obtained from r_{1a} and r_{0a} by deleting the last row in the respective matrices. We find that $S_{1a}^2(3) = S_1^2(1)$ and that $S_{0a}^2(3) = S_0^2(1)$. In Tables 2 and 3, we display numerical values of various measures of dispersion and dispersion effects for both factors B and C.

4   INTERPRETATION AND APPLICATION OF THE MEASURES

We now discuss the rationale for the proposed measures of dispersion effects and how they can be used in determining the optimum level of each control factor in view of reducing the process variability. We have

$$n(\text{MSPE}) = \left(\sum_{i=1}^{n} \delta_i\right) S_1^2(1) + \sum_{i=1}^{n}(1 - \delta_i)S_0^2(1)$$

Thus $S_1^2(1)$ and $S_0^2(1)$ are regarded as $(\text{MSPE})_1$ and $(\text{MSPE})_0$ in the notation of Section 2. If the first measure of the dispersion effect, $S_1^2(1)/S_0^2(1)$, is greater than 1, we then say that level 0 of the factor has less contribution to MSPE and therefore would be preferred to level 1 for stability against noise factors. If $S_1^2(1)/S_0^2(1)$ is less than 1, level 1 would be preferred to level 0. Notice that the comparison of the ratio $S_1^2(1)/S_0^2(1)$

can also be made with a critical value of the F distribution at the level of significance $\alpha = .05$ and the degrees of freedom $(\sum_{i=1}^{n} \delta_i)(r-1)$ and $(\sum_{i=1}^{n}(1-\delta_i))(r-1)$. We observe that

$$(N-p)\,\mathrm{MSE} = V_1 S_1^2(2) + V_0 S_0^2(2)$$

If the second measure of the dispersion effect, $S_1^2(2)/S_0^2(2)$, is larger than 1, we then conclude that level 0 of the factor has less contribution to MSE and therefore would be preferred to level 1 for reducing process variability due to noise factors. If $S_1^2(2)/S_0^2(2)$ is smaller than 1, level 1 would be preferred to level 0.

We notice that

$$(N-p)\,\mathrm{MSE} = V_1 S_1^2(3) + V_{0a} S_{0a}^2(3)$$
$$= V_{1a} S_{1a}^2(3) + V_0 S_0^2(3)$$
$$(N-p) = V_{1a} + V_0 = V_1 + V_{0a}$$

The third set of measures of the dispersion effect consists of four measures of the dispersion effect. As before, the numerical value of a measure greater than 1 indicates a preference of level 0 over level 1 and the numerical value less than 1 indicates a preference of level 1 over level 0 in terms of smaller dispersion. If the numerical values of all proposed measures are near 1, then both levels 1 and 0 are equally preferable in terms of dispersion.

Example 1 (cont.)

We find from Table 3 that numerical values of all measures of dispersion effects are less than 1 for both factors B and C. We thus prefer level 1 over level 0 or both factors. The best choice for level combination of control factors B and C is therefore $(1,1)$ for stability against noise. Notice that the numerical values of all measures are not only greater but also closer to 1 for the factor B than the factor C. The next choice for level combination of B and C is therefore $(0,1)$ for stability against noise. Our choices for the best and the next-best level combinations are also supported by the numerical values of $\sum_{j=1}^{r}(y_{ij} - \bar{y}_i)^2/(r-1)$ in Table 1. This is, of course, very natural because the inner array T in this example consists of all runs of a 2^2 factorial experiment. In practice, we would have a fractional factorial instead of a complete factorial as the inner array in most situations. We would then not be able to calculate $\sum_{j=1}^{r}(y_{ij} - \bar{y}_i)^2/(r-1)$ for all runs but only for those runs in the fractional factorial inner array. The best choice

of level combination using the methods described in this chapter may or
may not be present in the inner array.

5 INNER ARRAY INFLUENCE

We now discuss the influence of the inner array on the proposed measures
of dispersion effects. We first consider the situation in which the number of
runs in the inner array equals the number of parameters in β of (1) or, in
other words, $n = p$. The inner array is then a saturated design. The class
of designs with $n = p$ includes the known Plackett and Burman designs
(Plackett and Burman, 1946). For an inner array with $n = p$, we have
$S_u^2(1) = S_u^2(2)$, $u = 0$, 1, and vice versa. We observe that, for an inner
array with $n = p$, the first two measures of dispersion effects are identical.

We next study the measures in two extreme situations: (1) Two vectors
of residuals $\mathbf{y}_1 - \hat{\mathbf{y}}_1 = r_1\mathbf{y}$ and $\mathbf{y}_0 - \hat{\mathbf{y}}_0 = r_0\mathbf{y}$ at levels 1 and 0 of the factor
are uncorrelated under the standard linear model, i.e., $r_{01} = Ar_{11}$ for some
matrix A. In situation 1, we have $S_u^2(3) = S_u^2(2) = S_{ua}^2(3)$, $u = 0$, 1.
This, in turn, implies that for each u, $u = 0$, 1, all measures of dispersion
effects in the third set are identical to one another, and identical to the
second measure. We thus see that, in situation 1, there is no need for
the adjustment of residuals and for the third set of measures of dispersion
effects. In situation 2, we have $r_{0a} = 0$, $V_{0a} = 0$ and $\mathrm{SS}(r_{0a1}\mathbf{y}) = 0$. We
thus notice in situation 2 that level 1 of the factor makes all contribution
to SSE and that level 0 does not make any additional contribution to SSE.
In case $V_{0a} = V_{1a} = 0$, we have $V_0 = V_1 = (N - p)$, $r_{01} = Ar_{11}$, and
A is nonsingular. This is a situation in which levels 0 and 1 have equal
measures of dispersions. This is purely due to the inner array influence.
We have $V_{1a} \geq (\sum_{i=1}^n \delta_i)(r - 1)$ and $V_{0a} \geq [\sum_{i=1}^n (1 - \delta_i)](r - 1)$ for all
inner arrays. If $V_{1a} = (\sum_{i=1}^n \delta_i)(r - 1)$, then $S_{1a}^2(3) = S_1^2(1)$; and if $V_{0a} =
[\sum_{i=1}^n (1 - \delta_i)](r - 1)$, then $S_{0a}^2(3) = S_0^2(1)$. We note that V_{0a} and V_{1a} are
both nonzero for $r > 1$. [We assume naturally that there is at least one
$\delta_i = 1$ and at least one $(1 - \delta_i) = 1$.]

We now present an example of an inner array for which $S_1^2(2)$ and
$S_0^2(2)$ are uncorrelated for one factor but for which $S_1^2(2)$ and $S_0^2(2)$ are
correlated for all other factors under the standard linear model. This example is remarkable in displaying contrasting influence of the inner array
in measuring $S_u^2(2)$, $u = 0$, 1, and in the need for adjustment of residuals
for different factors.

Example 2

We consider a 2^5 factorial experiment, i.e., $m = 5$. We thus have five control factors each at two levels. Let the inner array $T(8 \times 5)$ be

$$T = \begin{bmatrix} 0 & 0 & 0 & 0 & 0 \\ 1 & 1 & 0 & 0 & 0 \\ 0 & 0 & 0 & 1 & 1 \\ 1 & 1 & 0 & 1 & 1 \\ 0 & 1 & 1 & 1 & 0 \\ 1 & 0 & 1 & 1 & 0 \\ 0 & 1 & 1 & 0 & 1 \\ 1 & 0 & 1 & 0 & 1 \end{bmatrix}$$

We consider a situation in which the model for a main effect plan fits the data adequately; i.e., there is no significant lack of fit under the standard linear model. Therefore, $n = 8$ and $p = 6$. The first column of X has all entries unity. The distinct rows of the remaining columns of X are obtained from T replacing 0 by (-1), and each distinct row is replicated r times. We denote an $(r \times r)$ matrix with all elements unity by J and

$$G = \begin{pmatrix} 6J & 2J & 2J & -2J \\ 2J & 6J & -2J & 2J \\ 2J & -2J & 6J & 2J \\ -2J & 2J & 2J & 6J \end{pmatrix}$$

It can be easily seen that

$$H = X(X'X)^{-1}X' = \frac{1}{8r} \begin{pmatrix} G & 0 \\ 0 & G \end{pmatrix}$$

$$R = \frac{1}{8r} \left(\begin{array}{c|c} 8rI_{4r} - G & 0 \\ \hline 0 & 8rI_{4r} - G \end{array} \right)$$

It now follows that $r_1 r_0' = 0$ for the factor 3 but that $r_1 r_0' \neq 0$ for the factors 1, 2, 4, and 5. Thus, $S_1^2(2)$ and $S_0^2(2)$ are uncorrelated for the factor 3 and are correlated for factors 1, 2, 4, and 5 under the standard linear model.

6 PROPERTIES

We now state some properties of the sample dispersions under the standard linear model and under the model (1–3). We first observe that the sample dispersions $S_1^2(1)$ and $S_0^2(1)$ do not depend on the fitted model and that all other sample dispersions depend on the fitted model. The sample dispersions $S_1^2(1)$ and $S_0^2(1)$ are always uncorrelated under the assumed model (1–3). The measures $S_1^2(2)$ and $S_0^2(2)$ may, however, be correlated. They are uncorrelated under the standard linear model if and only if $r_1 r_0' = 0$. Similarly, $S_1^2(3)$ and $S_0^2(3)$ are generally correlated, and they are uncorrelated under the standard linear model if and only if $r_{11} r_{01}' = 0$ or, equivalently, $r_1 r_0' = 0$.

The dispersion measures $S_1^2(3)$ and $S_{0a}^2(3)$, $S_{1a}^2(3)$ and $S_0^2(3)$, $S_{1a}^2(3)$ and $S_{0a}^2(3)$ are all uncorrelated under the standard linear model and under the assumption of normality. Therefore, under equality of pertinent dispersions, $S_1^2(1)/S_0^2(1)$, $S_1^2(3)/S_{0a}^2(3)$. $S_{1a}^2(3)/S_0^2(3)$, and $S_{1a}^2(3)/S_{0a}^2(3)$ have the central F distribution with appropriate degrees of freedom. The measures $S_1^2(2)/S_0^2(2)$ and $S_1^2(3)/S_0^2(3)$ have the central F distribution if and only if $r_1 r_0' = 0$. We question the use of measures of the dispersion effect $S_1^2(2)/S_0^2(2)$ and $S_1^2(3)/S_0^2(3)$ unless $r_1 r_0' = 0$.

7 POPULATION DISPERSION EFFECTS

In this section, we present the population dispersion effects corresponding to the sample measures proposed in Section 3. For the chosen factor, the parameters $\delta_i \sigma_i^2$, with $\delta_i = 1$ for $i = 1, \ldots, n$, are called the population dispersions of the chosen factor at level 1. The parameters $(1 - \delta_i)\sigma_i^2$, with $\delta_i = 0$ for $i = 1, \ldots, n$, are called the population dispersions of the chosen factor at level 0. It can be seen that

$$E(S_1^2(1)) = \frac{\sum_{i=1}^n \delta_i \sigma_i^2}{\sum_{i=1}^n \delta_i}, \qquad E(S_0^2(1)) = \frac{\sum_{i=1}^n (1 - \delta_i)\sigma_i^2}{\sum_{i=1}^n (1 - \delta_i)} \qquad (11)$$

Hence, $S_1^2(1)$ and $S_0^2(1)$ measure without bias the average dispersions at levels 1 and 0 of the chosen factor. The first sample dispersion effect measure is thus a measure of a population dispersion effect that is the ratio of the average dispersions at levels 1 and 0 of the chosen factor.

Let Σ_u be the submatrix of Σ corresponding to the runs with the chosen factor at level u, $u = 1, 0$. It can be checked that, for $u = 0, 1$

$$E(V_u S_u^2(2)) = \mathrm{Tr}(r_u r_u' \Sigma_u r_u r_u' + r_u r_{(1-u)}' \Sigma_{(1-u)} r_u') \tag{12}$$

Clearly, $E(V_u S_u^2(2)) = \mathrm{Tr}\, r_u r_u' \Sigma_u r_u r_u'$ if and only if $r_1 r_0' = 0$. In case $r_1 r_0' = 0$, the second sample dispersion effect is thus a measure of a population dispersion effect that is the ratio of a linear function of the level 1 dispersions and a linear function of the level 0 dispersions. Notice that, in case $r_1 r_0' = 0$, $S_u^2(3) = S_u^2(2) = S_{ua}^2(3)$, $u = 0, 1$. Therefore, the measures in the third set, $S_1^2(3)/S_0^2(3)$, $S_1^2(3)/S_{0a}^2(3)$, $S_{1a}^2(3)/S_0^2(3)$, are identical to the second sample dispersion effect.

In case $r_1 r_0' \neq 0$, considering $S_{ua}^2(3)$ for $u = 0, 1$, we have

$$E(V_{ua} S_{ua}^2(3)) = \mathrm{Tr}(r_{ua1} r_{ua1}')^{-1} r_{ua1} \Sigma_u r_{ua1}' \tag{13}$$

The sample dispersion effect $S_{1a}^2(3)/S_{0a}^2(3)$ is thus a measure of a population dispersion effect that is the ratio of a linear function of the level 1 dispersions and a linear function of the level 0 dispersions.

8   CONCLUSIONS

In industrial experiments for quality improvement, dispersion effects of factors play an important role. They are instrumental in the choice of an optimum combination of levels of control factors. This article presents both the sample and the corresponding population measures of dispersion effects. The sample dispersion effects are simple, meaningful, and unbiased estimators of the population measures.

APPENDIX

We now present the proofs of many statements we have made in the main body of the paper. We also present some valuable technical results in the investigation.

Let $D_1(N \times N)$ be a diagonal matrix with n sets of diagonal elements, and the elements in the ith ($i = 1, \ldots n$) set are equal to δ_i. We define $D_0 = I - D_1$. It can be seen that $D_1 D_0 = 0$ and both D_1 and D_0 are idempotent matrices. We have $R = D_1 R + D_0 R$. The matrices r_1 and

r_0 defined in Section 3 are in fact nonnull row vectors of $D_1 R$ and $D_0 R$, respectively. It can be seen that $RD_u R = r'_u r_u$, $V_u S_u^2(2) = \mathbf{y}'RD_u R\mathbf{y}$, and $\text{SSE} = \mathbf{y}'R\mathbf{y} = \mathbf{y}'RD_1 R\mathbf{y} + \mathbf{y}'RD_0 R\mathbf{y}$, $u = 1, 0$.

We now investigate the situation where $S_u^2(1) = S_u^2(2)$, $u = 1, 0$. In other words, we characterize the inner arrays for which $S_u^2(1) = S_u^2(2)$, $u = 1, 0$. We denote the row of the matrix X corresponding to the run i by $x'_i(1 \times p)$. Note that, for each i, $i = 1, \ldots, n$, the row x'_i is repeated r times in X. Let $X^*(n \times p)$ be a matrix whose ith row is x'_i. Notice that rows of X^* are in fact distinct rows of X. We have $X'X = r(X^{*\prime}X^*)$.

Theorem 1   We have $S_u^2(1) = S_u^2(2)$, $u = 1, 0$ if and only if $n = p$.

Proof   Note that $S_u^2(1) = S_u^2(2)$, $u = 1, 0$, hold if and only if $\hat{y}_i = \bar{y}_i$, $i = 1, \ldots, n$. The condition $\hat{y}_i = \bar{y}_i$, $i = 1, \ldots, n$, holds if and only if

$$\mathbf{x}'_{i_1}(X'X)^{-1}\mathbf{x}'_{i_2} = \begin{cases} \dfrac{1}{r} & \text{for } i_1 = i_2 \\ 0 & \text{for } i_1 \neq i_2; i_1, i_2 \in \{1, \ldots, n\}. \end{cases}$$

The above condition may be expressed as $X^*(X'X)^{-1}X^{*\prime} = (1/r)I_n$, or, equivalently, $X^*(X^{*\prime}X^*)^{-1}X^{*\prime} = I_n$, which is true if and only if $n = p$. This is because $\text{Rank}(I_n - X^*(X^{*\prime}X^*)^{-1}X^{*\prime}) = n - p$. This completes the proof.

When $n = p$, $X^*(n \times n)$ satisfies the condition $X^*(X^{*\prime}X^*)^{-1}X^{*\prime} = I_n$. Therefore, for $n = p$, we get $S_u^2(1) = S_u^2(2)$.

We now present results showing the influence of the inner array on the measures of dispersion $S_u^2(3)$, $S_{ua}^2(3)$, $S_u^2(2)$, and $S_u^2(1)$, $u = 1, 0$.

Theorem 2

$$1.\ V_{1a} \geq \left(\sum_{i=1}^{n} \delta_i\right)(r - 1), \qquad V_{0a} \geq \left(\sum_{i=1}^{n}(1 - \delta_i)\right)(r - 1)$$

$$2.\ V_{1a}S_{1a}^2(3) \geq \left(\sum_{i=1}^{n} \delta_i\right)(r - 1)S_1^2(1)$$

$$V_{0a}S_{0a}^2(3) \geq \left(\sum_{i=1}^{n}(1 - \delta_i)\right)(r - 1)S_0^2(1)$$

3. If $V_{1a} = \left(\sum_{i=1}^{n} \delta_i \right) (r-1)$, then $S_{1a}^2(3) = S_1^2(1)$

4. If $V_{0a} = \left(\sum_{i=1}^{n} (1 - \delta_i) \right) (r-1)$, then $S_{0a}^2(3) = S_0^2(1)$

Proof It can be checked that $\text{Cov}(y_{ij}, \bar{y}_u - \hat{y}_u) = \text{Cov}(\bar{y}_i, \bar{y}_u - \hat{y}_u)$ and, therefore, $\text{Cov}(y_{ij} - \bar{y}_i, \bar{y}_u - \hat{y}_u) = 0$. Moreover, $\text{Cov}(y_{ij} - \bar{y}_i, y_{uw} - \bar{y}_u) = 0$, $i \neq u$. It now follows that any contrast of $(y_{ij} - \bar{y}_i)$, $j = 1, \ldots, r$ for a fixed i with $\delta_i = 1$ is orthogonal to any contrast of $(y_{uw} - \bar{y}_u)$, $w = 1, \ldots$, r for a fixed u with $(1 - \delta_u) = 1$. Furthermore, any contrast of $(\bar{y}_i - \hat{y}_i)$ for all i with $\delta_i = 1$ is orthogonal to any contrast of $(y_{uw} - \bar{y}_u)$, $w = 1, \ldots$, r, for a fixed u with $(1 - \delta_u) = 1$. The results (1–4) follow immediately from the above facts, the relationship between the rank and the number of orthogonal contrasts, and the fact that the sum of squares is equal to the sum of sums of squares of orthogonal contrasts.

We now study the measures in two extreme situations: (1) $r_1 \mathbf{y}$ and $r_0 \mathbf{y}$ are uncorrelated, i.e., $r_1 r_0' = 0$, (2) $r_1 \mathbf{y}$ and $r_0 \mathbf{y}$ are completely correlated; i.e., $r_{01} = A r_{11}$ for some matrix A.

Theorem 3 Consider the situation $r_1 r_0' = 0$. Then, $S_u^2(3) = S_u^2(2) = S_{ua}^2(3)$, $u = 1, 0$.

Proof We first show that $S_1^2(3) = S_1^2(2)$ or, in other words, $\text{SS}(r_{11}\mathbf{y}) = \mathbf{y}' r_1' r_1 \mathbf{y}$. We observe that

$$\mathbf{y}' r_{11}' [r_{11} r_{11}']^{-1} r_{11} \mathbf{y}$$

$$= \mathbf{y}' r_1' \begin{bmatrix} (r_{11} r_{11})^{-1} & 0 \\ 0 & 0 \end{bmatrix} r_1 \mathbf{y}$$

$$= \mathbf{z}' r_1 \mathbf{y}, \qquad \text{where } r_1 r_1' \mathbf{z} = r_1 \mathbf{y}$$

$$= \mathbf{z}' r_1 r_1' \mathbf{z} = (r_1 r_1' \mathbf{z})'(r_1 r_1' \mathbf{z})$$

$$= (r_1 \mathbf{y})'(r_1 \mathbf{y}) = \mathbf{y}' r_1' r_1 \mathbf{y}$$

It follows from the representations of MSE in terms of $(S_1^2(2), S_0^2(2))$ and $(S_1^2(3), S_{0a}^2(3))$ that $V_{0a} S_{0a}^2(3) = V_0 S_0^2(2)$. The condition $r_1 r_0' = 0$ implies that $V_{0a} = V_0$. Thus, $S_{0a}^2(3) = S_0^2(2)$. The rest is similar. This completes the proof.

Theorem 4 If $r_{01} = Ar_{11}$, then we have $r_{0a} = 0$, $V_{0a} = 0$, and $SS(r_{0a1}\mathbf{y}) = 0$.

Proof We write $r_0 = A_0 r_{11}$ for a matrix A_0, whose independent rows are rows of A. This implies that $r_0 r_{11}' = A_0 r_{11} r_{11}'$ and, thus, $A_0 = r_0 r_{11}'(r_{11} r_{11}')^{-1}$. Hence, from (6), we get $r_{0a} = 0$. The rest is clear. This completes the proof.

We now present some results that are useful in establishing properties of the proposed measures in the paper.

Theorem 5 Suppose that $\mathbf{y} \sim N(X\beta, \sigma^2 I)$. A necessary and sufficient condition that (1) $(\mathbf{y}'r_1'r_1\mathbf{y}/\sigma^2) \sim$ central χ^2 with df $= \mathrm{Tr}\, r_1' r_1$, (2) $(\mathbf{y}'r_0'r_0\mathbf{y}/\sigma^2) \sim$ central χ^2 with df $= \mathrm{Tr}\, r_0' r_0$, and (3) furthermore, (1) and (2) are statistically independent, is that $r_1 r_0' = 0$.

It can be seen (see Rao, 1973) that a necessary and sufficient condition for (1), (2) and (3) to be true is that $r_1' r_1 r_0' r_0 = 0$. The condition is equivalent to $r_1 r_0' = 0$.

Theorem 6

1. $\mathbf{j}'r_u = 0$ and $\mathbf{j}'r_{ua} = 0$, $u = 1, 0$, where $\mathbf{j}'$ is a vector with all elements unity,
2. If $r_1 r_0' = 0$, then, for $u = 1, 0$:

   a. $r_u r_u'$ is an idempotent matrix.
   b. $(\mathbf{y}_u - \hat{\mathbf{y}}_u) = r_u r_u' \mathbf{y}_u$,
   c. $X_u' r_u r_u' = 0$,
   d. $\sum_{i=1}^{n} \sum_{j=1}^{r} \delta_i \hat{y}_i (y_{ij} - \hat{y}_i) = \sum_{i=1}^{n} \sum_{j=1}^{r} (1 - \delta_i) \hat{y}_i (y_{ij} - \hat{y}_i) = 0$

Proof The result 1 follows by considering the columns of X for the general mean and the factor chosen and from the fact that $X'R = 0$. The results 2a and 2b follow directly from the structure of R and the fact that R is an idempotent matrix. The result 2c follows from $X'R = 0$. From 2c, we get $\hat{\beta}' X_u' r_u r_u' \mathbf{y}_u = 0$; i.e., $\hat{\mathbf{y}}_u' r_u r_u' \mathbf{y}_u = 0$. The result 2b implies that $\hat{\mathbf{y}}_u'(\mathbf{y}_u - \hat{\mathbf{y}}_u) = 0$ and, hence, the result 2d is true. This completes the proof.

Theorem 7 For r_{0a} and r_{1a} in formulas (6) and (7), equation (8) holds.

Proof We prove the equation (8) for $r_{0a}\mathbf{y}$. The proof for $r_{1a}\mathbf{y}$ is similar. The fact $RX = 0$ implies that $r_0\hat{\mathbf{y}} = 0$, $r_{11}\hat{\mathbf{y}} = 0$ and, hence, $r_{0a}\hat{\mathbf{y}} = 0$. Since $\mathbf{y} - \hat{\mathbf{y}} = R\mathbf{y}$ and $R' = R$, we have $\mathbf{y} = \hat{\mathbf{y}} + R'\mathbf{y} = \hat{\mathbf{y}} + r_1'\mathbf{y}_1 + r_0'\mathbf{y}_0$. The independent rows of r_1 are in r_{11}, and the other rows of r_1 can be written as Qr_{11} for some matrix Q. The fact $(I - r_{11}'(r_{11}r_{11}')^{-1}r_{11})r_{11}' = 0$ implies that $(I - r_{11}'(r_{11}r_{11}')^{-1}r_{11})r_1' = 0$ and, hence, $r_{0a}r_1' = 0$. Hence, $r_{0a}\mathbf{y} = r_{0a}r_0'\mathbf{y}_0$. This completes the proof.

ACKNOWLEDGMENT

The work of the first author is sponsored by the Air Force Office for Scientific Research under AFOSR grant 87-0048.

REFERENCES

Box, G. E. P. and Meyer, R. D. (1986). Dispersion effects from fractional designs, *Technometrics*, *28*: 19–27.

Box, G. E. P. (1988). Signal-to-noise ratios, performance criteria and transformations, *Technometrics 30 (1)*: 1–40 (with discussions).

Carroll, R. J. and Ruppert, D. (1988). *Transformation and Weighting in Regression*, Chapman and Hall, New York.

Ghosh, S. (1987). Non-orthogonal designs for measuring dispersion effects in sequential factor screening experiments using search linear models, *Communications in Statistics, Theory and Methods*, *16 (10)*: 2839–2850.

Kacker, R. N. (1985). Off-line quality control, parameter design and the Taguchi Method, *Journal of Quality Technology*, *17*: 175–246 (with discussion).

León, R., Shoemaker, A., and Kacker, R. N. (1987). Performance measures independent of adjustment, *Technometrics*, *29 (3)*: 253–285 (with discussions).

Nair, V. N. (1986). Industrial experiments with ordered categorical data, *Technometrics*, *28*: 283–311 (with discussion).

Nair, V. N. and Pregibon, D. (1988). Analyzing dispersion effects from replicated factorial experiments, *Technometrics*, *30 (3)*: 247–257.

Phadke, M. S., Kacker, R. N., Speeney, D. V., and Grieco, M. J. (1983). Off-line quality control in integrated circuit fabrication using experimental design, *Bell System Technical Journal*, *62*: 1273–1310.

Plackett, R. L. and Burman, J. P. (1946). The design of optimum multifactorial experiments, *Biometrika*, *33*: 305–325.

Rao, C. R. (1973). *Linear Statistical Inference and Its Applications*, 2nd ed., Wiley, New York.

Scheffé, H. (1959). *The Analysis of Variance*, Wiley, New York.

Srivastava, J. N. (1975). *Designs for Searching Non-negligible Effects. A Survey of Statistical Designs and Linear Models* (J. N. Srivastava, ed.), North-Holland, Amsterdam, pp. 507–519.

Srivastava, J. N. (1987). On the inadequacy of the customary orthogonal arrays in quality control and general scientific experimentation and the need of probing designs of higher revealing power, *Communications in Statistics, Theory and Methods, 16 (10)*: 2901–2941.

Taguchi, G. and Wu, Y. (1985). *Introduction to Off-line Quality Control*, Central Japan Quality Control Association, Tokyo.

17

Designing Factorial Experiments: A Survey of the Use of Generalized Cyclic Designs

Angela M. Dean   The Ohio State University, Columbus, Ohio

1   INTRODUCTION

Experimentation is extremely common in industrial research, especially in the areas of product development and quality improvement. The characteristics of a finished product depend upon the ingredients or components in the product, the conditions under which the product was made, and the conditions under which it is to be stored or used. The development of a product usually involves the comparison of a number of experimental products which are made by varying the ingredients (components) and conditions. The usual objective of the experimentation is to study the effect of varying these factors or to determine the best combination of ingredients (components).

Any ingredient, component or condition which is to be varied in the course of the experiment is called a *treatment factor*, and its different forms are called the *levels* of the treatment factor. Most industrial experiments involve the simultaneous study of two or more treatment factors and therefore every experimental observation is a response obtained from one of the possible combinations of their experimental levels. These combinations are called the *treatment combinations*. Varying the levels of the factors simultaneously rather than one at a time is efficient in terms of time and cost considerations, and also allows the interactions between the factors to be monitored. Such an experiment is called a *factorial experiment*.

Ideally every observation should be taken under identical experimental conditions (other than those conditions which are being varied as part of the experiment). Sometimes, however, this ideal situation cannot be achieved. An experiment may be carried out over several days with large variations in temperature and humidity. Observations may be collected by different technicians, whose experimental procedures may differ, or collected in different laboratories, whose equipment may not be identical. Two examples of such experiments are described in detail in Section 2. Observations which are collected under similar experimental conditions are said to be in the same *block*. Experimental conditions may be quite different for observations in different blocks. Adjustments are then made for these differences during the analysis of the experimental data. If every block contains an observation on every treatment combination, the experiment is said to be arranged as a *complete block design*. Frequently, the blocks are not large enough to accommodate an observation on every treatment combination, in which case a decision must be made as to which treatment combinations will be observed in which block. The plan of such an experiment is called an *incomplete block design*.

Over the past forty years, various authors have described the use of incomplete block designs for factorial experiments. The use of generalized cyclic designs for this purpose was first discussed by John (1973). John showed that this class of designs provides efficient plans for factorial experiments and a straightforward analysis. Over the past fifteen years, numerous results concerning factorial experiments in generalized cyclic designs have been published. The purpose of this article is to draw together these results for readers with little algebraic background and little knowledge of the subject matter.

Generalized cyclic designs are described in detail in Section 3, with particular reference to their use for factorial experiments. Combinatorial properties of the designs are investigated in Section 6. In Section 4, the analysis of factorial experiments in incomplete block designs is presented. Expressions for efficiency factors for factorial experiments are given in Section 5, together with their upper bounds. Section 7 deals with the case of single replicate experiments where each treatment combination is observed exactly once, and with the case of fractional factorial experiments where only a fraction of the treatment combinations can be observed. In Section 8, a brief survey is given on other types of designs for factorial experiments.

2 FACTORIAL EXPERIMENTS IN INDUSTRY

In this section two industrial examples are described which involve factorial experiments arranged as incomplete block designs. The examples have been selected to illustrate experiments in two very different industries. In both cases, the designs are members of the class of generalized cyclic designs. The studies were originally published in order to advocate the types of designs discussed in this article.

The first example describes a 1956 experiment in the chemical industry, dealing with the decontamination of radioactive waste. The second example discusses a 1980 experiment on cake quality improvement in the food manufacturing industry.

Example 2.1

Barnett and Mead (1956) describe an experiment which was run in a laboratory operated by the Monsanto Chemical Company for the United States Atomic Energy Commission. The objective of the experiment was to determine the efficiency of a decontamination process for removing radioactive isotopes from liquid waste. In particular, the efficiency of the decontamination process was thought to depend on the amount of four process chemicals added to the liquid waste. The nature of the dependence of the process efficiency on these chemicals was to be determined. Efficiency was defined as the activity of alpha and beta particles remaining in the waste after decontamination. The four process chemicals were each observed at two levels as follows: barium chloride and aluminum sulfate each at 0.4 and 2.5 g/l, carbon at 0.08 and 0.4 g/l, sufficient sodium hydrochloric acid to bring the final pH to 6 or 10. Thus the experiment involved four treatment factors each at two levels, ($2 \times 2 \times 2 \times 2$ or a 2^4 factorial experiment). Two observations were taken on each of the 16 treatment combinations. Only eight observations could be taken in a single day, and the experiment was run over four days. No reason was given in the published study for suspecting a large day to day variability in the results of the experiment. However, it was noted that, at the end of the first day's observations, an unforeseen change of operators became necessary. Thus the block differences involved not only day to day differences but also operator differences. The data showed that the block differences were fairly large and could have biased the results of the experiment if they had been ignored. Further details of the experiment and its results are given in Barnett and Mead (1956).

Example 2.2

An experiment conducted by Spillers Ltd. was described by Lewis and Dean (1980) as follows. The company wanted to study the effects on cake quality of various amounts of glycerol and tartaric acid added to the cake mix. Three different amounts of glycerol and four different amounts of tartaric acid were selected by the researchers for study. The experiment therefore involved two treatment factors, glycerol and tartaric acid, at three and four levels, respectively (a 3×4 factorial experiment). The experiment was to be conducted during the normal baking routine which limited the total number of observations that could be taken. Three ovens were available and each could be used twice during the day set aside for the experiment. Since the temperature in different ovens cannot be presumed to be exactly the same, and the temperature in the same oven may vary for different bakings, the experiment involved six blocks. Each oven held six trays of cakes and an entire tray was needed to assess any one cake mix. Therefore only six of the twelve treatment combinations could be observed in each block.

It was thought advisable to assign the treatment combinations to the blocks in such a way that all twelve could be observed in each oven, six on the first run and six in the second. This precaution was taken so that if one of the ovens failed on the day of the experiment, leaving only two ovens available for experimentation, the treatment combinations would still be observed the same number of times. Such an assignment also allows oven differences to be monitored if desired.

Whenever blocks can be grouped into sets of blocks so that each set of blocks contains every treatment combination exactly once, the design is called *resolvable*. The experiment described in Example 2.1 was also run as a resolvable design. The paper of Lewis and Dean (1980) gives a list of efficient resolvable generalized cyclic designs for two treatment factors. The construction of such designs for general factorial experiments is discussed in Section 6.2.

3   DEFINITION AND CONSTRUCTION OF
GENERALIZED CYCLIC DESIGNS

This section begins with a discussion of the construction of *cyclic designs*, which are incomplete block designs suitable for experiments involving only one treatment factor. The construction method is then extended to produce generalized cyclic designs which are ideal block designs for experiments with several treatment factors. The analysis of these designs is discussed in Section 4.

Consider, first, an experiment involving only one treatment factor having v levels. The levels are chosen by the experimenter and, for simplicity, are coded as 0, 1, 2, ..., $v - 1$. The code 0 does not necessarily signify the absence of the treatment factor. In fact, codes 1, 2, ..., v could just as well have been used, but these are less convenient from a mathematical viewpoint when the properties of designs are studied. If the factor is quantitative, its levels are often arranged in increasing order and then coded 0, 1, ..., $v - 1$ in the same increasing order. However, this is not necessary and, in fact, it has been shown that other orders of coding are sometimes better (see Bailey, 1982). Except where stated, no particular order is implied in this chapter.

The codes form a *code cycle* in the sense that code 0 is followed by 1, code 1 is followed by 2, ..., code $v - 2$ is followed by $v - 1$, code $v - 1$ is followed by 0 and so on. This can be depicted as:

$$\boxed{\longrightarrow 0 \rightarrow 1 \rightarrow 2 \rightarrow \cdots \rightarrow v - 2 \rightarrow v - 1 \rightarrow}$$

Suppose that the experiment is to be run in blocks of size k (that is, k observations can be taken in each block). A cyclic design is constructed as follows. Select any k codes, hereafter called treatment labels, to be observed in any one block. Conventionally, label 0 is included among the selected labels, and the block to which these labels are assigned is called the *initial* or *principal* or *generating* block of the cyclic design. The other blocks are then determined as follows. The second block is obtained by replacing each treatment label in the initial block by the next label in the code cycle. The third block is obtained by replacing each label in the second block by the next label in the cycle, and so on. The cycling process stops as soon as the initial block reappears. Only one copy of the initial block is kept in the design. The design will consist of at most v distinct blocks, where v is the number of treatment labels. If a larger design is required, a second initial block can be selected and the cycling process repeated.

Example 3.1

Consider an experiment with one treatment factor having $v = 5$ levels coded 0, 1, 2, 3, 4. Suppose that the block size is $k = 3$ and treatment labels 0, 2, 3 are selected for the initial block. The second block then contains treatment labels 1, 3, 4 since, in the code cycle, label 0 is followed by 1, label 2 by 3, and label 3 by 4. The third block then contains 2, 4, 0 since label 1 is followed by 2, label 3 by 4, and label 4 is followed by label 0 in the cycle. The other blocks are obtained in a similar way. The design contains five distinct blocks since the sixth block would be a repeat of the initial block. The incomplete block design is written as follows, although the row headings are usually omitted.

Block 1	0	2	3
Block 2	1	3	4
Block 3	2	4	0
Block 4	3	0	1
Block 5	4	1	2

In a cyclic design, every treatment label occurs the same number of times in the design. This number is denoted by r (for replication) and the number of blocks is denoted by b. In the above example $r = k = 3$ and $b = v = 5$. If k and v are both divisible by an integer $d > 1$, then it is possible that the initial block will reappear in v/d cycles. In this case $b = v/d$ and $r = k/d$. The reason for this is discussed in Section 6.1 in the context of generalized cyclic designs.

An alternative method of obtaining the blocks of a cyclic design involves addition of treatment labels. Addition of two labels, c and d, is achieved by locating c in the code cycle and moving d positions along the cycle (or equivalently, locating d and moving along c positions). For illustration, consider the following code cycle for five levels of the treatment factor,

$$\longrightarrow 0 \rightarrow 1 \rightarrow 2 \rightarrow 3 \rightarrow 4 \rightarrow$$

Addition of treatment labels 2 and 4, for example, gives treatment label 1, as can be seen by locating 2 in the code cycle and moving along four positions, or locating 4 and moving along two positions. This method of addition is known as "addition modulo v" where v is the length of the code cycle. Looking back at Example 3.1, it can be seen that block 4, for instance, can be obtained by adding the fourth treatment label, which has code 3, to each label in block 1 (the initial block). In general, the sth block of a cyclic design can be obtained by adding (modulo v) the sth treatment label to every treatment label in the initial block.

All the properties of a cyclic design are determined by the selection of treatment labels for the initial block. Further details and lists of efficient designs can be found in John, Wolock and David (1972) and John (1981a, 1987). The properties will not be discussed here since they can be deduced from those of generalized cyclic designs.

In a factorial experiment, there are $n > 1$ treatment factors which may or may not have the same number of levels. Suppose that the jth factor has m_j levels which are coded 0, 1, 2, ..., $m_j - 1$. These m_j codes form exactly the same type of cycle as the codes for the treatment labels in a cyclic design. In a factorial experiment with n treatment factors, there are $v = m_1 \times m_2 \times \cdots \times m_n$ treatment combinations, each of which can be coded as an n-tuple $i_1 i_2 \ldots i_n$, where i_j is one of the codes 0, 1, ..., $m_j - 1$ $(j = 1, \ldots, n)$. The n-tuples are the treatment labels representing the treatment combinations. A *generalized cyclic design* is constructed in the same way as a cyclic design, by selecting k treatment labels to form the initial block, and then adding the sth treatment label to the labels in the initial block to obtain the sth block $(s = 2, 3, \ldots, b)$. Addition of two treatment labels involves the addition of the jth digits modulo m_j $(j = 1, \ldots, n)$ as shown in Example 3.2. The n-digit treatment labels are usually written in a standard order so that the "sth treatment label" is unambiguous. The standard order is that obtained when the treatment labels are regarded as n-digit numbers and written in ascending order. A generalized cyclic design with n-digit treatment labels is sometimes called a GC/n *design* or n-*cyclic design* for short.

Example 3.2

Consider an experiment, such as that in Example 2.2, with two factors, the first factor having $m_1 = 3$ levels (coded 0, 1, 2) and the second factor having $m_2 = 4$ levels (coded 0, 1, 2, 3). There are $v = 12$ treatment combinations whose coded labels can be written in standard order as follows:

00 01 02 03 10 11 12 13 20 21 22 23

Suppose the $k = 6$ treatment labels 00 02 11 12 20 23 are selected for the
initial block. The second block is obtained by adding the second treatment
label, 01, to each treatment label in the first block. The second block
therefore contains $00 + 01$, $02 + 01$, $11 + 01$, $12 + 01$, $20 + 01$ and $23 + 01$.
Similarly the seventh block, for example, is obtained by adding the seventh
label to the initial block and therefore contains $00 + 12$, $02 + 12$, $11 + 12$,
$12 + 12$, $20 + 12$, $23 + 12$. Addition is carried out separately on the two
digits. The code cycles for the first and second digits are

Thus, for example, $23 + 12$ involves adding $2 + 1$ using the first cycle and
$3 + 2$ using the second cycle. Consequently $23 + 12 = 01$. The design has
twelve distinct blocks, which are shown below with each row denoting the
treatment combinations in the same block

$$
\begin{array}{cccccc}
00 & 02 & 11 & 12 & 20 & 23 \\
01 & 03 & 12 & 13 & 21 & 20 \\
02 & 00 & 13 & 10 & 22 & 21 \\
03 & 01 & 10 & 11 & 23 & 22 \\
10 & 12 & 21 & 22 & 00 & 03 \\
11 & 13 & 22 & 23 & 01 & 00 \\
12 & 10 & 23 & 20 & 02 & 01 \\
13 & 11 & 20 & 21 & 03 & 02 \\
20 & 22 & 01 & 02 & 10 & 13 \\
21 & 23 & 02 & 03 & 11 & 10 \\
22 & 20 & 03 & 00 & 12 & 11 \\
23 & 21 & 00 & 01 & 13 & 12 \\
\end{array}
$$

As can be seen in the design in Example 3.2, every treatment label in a
generalized cyclic design occurs r times, and if $b = v$, then $r = k$. However,
if v and k are both divisible by an integer $d > 1$, then it is possible to
obtain a design with $b = v/d$ blocks, in which case $r = k/d$. The number
of distinct blocks in a design having a single initial block is discussed in
Section 6.1. Larger designs can be obtained by selecting a second initial
block and repeating the construction process. The choice of treatment
labels for each initial block is an art and is discussed in Section 5.

4 ANALYSIS OF FACTORIAL EXPERIMENTS IN BLOCK DESIGNS

4.1 The Model

For simplicity, consider first an experiment with two treatment factors, F_1 and F_2, having m_1 and m_2 levels, respectively, run as a block design with b blocks of size k (where $k \leq v = m_1 m_2$). Suppose that r observations are taken on each treatment combination and no treatment combination is observed more than once in a block. An appropriate model for such an experiment is often assumed to be of the form

$$y_{i_1 i_2 s} = \mu + \beta_s + \alpha_{i_1}^{10} + \alpha_{i_2}^{01} + \alpha_{i_1 i_2}^{11} + e_{i_1 i_2 s} \tag{1}$$

where $y_{i_1 i_2 s}$ are the responses from those treatment combinations $i_1 i_2$ observed in block s ($i_1 = 0, 1, \ldots, m_1 - 1; i_2 = 0, 1, \ldots, m_2 - 1; s = 1, \ldots, b$); $e_{i_1 i_2 s}$ are the corresponding error variables which are assumed to be independent and have identical normal distributions with mean zero and variance σ^2; μ is a constant; β_s is the effect of the sth block (that is, β_s is a parameter which is positive if the observations in the sth block tend to be inflated above the average and negative if they tend to be depressed); $\alpha_{i_1}^{10}$ is the effect of factor F_1 at level i_1, and is positive or negative depending on whether observations on level i_1 tend to be above or below the average; similarly, $\alpha_{i_2}^{01}$ is the effect of factor F_2 at level i_2; $\alpha_{i_1 i_2}^{11}$ is the joint effect of F_1 at level i_1 and F_2 at level i_2 over and above (or below) the sum of their individual effects.

If the values of $\alpha_0^{10}, \alpha_1^{10}, \ldots, \alpha_{m_1-1}^{10}$ differ substantially, then the different levels of factor F_1 produce significantly different responses, and we say that there is a significant *main effect* of F_1. Otherwise F_1 is said to be negligible. Similar comments apply to the main effect of F_2. If the values of $\alpha_{00}^{11}, \ldots, \alpha_{(m_1-1)(m_2-1)}^{11}$ differ substantially, then the difference in the response obtained when factor F_1 is at level i_1 from that obtained at level i_2 depends also upon the level of factor F_2. We say that the factors F_1 and F_2 *interact* or that there is a substantial *interaction* between F_1 and F_2, denoted by $F_1 F_2$. Otherwise the interaction $F_1 F_2$ is said to be negligible.

The notation can easily be extended to cover $n > 2$ factors. For example, if $x = x_1 x_2 \ldots x_n$, then $F^x = F_1^{x_1} F_2^{x_2} \ldots F_n^{x_n}$ represents the interaction between those factors for which $x_i = 1$ (that is, the joint effect of those factors over and above the sum of their individual effects and the interactions between subsets of those factors). Thus, in an n-factor experiment, if

$x = 110 \ldots 01$, then $F^x = F_1 F_2 F_n$ represents the interaction between factors F_1, F_2 and F_n. The effect of each interaction F^x can be represented in the model by $\alpha^x = \alpha^{x_1 x_2 \cdots x_n}$ with subscripts as appropriate.

It is sometimes reasonable to assume that some or all of the high order interactions are negligible, in which case their effects are omitted from the model. The first task in the analysis of a factorial experiment is to determine whether or not the interactions which are included in the model are also negligible. If there are no significant interactions between the factors, then the main effect of each factor can be analyzed separately. Typically, this involves comparing the effects of the different levels of each factor or, if a factor is quantitative, estimating the linear and quadratic trends in the response as the level of the factor increases. If, however, there are sizable interactions, then analysis of main effects may or may not be of interest. Instead, the experimenter may wish to compare the effects of the individual treatment combinations or, in the case of quantitative factors, to estimate a response surface (see Box and Draper, 1987) and to determine the trends in the response that can be ascribed to the interactions (see for example, Montgomery, 1984, Chapter 9.3).

4.2 Analysis of Variance

When the factorial experiment has n factors and every treatment combination is observed exactly once in every block, the analysis of variance table for testing the significance of each of the main effects and the interactions is the usual $(n+1)$-way cross-classification table (n treatment factors and one block factor). The treatment-block interactions are usually assumed to be negligible and are used to estimate error. The more common case of small block sizes requires that the estimated effect of each treatment combination be adjusted for those blocks in which it is observed (see for example, John, 1987, Chapter 1). The adjustment is done automatically by any standard analysis of variance computer package that can handle unbalanced data, provided that the block parameter is the first term entered into the model.

Once the effects of the treatment combinations have been adjusted for blocks, two problems can arise. First, it may no longer be possible to estimate one or more of the factorial effects. If there are $r > 1$ observations on each treatment combination, this problem can usually be avoided by using a *connected* design (see Section 6.3). If there is at most one observation on each treatment combination, the problem cannot be avoided and extreme care needs to be taken over the construction of the incomplete block design and the analysis of the data (see Section 7). The second problem that may occur is that the estimates of the different factorial effects (main effects and interactions) may be correlated, in which case, interpretation of the analysis becomes difficult. However, this problem can often be avoided since there are a large number of block designs in which the estimates do remain uncorrelated. Such designs are said to have *orthogonal factorial structure* (OFS). All complete block designs, balanced incomplete block designs, group divisible designs and generalized cyclic designs have orthogonal factorial structure (see, for example, John, 1987, Chapter 7). The analysis of a factorial experiment run as a generalized cyclic design is illustrated in Example 4.1.

Example 4.1

A generalized cyclic design suitable for the baking experiment of Example 2.2 is the design constructed in Example 6.3 and reproduced in Table 1, together with some hypothetical data. The treatment combinations have been randomly assigned to shelves of each oven and are listed in order from the top to the bottom shelf.

The analysis of variance table obtained from a standard analysis of variance package is shown in Table 2. The parameter representing the block effect has been entered into the model first. Consequently, the block sum of squares has not been adjusted for the effect of the different treatment

Table 1 A Generalized Cyclic Design and Hypothetical Data for the Baking Experiment of Section 2

Oven 1	Run 1	13	21	02	00	23	11
		12.1	18.4	17.9	10.2	10.1	15.6
	Run 2	20	22	03	01	10	12
		13.1	31.2	12.2	16.7	11.9	21.0
Oven 2	Run 1	23	12	21	03	10	01
		12.5	26.9	15.8	18.2	11.1	18.2
	Run 2	22	13	02	20	00	11
		31.5	17.7	22.4	18.8	11.8	14.6
Oven 3	Run 1	22	13	01	20	11	03
		30.4	17.1	20.0	14.5	19.9	16.0
	Run 2	12	02	23	00	21	10
		32.8	25.2	10.8	12.6	18.8	19.6

Table 2 Analysis of Variance Table for the Experiment in Table 1

Source of variation	Sum of squares	Degrees of freedom	Mean square	F ratio
Blocks (before fitting treatment factors)	137.4588	5	—	—
F_1 (after fitting blocks)	18.4838	2	9.24	1.60
F_2 (after fitting blocks)	1006.8594	3	335.62	58.21
$F_1 F_2$ (after fitting blocks and main effects)	155.5445	6	25.92	4.50
Error	109.5521	19	5.77	
Total	1427.8988	35		

combinations occurring in the different blocks, and it should not be used. The main effect parameters of the two factors have been entered into the model following the block effect, and finally the interaction parameter has been entered. The main effects could have been entered into the model in either order (following the block effect) without altering the sums of squares since the design has orthogonal factorial structure. The sums of squares can be calculated by hand using a matrix formulation as shown by John (1981b, and 1987, Section 7.4).

No parameter representing the shelf effect was used in the model since this effect was known to be very small before the experiment took place. The analysis of variance indicates that there is an interaction between the two factors F_1 (glycerol) and F_2 (tartaric acid). This will be investigated further in Example 4.2.

4.3 Estimation of Factorial Effects

Model (1) can be rewritten as

$$y_{is} = \mu + \beta_s + \tau_i + e_{is} \tag{2}$$

where $i = i_1 i_2$ and $\tau_i = \tau_{i_1 i_2} = \alpha_{i_1}^{10} + \alpha_{i_2}^{01} + \alpha_{i_1 i_2}^{11}$ is the effect of the ith treatment combination (using the standard order defined in Section 3). Some computer packages, such as SAS, will print out least squares estimates, $\hat{\tau}_i$, of the parameters, τ_i. These are shown in Table 3 for the data in Example 4.1. They were obtained from the PROC GLM option LSMEANS in SAS, using model (2). The least squares estimates are not unique and a different computer package may give a different set of estimates. However, if the design is connected (see Section 6.3), then all possible contrasts $\sum_i c_i \tau_i$ with $\sum_i c_i = 0$ can be estimated uniquely and the corresponding least squares estimates are given by $\sum_i c_i \hat{\tau}_i$.

Contrasts of particular interest in factorial experiments are those which have a product structure, that is, $c_i = c_{i_1 i_2} = c_{i_1}^{10} c_{i_2}^{01}$. When $\sum_{i_1} c_{i_1}^{10} = \sum_{i_2} c_{i_2}^{01} = 0$, then $\sum_i c_i \tau_i$ is the interaction contrast $\sum_{i_1} \sum_{i_2} c_{i_1}^{10} c_{i_2}^{01} \alpha_{i_1 i_2}^{11}$; when $\sum c_{i_1}^{10} = 0$ and $c_{i_2}^{01} = m_2^{-1}$ for $i_2 = 0, \ldots, m_2 - 1$, the contrast $\sum_i c_i \tau_i$ is the main effect contrast $\sum_{i_1} c_{i_1}^{10} (\alpha_{i_1}^{10} + \bar{\alpha}_{i_1 .}^{11})$; and when $\sum_{i_2} c_{i_2}^{01} = 0$ and $c_{i_1}^{10} = m_1^{-1}$ for all $i_1 = 0, \ldots, m_1 - 1$, then $\sum_i c_i \tau_i$ is the main effect contrast $\sum_{i_2} c_{i_2}^{01} (\alpha_{i_2}^{01} + \bar{\alpha}_{.i_2}^{11})$. Notice that main effect contrasts are always averaged over the interaction effects and consequently they may not be of interest if the interaction is non-negligible.

Table 3 Least Squares Estimates for τ_i in Example 4.1

Treatment combination $i_1 i_2$	Number in standard order i	Least Squares estimate $\hat{\tau}_i$
00	1	10.95
01	2	18.88
02	3	21.25
03	4	16.05
10	5	13.98
11	6	16.92
12	7	26.68
13	8	15.85
20	9	13.71
21	10	17.09
22	11	31.61
23	12	10.55

All of the above ideas extend in the obvious way to experiments with more than two factors

Example 4.2

In Example 4.1, the analysis of variance table showed that there was an interaction between the two treatment factors F_1 and F_2. The design is connected (see Example 6.4) and therefore all interaction contrasts of the above form can be estimated. The least squares estimates in Table 3 suggest that, as the coded level of factor 2 increases, a quadratic trend is produced in the response and that this quadratic trend differs slightly for different levels of the first factor. The two factors are quantitative, and if their actual levels correspond to their coded levels in increasing order, the above observations suggest that a linear F_1 × quadratic F_2 contrast might be of interest. If the actual levels of the two factors are equally spaced, then the values of $c_{i_1}^{10}$ and $c_{i_2}^{01}$ which give such a contrast can be obtained from a table of orthogonal polynomial coefficients (see, for example, Montgomery, 1984, Section 7.6), that is,

$$c_0^{10} = -1, \qquad c_1^{10} = 0, \qquad c_2^{10} = 1,$$
$$c_0^{01} = -1, \qquad c_1^{01} = 1, \qquad c_2^{01} = 1, \qquad c_3^{01} = -1$$

The least squares estimate of the linear F_1 × quadratic F_2 contrast is $\sum_{i_1} \sum_{i_2} c_{i_1}^{10} c_{i_2}^{10} \hat{\tau}_{i_1 i_2} = 11.30$. The standard error of this estimate is 3.92 (obtained from the ESTIMATE statement of PROC GLM in SAS), and the significance of the contrast can be tested using Scheffé's method of multiple comparisons. A t-test should be used only when a contrast is specified before the data is examined.

4.4   Average Efficiency Factors

The standard error of the least squares estimate of any contrast depends upon the particular block design being used. It is generally the case that no one block design gives the smallest standard errors for *all* contrasts of interest. Compromises have to be made. One of the most common compromises is to select the design that has the minimum average variance for the estimates of the pairwise comparisons $\tau_1 - \tau_j$. This is known as the *A-optimality* criterion.

For a given block design d, with r observations on each of the v treatment combinations, the *average efficiency factor* E_d of the design is defined as the "ratio of the average variance of the pairwise comparison estimators $\hat{\tau}_i - \hat{\tau}_j$ in d and the corresponding average variance for a randomized (complete) block design (RBD) with r blocks of size v, multiplied by σ_d^2/σ_R^2," where σ_d^2 is variance of the error variables in (1) when the block design d is used, and σ_R^2 is the variance of the error variables that would result if the randomized block design could be used. Thus,

$$E_d = \frac{\text{Average var}(\hat{\tau}_i - \hat{\tau}_j) \text{ in RBD}}{\text{Average var}(\hat{\tau}_i - \hat{\tau}_j) \text{ in } d} \times \frac{\sigma_d^2}{\sigma_R^2} \qquad (3)$$

For a randomized block design, $\text{Var}(\hat{\tau}_i - \hat{\tau}_j) = 2r^{-1}\sigma_R^2$ for every pairwise comparison $\tau_i - \tau_j$. Consequently, (3) reduces to

$$E_d = 2r^{-1}a_d^{-1} \qquad (4)$$

where $a_d = [\text{Average Var}(\hat{\tau}_i - \hat{\tau}_j) \text{ in } d]/\sigma_d^2$.

The calculation of a_d for general incomplete block designs is more complicated (see John, 1987, Section 2.1). It is equal to the average of the non-zero eigenvalues of the variance-covariance matrix of $\hat{\tau}_1, \ldots, \hat{\tau}_v$ multiplied by $2r\sigma_d^{-2}$ as shown in (9). A specific calculation of (9) for generalized cyclic designs is given in Section 5.

The A-optimal design has the maximum achievable value of E_d. An upper bound for E_d given by Williams, Patterson and John (1976), is

$$U_d = 1 - \frac{k^*(v - k^*)}{k^2(v - 1)} \qquad (5)$$

where k is the block size and k^* is k reduced modulo v (that is start at 0 in the code cycle of length v and move along k positions). The upper bound is achieved by all balanced block designs, and these are therefore A-optimal.

For factorial experiments it is preferable to consider separately the average variance of the estimates of a set of contrasts for each factorial effect. The average efficiency factor for a factorial effect F^x is defined as follows. Let $\sum_i c_{qi}^x \tau_i$, $q = 1, \ldots, w^x$, where $w^x = \prod_{j=1}^n (m_j - 1)^{x_j}$, be a set of orthogonal contrasts for the factorial effect F^x (that is $\sum_i c_{qi}^x c_{ti}^x = 0$ for $q \neq t$ and $= 1$ for $q = t$). Then the average efficiency factor E_{dx} of the factorial effect F^x in the design d is

$$E_{dx} = \frac{\text{Average var}(\sum_i c_{qi}^x \hat{\tau}_i) \text{ in RBD}}{\text{Average var}(\sum_i c_{qi}^x \hat{\tau}_i) \text{ in } d} \times \frac{\sigma_d^2}{\sigma_R^2} \qquad (6)$$

where the average is taken over the w^x contrasts $\sum_i c_{qi}^x \tau_i$, $q = 1, \ldots, w^x$.

For a randomized block design, $\text{Var}(\sum_i c_{qi}^x \hat{\tau}_i) = r^{-1}\sigma_R^2$ for every normalized contrast $\sum_i c_{qi}^x \tau_i$. Consequently (6) reduces to

$$E_{dx} = r^{-1} a_{dx}^{-1} \tag{7}$$

where $a_{dx} = \sigma_d^2 \times$ Average $\text{Var}(\sum_i c_{qi}^x \hat{\tau}_i)$ in d. The calculation of a_{dx} is discussed by John (1987, Chapter 7). An alternative method for calculating E_{dx} for generalized cyclic designs is given in Section 5.

Lewis and Dean (1984) showed that, when d has orthogonal factorial structure, the relationship between the factorial average efficiency factors E_{dx} and the design average efficiency factor E_d is

$$(v-1)E_d^{-1} = \sum_t \sum_{A(x,t)} w^x E_{dx}^{-1} \tag{8}$$

where $A(x,t)$ is the set of all index vectors x which satisfy $\sum_j x_j = t$, and w^x is defined above.

If d has orthogonal factorial structure, (5) provides an upper bound for the average efficiency factor of the ith effect when v is replaced by m_i (see Lewis and Dean, 1985). Upper bounds for average efficiency factors of the interactions are more complicated since (8) must be satisfied (see Example 5.1, and Lewis and Dean, 1984, or Bailey, 1985a). In section 5, a simple method is described for obtaining high factorial average efficiency factors in generalized cyclic designs.

5   FACTORIAL EFFICIENCY FACTORS IN GENERALIZED CYCLIC DESIGNS

John (1973) showed that the efficiency of the main effect of the hth factor F_h in a generalized cyclic design depends upon the frequency with which each code occurs as the hth digit of the treatment labels in the initial block ($h = 1, \ldots, n$). In particular, he showed by example that the highest efficiencies seem to be obtained when the frequencies of occurrence of the codes are as equal as possible. For example, the design of Table 1 has codes 0, 1, 2 for the first factor, each occurring twice in the first block, and for the second factor codes 0 and 2, each occurring once and, for codes 1 and 3, each occurring twice. In both cases the frequencies are as equal as possible. The efficiency of the second main effect is therefore unlikely to be increased by altering how many codes occur with each frequency. However, it is possible that the efficiency could be increased a little by changing *which* of the codes occur once and which occur twice. Such a change can also alter the number of distinct blocks in the design (see Section 6.1).

The frequency argument also applies to the efficiency of t-factor interactions F^x, $(\sum x_h = t, t \leq n)$, as follows. Delete all the digits of the treatment labels in the initial block corresponding to $x_h = 0$ ($h = 1, \ldots, n$). Each label then contains the t codes corresponding to the factors in the interaction F^x. A high value of the average efficiency factor E_{dx}, defined in (5), tends to occur when the frequencies of occurrence of the t-digit codes are as equal as possible. If there is more than one initial block in the design, then all the initial blocks should be considered as a single block when the frequency count is made.

Due to the construction method of a generalized cyclic design and the cycling property of the codes, the requirement that the frequency of occurrence of each code (or t-digit code) should be as equal as possible in the initial block(s) is the same as the requirement that each code (or t-digit code) should occur together with the zero code (or t-digit code) in the same, or almost the same, number of blocks. This, in turn, is the same as the requirement that all pairs of codes (or t-digit codes) occur together in the same block an equal, or almost equal, number of times, and this seems to ensure high average efficiency factors. This frequency argument has been supported by the work of many authors, but it has not yet been proved mathematically (see John, 1987, Section 2.5).

For the hth main effect ($1 \leq h \leq n$), let p^h_{ij} be the number of times that code i occurs as the hth digit in the treatment labels in block j, ($i = 0, \ldots, m_h - 1; j = 1, \ldots, b$). Let λ^h_i be the number of blocks in which code i and code 0 both occur as the hth digit; then $\lambda^h_i = \sum_{j=1}^b p^h_{0j} p^h_{ij}$. The highest average efficiency factor for the hth main effect tends to occur in designs for which the λ^h_i, $i = 1, \ldots, m_h$ are as equal as possible.

Similarly, for the t-factor interaction F^x, let $i = i_1 i_2 \ldots i_t$ denote the t digits of the treatment labels corresponding to those factors for which $x_h = 1$ ($h = 1, \ldots, n$). Let p^x_{ij} be the number of times that the t-digit code i occurs in block j, and let $\lambda^x_i = \sum_{j=1}^b p^x_{0j} p^x_{ij}$. The highest factorial average efficiency factors E_{dx} tend to occur in designs for which the λ^x_i are as equal as possible.

For generalized cyclic designs, (4) and (7) can be calculated from the λ^y_i with $y = 11 \ldots 1$ and $i = i_1 i_2 \ldots i_n$. In general, for a connected design,

(4) can be re-expressed as

$$E_d = (v-1)\left(\sum_{u=1}^{v-1} e_u^{-1}\right)^{-1}$$

$$u = u_1 u_2 \ldots u_n; \qquad u_h = 0, \ldots, m_h - 1; \qquad h = 1, \ldots, n \qquad (9)$$

where e_1^{-1}, e_2^{-1}, $\ldots$, e_{v-1}^{-1} are called the *canonical efficiency factors* and are the non-zero eigenvalues of $r\sigma^{-2}V$, where V is the variance-covariance matrix of $\hat{\tau}_1, \ldots, \hat{\tau}_v$. For generalized cyclic designs, e_1, e_2, $\ldots$, e_v are functions of the $\lambda_{i_1 i_2 \ldots i_n}^{1 \; 1 \; \ldots \; 1} = \lambda_{i_1 i_2 \ldots i_n}$ as follows,

$$e_u = e_{u_1 u_2 \ldots u_n}$$

$$= 1 - (rk)^{-1} \sum_{i_1=0}^{m_1-1} \sum_{i_2=0}^{m_2-1} \cdots \sum_{i_n=0}^{m_n-1} \lambda_{i_1 i_2 \ldots i_n} \cos\left[2\pi \sum_{h=1}^{n} u_h i_h / m_h\right] \qquad (10)$$

The factorial average efficiency factors E_{dx} given in (7) can re-expressed as

$$E_{dx} = w^x \left(\sum_{u=1}^{v} \delta(u, x) e_u^{-1}\right)^{-1} \qquad (11)$$

where $w^x = \prod_{h=1}^{v}(m_h - 1)^{x_h}$ and where $\delta(u, x) = 1.0$ if $u_h = 0$ when $x_h = 0$ and if $u_h \neq 0$ when $x_h = 1$, and $\delta(u, x) = 0.0$ otherwise (see Williams, 1975, John and Lewis, 1983, and John, 1987, Section 7.6).

Example 5.1

The generalized cyclic design in Table 1 is listed by Lewis and Dean (1980) as having factorial average efficiency factors $E_{10} = 1.0$, $E_{01} = 0.96$ and $E_{11} = 0.79$. These can be calculated from (10) and (11) as shown below.

The values of λ_i can be obtained from Table 1 by counting the number of times that $i = i_1 i_2$ occurs in the same block as 00. Consequently, $\lambda_{00} = r = 3$, $\lambda_{01} = \lambda_{03} = 0$, $\lambda_{02} = 3$, $\lambda_{10} = \lambda_{12} = \lambda_{20} = \lambda_{22} = 1$, $\lambda_{11} = \lambda_{13} = \lambda_{21} = \lambda_{23} = 2$, giving $e_{01} = e_{03} = e_{10} = e_{11} = e_{13} = e_{20} = e_{21} = e_{23} = 1.0$, $e_{12} = e_{22} = 5/9$, $e_{02} = 8/9$.

Thus, $E_{10} = 2(e_{10}^{-1} + e_{20}^{-1})^{-1} = 1.0$, $E_{01} = 3(e_{01}^{-1} + e_{02}^{-1} + e_{03}^{-1})^{-1} = 0.96$ and $E_{11} = 6(e_{11}^{-1} + e_{12}^{-1} + e_{13}^{-1} + e_{21}^{-1} + e_{22}^{-1} + e_{23}^{-1})^{-1} = 0.7895$. The average efficiency factor E_d is $11(\sum_{u_1} \sum_{u_2} e_{u_1 u_2}^{-1})^{-1} = 0.8644$. The upper bound for E_d is given by (5), namely, $U_d = 1 - 6(12 - 6)/36(12 - 1) = 0.9091$. Therefore, in terms of the average efficiency of the design, it is likely that better designs exist. However, the design is good for estimating contrasts in the main effects, since $E_{10} = 1.0$, $E_{01} = 0.96$, and for E_{01} the upper bound (5) gives $U_{01} = 1 - 2(4 - 2)/36(4 - 1) = 0.9630$, which leaves little room for improvement. For the interaction, (8) gives

$$6E_{11}^{-1} = 11E_d^{-1} - 2(1.0)^{-1} - 3(0.96)^{-1} \geq 11U_d^{-1} - 2.0 - 3.125 = 6.9749$$

Hence $E_{11} \leq 0.8602$. This bound can be tightened a little using (3.3) of Bailey (1985) which gives $U_{11} = 0.8533$. Since $E_{11} = 0.79$ for the design of Table 1, it is possible that a better design could be found. However, the requirement of resolvability may decrease the achievable efficiency.

In summary, the construction of efficient generalized cyclic designs is best achieved by monitoring the frequencies of the t-digit codes (for all $t = 1, \ldots, n$) in the initial block(s). The selection of one-digit codes is facilitated by consulting the tables of John, Wolock and David (1972) which list the most efficient selection. Orthogonal factorial structure is guaranteed for all generalized cyclic designs. Other properties such as the number of distinct blocks, resolvability, and connectivity are discussed in detail in Section 6. A list of efficient resolvable designs is given by Lewis and Dean (1980). A list of designs with block size $k = 2$ (paired comparison designs) is given by Lewis and Tuck (1985).

6 COMBINATORIAL PROPERTIES OF GENERALIZED CYCLIC DESIGNS

All properties of generalized cyclic designs depend upon the selection of the treatment labels for the initial block. In this section, it is shown how the number of distinct blocks and the resolvability and the connectivity of the design are determined from the initial block.

6.1 Number of Distinct Blocks

To construct a generalized cyclic design with at most v/d distinct blocks, $d > 1$, the treatment labels must first be partitioned into v/d sets of d labels. The procedure for forming the sets is described below and illustrated in Example 6.1. The initial block of the design consists of a selection of k/d of these sets. The other blocks of the design are then found in the usual way. Notice that d must divide both v and k.

To partition the treatment labels into sets of size d, first, find the lowest common multiple, μ, of $m_1, m_2, \ldots, m_n$ (that is the smallest integer that is divisible by every m_h, $h = 1, \ldots, n$). Now d may or may not divide μ. Consider first the case where d divides μ, then $d = \mu/t$ for some integer t. Select a treatment label $i_1 i_2 \ldots i_n \neq 00 \ldots 0$ which satisfies

$$t = \mathrm{HCF}(\mu, i_1 \mu/m_1, i_2 \mu/m_2, \ldots, i_n \mu/m_n) \qquad (12)$$

where selection of $i_h = 0$ is the same as selection of $i_h = m_h$ due to the code cycle, and where HCF stands for highest common factor. If the selected treatment label $i_1 i_2 \ldots i_n$ is added to itself repeatedly until $00 \ldots 0$ is reached, a set S of d distinct treatment labels is formed. (This is the reverse of the procedure detailed by Dean and John, 1975, where $i_1 i_2 \ldots i_n$ is selected and t and d are then calculated.) The resulting set S of d labels is said to have been *generated* by $i_1 i_2 \ldots i_n$ and forms an algebraic structure called a *subgroup of the group of treatment labels*. The remaining treatment labels are partitioned into sets of size d by adding each treatment label to those labels in the subgroup, S. In all, v/d distinct sets are formed (including S) and these are known as the *cosets* of S. To construct the design, k/d cosets (including S) are selected to form the initial block. When the other blocks of the design are obtained by adding each treatment label in turn to the labels in the initial block, at most v/d distinct blocks will result, and each of these blocks will contain a selection of cosets of S. Generally the number of distinct blocks will be exactly v/d, but occasionally the cosets selected for the initial block are able to be combined into a larger subgroup (say S^* of size $d^* > d$), and some of its cosets (see Example 6.1). In this case the design would only have v/d^* distinct blocks.

If d does not divide μ, then the above procedure must be modified slightly. This is discussed after Example 6.1.

Example 6.1

Suppose that a 3×6 factorial experiment is to be run in six blocks, and six observations can be taken per block. Consider using a generalized cyclic design with $v = 3 \times 6 = 18$ treatment labels, and $b = 6 = 18/3$ blocks of size $k = 6$. Now $\mu = 6$ and $d = 3$, therefore $d = \mu/t$ when $t = 2$. From (12), treatment label $i_1 i_2 \neq 00$ must be selected so that $\mathrm{HCF}(6, 2i_1, i_2) = 2$. There are eight possibilities since any one of the choices $i_1 i_2 = 02, 04, 10,$ 12, 14, 20, 22 or 24 satisfies the requirement.

Suppose that $i_1 i_2 = 12$ is selected; then the subgroup of size $d = 3$ is $S = (12, 24, 00)$ and the other treatment labels are divided into 5 cosets of

S as follows. The sixth coset is S itself.

$$01 + S = (01 + 12, 01 + 24, 01 + 00) = (13, 25, 01)$$
$$02 + S = (02 + 12, 02 + 24, 02 + 00) = (14, 20, 02)$$
$$03 + S = (03 + 12, 03 + 24, 03 + 00) = (15, 21, 03)$$
$$04 + S = (04 + 12, 04 + 24, 04 + 00) = (10, 22, 04)$$
$$05 + S = (05 + 12, 05 + 24, 05 + 00) = (11, 23, 05)$$

The required design has an initial block consisting of $k/d = 2$ cosets one of which, by convention, should be S. If the coset (15,21,03) is selected, the initial block becomes $H = (00, 12, 24, 15, 21, 03)$. This is an unfortunate choice because H happens to be a subgroup of size 6, and only $v/6 = 3$ distinct blocks will result. This can be checked by setting $t = 1$ and $i_1 i_2 = 15$. Since such circumstances are not easy to foresee, it is recommended that the selection of cosets is changed whenever fewer than the desired number of distinct blocks results. There are thus four possible selections for the initial block. It was suggested in Section 5 that the frequencies of the codes in the initial block should be as equal as possible, in which case $(14, 20, 02)$ and $(10, 22, 04)$ should be avoided. This leaves two possibilities, namely $(13, 25, 01)$ and $(11, 23, 05)$. Arbitrarily selecting the first of these gives the following design in six blocks of size 6.

00	12	24	13	25	01
01	13	25	14	20	02
02	14	20	15	21	03
03	15	21	10	22	04
04	10	22	11	23	05
05	11	23	12	24	00

Dean and Lewis (1980) used the following shorthand notation for an initial block of a design such as that in Example 6.1. The initial block, H, is formed from S and the coset $01 + S$, that is, $H = (S, S + 01)$. This can be represented as $H = S [+] R$ where $R = (00, 01)$ and $[+]$ means add every treatment label in R to every treatment label in S.

It is not always the case that d divides μ; however, d can always be written as $d = d_1 d_2 \ldots d_q$, where $d_j = \mu/t_j$, $1 < t_j < \mu$, $j = 1, \ldots q$, for some integer q. This representation can sometimes be written in more than one way, resulting in different selections of designs. The subgroup

S is generated iteratively. First, find a subgroup S_1, of size $d_1 = \mu/t_1$ using the above procedure. Repeat the procedure to obtain a subgroup S_2 of size $d_2 = \mu/t_2$ so that S_1 and S_2 have only the label $00 \ldots 0$ in common. This *can* always be done, but S_1 may need to be changed or a different representation of d used in order to achieve it. Combine S_1 and S_2 into a subgroup S_{12} of size $d_1 d_2$ by adding every label in S_1 to every label in S_2. If $q \geq 3$, then find a subgroup S_3 of size $d_3 = \mu/t_3$ so that S_3 and S_{12} have only the label $00 \ldots 0$ in common. Combine S_{12} and S_3 into a subgroup S_{123} of size $d_1 d_2 d_3$. If $q \geq 4$, repeat the procedure until eventually $S = S_{13\ldots q}$ which is of the required size. An illustration is given for $q = 2$ in Example 6.2.

Example 6.2

Consider an experiment with two treatment factors, each having 6 levels, which is to be run as an incomplete block design with $b = 4$ blocks of size $k = 18$. Now $b = v/d = 36/9$. So $d = 9$ and $\mu = 6$. Thus d cannot be written as μ/t but it can be written as $d_1 d_2$ where $d_1 = d_2 = 6/2$, giving $t_1 = t_2 = 2$. To find S_1, choose $i_1 i_2$ so that $\text{HCF}(6, i_1, i_2) = 2$. The possibilities are $i_1 i_2 = 02,20,22$. Let $i_1 i_2 = 22$, then $S_1 = (22, 44, 00)$. Repeating the procedure for S_2 gives, for example, $S_2 = (20, 40, 00)$. Combining these two subgroups gives a subgroup of size $d = 9$, namely, $S = S_{12} = (42, 02, 22, 04, 24, 44, 20, 40, 00)$. There are three other cosets $S + 01$, $S + 10$ and $S + 11$. The initial block consists of S and preferably the coset $S + 11$ (in order to have as many different codes as possible in the initial block).

6.2   Resolvability

Resolvable designs are block designs whose b blocks can be divided into r sets of v/k blocks so that the blocks within a set contain all the treatment labels exactly once. Resolvable designs are frequently used as a safeguard against unforeseen termination of an experiment, or against loss of data from an entire section of the experiment. The second experiment discussed in Section 2 is an example of a situation where such a precaution is advisable.

The class of generalized cyclic designs is rich in resolvable designs, some of which are listed by Lewis and Dean (1980). The specific algebraic requirements on the initial block for obtaining a resolvable design are discussed in detail by Dean and Lewis (1980). Clearly, a resolvable design cannot exist if v/k is not an integer. If it is an integer, then the resolvability depends on the difference set D_H of the initial block H. The difference set D_H contains the $k(k-1)$ differences between pairs of treatment labels in H, together with the label $00\ldots0$. The difference between the pair of treatment labels $a_1 a_2 \ldots a_n$ and $b_1 b_2 \ldots b_n$ is

$$a_1 a_2 \ldots a_n - b_1 b_2 \ldots b_n = (a_1 - b_1)(a_2 - b_2) \ldots (a_n - b_n)$$

where $(a_j - b_j)$ is calculated from the appropriate code cycle, starting at a_j and moving b_j steps in the reverse direction. For example, if the jth treatment factor has $m_j = 4$ levels, the code cycle is

$$\boxed{\longrightarrow 0 \to 1 \to 2 \to 3 \to}$$

so $1 - 2 = 3$. An alternative method of calculation makes note of the fact that 1 is the same as 5 in the code cycle so that $1 - 2 = 5 - 2 = 3$.

To calculate the difference set, D_H, the $k(k-1)$ differences must be calculated. However, a shortcut method can be found when H is expressed as $S\,[+]\,R$, (see Section 6.1), since $D_H = S\,[+]\,D_R$ where D_R is the difference set of R. This method involves calculating only $k(k-d)/d^2$ differences. Let the set $\bar{D}_H$ contain all the treatment labels not in D_H plus the label $00\ldots0$. The design is resolvable if (and only if) a set C of v/k treatment labels can be selected from $\bar{D}_H$ with the property that the difference set D_C of C includes only labels in $\bar{D}_H$. The first set of blocks in the resolvable design is obtained by adding the treatment labels in C in turn to the initial block. The rest of the design is obtained in the usual way. The procedure is illustrated in Example 6.3. It is possible that no set C can be chosen from $\bar{D}_H$ with the required properties. In this case, the design with initial block H is not resolvable.

Example 6.3

The second experiment in Section 2 was run as a generalized cyclic design with six blocks of size 6. There were $v = m_1 m_2 = 3 \times 4 = 12$ treatment combinations in total; thus a resolvable design would consist of three sets of two blocks, each set containing all twelve treatment labels. Since only $b = 6 = 12/2$ blocks are required, the initial block must contain a subgroup of size 2 and two other cosets. The only subgroup of size 2 is $S = (00, 02)$ since $6 = \text{HCF}(12, 4i_1, 3i_2)$ must be satisfied (see Section 6.1). Suppose that the two cosets (11,13) and (21,23) are selected for the initial block;

then,

$$H = (00, 02, 11, 13, 21, 23) = (00, 02)\,[+]\,(00, 11, 21)$$

Thus

$$R = (00, 11, 21)$$

and

$$D_R = (00 - 11, 00 - 21, 11 - 00, 11 - 21, 21 - 00, 21 - 11, 00)$$
$$= (23, 13, 11, 20, 21, 10, 00).$$

Then, ignoring repeated treatment labels,

$$D_H = (00, 02)\,[+]\,(23, 13, 11, 20, 21, 10, 00)$$
$$= (23, 21, 13, 11, 20, 22, 10, 12, 00, 02)$$

and

$$\bar{D}_H = (00, 01, 03).$$

Note that D_H can also be obtained by calculating the differences of the 30 pairs of labels in H. A set of C of size $v/k = 2$ must be selected from $\bar{D}_H$ so that D_C is contained in $\bar{D}_H$. Either $C = (00, 01)$ or $C = (00, 03)$ fulfills this requirement. Taking $C = (00, 01)$, the resolvable design is

$$\text{Set 1} \begin{cases} \text{Block 1} & 00\ 02\ 11\ 13\ 21\ 23 \\ \text{Block 2} & 01\ 03\ 12\ 10\ 22\ 20 \end{cases}$$

$$\text{Set 2} \begin{cases} \text{Block 3} & 10\ 12\ 21\ 23\ 01\ 03 \\ \text{Block 4} & 11\ 13\ 22\ 20\ 02\ 00 \end{cases}$$

$$\text{Set 3} \begin{cases} \text{Block 5} & 20\ 22\ 01\ 03\ 11\ 13 \\ \text{Block 6} & 21\ 23\ 02\ 00\ 12\ 10 \end{cases}$$

This is the design listed in Table 3 of Lewis and Dean (1980), and discussed in Example 5.1. If the cosets (11, 13) and (10, 12) had been selected for the initial block, then $\bar{D}_H$ would equal (00) and the design would not be resolvable.

6.3 Connectivity

As mentioned in Section 4, a connected design is a design which allows every comparison of the treatment combinations to be estimated. An easy way to check for connectivity in a generalized cyclic design is as follows. Write the initial block as $H = S\,[+]\,R$. Generate a subgroup, T, from the treatment labels in R in a similar way to that described in the paragraph preceding Example 6.2. That is, generate the subgroup T_1 using the first nonzero treatment label listed in R. Choose the next treatment label in R that is not in T_1 and generate the subgroup T_2. Combine T_1 and T_2 into the subgroup T_{12}. Choose the next treatment label in R that is not in T_{12} and generate the subgroup T_3. Combine T_{12} and T_3 into the subgroup T_{123}. Continue in this manner until all the labels in R have been used or passed over. Let the resulting subgroup be called T. Now, combine the subgroups T and S into a subgroup K. If K contains all v treatment labels, then the design is connected (see Dean and Lewis, 1986).

Example 6.4

The design of Example 6.3 has initial block $H = (00\ 02)\,[+]\,(00\ 11\ 21)$. Use label 11 to generate T_1, that is $T_1 = (11\ 22\ 03\ 10\ 21\ 02\ 13\ 20\ 01\ 12\ 23\ 00)$. Since T_1 contains all the treatment labels, so does K, and the design is connected.

In a factorial experiment, it is not always necessary to estimate every comparison of the treatment combinations since interactions between several treatment factors can often be assumed to be negligible. If K contains v/q distinct treatment labels, then $q - 1$ orthonormal treatment contrasts are non-estimable. The method of determining non-estimable contrasts is identical to the method described in Section 7 for single replicate designs.

7 SINGLE REPLICATE AND FRACTIONAL FACTORIAL EXPERIMENTS

7.1 Confounding in Single Replicate Experiments

A single replicate factorial experiment is an experiment in which exactly one observation is taken on each treatment combination. If the experiment is run as incomplete block design, the design is disconnected and, inevitably, some of the factorial contrasts are not estimable after adjusting for blocks. The non-estimable contrasts are said to be *confounded* or *confounded with blocks.* It is often possible to select a design with the property that the confounded contrasts belong to a few specified factorial effects, preferably the high order interactions.

The problem of constructing a design which confounds certain specified contrasts and allows all other contrasts to be estimated has engaged the attention of researchers in experimental design since the 1930's. Much of the early work was based on Galois fields and orthogonal and balanced arrays (see for example Bose and Kishen, 1940, and Nair and Rao, 1948). Probably the best-known confounding method, which is applicable when all factors have the same prime number, m, of levels involves solving a set of linear equations modulo m. This "classical" method dates back to Fisher (1942) and Bose (1947), and the details are given in most standard textbooks (e.g., Kempthorne, 1952, Chapters 14, 16, 17; Montgomery, 1984, Chapter 10).

Over the past twenty years various methods have been published which extend the classical method of confounding to more general situations (see Voss, 1986a; Lin, 1987; and their references). Voss (1986a) shows that almost all of these methods, within the realm of their applicability, generate the same class of designs, and this he calls the *general classical* class of designs.

John and Dean (1975) and Dean and John (1975) show that the class of single replicate generalized cyclic designs can also be used to obtain suitable confounding schemes. Single replicate designs for asymmetrical experiments (where at least two factors have a different number of levels) are listed by Dean and John (1975) and by Lewis (1982) together with details of the confounded degrees of freedom. The designs obtained by the general classical method of confounding are all included within the class of generalized cyclic designs (see Voss and Dean, 1987). Furthermore, the theory of confounding described by Bailey (1977) is directly applicable to these designs and can be used to guide the selection of treatment labels for the initial block (see Examples 7.2 and 7.3).

A single replicate generalized cyclic design is constructed as described in Section 6.1. Since $b = v/k$ blocks are required, where k is the block size, the initial block should consist of a subgroup S of size $d = k$. Following Section 6.3, $K = S$ and therefore $(v/k) - 1 = b - 1$ treatment contrasts are confounded. The subgroup structure of the initial block ensures that the confounded contrasts belong to individual factorial effects (rather than being a combination of contrasts from two or more factorial effects).

Example 7.1

Consider a factorial experiment with three factors having 2, 4 and 6 levels, respectively, which is to be run in six blocks of size 8. Now $d = 8$ since $b = 6 = 48/8$, and $d = d_1 d_2 = (\mu/t_1)(\mu/t_2) = (12/3)(12/6)$. Following the procedure of Section 6.1, first, find a treatment label $i_1 i_2 i_3$ that satisfies (12) with $t = t_1 = 3$, that is,

$$3 = \text{HCF}(12, 6i_1, 3i_2, 2i_3)$$

There are eight possible choices, namely, 010, 013, 030, 033, 110, 113, 130, 133. Remembering the advice in Section 5 to include as many codes as possible in the initial block, 113 might be selected, giving $S_1 = (000, 113, 020, 133)$. Now, find a second treatment label satisfying (12) with $t = t_2 = 6$, that is,

$$6 = \text{HCF}(12, 6i_1, 3i_2, 2i_3)$$

There are seven possibilities, namely, 003, 020, 023, 100, 103, 120, 123. However, 020 is contained in S_1 and is therefore discarded. Selecting 123 gives $S_2 = (000, 123)$ and $S_{12} = (000, 113, 020, 133, 123, 030, 103, 010)$. The resulting design is given below with rows representing blocks.

000	010	020	030	103	113	123	133
001	011	021	031	104	114	124	134
002	012	022	032	105	115	125	135
003	013	023	033	100	110	120	130
004	014	024	034	101	111	121	131
005	015	025	035	102	112	122	132

The generalized cyclic design in Example 7.1 was constructed without regard to which factorial contrasts are confounded. Since the design has $b = 6$ blocks, a total of $b - 1 = 5$ factorial contrasts must be nonestimable. It seems likely that some of the confounded contrasts belong to the third main effect F^{001} since only two of the codes for the third factor occur in each block. Similarly the interaction between the first and third factors F^{101} is likely to be confounded. It is not immediately obvious whether any of the other factorial effects contain confounded contrasts. However, the confounding theory of Bailey (1977) can be used to provide the answer as follows.

First, note that every main effect or interaction F^x has $w^x = \prod_{h=1}^{n}(m_h - 1)^{x_h}$ degrees of freedom associated with it. Each degree of freedom corresponds to a particular contrast and can be labelled as $A^{y_1} B^{y_2} \ldots N^{y_n}$, where $1 \leq y_h \leq m_h - 1$ (and $y_h = 0$ if the nth factor is not in the interaction). Thus, for example, in an experiment with two factors at three levels, the four degrees of freedom for the interaction can be labelled as AB, $A^2 B^2$, AB^2, $A^2 B$. In standard textbooks, AB represents the pair of degrees of freedom $(AB, A^2 B^2)$ while AB^2 represents the pair $(AB^2, A^2 B)$. For the present purpose it is more convenient to refer to each degree of freedom separately.

Suppose that the initial block S of a single replicated generalized cyclic design is obtained by combining q subgroups S_1, S_2, $\ldots$, S_q as shown in Section 6.1, and suppose that S_j is generated by the treatment label $a_1^j a_2^j \ldots a_n^j$, $(j = 1, \ldots, q)$. Then, following Bailey (1977), the confounded degrees of freedom are those F^y which satisfy

$$\frac{\mu}{m_1} a_1^j y_1 + \frac{\mu}{m_2} a_2^j y_2 + \cdots + \frac{\mu}{m_n} a_n^j y_n = 0 \bmod \mu \tag{13}$$

for all $j = 1, 2, \ldots, q$.

Example 7.2

The initial block of the design in Example 7.1 is obtained by combining two subgroups which are generated by the treatment labels $i_1^1 i_2^1 i_3^1 = 113$ and $i_1^2 i_2^2 i_3^2 = 123$, respectively. The confounded degrees of freedom are those $F^y = A^{y_1} B^{y_2} C^{y_3}$ which satisfy (7.1), that is,

$$6y_1 + 3y_2 + 6y_3 = 0 \bmod 12$$

and

$$6y_1 + 6y_2 + 6y_3 = 0 \bmod 12$$

$$y_1 = 0, 1; \qquad y_2 = 0, 1, 2, 3; \qquad y_3 = 0, 1, 2, 3, 4, 5.$$

So the five confounded degrees of freedom are those for which $y_1 y_2 y_3 = 002, 004, 101, 103, 105$, namely C^2, C^4, AC, AC^3 and AC^5. Thus the confounded contrasts belong to two factorial effects, namely F^{001} (two degrees of freedom) and F^{101} (three degrees of freedom). All other factorial effects are fully estimable.

The design in Examples 7.1 and 7.2 may be regarded as unsatisfactory in practice due to the fact that the third main effect is confounded. The lists of Dean and John (1975) and Lewis (1982) do not mention designs which confound main effects. However, no alternative design is offered by either source and, in fact, no generalized cyclic design exists which avoids confounding this main effect. It is, however, possible to confound the three-factor interaction instead of the two-factor interaction (see Example 7.3).

In practice, an experimenter would wish to choose the confounding scheme and then generate the design. For single replicate generalized cyclic designs, this can be achieved by reversing the roles of (12) and (13) (see Bailey, 1977, Theorem 4.2). Thus, if $d - 1 = b - 1$ degrees of freedom are to be confounded, first, write d as $\prod_{i=1}^{q}(\mu/t_i)$. Then, select $y_1^j y_2^j \ldots y_n^j$ to satisfy

$$t_j = \mathrm{HCF}(\mu, y_1^j \mu/m_1, y_2^j \mu/m_2, \ldots, y_n^j \mu/m_n) \qquad (14)$$

$j = 1, \ldots, q$. Each selected degree of freedom generates a subgroup of degrees of freedom. The subgroups are then combined as in Section 6.1 to form a subgroup S_D of degrees of freedom to be confounded. The treatment labels $i_1 i_2 \ldots i_n$ in the initial block of the design that confounds the degrees of freedom in S_D are those which satisfy (13), that is,

$$i_1 y_1^j \mu/m_1 + i_2 y_2^j \mu_2/m_2 + \cdots + i_n y_n^j \mu_n/m_n = 0 \bmod \mu \qquad (15)$$

for all $j = 1, \ldots, q$.

Example 7.3

Consider the experiment of Example 7.1, namely, a $2 \times 4 \times 6$ experiment in $b = 6$ blocks of size $k = 8$. A total of $6 - 1 = 5$ degrees of freedom must be selected for confounding. Let $d = b$ then, since $\mu = 12$, $d = \mu/t$ for $t = 2$. Select a degree of freedom $y_1 y_2 y_3$ that satisfies (14) namely,

$$2 = \mathrm{HCF}(12, 6y_1, 3y_2, 2y_3).$$

The possible choices are $y_1 y_2 y_3 = 001, 005, 021, 022, 024, 025, 101, 102, 104, 105, 121, 122, 124, 125$. In order to confound the three-factor interaction, one of 121, 122, 124, 125 should be selected. Consider selecting 121, that is AB^2C; then the entire set of confounded degrees of freedom S_D is generated by 121. Thus the confounded degrees of freedom are $S_D = (121, 002, 123, 004, 125)$ or AB^2C, C^2, AB^2C^3, C^4 and AB^2C^5. It can be seen that two degrees of freedom from F^{001} and three degrees of freedom from F^{111} have been selected for confounding. The treatment labels $i_1 i_2 i_3$ in the initial block of the generalized cyclic design satisfy (15) with $y_1 y_2 y_3 = 121$, namely,

$$6i_1 + 6i_2 + 2i_3 = 0 \bmod 12$$

giving $S = (000, 013, 020, 033, 103, 110, 123, 130)$. Notice that this design could have been selected in Example 7.1 if the initial choice of generators had been 013 and 123 instead of 113 and 123. Notice also that $d = b = 6$ can be represented as $d = d_1 d_2 = (\mu/t_1)(\mu/t_2)$ with $t_1 = 4$ and $t_2 = 6$. Sometimes a different representation can lead to alternative confounding schemes.

Explicit formulas are given by Collings (1984), El Mossadeq, Kobilinsky and Collombier (1985) and Voss and Dean (1988) for obtaining the initial block from a set of confounded degrees of freedom. However, for small designs, an exhaustive search is not difficult.

If a factor has a non-prime number, m, of levels, it can be replaced by two or more pseudofactors as described by Kempthorne (1952, Chapter 18). However, unless m is divisible by a square of a prime number, the designs obtained by using pseudofactors and those obtained without pseudofactors are equivalent in terms of the number of degrees of freedom confounded from each factorial effect (see Voss and Dean, 1987, Theorem 3).

7.2 Aliasing in Fractional Factorial Experiments

A fractional factorial experiment is a factorial experiment in which only a fraction of the v treatment combinations is observed. The simplest method of selecting k treatment combinations when k divides v is to choose one block of a single replicate design which has v/k blocks of size k. This is the only method that will be considered in this section.

The set S_D of confounded degrees of freedom in the block design are, of course, non-estimable in the fractional experiment. This set of degrees of freedom is called the *defining relation* for the experiment. Thus, for example, if one block was to be selected from the design in Example 7.3, the fractional factorial experiment would involve observations on just $k = 8$ treatment combinations, and would be called a 1/6th fraction. The defining relation for the fraction would be the set of degrees of freedom S_D and would be written as

$$I = C^2 = C^4 = AB^2C = AB^2C^3 = AB^2C^5$$

Contrasts belonging to various factorial effects are confounded with each other in fractional factorial experiments and cannot be estimated separately (see, for example, Montgomery, 1984, Chapter 11). Such factorial contrasts are said to be *aliased*. In a generalized cyclic design, the sets of aliased contrasts are identified with sets of aliased degrees of freedom. The list of aliased degrees of freedom is called the aliasing scheme for the design. The aliasing scheme is obtained by multiplying each degree of freedom F^y by the degrees of freedom in the defining relation. For the above fractional factorial experiment, the aliasing scheme consists of the following distinct sets of aliased degrees of freedom.

$$
\begin{aligned}
I &= C^2 &&= C^4 &&= AB^2C &&= AB^2C^3 &&= AB^2C^5 \\
A &= AC^2 &&= AC^4 &&= B^2C &&= B^2C^3 &&= B^2C^5 \\
B &= BC^2 &&= BC^4 &&= AB^3C &&= AB^3C^3 &&= AB^3C^5 \\
B^2 &= B^2C^2 &&= B^2C^4 &&= AC &&= AC^3 &&= AC^5 \\
B^3 &= B^3C^2 &&= B^3C^4 &&= ABC &&= ABC^3 &&= ABC^5 \\
C &= C^3 &&= C^5 &&= AB^2C^2 &&= AB^2C^4 &&= AB^2 \\
AB &= ABC^2 &&= ABC^4 &&= B^3C &&= B^3C^3 &&= B^3C^5 \\
AB^3 &= AB^3C^2 &&= AB^3C^4 &&= BC &&= BC^3 &&= BC^5
\end{aligned}
$$

The number of rows in the aliasing scheme is always equal to the number of treatment combinations observed in the fractional factorial experiment. The above fraction is unlikely to be used in practice due to the severe loss of information in the third main effect. Not only is it impossible to estimate the contrasts corresponding to C^2 and C^4, but also those corresponding to C, C^3, C^5 and AB^2 cannot be estimated separately. Ideally, the defining relation should contain only degrees of freedom $F^y = A^{y_1}B^{y_2}\ldots N^{y_n}$ with at least five of $y_1, y_2, \ldots, y_n$ being non-zero, in which case main effects are only aliased with interactions involving four or more factors, and two-factor interactions are only aliased with interactions involving three or more factors. The minimum number of non-zero y_i in the defining relation is called the *resolution number* of the fraction. A list of defining relations for resolution 3–6 fractions, together with the generators for the corresponding initial block of the generalized cyclic design, is provided by Lewis (1982) for asymmetrical experiments.

Fractional factorial experiments can be run as block designs by a further application of (14) and (15). Cochran and Cox (1957, Chapter 6A) list a number of fractional factorial 2^n experiments, some of which are in blocks. All of their designs can be obtained by the methods of this article, as illustrated in Example 7.4.

Example 7.4

Consider a fraction of a 2^6 factorial experiment run in four blocks of size 8. Since only 32 observations are to be taken and there are $2^6 = 64$ treatment combinations, a 1/2 fraction is required.

Start by dividing the treatment combinations into two blocks of size 32, confounding one degree of freedom. Now $\mu = 2$, therefore $d = 2 = \mu/t$ where $t = 1$, and (14) gives

$$1 = \text{HCF}(2, y_1, y_2, \ldots, y_6)$$

Let $y_1 y_2 \ldots y_6 = 111111$, then $S_D = (000000\ 111111)$ giving a resolution 6 design with defining relation $I = ABCDEF$. The aliasing scheme can be obtained using the method illustrated above. The 32 treatment combinations selected for the design are obtained from one of the blocks of the generalized cyclic design whose initial block satisfies (15) namely,

$$i_1 + i_2 + i_3 + i_4 + i_5 + i_6 = 0 \bmod 2 \qquad (16)$$

It is simplest, but not necessary, to select the initial block itself, as shown in Table 4. To divide these 32 treatment combinations into four blocks of size 8 confounding three degrees of freedom and their aliases, (14) is applied again with $d = b = 4 = \mu^2/t_1 t_2$, where $t_1 = t_2 = 1$. Degrees of freedom F^y must be selected which satisfy

$$t_j = 1 = \text{HCF}(2, y_1^j, y_2^j, \ldots, y_6^j), \qquad j = 1, 2$$

Let $y_1^j y_2^j \ldots y_6^j = 111000$, giving $S_{D1} = (000000\ 111000)$, and let $y_1^2 y_2^2 \ldots y_6^2 = 110100$, giving $S_{D2} = (000000\ 110100)$. Then $S_D = (000000\ 111000\ 110100\ 001100)$. The three degrees of freedom, together with their aliases, selected for confounding are therefore $ABC = DEF$, $ABD = CEF$, $CD = ABEF$. The treatment combinations in the initial block of the generalized cyclic fraction satisfy (15) using $y_1 y_2 \ldots y_6 = 111000$ and 110100 (that is, the generators of the two subgroups S_{D1} and S_{D2}), namely,

$$i_1 + i_2 + i_3 = 0 \bmod 2$$
$$i_1 + i_2 + i_4 = 0 \bmod 2$$

Table 4   A 1/2 replicate of a 2^6 experiment; Resolution 6

000000	000011	110000	110011	101101	101110	011101	011110
001100	001111	111100	111111	100001	100010	010001	010010
011000	011011	101000	101011	110101	110110	000101	000110
010100	010111	100100	100111	111001	111010	001001	001010

Remembering that the blocks of the generalized cyclic design to be included in the fraction are only those whose treatment combinations satisfy (16), the resulting design is identical to that given by Cochran and Cox (1957, plan 6A.6), and is shown in Table 4, where rows denote blocks.

If all three-, four-, five- and six-factor interactions can be assumed to be negligible, then all main effects and two-factor interactions with the exception of $CD = F^{001100}$ can be estimated. There are eight degrees of freedom for error. These correspond to the negligible contrasts of the ten aliased pairs of three-factor interactions apart from the two confounded pairs $(ABC = DEF,\ ABD = CEF)$.

8   OTHER DESIGNS FOR FACTORIAL EXPERIMENTS

Gupta (1983, 1985) and Mukerjee (1984, 1986) discuss alternative methods of constructing multi-replicate block designs with orthogonal factorial structure. These are based on the Kronecker product of the incidence matrices obtained by considering each factor separately. These authors have published several results on the efficiencies of these designs, all of which are summarized in their recent monograph (Gupta and Mukerjee, 1989).

Patterson (1976) discusses the use of an algorithm, which he calls the DSIGN algorithm, for constructing designs with various types of blocking structures for factorial experiments. Those designs which have orthogonal factorial structure and a single blocking factor as discussed in this article are generalized cyclic designs. Patterson and Bailey (1978) give examples of designs with more complicated blocking structures.

Generalized cyclic designs have also been used to provide efficient designs for multi-replicate factorial experiments in the presence of several crossed blocking factors (see John and Lewis, 1983; Bailey, 1985b; Lewis, 1986; Dean and Lewis, 1988).

Non-equireplicate designs with orthogonal factorial structure are obtained by Puri and Nigam (1978) and Gupta and Mukerjee (1988) by merging of treatment labels, and by Mukerjee and Bose (1988) using the Kronecker product method of construction.

Chauhan and Dean (1986) introduce the idea of partial orthogonal factorial structure, where a specified subset of factorial effects can be estimated independently of all other factorial effects, but not all of the factorial effects are independent of each other. Examples of designs with this property are obtained by modifying generalized cyclic designs by merging or dividing treatment labels. Further examples and results are given by Gupta (1987).

Some of the single replicate factorial designs obtained by Patterson (1976) using the DSIGN algorithm are single replicates from resolvable generalized cyclic designs. These designs do not have the property that the confounded contrasts belong to individual factorial effects. However, they can be useful for factorial experiments since many allow low order contrasts to be estimated independently or almost independently of blocks (see Lewis, Dean and Lewis, 1983). Other useful single replicate designs can be obtained by deleting treatment combinations from a starting design (e.g., Sardana and Das, 1965; Voss, 1986b).

There is a huge literature dealing with the construction of designs for fractional factorial experiments under various criteria for good designs. Many of these designs are based on orthogonal and balanced arrays and differ from those in Section 7.2. For summaries of available designs and construction methods, see Srivastava (1978), Raktoe, Hedayat and Federer (1980) and Dey (1985).

9   CONCLUSION

The class of generalized cyclic designs provides a large source of designs with orthogonal factorial structure for multiple-replicate, single-replicate and fractional factorial experiments. The construction of efficient multiple-replicate designs is straightforward, as is the construction of designs for single-replicate and fractional factorial experiments with specified confounding patterns. For single-replicate or fractional experiments with all factors having the same prime number of levels, the generalized cyclic construction method is identical to the well-known classical method.

The class of generalized cyclic designs is large enough to allow a selection of efficient designs for most experiments. The designs are easy to specify and their analysis is straightforward.

ACKNOWLEDGMENTS

I am extremely grateful to Rosemary Bailey, Chand Chauhan, Susan Lewis, Jeffrey Nunemacher, Donald Preece and Daniel Voss for reading this article in great detail and for providing me with many valuable comments and suggestions. I am also grateful to Dolores Wills for typing the article accurately, quickly and at very short notice.

REFERENCES

Bailey, R. A. (1977). Patterns of confounding in factorial designs. *Biometrika*, 64, 597–603.

Bailey, R. A. (1982). The decomposition of treatment degrees of freedom in quantitative factorial experiments. *J. R. Statist. Soc.* B, 44, 63–70.

Bailey, R. A. (1985a). Balance, orthogonality and efficiency factors in factorial design. *J. R. Statist. Soc.* B, 47, 453–458.

Bailey, R. A. (1985b). Factorial design and abelian groups. *Linear Algebra Appl.*, 70, 349–368.

Barnett, M. K. and Mead, F. C., Jr. (1956). A 2^4 factorial experiment in four blocks of eight: A study in radioactive decontamination. *Appl. Statist.*, 5, 122–131.

Bose, R. C. (1947). Mathematical theory of the symmetrical factorial design. *Sankhyā*, 8, 107–166.

Bose, R. C. and Kishen, K. (1940). On the problem of confounding in general symmetrical factorial design. *Sankhyā*, 5, 21–36.

Box, G. E. P. and Draper, N. R. (1987). *Empirical Model Building and Response Surfaces*. John Wiley, New York.

Chauhan, C. K. and Dean, A. M. (1986). Necessary and sufficient conditions for orthogonal factorial structure. *Ann. Statist.* 14, 743–752.

Cochran, W. G. and Cox, G. M. (1957). *Experimental Designs*, Second Edition, John Wiley, New York.

Collings, B. J. (1984). Generating the intrablock and interblock subgroups for confounding in general factorial experiments. *Ann. Statist.*, 12, 1500–1509.

Dean, A. M. and John, J. A. (1975). Single replicate factorial experiments in generalized cyclic designs: II. Asymmetrical arrangements. *J. R. Statist. Soc.* B, 37, 72–76.

Dean, A. M. and Lewis, S. M. (1980). A unified theory for generalized cyclic designs, *J. Stat. Plan. Inf.*, 4, 13–23.

Dean, A. M. and Lewis, S. M. (1986). A note on the connectivity of generalized cyclic designs. *Comm. Statist.*, 15(11), 3429–3433.

Dean, A. M. and Lewis, S. M. (1988). Towards a general theory for the construction of multi-dimensional block designs. *Utilitas Math.*, 34, 33–44.

Dey, A. (1985). *Orthogonal Fractional Factorial Designs.* Wiley, Eastern, New Delhi.

El Mossadeq, A., Kobilinsky, A. and Collombier, D. (1985). Construction d'orthogonaux dans les groupes abéliens finis et confusions d'effets dans les plans factoriels. *Linear Algebra Appl.*, 70, 303–320.

Fisher, R. A. (1942). The theory of confounding in factorial experiments in relation to the theory of groups. *Ann. Eugenics*, 11, 341–353.

Gupta, S. C. (1983). Some new methods for constructing block designs having orthogonal factorial structure, *J. R. Statist, Soc.* B 45, 297–307.

Gupta, S. C. (1985). On Kronecker block designs for factorial experiments. *J. Statist. Plan. Inf.*, 11, 227–236.

Gupta, S. C. (1987). On designs for factorial experiments derivable from generalized cyclic designs. *Sankhyā*, 49, 90–96.

Gupta, S. C. and Mukerjee, R. (1988). Non-equireplicate factorial designs through merging of treatments (submitted for publication).

Gupta, S. C. and Mukerjee, R. (1989). A calculus for factorial arrangements. *Lecture Notes in Statistics*, 59, Springer-Verlag.

John, J. A. (1973). Generalized cyclic designs in factorial experiments. *Biometrika*, 60, 55–63.

John, J. A. (1981a). Efficient cyclic designs. *J. R. Statist. Soc.* B, 43, 76–80.

John, J. A. (1981b). Factorial-balance and the analysis of designs with factorial structure. *J. Statist. Plan. Inf.* 5, 99–105.

John, J. A. (1987). *Cyclic Designs.* Chapman and Hall, New York.

John, J. A. and Dean, A. M. (1975). Single replicate factorial experiments in generalized cyclic designs: I. Symmetrical arrangements. *J. R. Statist. Soc.*, B, 37, 1, 63–71.

John, J. A. and Lewis, S. M. (1983). Factorial experiments in generalized cyclic row-column designs. *J. R. Statist. Soc.* B., 45, 245–251.

John, J. A., Wolock, F. W. and David, H. A. (1972). Cyclic designs. *Nat. Bur. Standards Appl. Math. Ser.* 62.

Kempthorne, O. (1952). *The Design and Analysis of Experiments.* Wiley, New York.

Lewis, S. M. (1982). Generators for asymmetrical factorial experiments. *J. Statist. Plan. Inf.*, 6, 59–64.

Lewis, S. M. (1986). Composite generalized cyclic row-column designs. *Sankhyā*, B 48, 373–379.

Lewis, S. M. and Dean, A. M. (1980). Factorial experiments in resolvable generalized cyclic designs. *B.I.A.S.* 7, 2, 159–167.

Lewis, S. M. and Dean, A. M. (1984). Upper bounds for factorial efficiency factors. *J. R. Statist. Soc.*, B, 46, 273–278.

Lewis, S. M. and Dean, A. M. (1985). A note on efficiency-consistent designs. *J. R. Statist. Soc.*, B, 47, 261–262.

Lewis, S. M., Dean, A. M. and Lewis, P. H. (1983). Efficient single replicate designs for two-factor experiments. *J. R. Statist. Soc.*, B, 45, 224–227.

Lewis, S. M. and Tuck, M. G. (1985). Paired comparison designs for factorial experiments. *Appl. Statist.*, 34, 227–234.

Lin, P. K. H. (1987). Confounding in mixed factorial experiments through isomorphisms. *Comm. Statist.*, 16, A, 421–429.

Montgomery, D. C. (1984). *Design and Analysis of Experiments*, Second Edition. John Wiley, New York.

Mukerjee, R. (1984). Applications of some generalizations of Kronecker product in the construction of factorial designs. *J. Indian Soc. Agric. Statist.*, 36, 38–46.

Mukerjee, R. (1986). Construction of orthogonal factorial designs controlling interaction efficiencies. *Comm. Statist.*, 15, 1535–1548.

Mukerjee, R. and Bose, M. (1988). Non-equireplicate Kronecker factorial designs. *J. Statist. Plan. Inf.*, 19, 261–267,

Nair, K. R. and Rao, C. R. (1948). Confounding in asymmetrical factorial experiments. *J. R. Statist. Soc.* B, 10, 109–131.

Patterson, H. D. (1976). Generation of factorial designs. *J. R. Statist. Soc.*, B, 38, 175–179.

Patterson, H. D. and Bailey, R. A. (1978). Design keys for factorial experiments. *Appl. Statist.*, 27, 335–343.

Puri, P. D. and Nigam, A. K. (1978). Balanced factorial experiments II. *Commun. Statist.*, A, 7, 591–605.

Raktoe, B. L., Hedayat, A. and Federer, W. T. (1980). *Factorial Designs*. John Wiley, New York.

Sardana, M. G. and Das, M. N. (1965). On the construction and analysis of some confounded asymmetrical factorial designs. *Biometrics*, 21, 948–956.

Srivastava, J. N. (1978). A review of some recent work on discrete optimal factorial designs for statisticians and experimenters. *Developments in Statistics*, 1, 267–329. Academic Press, New York.

Voss, D. T. (1986a). On generalizations of the classical method of confounding to asymmetric factorial experiments. *Comm. Statist. A*, 15, 1299–1314.

Voss, D. T. (1986b). First order deletion designs and the construction of efficient nearly orthogonal factorial designs in small blocks. *J. A. S. A.*, 81, 813–818.

Voss, D. T. and Dean, A. M. (1987). Equivalence of methods of confounding in asymmetrical single replicate factorial designs. *Ann. Statist.*, 15, 376–384.

Voss, D. T. and Dean, A. M. (1988). On confounding in single replicate factorial experiments. *Utilitas Math.*, 33, 59–64.

Williams, E. R. (1975). A new class of resolvable block designs. *Ph.D. thesis. Edinburgh University.*

Williams, E. R., Patterson, H. D. and John, J. A. (1976). Resolvable designs with two replications. *J. R. Statist. Soc.*, B, 38, 296–301.

18

Crossover Designs in Industry

Damaraju Raghavarao Temple University, Philadelphia, Pennsylvania

1 INTRODUCTION

The topic of experimental designs has several areas of application; however, the terminology used here is borrowed from agriculture experiments, where it was initially used. In this context, *treatments* are the ones intentionally introduced by the experimenter to study their relative effects, and *experimental units* (eu) are the ones on which the treatments are applied to get the required response. The eu depends on the nature of the experiment. The eu may be a piece of equipment, a worker, a manufacturing company, etc., in industrial experiments.

In experiments, depending on the problem and/or eu, one or several treatments can be given to each eu. *Crossover designs* are that class of designs in which a sequence of all or a subset of treatments are given to each eu over different time periods. Crossover designs have been used for many years in a broad spectrum of research areas, including agriculture experiments (Cochran, 1939), dairy husbandry (Cochran et al., 1941), bioassay procedures (Finney, 1956), clinical trials (Grizzle, 1965), psychological experiments (Keppel, 1973), and weather modification experiments (Mielke, 1974). These designs, somehow, have not been used in industrial settings, and the purpose of this chapter is to acquaint those researchers with crossover designs in general, and with two-treatment, two-period designs in particular.

The advantage of a crossover design is the economic use of experimental material and the more sensitive treatment comparisons that result from the elimination of interunit variation. Furthermore, in some experimental settings, crossover designs may be the natural choice for organizing the study.

When different treatments are applied to each eu, every treatment produces two types of effects: (1) *direct effect* (DE) and (2) *residual effect* (RE). Direct effect of the treatment is the effect of the treatment in the period of its application, and residual effect is the effect of the treatment in the periods subsequent to the period of application of the treatment. One distinguishes different residual effects of a treatment by defining the *ith-order residual effect* as the residual effect of the treatment in the ith period after its application, $i = 1, 2, \ldots$. Usually the ith-order residual effect will be greater than the $(i+1)$th-order residual effect, and most of the crossover designs in the literature consider only the first-order residual effects. The crossover designs accounting for residual effects may be called *crossover designs with residual effects (CODWR)*. If the experimenter does not want information on residual effects, then enough washout period should be used between periods while switching treatments, and crossover designs ignoring residual effects may be called *crossover designs without residual effects* (CODWOR). It is worthwhile to note that even if the experimenter is not interested in residual effects, in some experimental situations, CODWR is the only appropriate design.

Latin square, Youden square, and lattice square designs are some of the commonly used CODWOR and are discussed in standard experimental design books. The literature on CODWR is present mainly in research papers and, in this chapter, a brief summary of this work useful to practitioners will be given.

Example 1.1

Suppose a company having four plants is interested in studying the production and accident rates for different time intervals between rotation of shifts. Let weekly rotation of shifts be denoted A, biweekly B, triweekly C, and monthly D. Furthermore, let each kind of rotation be tried for three months before changing the rotation pattern. A convenient plan for organizing this study is outlined in Table 1.

Table 1   Rotation Plan
for Shifts

Periods (months)	Plants			
	I	II	III	IV
0–3	A	B	C	D
3–6	B	C	D	A
6–9	D	A	B	C
9–12	C	D	A	B

It is not feasible to introduce washout periods between changes of rotating shifts and the statistical analysis must use residual effects in the linear model. In plant I, the sequence of rotation of shifts will be A, B, D, and C for the four quarters in a year; in plant II the sequence of rotation of shifts will be B, C, A, and D for the four quarters in a year; etc.

When the plant quarterly production is used as a response variable, the normal theory analysis is valid. For the plant quarterly accident rate, nonparametric methods are required, and they are known in literature for a two-treatment, two-period design.

2   WILLIAMS' DESIGNS AND ANALYSIS

The commonly used CODWR are the Williams' designs (cf. Williams, 1949), which consist of one or two latin squares, where the number of treatments v is even or odd, such that every ordered pair of distinct treatments (θ, ϕ), $\theta \neq \phi$, occurs together equally, often in successive rows. The design given in Section 1 is Williams' CODWR for $v = 4$, and it can easily be verified that every ordered pair of treatments occurs in successive rows in exactly one column.

For $v = 5$, one needs to use two latin squares as given below:

$$
\begin{array}{ccccc}
A & B & C & D & E \\
B & C & D & E & A \\
E & A & B & C & D \\
C & D & E & A & B \\
D & E & A & B & C
\end{array}
\qquad
\begin{array}{ccccc}
D & E & A & B & C \\
C & D & E & A & B \\
E & A & B & C & D \\
B & C & D & E & A \\
A & B & C & D & E
\end{array}
$$

In the above design, every ordered pair of distinct treatments occurs in successive rows in exactly two columns. Here, also, the rows represent the periods and the columns, the units or sequences.

The statistical property possessed by Williams' CODWR is that the variances of the estimated contrasts of direct effects (DE) of treatments are all equal and the variances of the estimated contrasts of first-order residual effects (RE) of treatments are all equal. In this sense, these designs are balanced. It may be noted that the Williams' designs under discussion here consider only first-order residual effects.

Let $d(i, j)$ be the treatment in the ith row and the jth column of the design. For example, in the illustrated design for $v = 5$, $d(2, 4) = E$, $d(4, 7) = C$. The linear model assumed for the analysis is

$$Y_{ij} = \mu + \rho_i + \gamma_j + \tau_{d(i,j)} + \delta_{d(i-1,j)} + e_{ij}$$

where Y_{ij} is the response variable in the ith row and the jth column, μ the general mean, ρ_i the ith row effect, γ_j the jth column effect, $\tau_{d(i,j)}$ the DE of the treatment $d(i, j)$, $\delta_{d(i-1,j)}$ the RE (first-order) of the treatment $d(i-1, j)$ (note $\delta_{d(0,j)} \equiv 0$); e_{ij} are random errors assumed to be normally and independently distributed with mean zero and variance σ^2.

Let T_α be the total response for the αth treatment, R_α the total responses in the immediately succeeding periods after using the αth treatment, P_i the total response for the ith period, C_j the total response for the jth column, $C^{(\alpha)}$ the sum of the column totals for those columns where the treatment α occurs in the last row, and G the grand total. Further, let $t = 1$ or 2 depending on whether v is even or odd. A solution for direct effects (DE) and residual effects (RE) from the normal equations is

$$\hat{\tau}_\alpha = \left\{ tv(v^2 - v - 2) \right\}^{-1} \left\{ (v^2 - v - 1)T_\alpha + vR_\alpha + C^{(\alpha)} + (P_1 - vG) \right\}$$

$$\hat{\delta}_\alpha = \left\{ tv(v^2 - v - 2) \right\}^{-1} \left\{ v^2 R_\alpha + vT_\alpha + vC^{(\alpha)} + vP_1 - (v + 2)G \right\}$$

where $\hat{\tau}_\alpha$ and $\hat{\delta}_\alpha$ are estimators of τ_α and δ_α, respectively. The ANOVA table is then given in Table 2.

The variances of the elementary contrasts can be evaluated and are given by

$$\mathrm{Var}(\hat{\tau}_\alpha - \hat{\tau}_\beta) = [2(v^2 - v - 1)/\left\{ tv(v^2 - v - 2) \right\}]\sigma^2$$

$$\mathrm{Var}(\hat{\delta}_\alpha - \hat{\delta}_\beta) = [2v^2/\left\{ tv(v^2 - v - 2) \right\}]\sigma^2$$

Table 2 ANOVA Table for William's Designs

Source	df	SS	MS	F
Periods	$v - 1$	$\sum_{i=1}^{v} \dfrac{P_i^2}{tv} - \dfrac{G^2}{tv^2}$		
Units	$tv - 1$	$\sum_{j=1}^{tv} \dfrac{C_j^2}{v} - \dfrac{G^2}{tv^2}$		
DE (ignoring RE)	$v - 1$	$\sum_{l=1}^{v} \dfrac{T_l^2}{tv} - \dfrac{G^2}{tv^2}$		
RE (eliminating DE)	$v - 1$	$\sum_{h=1}^{v} \dfrac{t(v^2 - v - 2)\hat{\delta}_h^2}{v}$	MS_r	MS_r / MS_e
DE (eliminating RE)	$v - 1$	$\sum_{l=1}^{v} \dfrac{tv(v^2 - v - 2)\hat{\tau}_l^2}{(v^2 - v - 1)}$	MS_d	MS_d / MS_e
RE (ignoring DE)	$v - 1$	a		
Error	$(v - 1)(tv - 3)$	b	MS_e	
Total	$tv^2 - 1$	$\sum_{i,j} Y_{ij}^2 - \dfrac{G^2}{tv^2}$		

[a] = DE (ignoring RE) SS + RE (eliminating DE) SS − DE (eliminating RE) SS
[b] = Total SS − periods SS − units SS − DE (ignoring RE) SS − RE (eliminating DE) SS

where Var( ) is the variance of the quantity in the parenthesis for $\alpha, \beta = 1$, 2, ..., v; $\alpha \neq \beta$. Clearly, $\text{Var}(\hat{\delta}_\alpha - \hat{\delta}_\beta) > \text{Var}(\hat{\tau}_\alpha - \hat{\tau}_\beta)$, and these designs are less precise in estimating residual effects contrasts. To reduce the disparity in these variances, extra-period crossover designs are introduced in the literature (cf. Patterson and Lucas, 1959), in which the last row of the original Williams' design is repeated into an extra period. From Table 2 and estimated variances, inferences on the direct and residual effects can be drawn by standard procedure and will be illustrated numerically.

Consider the study discussed in Example 1.1. Let artificial production data (in 10,000 units/quarter) be given in parentheses against each treatment (see Table 3).

Table 3   Plant Study Data

| Period | Plant | | | | Quarter |
(months)	I	II	III	IV	totals
0–3	$A(5)$	$B(8)$	$C(6)$	$D(1)$	20
3–6	$D(2)$	$A(4)$	$B(7)$	$C(5)$	18
6–9	$B(10)$	$C(7)$	$D(2)$	$A(5)$	24
9–12	$C(6)$	$D(3)$	$A(8)$	$B(9)$	26
Plant totals	23	22	23	20	88

The calculations leading to Table 5 are done as follows:

$$\text{Correction factor (CF)} = \frac{G^2}{tv^2} = \frac{88^2}{4^2} = 484$$

$$\text{Total SS} = 5^2 + 2^2 + \cdots + 9^2 - \text{CF} = 104$$

$$\text{Periods (rows) SS} = \frac{20^2 + 18^2 + 24^2 + 26^2}{4} - \text{CF} = 10$$

$$\text{Plants (columns) SS} = \frac{23^2 + 22^2 + 23^2 + 20^2}{4} - \text{CF} = 1.5$$

$$40\hat{\tau}_\alpha = 11T_\alpha + 4R_\alpha + C^{(\alpha)} + (P_1 - 4G)$$

$$40\hat{\delta}_\alpha = 4T_\alpha + 16R_\alpha + 4C^{(\alpha)} + 4P_1 - 6G$$

The calculation of $\hat{\tau}_\alpha$ and $\hat{\delta}_\alpha$ will be facilitated by Table 4.

Since $d(1,1) = d(2,2) = d(3,4) = d(4,3) = A$, $t_A = y_{11} + y_{22} + y_{34} + y_{43} = 5 + 4 + 5 + 8 = 22$. Furthermore, the RE of A is accounted in the

Table 4

Treatment	T_α	R_α	$C^{(\alpha)}$	$\hat{\tau}_\alpha$	$\hat{\delta}_\alpha$
A	22	18	23	0.125	0.5
B	34	12	20	2.750	−1.0
C	24	15	23	0.375	−0.5
D	8	23	22	−3.250	1.0
Total	88	66	88	0	0

responses y_{21}, y_{32}, and y_{44}, and hence $R_A = y_{21} + y_{32} + y_{44} = 2+7+9 = 18$. Since treatment A occurs in the last row of column 3, $C^{(A)} = C_3 = 23$. The terms $\hat{\tau}_A$ and $\hat{\delta}_A$ are calculated from formulae on p. 522. The other rows of the Table 4 can be accordingly filled. Note that

$$\sum_{\alpha=1}^{v} T_\alpha = G, \qquad \sum_{\alpha=1}^{v} R_\alpha = G - P_1, \qquad \sum_{\alpha=1}^{v} C^{(\alpha)} = G,$$

$$\sum_{\alpha=1}^{v} \hat{\tau}_\alpha = 0 = \sum_{\alpha=1}^{v} \hat{\delta}_\alpha$$

Now,

$$\text{DE (ignoring RE) SS} = \frac{22^2 + 34^2 + 24^2 + 8^2}{4} - \text{CF} = 86$$

RE (eliminating DE) SS

$$= \frac{(4^2 - 4 - 2)}{4} \left\{ (0.5)^2 + (-1.0)^2 + (-0.5)^2 + (1.0)^2 \right\}$$

$$= 6.25$$

$$\text{ESS} = 104 - 10 - 1.5 - 86 - 6.25$$

$$= 0.25$$

DE (eliminating RE) SS

$$= \frac{4(4^2 - 4 - 2)}{4} \left\{ (0.125)^2 + (2.75)^2 + (0.375)^2 + (-3.25)^2 \right\}$$

$$= 66.48$$

RE (ignoring DE) SS

$$= 86 + 6.25 - 66.48 = 25.77$$

The above information can be summarized in Table 5.

The null hypothesis of equality of direct effects can be tested from the F ratio of direct effects (eliminating RE) MS/error MS, with $v - 1$ numerator and $(v - 1)(tv - 3)$ denominator degrees of freedom (df). This F in the present example is

$$F = 277.0 \qquad \text{with } (3, 3) \text{ df}$$

Table 5   ANOVA Table for the Artificial Data

Source	df	SS	MS	F
Periods (quarters)	3	10.00		
Columns (plants)	3	1.50		
DE (ignoring RE)	3	86.00		
RE (eliminating DE)	3	6.25	2.08	26.0
DE (eliminating RE)	3	66.48	22.16	277.0
RE (ignoring DE)	3	25.77		
Error	3	0.25	0.08	

The critical F value for three numerator and three denominator degrees of freedom at 0.05 level of significance is 9.8. The observed F value far exceeds the critical value and, hence, significant differences in the production for different shift rotation schedules will be concluded.

The null hypothesis of equality of residual effects can be tested from the F ratio of residual effects (eliminating DE) MS/error MS, with $v - 1$ numerator and $(v - 1)(tv - 3)$ denominator degrees of freedom. This F in the present example is

$$F = 26.0 \qquad \text{with } (3, 3) \text{ df}$$

and exceeds the 0.05-level critical value 9.28. Significant differences in the residual effects will be concluded, thereby implying that the impact of shift rotations are not equally felt for the four schedules after a change occurs.

Suppose the primary interest of the investigator is to compare the production rates with weekly rotations and triweekly rotations. In other words, the experimenter intends to make inferences on τ_A and τ_C. To this end, we note that

$$\mathrm{var}(\hat{\tau}_A - \hat{\tau}_C) = [2(4^2 - 4 - 1)/\{4(4^2 - 4 - 2)\}]\sigma^2 = 11\sigma^2/20$$

and that it is estimated by $11(0.08)/20 = 0.044$. Note that σ^2 is estimated by the error MS of the ANOVA table.

A 95% confidence interval on $\tau_A - \tau_C$ is

$$\hat{\tau}_A - \hat{\tau}_C \pm \{t_{0.025}[(v - 1)(tv - 3)]\} \left\{ \sqrt{\widehat{\mathrm{var}}(\hat{\tau}_A - \hat{\tau}_C)} \right\}$$

where $\widehat{\text{var}}()$ is the estimated variance and $t_\alpha(\nu)$ is the upper 100α percentile of a Student's t distribution with ν degrees of freedom. Substituting the numerical values

$$0.125 - 0.375 \pm (3.182)(0.21) = -0.25 \pm 0.67$$

Thus, $-0.92 \leq \tau_A - \tau_B \leq 0.42$ and the null hypothesis $H_0 : \tau_A = \tau_B$ against the alternative hypothesis $H_A : \tau_A \neq \tau_B$ is tenable at the 0.05-level of significance.

Inferences on other contrasts of interest of direct or residual effects can similarly be performed.

3 TWO-TREATMENT, TWO-PERIOD DESIGNS— PARAMETRIC APPROACH

In this section, we restrict ourselves to using only two treatments in two periods. Let n experimental units be available, and let A and B be the two treatments. The treatment sequence AB (A followed by B) will be given to n_1 randomly selected units form the available n units, and the sequence BA (B followed by A) will be given to the remaining $n - n_1 = n_2$ units. The experimental plan then looks as follows:

$$
\begin{array}{ccccccc}
A & A & \cdots & A & B & B & \cdots & B \\
B & B & \cdots & B & A & A & \cdots & A \\
\end{array}
$$

$$n_1 \text{ units} \qquad n_2 \text{ units}$$

Let Y_{ijk} be the response in the kth period on the jth experimental unit in the ith sequence; $i = 1, 2; j = 1, 2, \ldots, n_i; k = 1, 2$. Analogously to the model considered in Section 2, we assume that

$$Y_{ijk} = \mu + \rho_k + \gamma_{ij} + \tau_{d(k,ij)} + \delta_{d(k-1,ij)} + e_{ijk}$$

where ij denotes the jth unit in the ith sequence, the symbols denoting the same effects as defined earlier. Using normal distribution theory or under nonparametric approach, it is necessary to assume γ_{ij} to be random effects in order to estimate the direct and residual effects contrasts of the two treatment effects. In view of this, we further assume that γ_{ij} are independently and normally distributed with mean zero and variance σ_u^2; e_{ijk}

are independently and normally distributed with mean zero and variance σ^2. Further, we assume that γ_{ij} and e_{ijk} are independently distributed.

Put $Y_{ij} = Y_{ij1} + Y_{ij2}$. Then $\{Y_{1j}\}$ and $\{Y_{2j}\}$ are two independent samples with

$$E(Y_{1j}) = 2\mu + \rho_1 + \rho_2 + \tau_A + \tau_B + \delta_A$$
$$E(Y_{2j}) = \mu + \rho_1 + \rho_2 + \tau_A + \tau_B + \delta_B$$
$$\mathrm{var}(Y_{1j}) = \mathrm{var}(Y_{2j}) = 2(\sigma^2 + 2\sigma_u^2)$$

Under the null hypothesis $H_0 : \delta_A = \delta_B$, the two samples $\{Y_{1j}\}$ and $\{Y_j\}$ have the same mean and same variance, and the hypothesis can be tested by the standard two-sample t test.

Again, put $Z_{ij} = Y_{ij1} - Y_{ij2}$. The $\{Z_{1j}\}$ and $\{Z_{2j}\}$ are two independent samples with

$$E(Z_{1j}) = \rho_1 - \rho_2 + \tau_A - \tau_B - \delta_A$$
$$E(Z_{2j}) = \rho_1 - \rho_2 - \tau_A + \tau_B - \delta_B$$
$$\mathrm{var}(Z_{1j}) = \mathrm{var}(Z_{2j}) = 2\sigma^2$$

When the previous hypothesis $\delta_A = \delta_B$ is retained, under the additional hypothesis $H_0 : \tau_A = \tau_B$, the two independent samples $\{Z_{1j}\}$ and $\{Z_{2j}\}$ are taken from populations having the same mean and same variance, and this hypothesis can also be tested by the standard two sample t test.

Since the test for $H_0 : \delta_A = \delta_B$ is used as a preliminary test for testing the hypothesis of equal treatment effects, it should be tested at a level of significance higher than the conventional level. Grizzle (1965) recommends the use of 0.1 as the level of significance for testing $H_0 : \delta_A = \delta_B$.

On the other hand, when the hypothesis $\delta_A = \delta_B$ is rejected, under the null hypothesis $H_0 : \tau_A = \tau_B$, the two independent samples $\{Y_{1j1}\}$ and $\{Y_{2j1}\}$ are taken from populations having the same mean and same variance, and the hypothesis can be tested by a two-sample t test based on the samples $\{Y_{1j1}\}$ and $\{Y_{2j1}\}$. In this case, the second-period data are wasted in drawing inferences on direct effects and results in a loss of power associated with this test.

Alternatively, one may perform a split-plot design analysis (see Cochran and Cox, 1957, Chapter 7) in the present setting by considering sequences as main plot treatments and periods as subplot treatments. (See Table 6).

Table 6   A skeleton ANOVA

Source	df
Sequences	1
Units/sequences (Error a)	$n-2$
Periods	1
Sequences × Periods	1
Error (b)	$n-2$
Total	$2n-1$

In the context of Grizzle's model, the difference in carryover effects can also be interpreted as the difference in the two sequences and, hence, the null hypothesis $H_0 : \delta_A = \delta_B$ can be tested using the F ratio of sequences MS/Error (a) MS with 1 numerator and $n-2$ denominator degrees of freedom. If this null hypothesis is retained, one may test $H_0 : \tau_A = \tau_B$ from the F ratio of sequences × periods MS/Error (b) MS with 1 numerator and $n-2$ denominator degrees of freedom. Note that, in Grizzle's model, the difference in direct effects can be interpreted as the sequences X periods interaction. If $\delta_A = \delta_B$ is not valid, one may test $H_0 : \tau_A = \tau_B$ using an approximate t test based on the first-period mean differences of the two sequences as described by Cochran and Cox (1957, pp. 298–299). This approximate test uses error(a) and (b) MS, and entire data provides information.

Since readers are familiar with t tests and split-plot analysis, no numerical example is provided.

Patel (1983) proposed that two baseline measurements be obtained, whenever feasible, one before each treatment period, and he supplied procedures to analyze data after taking into account the baseline measurements. (Also, see Wallenstein and Fleiss, 1988.)

4   TWO TREATMENT, TWO-PERIOD DESIGN— NONPARAMETRIC APPROACH

Koch (1972) discussed the use of non-parametric methods in the statistical analysis of these designs. Continuing the same setting and model as in Section 3 except that the normal distributional assumption is dropped on γ_{ij} and e_{ijk}, under the null hypotheses $H_0 : \delta_A = \delta_B$, the samples $\{Y_{1j}\}$

and $\{Y_{2j}\}$ come from the same population, and the H_0 can be tested by the Wilcoxon test (cf. Walpole and Myers, 1978, pp. 474–477)].

If $\delta_A = \delta_B$ is valid, the null hypothesis $H_0; \tau_A = \tau_B$ can be tested once again by the Wilcoxon two-sample test using the two independent samples $\{Z_{1j}\}$ and $\{Z_{2j}\}$.

To test the null hypothesis $H_0 : \tau_A = \tau_B$ for any δ_A and δ_B, one ignores the second-period data and uses the Wilcoxon two-sample test on the two independent samples $\{Y_{1j1}\}$ and $\{Y_{2j1}\}$.

Finally, to test the null hypothesis $H_0 : \tau_A = \tau_B$, $\delta_A = \delta_B$, one uses the bivariate Wilcoxon test. Let R_{ijk} denote the rank of Y_{ijk} for $i = 1, 2$; $j = 1, 2, \ldots, n_i$ for each $k = 1, 2$. Let $\bar{R}_{i.k} = \sum_j R_{ijk}/n_i$, $M = (n+1)/2$, $\mathbf{U}_i = (\bar{R}_{i.1} - M, \bar{R}_{i.2} - M)$ and

$$ S = \sum_i \sum_j \begin{bmatrix} (R_{ij1} - M)^2 & (R_{ij1} - M)(R_{ij2} - M) \\ (R_{ij1} - M)(R_{ij2} - M) & (R_{ij2} - M)^2 \end{bmatrix} $$

The required test statistic is

$$ L = (n - 1) \sum_i n_i U_i' S^{-1} U_i $$

and is distributed as a χ^2 variable with 2 df under the null hypothesis $H_0 : \tau_A = \tau_B$, $\delta_A = \delta_B$.

Consider an example in which weekly rotation shifts (A) and triweekly rotation shifts (C) are compared in a crossover design using quarterly time period for each treatment in eight plants (see Table 7). Let the artificial data on the number of accidents during that study be shown in parenthesis:

The ranks for the sums for the eight plants are 4, 1, 3, 5, 6, 2, 7, and 8, respectively. The sum of the ranks for the first sample is $w_1 =$

Table 7   Number of Accidents in Plant Study

Period (months)	Plant							
	I	II	III	IV	V	VI	VII	VIII
0–3	A(2)	A(1)	A(4)	A(1)	B(7)	B(4)	B(2)	B(10)
3–6	B(5)	B(2)	B(2)	B(7)	A(2)	A(0)	A(9)	A(2)
Sum	7	3	6	8	9	4	11	12
Difference	−3	−1	2	−6	5	4	−7	8

$4 + 1 + 3 + 5 = 13$ and, for the second sample, $w_2 = 36 - 13 = 23$. Thus, $u_1 = 13 - 4(5)/2 = 3$, $u_2 = 23 - 4(5)/2 = 13$, and one takes $u = 3$. From Table XVI of Walpole and Myers (1978), $P(U \leq a \mid H_0) = 0.1$ and this probability is larger than $\alpha = 0.05$. Thus, the equality of residual effects is retained at 0.05-level of significance. The ranks for the differences for the eight plants are, respectively, 3, 4, 5, 2, 7, 6, 1, and 8, and the sum of the ranks for the first sample is $w_1 = 3 + 4 + 5 + 2 = 14$ and, thus, $w_2 = 22$, $u_1 = 4$, and $u_2 = 12$. Taking $u = 4$, $P(U \leq u \mid H_0) = 0.171$. Thus, the null hypothesis $H_0 : \tau_A = \tau_B$ is also retained. In order to simultaneously test $H_0 : \tau_A = \tau_B$, $\delta_A = \delta_B$, one ranks period 1 data alone and period 2 data alone and calculates

$$\mathbf{u}_1' = [-1.5, 0.5], \qquad \mathbf{u}_2' = [1.5, -0.5],$$

$$S = \begin{bmatrix} 40.5 & -20.0 \\ -20.0 & 37.0 \end{bmatrix}$$

$$L = 3.234$$

The critical value of a χ^2 distribution with 2 df and at 0.05-level of significance is 5.991. Since observed L is smaller than the critical value, H_0: $\tau_A = \tau_B$, $\delta_A = \delta_B$ is retained at 0.05-level of significance.

REFERENCES

Cochran, W. G. (1939). Long-term agricultural experiments, *J. Roy. Stat. Soc.*, 6B: 104–148.

Cochran, W. G., Autrey, K. M., and Canon, C. Y. (1941). A double change-over design for dairy cattle feeding experiments, *J. Dairy Sci.*, 24: 937–951.

Cochran, W. G. and Cox, G. M. (1957). *Experimental Designs*, 2nd ed., Wiley, New York.

Finney, D. J. (1965). Cross-over designs in bio-assay, *Proc. Roy. Soc.*, 145B: 42–61.

Grizzle, J. E. (1965). The two-period change-over design and its use in clinical trials, *Biometrics*, 21: 467–480.

Keppel, G. (1973). *Design and Analysis: A Researcher's Handbook*, Prentice–Hall, Englewood Cliffs, N. J.

Koch, G. G. (1972). The use of non-parametric methods in the statistical analysis of the two-period change-over design, *Biometrics*, 28: 577–584.

Mielke, P. W. Jr. (1974). Squared rank test appropriate to weather modification cross-over design, *Technometrics*, 16: 13–16.

Patel, H. I. (1983). Use of baseline measurements in the two-period crossover design, *Comm. Statist.*, 12: 2693–2712.

Patterson, H. D. and Lucas, H. L. (1959). Extra-period change-over designs, *Biometrics*, 15: 116–132.

Wallenstein, S. and Fleiss, J. L. (1988). The two-period crossover design with baseline measurements. *Comm. Statist.*, 17: 3333–3344.

Walpole, R. E. and Myers, R. H. (1978). *Probability and Statistics for Engineers and Scientists*, Macmillan, New York.

Williams, J. (1949). Experimental designs balanced for the estimation of residual effects of treatments, *Australian J. Sci. Res.*, 2A: 149–168.

Index